DIGITAL TELEPHONY AND NETWORK INTEGRATION

Bernhard E. Keiser
and
Eugene Strange

VAN NOSTRAND REINHOLD COMPANY
New York

Library of Congress Catalog Card Number: 83–21698
ISBN: 0–442–24659–5

Manufactured in the United States of America

Published by Van Nostrand Reinhold Company Inc.
115 Fifth Avenue
New York, New York 10003

Van Nostrand Reinhold Company Limited
Molly Millars Lane
Wokingham, Berkshire RG11 2PY, England

Van Nostrand Reinhold
480 La Trobe Street
Melbourne, Victoria 3000, Australia

Macmillan of Canada
Division of Canada Publishing Corporation
164 Commander Boulevard
Agincourt, Ontario M1S 3C7, Canada

15 14 13 12 11 10 9 8 7 6 5 4 3 2

Library of Congress Cataloging in Publication Data
Keiser, Bernhard, 1928–
Digital telephony and network integration.

Includes index.
1. Digital telephone systems. I. Strange, Eugene.
II. Title.
TK6421.K45 1984 621.385 83–21698
ISBN 0–442–24659–5

To our wives, Evelyn Keiser and Elsie Strange, who gave unselfishlessly of vast amounts of their time in typing the manuscript.

Preface

What is "digital telephony"? To the authors, the term digital telephony denotes the technology used to provide a completely digital point-to-point voice communication system from end to end. This implies the use of digital technology from one end instrument through the transmission facilities and switching centers to another end instrument. Digital telephony has become possible only because of the recent and ongoing surge of semiconductor developments allowing microminiaturization and high reliability along with reduced costs.

This book deals with both the future and the present. Thus, the first chapter is entitled, "A Network in Transition." As baselines, Chapters 2, 3, and 10 provide the reader with the present status of telephone technology in terms of voice digitization as well as switching principles.

The book is an outgrowth of the authors' continuing engineering education course, "Digital Telephony," which they have taught since January, 1980, to attendees from business, industry, government, common carriers, and telephony equipment manufacturers. These attendees come from a wide variety of educational backgrounds, but generally have the equivalent of at least a bachelor's degree in electrical engineering.

The book has been written to provide both the engineering student and the practicing engineer a working knowledge of the principles of present and future voice communication systems based upon the use of the public switched network. Problems or discussion questions have been included at the ends of the chapters to facilitate the book's use as a senior level or first year graduate level course text.

Numerous clients and associates of the authors as well as hundreds of students have provided useful information and examples for the text, and the authors wish to thank all those who have so contributed either directly or indirectly.

The first chapter, which is a joint effort of both authors, provides an overview of the field. Chapters 2–4 deal with the subject of speech digitization, while Chapters 5 and 6 are devoted to the use of digital technology in the telephone network and for transmission in general. Chapters 7–9 treat three principal facility types for digital transmission: microwave radio, communication satellite systems, and fiber optics.

Chapters 10–12 begin with the present status of telephony switching systems and progress through a description of digital switching architecture and switching systems using stored program control. The evolution of the switched digital network is covered in Chapter 13, leading finally to the integrated services digital network (ISDN), which is the subject of Chapter 14.

Because of the rapidly changing nature of the subjects covered in this volume, the authors invite reader comments, questions and suggestions for future editions.

The authors acknowledge the useful comments and suggestions provided by the reviewers of the text.

Dr. Keiser thanks his daughter, Nancy, for drafting many of the illustrations, and Mr. Strange thanks Mr. Mike Fontana for his drafting support. The book would not have been possible without the contributions of these individuals.

BERNHARD E. KEISER
EUGENE STRANGE

Vienna, Virginia

Contents

DIGITAL TELEPHONY AND NETWORK INTEGRATION

1
A Network in Transition

1.1. INTRODUCTION

To say that telephony is in a state of transition is to describe the history of telephony. It always has been in a state of transition and will so continue for the foreseeable future. It is becoming increasingly difficult to distinguish one telephone network from another now that direct dialing is available in most countries, but we still do not have a single, worldwide network.

In this introductory chapter, we look at one network not only as a standalone network but also as a very significant portion of worldwide telephony. Since the authors are American and have had more extensive exposure to the network in the United States, it is only natural to focus on that network. Because the networks in the United States and Canada operate as a transparent, integrated network, the evolution of the network in the United States is considered to typify that of the Canadian network as well, and, to a certain extent, of other national networks throughout the world.

In order to focus on the extent of the evolution, this chapter highlights, in snapshot form, some statistics and characteristics of the American telephone network at three 20-year intervals—past, present, and future. The nature of the network 20 years from now, of course, is highly speculative, but some of the trends are readily apparent.

1.2. THE NETWORK YESTERDAY

Over twenty years ago (1964), all telephone calls in the public switched network were switched by operation of metallic elements to establish metallic paths through switching system networks. Some of those electromechanical switching systems had already been in service for over 40 years, and some manual switchboards were still in service. Direct dialing of toll calls was gaining in popularity as more local and toll switching systems were being equipped with the capability for direct distance dialing (DDD), although manual toll switchboards still were handling a high percentage of toll calls.

The development of multichannel transmission systems had been proceeding at a faster pace than that of switching systems. Between 1948 and 1964, there were few major improvements in switching system technology, the most significant being the introduction of stored program control in 1964. In 1941, the Type L1 carrier system was introduced, providing 480 voice channels over a pair of 0.69-cm (0.27-in.) coaxial tubes, later expanded to 600 voice channels over a pair of 0.95-cm (0.375-in.) coaxial tubes. In 1953, the L3 carrier system was placed in service, providing 1860 voice channels per pair of coaxial tubes. Initially, the coaxial cables were constructed with eight tubes, providing three working systems and one spare pair. In 1960, the tube capacity was increased to 12, enabling the five working systems to provide 9300 voice channels per cable.[1]

Microwave radio relay systems began spanning the country with the introduction of the TD-2 system in 1950 and had accumulated some 40 million circuit-kilometers by 1961, exceeding the cumulative circuit-distance of wire and cable systems by over 10%. The early TD-2 systems provided 480 voiceband channels, later increased to 600, or one black-and-white television signal, on each of five two-way radio channels operating in the 4-GHz frequency band. One radio channel was reserved as a protection channel. In the mid-1950s, the horn reflector antenna and its circular waveguide system were developed, enabling transmission of both vertically and horizontally polarized signals. This doubled the TD-2 route capacity to 6000 voiceband channels using ten working radio channels and two protection channels. The horn reflector operates in the 4-GHz, 6-GHz, and 11-GHz frequency bands simultaneously, permitting further increases in route capacity.[2]

Growth in the number of telephones in service, which had slowed considerably during the 1930s, increased substantially after the end of World War II, increasing from 23.5 million at year-end 1941 to 77.4 million at year-end 1961, a compounded annual growth rate of over 6%. During the same time period, the number of telephones in the world was increasing at about the same rate from 45.2 million to 150 million. During the 50 years from year-end 1911 to year-end 1961, compound annual growth rates of telephones in service for different areas were: United States, 4.57%; Europe, 5.47%; all other areas, 7.06%; and the aggregate, 5.11%.[3]

Technology had been advancing slowly. Although the transistor had been invented in 1948, all systems described above use older technology. The first transoceanic submarine cable, completed in 1956, provided 36 voice channels between North America and Europe in parallel one-way cables. A second such pair was laid to Europe in 1959. Yet, most transatlantic conversations still were conducted over high-frequency radio links. The first telephone call by satellite was made on August 12, 1960 when President Eisenhower's voice from California was bounced off Echo I, a 30-meter, aluminum-covered balloon, and was recovered by a ground station in Maine. Other laboratory research during the 1950s and early 1960s would emerge later to contribute to the rapid evolution of telephony technology that would occur during the ensuing 20-year period.[4]

Digital technology was introduced in 1962 when the first T1 digital carrier system was placed in service. The T1 enabled 24 voiceband channels to be multiplexed over two nonloaded exchange cable pairs, employing an early version of pulse code modulation (PCM).

As the number of telephones increased, so did the number of conversations per person. In the United States, the number of telephones per 100 population increased from about 16 in 1939 to about 38 in 1959. During the same period, the average number of calls per person per year increased from 222 in 1939 to 472 in 1959. Telephone usage in Canada led the world in 1959 with about 515 average number of calls per person.[5] Most of the calls, however, were "plain old telephone service" or POTS. Some data calls were being placed at very low data rates, rarely exceeding 600 or 1200 b/s.

1.3. THE NETWORK TODAY

The development of semiconductor technology, which effectively began with the transistor, has made possible the telecommunications explosion which has occurred during the last 20 years and is continuing. Solid state devices are employed in virtually every facet of the telephone network. The digital computer has revolutionized the way that business and financial transactions are conducted, and the marriage of computers and communications has made the telephone network of "yesterday" obsolete.

Transmission system capacities have multiplied. An underground transcontinental coaxial cable, hardened to protect against damage from natural causes or bomb damage except from a direct nuclear blast, was placed in service in late 1964. The 12-tube cable and L3 multiplex system provided a route capacity of 9300 voice channels. A 20-tube cable, combined with a solid state L4 multiplex system, increased coaxial route capacity to 32 400 in 1967 with 3600 voiceband channels per pair of tubes. The L5E multiplex system provides 13 200 voiceband channels per tube pair, further increasing the coaxial route capacity to 132 000 voiceband channels over a 22-tube cable, an increase of 73 to 1 over the L1 system.[6]

Solid state technology also increased the capacity of microwave routes while reducing space and power requirements. Digital microwave is in widespread use on short routes, but it is at a disadvantage on long-haul routes because of its relatively low spectrum efficiency compared with analog single-sideband systems. The extent of transmission capabilities that are available to customers can be illustrated by the fact that private facilities provided to the National Aeronautic and Space Administration for the Gemini space program in the mid-1960s included 56 voice channels, 2 TV channels, 6 teletypewriter channels, 4 wideband data channels, and 10 medium speed data channels between Texas and Florida.[7] Other advanced transmission systems which have been developed include high-capacity submarine cable systems, both analog and digital satellite relay systems, optical fiber cable systems, and digital coaxial systems.

Switching systems using stored-program computer control had been introduced into the network in 1964, but electromechanical systems continued to increase in number. The introduction of digital local and toll switching systems in 1976, using PCM coding, signaled a new era in telephony. Digital systems have become increasingly popular with local telephone companies because of their reliability, low maintenance costs, and space savings. Moreover, additional savings are possible when digital switching systems are interconnected by digital transmission systems. However, of the estimated 21 000 local and toll switching systems in service in the United States at the end of 1984, only some 4000 are believed to use digital technology. Additionally, about 3000 are analog space division systems with electronic common control, and the remaining 14 000 or so are electromechanical systems, of which roughly 50–65% are believed to employ step-by-step switching technology, first introduced into service before 1900. The replacement of older switching systems with digital systems is proceeding at a rapid pace numerically but much more slowly proportionately. Of major concern to common carriers are the economic and technical matters associated with operating a mixed network for the long period of transition.[8]

Services have proliferated along with the improvements in switching and transmission. As the number of computers has multiplied and terminals have become more "intelligent," the demand for more reliable data communication at higher speed has resulted in a wide variety of data services. Data speeds of up to 4800 b/s over public switched network connections and up to 9600 b/s (19.2 kb/s with parallel lines) over private lines are quite common, and data speeds of 56 kb/s and 1.544 Mb/s are being used increasingly. Packet-switched data networks provide high quality services for interactive data users, and users have a choice of several satellite services and facsimile networks. Teletypewriter Exchange Service and Telex are still popular services. More recently, videophone teleconferencing service has become available in a limited number of metropolitan areas. Long distance telephone services are offered to the general public by several common carriers and by many firms that lease private circuits from common carriers and resell message-rate services.

The network structure in the United States and network connections to foreign countries are undergoing radical changes. On January 1, 1984, the American Telephone and Telegraph Company (AT&T) divested all wholly owned Bell Operating Companies (BOCs) as a result of settlement of a government antitrust suit. The divested BOCs were grouped into seven independent regional holding companies. AT&T retained its long lines communication system, Western Electric Company, and Bell Laboratories. Under the terms of the consent decree, all installed customer premises equipment was allocated to AT&T, but the divested BOCs were permitted to reenter the terminal equipment business. In addition, AT&T was allowed to enter other competitive telecommunications markets.

The restructuring of the network following divestiture will evolve during a transition period of several years as competitive long distance systems become more fully implemented. Concurrently, competitive long distance carriers are beginning to provide transoceanic voice service as interconnection agreements are negotiated with foreign telephone administrations.

1.4. THE NETWORK TOMORROW

Will the changes in the network during the next 20 years take the form of evolution or revolution? Perhaps both. Evolution will characterize the conversion of the network from predominantly analog to predominantly digital. Services, however, are more likely to experience several progressive steps of revolution. The key technological developments that will make "tomorrow's" network possible are continuing advances in microprocessors and optical fiber transmission. As sufficient portions of the network become digital to enable common carriers to offer new services with the expectation of economic return, new services will materialize, and some will seem radically different from anything available today.

Near-term, an increasing fanfare of new data services will be available to businesses whose needs are sufficient to justify the cost. Videotex likely will become widely available in the very early portion of the period. When protocols for subscriber loops and network signaling become fully standardized, the Integrated Services Digital Network (ISDN), defined and described in Chapter 14, will evolve within the framework of the existing digital network characteristics with a channel limitation of 64 kb/s initially.

What form will the future network take? As optical fiber becomes more cost effective and more technologically advanced, it is likely to revolutionize subscriber loop technology. Wideband services, including color video, then will become technically possible between serving common carrier offices and business and residential locations. Such capability will create demand for new network services to use that capability. That, in turn, will drive the development of switching and transmission systems having greater capabilities to respond to the demand.

A wide variety of digital rate requirements will need a new generation of switching equipment to accommodate them. Excellent voice quality can be provided by digital techniques at data rates of less than the current 64 kb/s. Therefore, a variable bit-rate transmission system would seem to be the solution. Separate specialized networks may provide the different services to users,

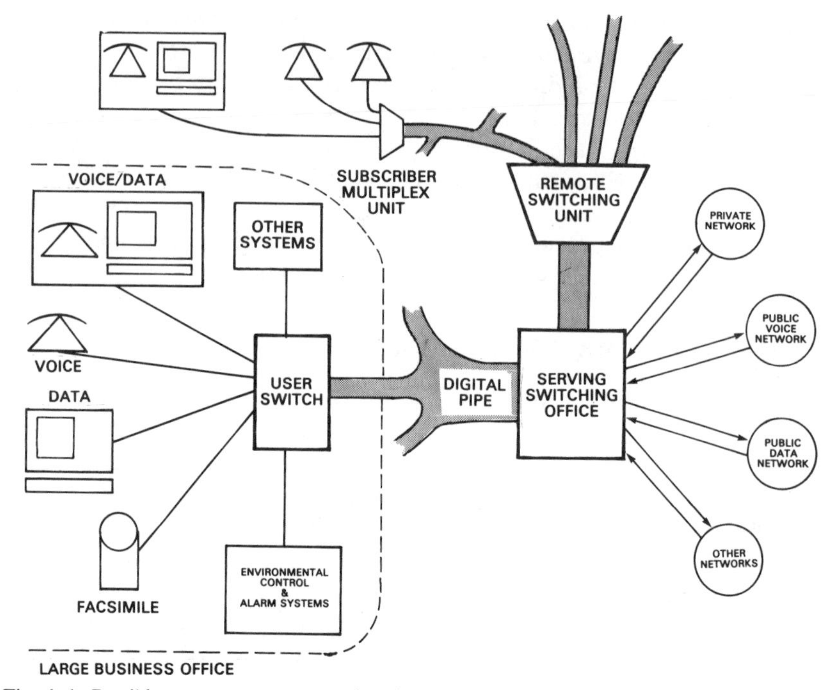

Fig. 1–1. Possible access arrangement for the Integrated Services Digital Network (ISDN). (Based on Fig. 1, Irwin Dorros, "Telephone nets go digital," *IEEE Spectrum,* April 1983, © 1983 IEEE.)

or a single advanced network could evolve. Whatever network characteristics evolve, the user will want transparency with a single access for all services. Will the network access resemble that depicted in Fig. 1–1, or will it take a different shape?[9] One thing is certain—it will be digital!

REFERENCES

1. *Engineering and Operations in the Bell System* (Bell Telephone Laboratories, 1977), 327–329.
2. Ibid, 333–336.
3. Elizabeth Wrenshall, "A Fabulous 50 Years," *Bell Telephone Magazine,* Spring 1963, 42–45.
4. "Bell Laboratories' Contributions To Space Technology," *Bell Telephone Magazine,* Autumn 1961, 17–27.
5. Elizabeth Wrenshall, "A Statistician Looks at the World's Telephones," *Bell Telephone Magazine,* Autumn 1960, 52–56.
6. *Telecommunications Transmission Engineering* (AT&T, 1977), 321.

7. "GLDS Gemini Launch Data System," *Bell Telephone Magazine,* Winter 1965–1966, 8–9.
8. There are no known firm statistics relative to the actual census of switching systems. The estimated figures quoted have been compiled from some actual statistics and from estimates of a variety of industry sources.
9. Irwin Dorros, "Telephone Nets Go Digital," *IEEE Spectrum* April 1983, 49.

2
Speech Digitization Fundamentals

2.1. INTRODUCTION

In digital telephony, speech is handled in digital form. Accordingly, the speech from the telephone transmitter must be converted from its analog waveform to a series of ones and zeros. A multitude of speech digitization methods has been devised; these methods can be categorized as waveform, parametric, and hybrid coding. A *waveform coder* takes the actual waveform and produces a series of ones and zeros representative of that waveform according to a set of rules. A *parametric coder,* sometimes called a *source coder,* attempts to detect certain characteristics of speech, such as pitch and amplitude, and produces a series of ones and zeros according to a set of rules that describe the detected speech characteristics rather than the waveform. A *hybrid coder* is one that uses both waveform and parametric principles in its production of a digital version of speech.

A speech waveform is a continuous function of time, but it or its parameters are to be converted into a series of digits which occur at a specific rate. Accordingly, the speech must be sampled periodically. Thus sampling is an important step in speech digitization.

Each sample conveys a magnitude based on the type of coding being used. This magnitude must be expressed as a series of digits. The magnitude may be a waveform amplitude or a speech parameter value at some sampling instant. The process of converting a waveform's magnitude at a given instant into a series of digits is called *quantization.*

The processes of speech digitization described in this chapter generally, but not always, result in the need for a bandwidth greater than the original analog speech bandwidth. The reasons for this increase are described in this chapter.

The relative performance capabilities of the various speech digitization methods are summarized at the close of the chapter, along with the advantages of speech digitization from the viewpoints of overall performance as well as system cost and reliability.

2.2. SPEECH CODING APPROACHES

Waveform and parametric coding constitute the two basic approaches to encoding a source which, in telephony, usually is thought of as producing a voice waveform, but which may produce various signaling and supervision tones such as dial tone, address signaling pulses or tones, or ringing and busy tones.

Waveform coding is performed by transforming the waveform itself into a series of digits. The rules used to achieve this transformation may be applied to the waveform itself, or to a differential version of the waveform which has been formed by subtracting a recent past value from the present value, as illustrated in Fig. 2–1. As can be seen from the figure, the differential waveform has a smaller amplitude, in general, than does the original waveform. This results from the fact that the original waveform is band-limited and thus its value does not change rapidly over a very short time interval. Recovery of the original waveform from the differential version is accomplished by integration. Chapter 4 describes this process in detail.

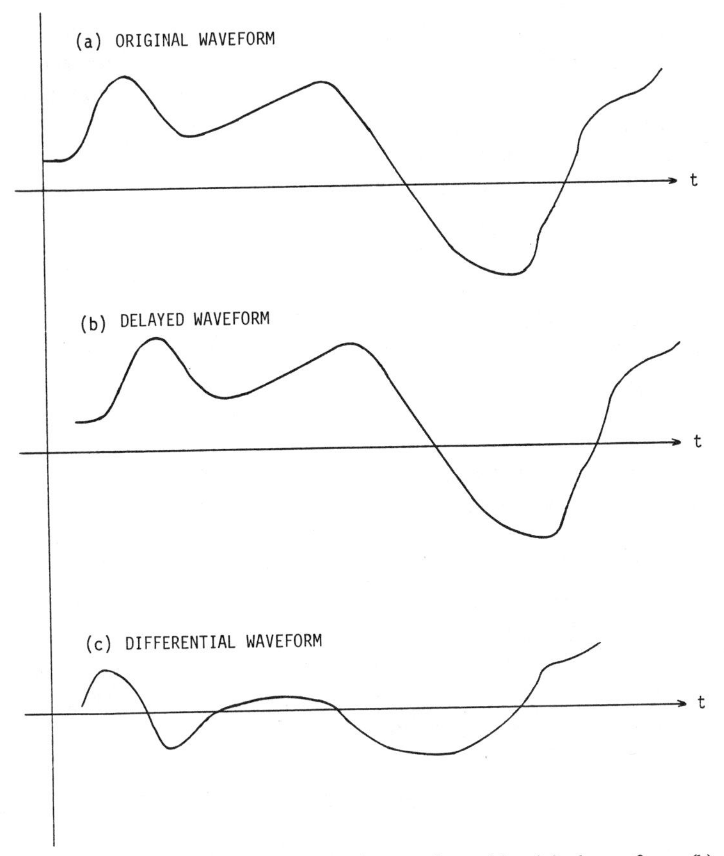

Fig. 2–1. Formation of a differential version of a waveform: (a) original waveform; (b) delayed waveform; (c) differential waveform.

The coding parameters may be varied during transmission with respect to their numerical values to accommodate different waveform characteristics. Such variable techniques are called *adaptive coding.* Adaptive coding is very useful in dealing with speech waveforms which may change, for example, from male to female voices, or with waveforms which may change their volume and pitch rapidly.

Parametric coders do not attempt to deal with the speech or signaling waveform at all. Instead, they function by extracting some characteristic of the waveform. For example, the waveform may be filtered into a number of subbands, each of whose amplitudes is transmitted, together with an indication that the wave is "voiced" at a given pitch, or that it is "unvoiced," i.e., is characterized by a random function. Another approach involves taking a Fourier or other transform of the wave and then transmitting the parameters of the transform.

The concept of *prediction* enters often into source coding. Many speech encoders predict the next digital value of the input and combine this predicted value with the actual next value to produce a quantized differential. The prediction is based upon a weighted sum of recent past values. At the receiver the differentials are summed to obtain the total value of the signal.

The speech coding technique commonly used in telephony today is a waveform coding scheme called pulse-code modulation (PCM), the subject of Chapter 3. The coding for PCM currently is done at the central office. However, such coders now can be implemented at very low cost on a single chip, thus allowing their placement within the telephone handset. However, for such techniques to be accepted and used on a widespread basis, further system developments are needed, as outlined in Chapter 4.

2.3. SAMPLING

If a waveform contains significant frequencies as high as W Hz, then, according to Nyquist's Theorem, that waveform can be reconstructed perfectly if it is sampled at a rate of $2W$ per second. Actually, if the lowest significant frequency is greater than zero, the sampling rate can be decreased to twice the actual bandwidth. The sampling theorem can be stated as follows:

If a signal that is band-limited is sampled at regular intervals at a rate at least twice the highest frequency in the band, then the samples contain all of the information of the original signal.

Because frequencies outside the band W may be reproduced by the receiving desampler inside the band W, it is important to precede a sampler by a filter to minimize such out-of-band energy.

To illustrate the operation of a sampler, let $f(t)$ be a band-limited signal which has no spectral components above W Hz. Multiple $f(t)$ by a uniform train of time impulse functions $\delta_T(t)$, as illustrated in Fig. 2–2.

The result of this multiplication is:

$$f_s(t) = f(t)\delta_T(t) = \sum_{n=-\infty}^{\infty} f(nT)\delta(t - nT) \tag{2–1}$$

The Fourier transform of $f(t)\delta_T(t)$, according to the frequency convolution theorem, is given by

$$2\pi f_s(t) \leftrightarrow F(\omega) * \omega_0 \delta_{\omega_0}(\omega)$$

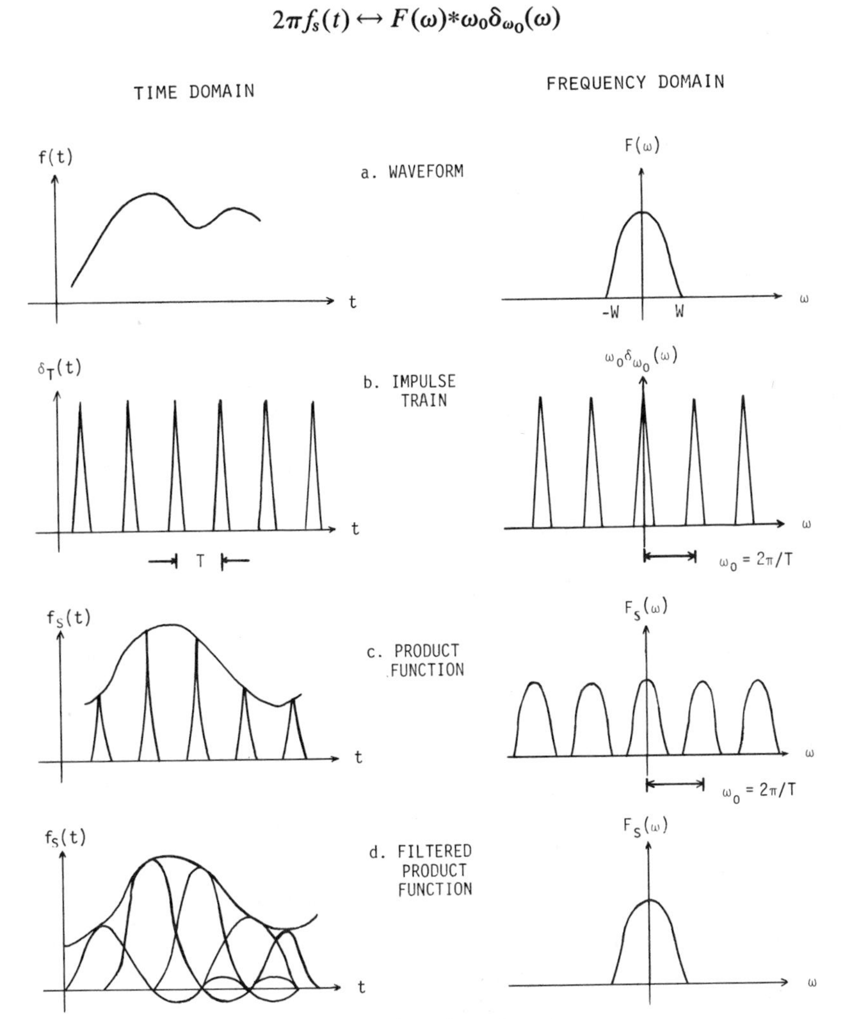

Fig. 2–2. Illustration of sampling theorem.

Letting $\omega_0 = 2\pi/T$,

$$f_s(t) \leftrightarrow (1/T)F(\omega)*\delta_{\omega_0}(\omega)$$

In the frequency domain, the corresponding function is a series of harmonics spaced $\omega_0 = 2\pi/T$. The product function $f(t)\delta_T(t)$ is a sequence of impulses located at regular intervals of T seconds each, and having amplitudes equal to the values of $f(t)$ at the corresponding instants. In other words, this is a pulse amplitude modulated (PAM) sequence. Note that the process of multiplication corresponds, in the frequency domain, to a replication of the original spectrum $F(\omega)$ every ω_0 radians/second.

If $\omega_0 > 2(2\pi W)$, the spectral representations of $F(\omega)$ do not overlap, i.e., $2\pi/T > 2(2\pi W)$, and thus $T \leqslant 1/2W$.

Therefore, if $f(t)$ is sampled at regular intervals less than $1/2W$ seconds apart, the sampled spectral density function $F_s(\omega)$ will be a periodic replica of the true $F(\omega)$. Thus $F_s(\omega)$ will contain all the information of $f(t)$. Removal of all but the spectrum around zero frequency by low-pass filtering then leaves the reconstructed function as the output.

2.4. QUANTIZATION

Each time a signal is sampled it is found to exhibit a value which must be expressed in digital form, i.e., as a series of ones and zeros. This is done by quantization, a process in which the continuous range of values of an input signal is divided into nonoverlapping subranges. The presence of the input signal in a particular subrange results in the production of a unique series of bits by the quantizer. In designing a quantizer, the level of waveform detail desired is first established, and then the quantization levels or bins are set up accordingly. Thus quantization may be said to remove details that the designer regards as irrelevant.

Figure 2–3 illustrates a typical quantizing characteristic. The fact that irrelevant details are removed results in a reconstructed output that can take on only specific values. Thus an error has been generated known as *quantizing noise.* A linearly increasing input (ramp function), for example, exhibits the quantizing noise indicated in Fig. 2–4.

For an oscillating input of the type that might be found in a speech waveform, the error exhibits the characteristics of the input near its peaks, as illustrated in Fig. 2–5.

The choice of the level spacing, or step size, may be a difficult one if the input is highly variable, as speech generally is. For example, when the speaker is talking loudly the waveform peaks are large, and large steps are desirable to prevent peak clipping and still not use an unreasonably large number of

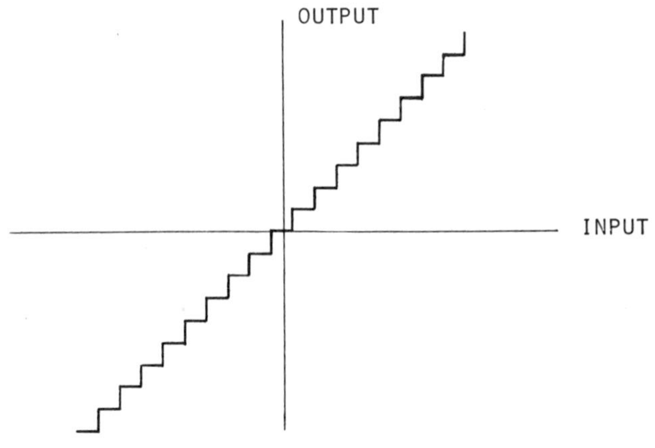

Fig. 2–3. Typical quantizing characteristic.

steps. On the other hand, when the speaker is talking softly, small steps are desirable. One solution to this dilemma is the use of adaptive quantization. The step size is adapted to the signal variance based upon a memory associated with the quantizer. Usually the step size is modified for each new input sample, based upon a knowledge of which quantizer slots were occupied by the previous samples. Adaptive quantization is discussed further in Chapter 3.

Another approach to the problem of preserving the large dynamic range characteristic of human speech is the use of nonuniform quantization, as illustrated in Fig. 2–6. Here the total number of steps remains the same as

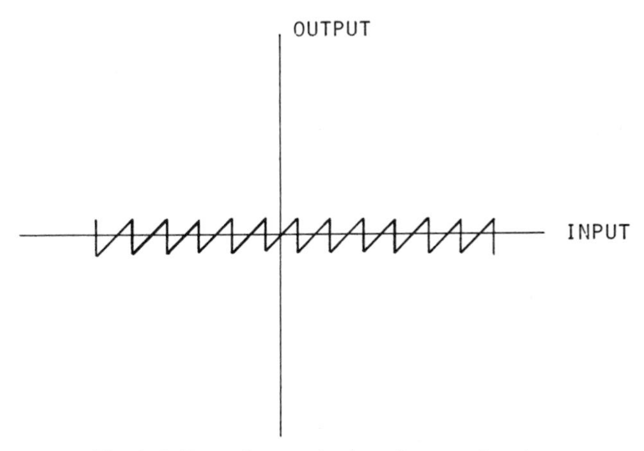

Fig. 2–4. Errors in quantization of a ramp function.

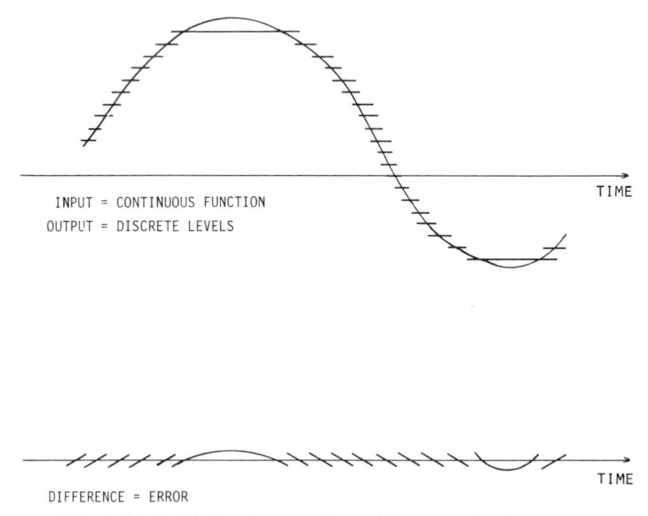

Fig. 2–5. Oscillating input and resulting quantization error.

in the case of uniform quantization, but low levels are represented by a large number of small steps and high levels are represented by a smaller number of large steps. Note that the use of nonuniform quantization is equivalent to passing the input waveform through an amplitude compressor, quantizing linearly, and then reconstructing the result by dequantizing linearly and

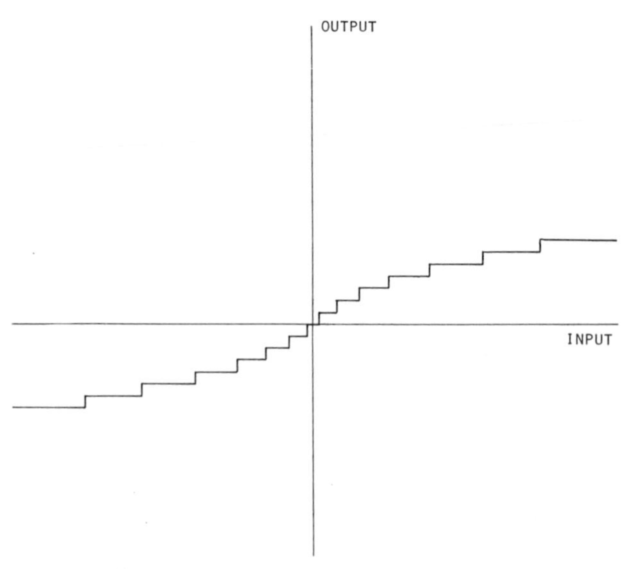

Fig. 2–6. Nonuniform quantizer characteristic.

passing the result through an amplitude expander. Nonuniform quantization is discussed further in Chapter 3.

2.5. EFFECT OF DIGITIZATION ON BANDWIDTH

The digitization of a waveform produces a bit stream whose rate depends upon the sampling rate and the number of bits per sample produced by the quantizer. Nyquist's theorem states that the sampling rate must be at least twice the maximum significant frequency contained in the input. To keep quantization noise low, quantization with a small step size is needed; this means a large total number of steps and thus many bits per sample.

For current telephone applications, speech is sampled 8000 times per second and quantized at 8 bits/sample, resulting in a 64 kb/s stream. If the keying is done on the basis of one bit per symbol, then a 64 kbaud stream is produced. Its bandwidth depends on the selectivity of the waveform shaping filters used, but probably will be on the order of 100 kHz. On the other hand, if a highly efficient keying technique providing 4 bits/symbol is used, then a 16-kbaud stream is produced, occupying a bandwidth on the order of 25 kHz. In practice, however, 24 telephone channels usually are time-division multiplexed together to produce a 1.544 Mb/s stream which then may be transmitted using a seven-level format resulting in a 386 kHz bandwidth for the 24 channels, or an equivalent of nearly 16 kHz per channel.

In general, the term b/s per Hz is used as a measure of spectrum efficiency for digital modulation techniques.

The foregoing discussion addressed the digitization of a *waveform.* If parametric coding is used instead, as will be discussed in detail in Chapter 4, the speech can be represented by a 2.4-kb/s stream. Since some data modems are capable of sending 9.6 kb/s or more over conditioned telephone facilities, one might say that the equivalent bandwidth of such digitized channel is only one-fourth that of a 3 kHz voice channel, or about 750 Hz.

2.6. SPEECH DIGITIZER PERFORMANCE

A wide variety of speech digitizers has been devised, ranging all the way from 64-kb/s pulse-code modulation (PCM) systems (see Chapter 3) to very low bit rate formant vocoders (see Chapter 4). The speech reproduced by these devices may be described as being in one of the following categories:

1. *Toll quality:* quality based[1] upon a laboratory test in which the signal-to-noise ratio exceeds 30 dB, the frequency response is 200 ~ 3200 Hz, and the harmonic distortion is less than 3%. This quality is accept-

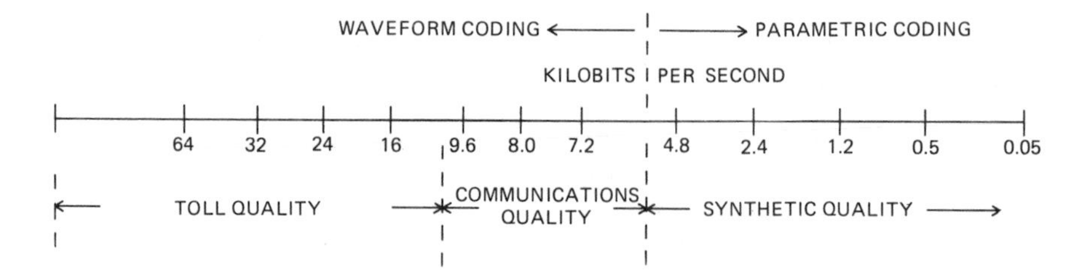

Fig. 2–7. Quality achievable in speech coding. (Courtesy J. L. Flanagan, Ref. 1, © IEEE, 1978.)

able to most members of the general public, but is not necessarily achieved on many connections.

2. *Communications quality:* quality which is acceptable to military, amateur, and citizens-band operators, often in a mobile environment.
3. *Synthetic quality:* the quality of computer-generated speech, which often lacks human naturalness.

Figure 2–7 indicates that waveform coding can provide toll quality at data rates on the order of 14 kb/s and higher. At lower rates, down to about 5 kb/s, the quality is judged to be communications quality. At rates below about 5 kb/s, parametric or "source" coders provide better intelligibility than waveform coders, but with synthetic quality.

The bit rate at which parametric coders perform better than waveform coders is not clearly established. Figure 2–8 summarizes the impressions expressed by large groups of listeners to different speech coders.[2] Here the parametric coders are called *vocoders,* a name commonly used for them.

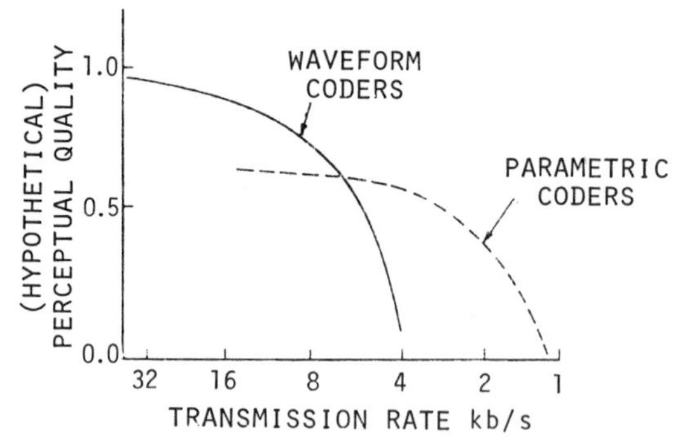

Fig. 2–8. Perceptual quality comparison of waveform and parametric coders. (Courtesy J. L. Flanagan et al., Ref. 2, © IEEE, 1979.)

Nevertheless, the superiority of waveform coders above about 6 to 8 kb/s is clearly evident.

Two general criteria are used in judging the performance of a speech coder. One is the ratio of signal to quantization noise, (SNR), which can be derived analytically. Since quantization noise is generated by the coder, not the channel, the SNR depends primarily on the type of coder used, as well as its bit rate. Alternatively, subjective and perceptual effects (How does it sound?) can be used to judge coder performance. The results can be evaluated only by actual listener tests. They include not only quantization noise effects but also the various types of distortions which a speech coder may produce and encounter.

2.7. SPEECH CODING ADVANTAGES

Section 2.5, "Effect of Digitization on Bandwidth," has shown that the digitization of speech increases its bandwidth if standard waveform coding is used. Offsetting this disadvantage, however, are a number of advantages:

1. The coded signal is easier for a detector to recognize than is an uncoded one. As a result, lower carrier-to-noise ratios can be tolerated than for analog signals. Only bits and bit combinations need be detected, rather than the exact values of a continuous waveform.
2. The coded signal can be regenerated efficiently. This allows the elimination of noise and distortion that may have entered between or at the repeaters of a transmission facility. However, severe noise levels may result in bit errors which are not eliminated or reduced unless special error correction techniques are used.
3. Coded signals are easily encrypted. This can be achieved using scrambling techniques of the type discussed in Chapter 5. Based on pseudonoise sequences, these techniques continually change the significance of each bit combination. Thus only a receiver with the proper "key" can recover the transmitted intelligence.
4. Coding makes possible the combination of transmission and switching functions. Since both speech and signaling are in the form of bit streams, such combinations become possible.
5. Coding provides a uniform format for different types of signals. This format is the same whether the input is dial tone, addressing, ringing, speech, or disconnect signal. Everything is in the form of a digital bit stream.

The foregoing technological advantages translate into economic advantages as well. Digitized speech allows the use of digital techniques throughout the network with their large scale economies. In addition, as the point of digitization moves toward the end user, a greater portion of network costs are incurred only when the user is ready to use and pay for them.[3] Moreover, digital techniques require less critical circuit adjustments, thus reducing maintenance costs.

PROBLEMS

2.1 Explain why parametric coding devised for speech will not accurately reproduce switching, signaling and data waveforms from the analog network.

2.2 Can a waveform coder be speech specific? Explain.

2.3 A corporation leases a conditioned voice grade line connecting its offices on opposite sides of the country. How many digital voice circuits at 2.4 kb/s can be accommodated on this line using 9.6-kb/s modems? Using 14.4-kb/s modems?

2.4 Explain why a digital decoder can be used only with a digital encoder design based on the same algorithm.

REFERENCES

1. Flanagan, J. L., "Opportunities and Issues in Digitized Voice," *IEEE-EASCON '78 Record,* pp. 709–712.
2. Flanagan, J. L., et al., "Speech Coding," *IEEE Trans. Comm.,* Vol. COM-27, No. 4, p. 729 (April, 1979).
3. Bellamy, J. C., *Digital Telephony,* John Wiley & Sons, Inc., New York, N.Y., 1982.

3
Pulse Code Modulation

3.1. INTRODUCTION

The term pulse code modulation (PCM) refers to the use of a specific set of rules for transforming a waveform into a stream of digits, and vice versa. To clarify the concept, however, the first point to be made is that PCM *is* coding, but it *is not* necessarily the modulation of the carrier. Modulation usually refers to an alteration of a periodic function (often a radio frequency sine wave) to cause it to convey intelligence. This alteration may be a change in amplitude or frequency or phase of the sine wave. These are the ways in which a carrier can be modulated. The PCM process, in contrast, is a coding technique. The resulting stream of ones and zeros can be used to modulate a carrier in any of the ways mentioned here, or special combinations of them.

The coding process known as PCM uses such techniques as sampling and quantization, already discussed in Chapter 2. It also involves the concept of synchronization. Synchronization refers to the timing with which information is transmitted. The receiver must know when to look for a new bit, and which bits constitute bytes or samples, etc. As applied to the transmission of a bit stream, communication may be accomplished either on an asynchronous or a synchronous basis.

Asynchronous communication refers to the transfer of data at nonuniform rates. A stop interval is required to guarantee that each character will begin with a one-to-zero transition, even if the preceding character was entirely zeros.

Synchronous communication is the technique used for PCM transmission, as well as for many other speech coding techniques. Synchronization can be performed on either a digit or a frame basis. *Digit synchronization* uses the timing information contained in the transmitted signal without altering the signal itself. In other words, the receiver simply derives its timing from changes in the bit stream itself. *Frame synchronization,* in contrast, uses a fixed synchronization signal once per frame, where "frame" refers to the block or group of digits transmitted as a unit, over which a coding procedure is applied for synchronization or error control purposes. This is the type of synchronization used in PCM transmission.

A major advantage of synchronous communication over asynchronous is that all, or nearly all, bits are used to transmit data, since stop intervals or bits are not needed for framing.

To determine which groups of bits constitute a character, a "sync" character is used, which is chosen so that its bit arrangement is significantly different from that of any regular character being transmitted. As will be seen in Chapter 5, in a PCM stream of bits a large but specific number of information bits is followed by a single "framing" bit, followed by more information bits, etc. If the framing bits were selected from the overall bit stream, they would constitute a much lower rate stream. The sync character is sent within this lower rate stream. For example, 24 PCM voice channels are transmitted at a rate of 1.544 Mb/s. Within that 1.544 Mb/s stream is a framing bit stream whose rate is 8000 b/s.

The receiver must recognize the sync character in order for it to determine which groups of bits constitute characters. Accordingly, when transmission begins, the receiver goes into its "sync search" mode. This also occurs if a data dropout occurs. In the sync search mode, bits are shifted into the receiver's shift register. The contents of this shift register then are compared continually with the stored sync character which resides in another receiver register. This process continues until a match occurs. When a match is found on two successive sync characters, the receiver raises a "character available" flag every eight bits and normal receiver operation proceeds.

3.2. BASIC PCM ENCODING

A continuous waveform representing speech or data can be transformed into a bit stream using PCM.[1] The procedure is as follows: The waveform is sampled at a rate of at least $2W$ Hz, where W is the highest significant frequency contained in the waveform. Usually higher frequencies will have been heavily attenuated by a filter so they do not result in spurious output components. The sampling rate can be less than $2W$ if the lowest frequency contained in the continuous waveform is nonzero. Standard PCM as used today on the public switched network operates with a sampling rate of 8000/second.

The sampling process produces a pulse train that is amplitude modulated, as illustrated in Fig. 2–2(c).

The continuous waveform now has been made discrete in the time domain by virtue of the sampling process. Next it must be made discrete in the amplitude domain as well. This is accomplished by the quantization process, in which a large number of possible amplitude levels for the waveform are established. This number is a multiple of 2. For example, a total of 256 possible levels, 128 positive and 128 negative, can be achieved by 8-bit quantization, since $2^8 = 256$. This 8-bit quantization is the standard used in all of the new PCM systems being implemented in the public switched network.

In general, the PCM process quantizes the amplitude of each signal sample into one of 2^B levels. Thus the amount of information is said to be B bits/sample. Since there are $2W$ samples/second, the information rate is $2WB$ bits/second.

The quantization process thus represents discrete amplitude levels by distinct binary words of length B. For example, if $B = 2$, there are four levels or 00, 01, 10 and 11.

The PCM decoding process maps the binary words back into amplitude levels. The initial result is an amplitude-time pulse sequence which is low-pass filtered using a filter with a cutoff frequency W.

Figure 3–1 illustrates the use of 8 bits in producing the 256 possible encoded words representing the continuous range of sample values. As shown, however, the use of this code results in the sending of a large number of zeros, since the most probable speech input amplitudes fall in the middle of the code range, where the zeros predominate. As will be discussed in Chapter 5, a digital zero corresponds to a zero level on the digital line. Since timing for systems such as the T1 and other standard digital arrangements is derived from the data signal, a long sequence of zeros means that there is no data signal from which to derive timing. Accordingly, a mixture of zeros and ones is needed. This mixture can be provided by inverting the code words

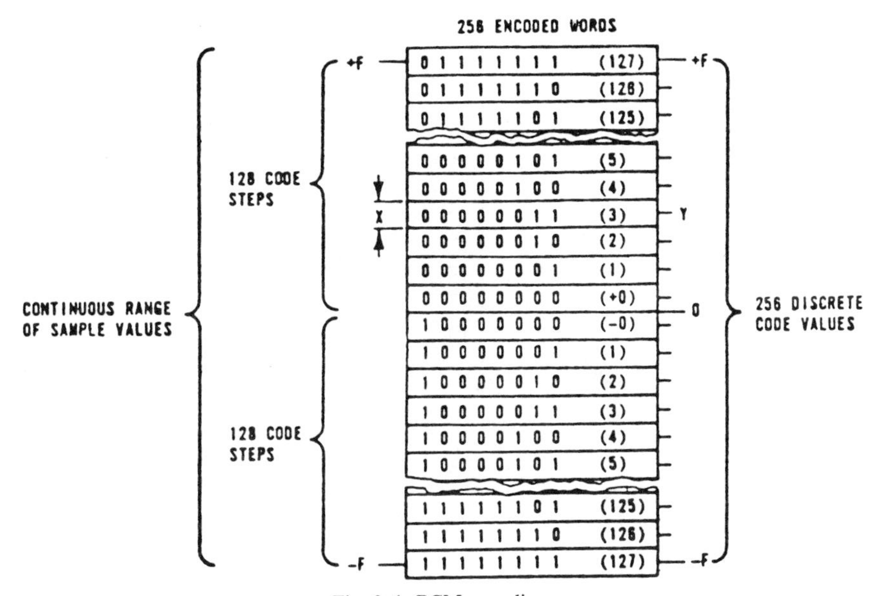

Fig. 3–1. PCM encoding.

thus using what is called *zero code suppression.* Figure 3–2 illustrates the use of this technique. The −127 level corresponds to a sample that produces the all zero code. When this level is reached, a one is placed in the next to least significant bit, and the word is transmitted as 00000010, corresponding to −125. The use of zero code suppression as illustrated here is important, since on a T1 repeatered line no more than 14 consecutive zeros can be transmitted.[2]

The foregoing discussion has described the 8-bit PCM parameters of the voice encoding system currently used on the public switched network. How well does it perform? Is there an incentive to change from $B = 8$ to some other value of B? An increase in B will increase the spectrum occupied, or require higher level modulation techniques (see Chapter 6). Accordingly, B should be kept as small as possible consistent with meeting performance requirements.

Small values of B (i.e., $B < 8$) mean coarser quantization and a resulting increased quantization error. If Δ is the quantizer step size, then the error E will be between -0.5Δ and $+0.5\Delta$ for any given sample. Moreover, the

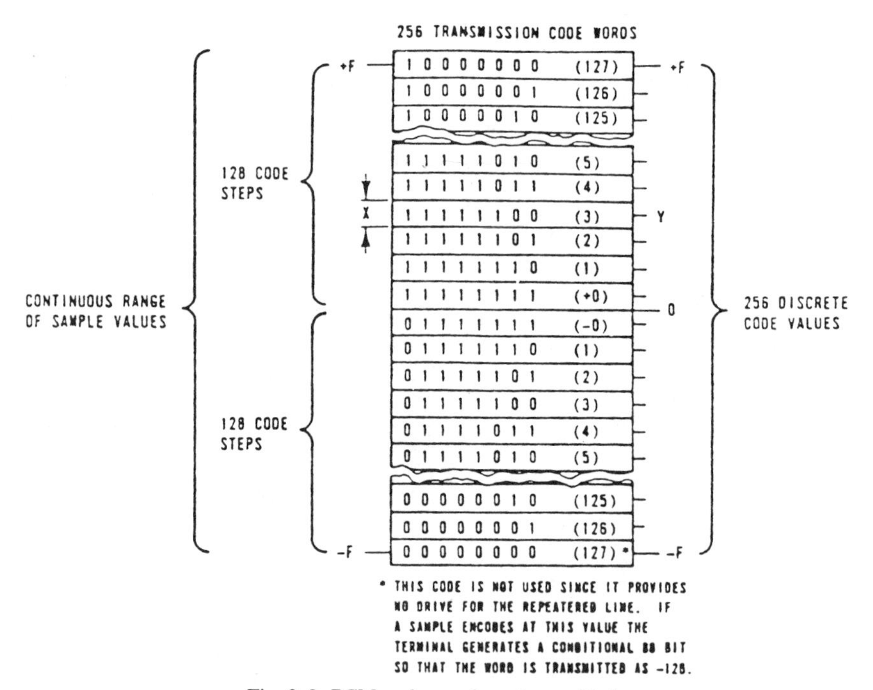

Fig. 3–2. PCM code words as transmitted.

error E is assumed to be uniformly distributed over the interval -0.5Δ to $+0.5\Delta$, so that $\Delta p(E) = 1$, or

$$p(E) = 1/\Delta \tag{3–1}$$

where $p(E)$ is the amplitude probability distribution of the error E. Eq. (3–1) is valid provided the signal does not overload the quantizer and provided the signal traverses a large number of steps. To determine the ratio of signal to quantization error (noise) on a power basis, the mean-square error must be calculated. It is

$$\int_{-0.5\Delta}^{0.5\Delta} E^2 p(E)\, dE = \Delta^2/12 \tag{3–2}$$

Then for an input level having a value X, the signal-to-quantization-noise ratio is

$$\mathrm{SNR} = X^2/(\Delta^2/12) \tag{3–3}$$

A formula can be developed for performance limited by quantization error, not channel noise. This formula is based upon the input signal being complex enough, or sufficiently finely quantized, that the quantization error has a uniform distribution throughout the interval -0.5Δ to $+0.5\Delta$. Moreover, the quantization is assumed to be fine enough that signal-correlated patterns do not appear in the error waveform, as was illustrated in Fig. 2–5. Thus the effect of errors must be measurable in terms of a noise power or error variance. Finally, the quantizer must be aligned with the amplitude range, which means that any peak clipping that occurs will be equal for both positive and negative peaks. In other words, the signal has a zero mean value, i.e., no dc component.

Since the number of quantization levels is 2^B, it follows that if the saturation limits of the quantizer are $\pm X_{pk}$, then the step size Δ is

$$\Delta = 2X_{pk}/2 \tag{3–4}$$

Letting A equal the peak amplitude of an input wave, $A < X_{pk}$ if clipping is to be avoided. Let $A/X_{pk} = m$, the fractional portion of the quantizer range occupied by the input waveform. For a sinusoidal input, the rms signal power is $(A/\sqrt{2})^2 = A^2/2$. Thus

$$\begin{aligned}\text{SNR} &= 10\log_{10}[(A^2/2)/\Delta^2/12] = 10\log_{10}[6A^2/\Delta^2] \\ &= 10\log_{10}[6m^2X_{\text{pk}}^2/\Delta^2] = 10\log_{10}[(3m^2/2)2^{2B}]\end{aligned}$$

As a result,

$$\text{SNR} = 1.76 + 20\,log\,m + 6B \tag{3–5}$$

This is the PCM performance equation for uniform encoding.

For example, suppose a system must be able to provide a 35 dB SNR for inputs whose magnitudes may differ as much as 40 dB. What number of quantization bits B must be used if the encoding is uniform? Using equation (3–5), one finds that $B = 12.2$. Thus to meet the stated requirements, 13-bit quantization would be required. Actual PCM telephony systems, however, use only 8 bit or less. This smaller number is feasible through the use of compression and nonuniform quantization, which are discussed in Section 3.3.

The foregoing SNR formulas are valid generally for toll quality speech links except for amplitude alignment. This problem results from the fact that a single encoder-decoder system must handle several speakers or circuits, all of which have nonstationary speech signals. The result is a deterioration of quantizer performance, i.e., lower SNR values than those predicted by the SNR equations. Solutions to this problem take two forms: nonuniform quantization and adaptive quantization. These two subjects are discussed in the next section and in Chapter 4, respectively.

To detect the presence or absence of a PCM (or other) pulse reliably requires a certain signal-to-channel-noise ratio. Two assumptions will be made to determine what this ratio must be. First, the detector is assumed to be ideal, i.e., no error is contributed by the detector. (Realistically, this simply means that detector error is negligible compared with the error produced by channel noise.) Second, the channel noise is assumed to have a uniform power spectrum, to have a zero mean value, and to have a gaussian amplitude distribution. (As far as the receiver is concerned, the noise is flat over the receiver passband and has the characteristics of thermal noise within this band.)

The input PCM pulses presented to the detector are assumed to be of short width (less than a time τ, to be defined), and to have been filtered by an ideal low-pass filter of bandwidth W. The pulse centered at time $t = m\tau$, where $\tau = 1/2W$, has the form

$$V = V_0[\sin(\pi/\tau)(t - m\tau)/(\pi/\tau)(t - m\tau)] \tag{3–6}$$

and will be zero at $t = k\tau$, where $k \neq m$, as illustrated in Fig. 3–3. However, by sampling the pulse at $t = m\tau$, only the pulse belonging to that time is

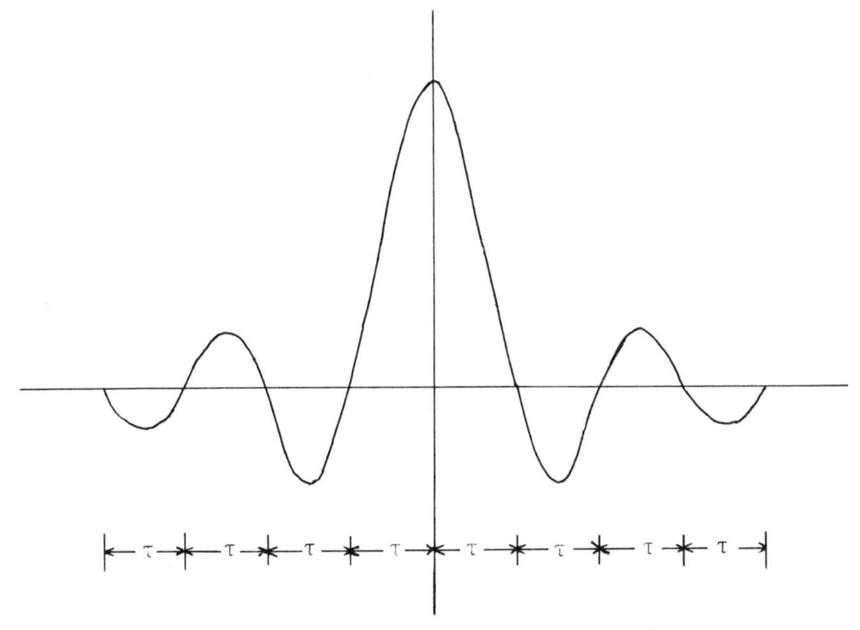

Fig. 3–3. Pulse of the form $V_0\,[\sin(\pi t/\tau)/(\pi t/\tau)]$.

seen, all others being zero at that instant, assuming the pulses occur at multiples of τ seconds.

As will be seen in Chapter 5, the actual transmission line code usually used is known as bipolar return to zero (BRZ) or alternate mark inversion (AMI).

For bipolar transmission, detection occurs as follows. If the signal when sampled has an output $>V_0/2$, a pulse is said to be present, whereas if the signal when sampled exhibits an output $<V_0/2$, no pulse is said to be present. Thus an error occurs for an instantaneous noise value $> V_0/2$ in the right direction at the sampling time. (The noise is assumed to have a zero mean value.)

Let σ = rms noise amplitude and P_s = signal "power" = V_0^2. The noise "power" in bandwidth W is $N = \sigma^2$. For unipolar transmission (V_0,0), the probability of an error occurring then is

$$p_e = \operatorname{erfc}\,(V_0/2\sigma) = \operatorname{erfc}\,\{(P_s/4N)^{1/2}\} \tag{3–7}$$

where

$$\operatorname{erfc}\,(x) = (1/\sqrt{2\pi}) \int_x^{\infty} e^{-\lambda^2/2}\, d\lambda \tag{3–8}$$

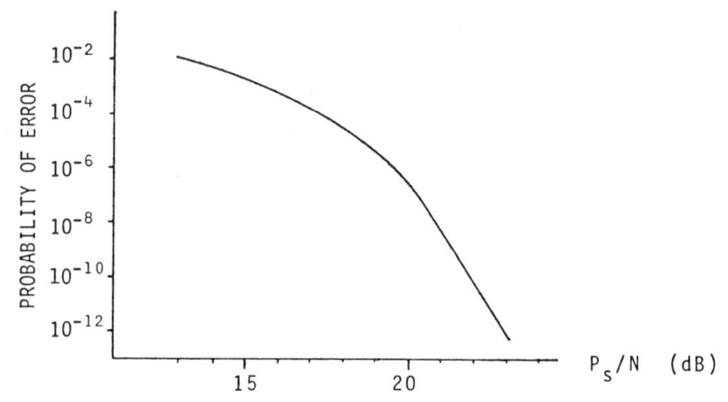

Fig. 3–4. Probability of detection error for pulse in random noise.

Eq. (3–8) is plotted in Fig. 3–4. It shows how the channel error rate varies with signal-to-channel-noise ratio for the case of simple pulse detection.

For bipolar transmission (V_0, 0, $-V_0$) the error rate is somewhat higher (about 40%) because both positive and negative noise pulses can cause unwanted threshold crossings when the zero-level signal is sent. By comparison, unipolar transmission (V_0, 0) is influenced only by positive noise pulses for the zero level signal and only by negative pulses when the V_0 level is sent.

If the AMI detector reproduces the unwanted pulse as a one, then the probability of error is doubled for transmitted zeros. As a result, the overall probability of error is increased by 50% if zeros and ones are equally likely.

3.3. COMPRESSION AND NONUNIFORM QUANTIZATION

As noted in the last section, the nonstationary nature of speech (variations in probability density and talker volume) results in an amplitude alignment problem and a corresponding ineffective use of a uniform quantizer's characteristic. Some speech inputs may use only the lowest levels of the characteristic while others may use primarily the higher levels, and even suffer clipping. Thus it is desirable to provide large end steps for the relatively infrequent large amplitude ranges, while providing a finer quantization for the lowest levels. Alternatively, encoding might be achieved at a lower bit rate for a given quality and dynamic range through the use of nonuniform quantization.

A compressor, as used in analog circuits, is a device that accepts a large dynamic range and reduces the dynamic range according to a predetermined compression law. A companion device, the expander, has the inverse characteristic. It takes the compressed signal and restores it to its original dynamic range. The overall process is called *companding.* It is useful in transmitting speech over channels whose dynamic range is limited. Companding may be syllabic or instantaneous, or somewhere between. Syllabic companding uses the fact that the speech power level varies only slightly from one 125 μs sample to the next. The gain of an amplifier is made to vary automatically to achieve increased dynamic range, whereas in the case of instantaneous companding, an amplifier with a nonlinear input-output characteristic is used. "Nearly instantaneous companding," a third technique, is discussed in Chapter 4.

For good voice reproduction, the signal-to-distortion ratio should be kept constant over a wide dynamic range. This means that the distortion should be proportional to the signal amplitude for any signal level. A logarithmic compression law achieves this result.

The use of a nonuniform quantizer is equivalent to the presentation of a compressed signal to a uniform quantizer, with subsequent expansion of the output. The compression law used in North American telephone networks is the μ-law. It is expressed by the equation

$$|\nu| = V \ln\left(1 + \frac{\mu|X|}{V}\right) \Big/ \ln(1+\mu), \qquad \mu > 0 \tag{3–9}$$

where

ν = output of speech compressor,
V = overload level of speech compressor designed for nonuniform quantization,
μ = dimensionless quantity in expression for logarithmic compression function,
X = input to speech compressor.

Figure 3–5 illustrates companding characteristics for various values of μ.

The $\mu = 0$ curve corresponds to uniform (linear) quantization. Based upon statistical observations of the dynamic range of speech signals, desirable values of μ are on the order of 100 or more.[3] The gain achieved over uniform quantization is shown in Fig. 3–6 for various values of μ. The gain is seen to be a function of the normalized input level. The areas under the curves measure how well quantizer performance changes as X_{rms} changes. The value $\mu = 255$ has been selected for 8-bit speech quantizers because it can be approximated closely by a set of eight straight line segments for either polarity (15 segments peak to peak). Where 7-bit quantizers are in use, usually in older designs, $\mu = 100$.

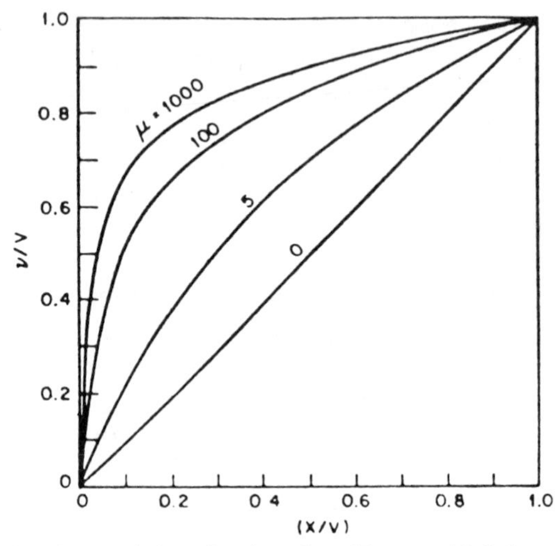

Fig. 3–5. Companding characteristics using the μ-law. (Courtesy N. S. Jayant, Ref. 3, © IEEE, 1974.)

The compression law used in European networks is called the A-law. It is expressed by the equation

$$|\nu| = V[(1 + \ln A|X|)/(1 + \ln A)] \quad \text{for} \quad 1/A \leqslant X \leqslant 1 \qquad (3\text{–}10a)$$

$$|\nu| = V[A|X|/(1 + \ln A)] \quad \text{for} \quad 0 \leqslant X \leqslant 1/A \qquad (3\text{–}10b)$$

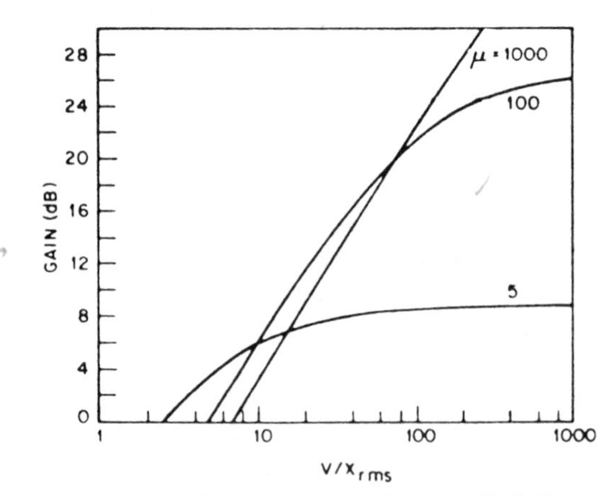

Fig. 3–6. Logarithmic quantizer performance curves. (Courtesy N. S. Jayant, Ref. 3, © IEEE, 1974.)

The logarithmic curve is approximated by a linear segment for small signals. The value of X/V after which the curve transitions smoothly to a true logarithmic form is $1/A$. The parameter A is usually set at 87.6. Segmented versions of A-law companding often use 13 segments.

Over the full input dynamic range, A-law companding provides a more nearly constant SNR than does μ-law, as illustrated in Fig. 3–7. However, μ-law provides somewhat more dynamic range.

The minimum step size for A-law is 2/4096, whereas for μ-law the minimum is 2/8159. Moreover, the A-law approximation used does not define a zero-level output for the first quantization interval. In other words, it uses what is known as a mid-riser quantizer. For low-level signals, the companding advantage, i.e., the ratio of the smallest step with and without companding, is 24 dB (4096/256) for A-law versus 30 dB (8192/256) for μ-law.

The term "optimum" quantizer refers to one that yields minimum mean-square error when matched to the variance and the amplitude distribution of the signal. However, the nonstationary nature of speech results in less than satisfactory results with optimum quantizers, especially if the number of quantization levels is small. If a quantization boundary is placed at the zero level, then during idle channel conditions, the output of the quantizer jumps back and forth between the lowest magnitude quantization levels. If these lowest quantization levels are greater than the amplitude of the background noise, the output noise can exceed the input noise. Thus such a boundary is undesirable in quantizer design.

Quantizers based on the μ-law produce lower idle channel noise than do

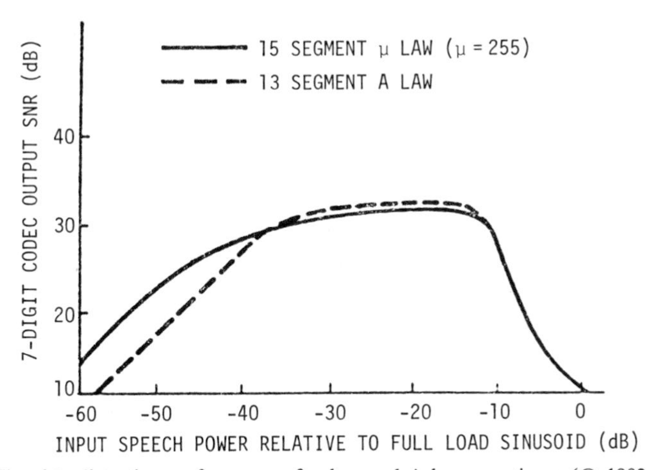

Fig. 3–7. Signal-to-distortion performance of μ-law and A-law quantizers. (© 1982, Bell Telephone Laboratories. Reprinted by permission.)

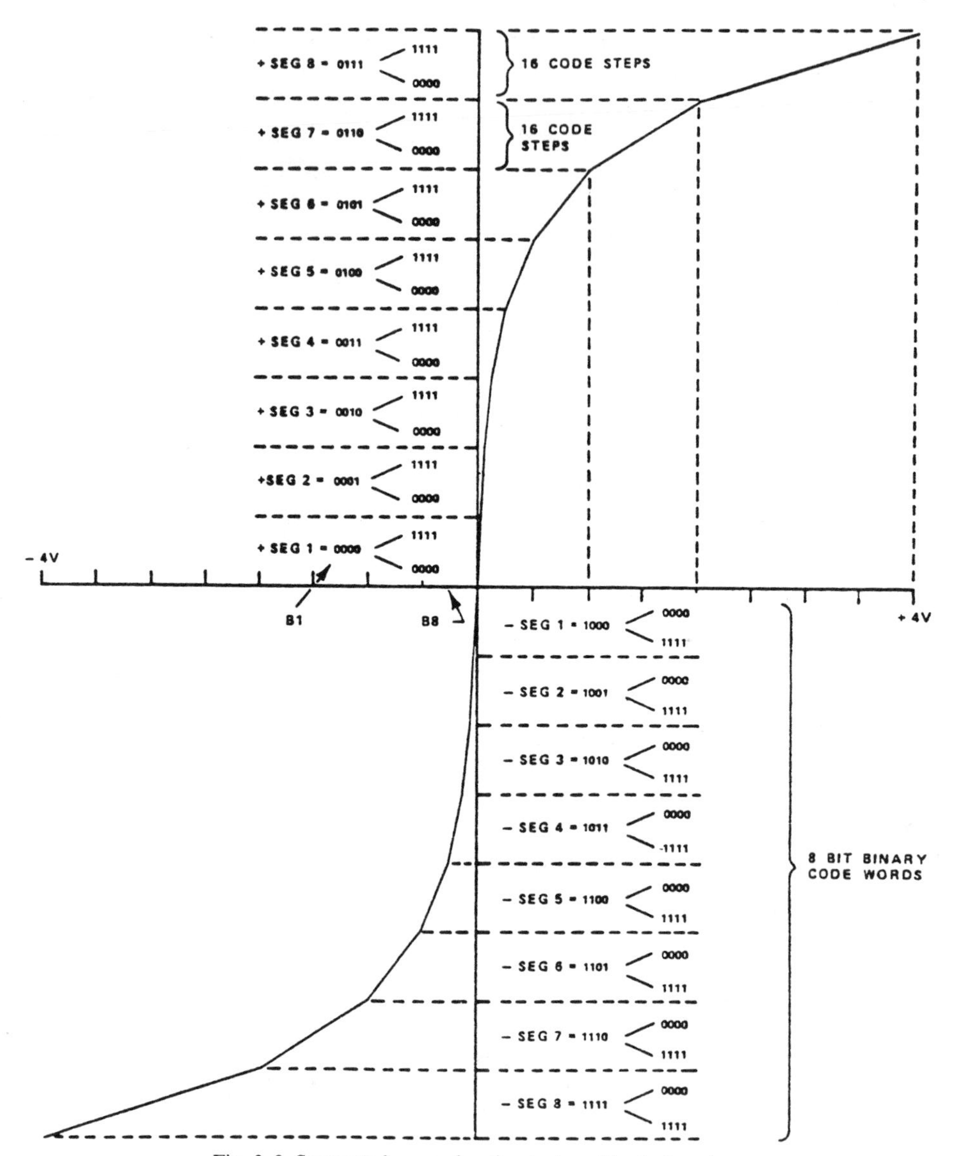

Fig. 3–8. Segmented approximation to logarithmic function.

optimum quantizers. For $\mu = 255$, the ratio of maximum to minimum output level is 4015 : 1.

Segmented approximations to the logarithmic functions are implemented so that each successive segment changes its slope by a factor of two. Each linear segment then contains an equal number of coding intervals, with Bit 1 being determined by the polarity (0 is +; 1 is −); Bits 2, 3, and 4 being determined by the segment of the encoder characteristic into which the sample value falls (binary progression); and Bits 5, 6, 7, and 8 being encoded using linear quantization within each segment. Figure 3–8 illustrates this coding scheme.

The advantages of nonuniform quantization increase with the crest factor (peak ÷ rms) of the signal. The values of μ cited relate to the standard 7- and 8-bit quantizations used in the public switched network.

The "optimum" quantizer (whether uniform or nonuniform) is one that maximizes the SNR, with the effect of overload errors included.

3.4. PCM PERFORMANCE

Two methods exist for the evaluation of a voice coding technique. One is the SNR, the ratio of signal to quantization noise. This is an objective rating, and one that can be calculated. The other is, How does the result sound? The response of a listener to the quality of a given coding technique is a subjective but a very important factor in the technique's acceptability.

The SNR has inadequacies as a performance measure because the quantization error sequence has components that depend on the signal, i.e., they are signal correlated. These components appear more like distortion than background noise. This distortion does not have the same annoyance value as does independent additive noise of equal variance, provided $B = 7$ or 8 or more, as is the case on the public switched network, as well as in high-fidelity digital recording. In Chapter 4, however, the reverse will be found to be true for $B < 6$, i.e., distortion will be more objectionable than background noise in these lower bit rate systems.

Table 3–1 summarizes the performance characteristics of PCM as used in the currently manufactured D4 channel banks[4] (8-bit quantization, $\mu = 255$). As can be seen from the table, PCM furnishes toll quality. Shown in Table 3–1 are the end-to-end system requirement (the design objective), the allocation to the coder-decoder (CODEC), and the allocation to the digital-to-analog converter (DAC). Figure 3–9 shows the transmit and receive filter frequency responses for the D4 channel bank.[5]

While PCM is designed primarily to handle voice, occasions arise in which a subscriber may use a voice line for data transmission via modems. Thus the ability to handle the tones produced by high-speed modems also is important. PCM has been found to pass modem signals satisfactorily at 4800 b/s for up to eight tandem A/D conversions, and at 9600 b/s for up to four tandem A/D conversions. At 14 400 b/s, only one A/D conversion generally can be tolerated. Accordingly, 4800 b/s usually is the maximum rate for switched lines, with higher rates generally using private conditioned lines.

Table 3–1. PCM Performance Characteristics.†

	1-KHZ LEVEL (DBM0)	SYSTEM REQUIREMENT (DB)	D4 CODEC OBJECTIVE (DB)	DAC OBJECTIVE (DB)
Signal-To-Distortion*	0	33	33	35
	−10	33	33	35
	−20	33	33	35
	−30	33	33	34
	−40	27	27	30
	−45	22	22	26
Gain Tracking*	+3	±0.5	±0.25	±0.08
	0 (Reference)	—	—	—
	−10	±0.5	±0.25	±0.08
	−20	±0.5	±0.25	±0.08
	−30	±0.5	±0.35	±0.12
	−40	±1.0	±0.50	±0.23
	−45	±1.0	±0.75	±0.35
Harmonic Distortion	1-kHz Level 0 dBm0	−40	−50	−55
Gain Stability (0°C–50°C)	1-kHz Level 0 dBm0	±0.25	±0.20	±0.05 (150 ppm °C) (0°C–75°C)
Crosstalk Coupling Loss	Level 0 dBm0 200 to 3400 Hz	>65	>75	Not applicable
Idle Channel Noise		23 dBrn C0	20 dBrnC0	Not applicable

* Includes crossover error.

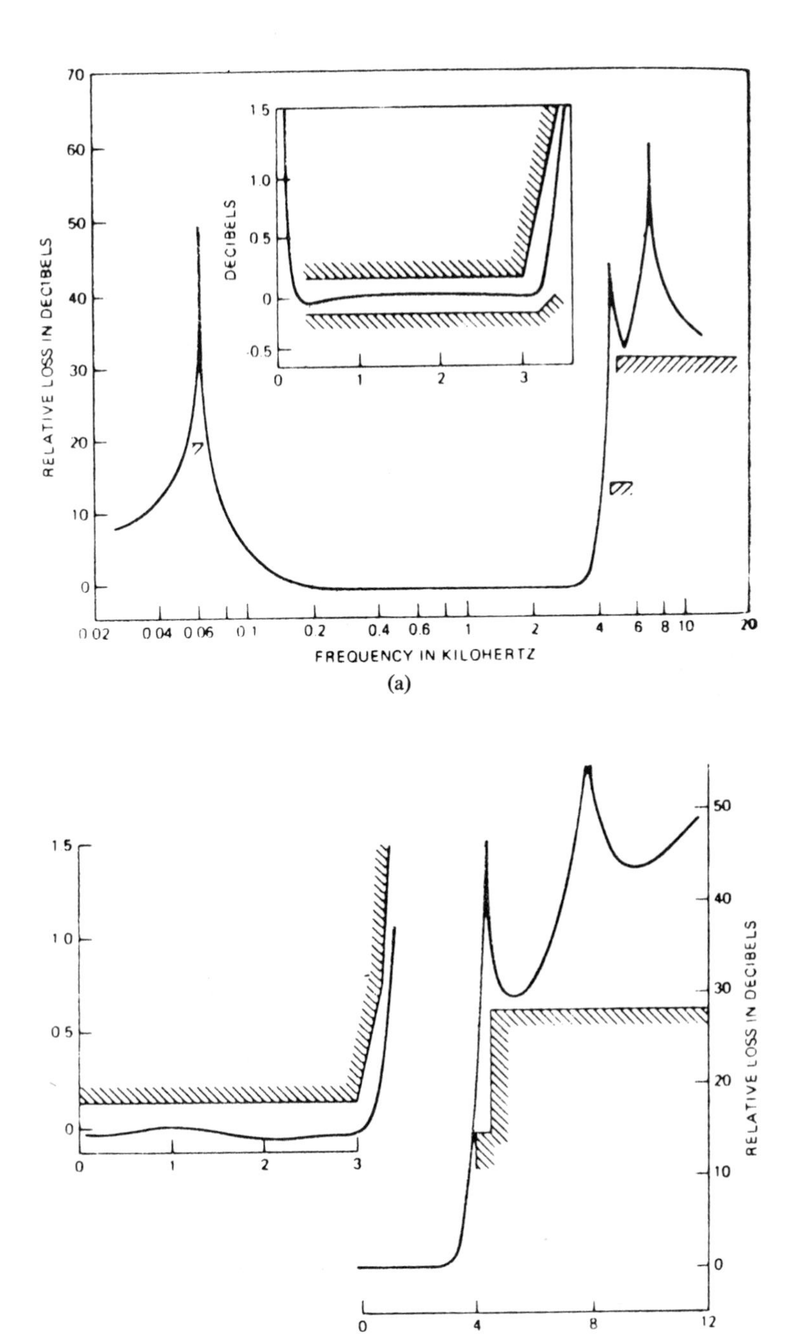

Fig. 3–9. (a) Transmit and (b) receive filter frequency responses of the D4 channel Bank (Courtesy American Telephone and Telegraph Company. Copyright 1982. Reprinted with permission.)

Applications information can be found in Bell System Technical Reference PUB41004[6] and in McNamara.[7]

Thus far, the effect of channel noise on bit error rate has been shown, but the tolerance of PCM to a given bit error rate has not been discussed. Digital facilities generally are rated in terms of error-free seconds and exhibit a bit error rate less than 10^{-7} (at worst, 10^{-5} where long-haul analog circuits may be involved in a tandem connection). PCM, however, works quite well at a bit error rate of 10^{-4}. This requires only a 17 dB signal-to-channel-noise ratio according to Fig. 3–4. Consequently, PCM is insensitive to the occasional errors that may exist on a facility.

In conclusion, PCM, as used on the public switched network and on equivalent private line facilities, provides toll quality voice transmission as well as transparency to other analog signals. Consequently, it can be regarded as providing a "universal" voice channel.

PROBLEMS

3.1 Calculate the number of quantization bits needed for a linear PCM system to provide a 30-dB SNR with a 48-dB dynamic range.

3.2 Explain why the advantages of nonuniform quantization increase with crest factor.

3.3 Explain the difference between asynchronous and synchronous communication. Why is synchronous communication said to be more efficient?

3.4 Explain the difference between bit and frame synchronization.

3.5 Why can a truly optimum fixed quantizer not be devised for speech?

3.6 If digital transmission is used on interoffice trunks for analog voice service, explain why end users still must use modems to transmit data.

REFERENCES

1. Oliver, B. M., Pierce, J. R., Shannon, C. E., "The Philosophy of PCM," *Proceedings IRE,* Vol. 36, pp. 1324–1331 (Nov. 1948).
2. *Transmission Systems for Communications,* Bell Telephone Laboratories, Inc., 1971.
3. Jayant, N. S., "Digital Coding of Speech Waveforms: PCM, DPCM, and DM Quantizers," *Proceedings IEEE,* Vol. 62, pp. 611–632 (May, 1974).
4. Crue, C. R., et al., "The Channel Bank," *Bell System Technical Journal,* Vol. 61, pp. 2611–2664, (Nov. 1982).
5. Adams, R. L., et al., "Thin Film Dual Active Filter for Pulse Code Modulation Systems," *Bell System Technical Journal,* Vol. 61, pp. 2815–2838, (Nov. 1982).
6. "Data Communications Using Voiceband Private Line Channels," Bell System Technical Reference PUB 41004, American Telephone and Telegraph Company, 1973.
7. McNamara, J. E., *Technical Aspects of Data Communication,* Digital Equipment Corporation, 1977.

4
Efficient Speech Coding Techniques

4.1. INTRODUCTION

The desire to conserve spectrum is a very widespread one, prevalent not only with respect to the radiated spectrum, but also significant in terms of the conducted spectrum that exists on any cable transmission facility. Numerous multilevel modulation techniques (see Chapter 6) have been devised to increase the number of bits per second that can be transmitted per Hertz of bandwidth. It should be no surprise, therefore, that significant efforts are being made to encode voice at lower information rates (b/s) than required for standard 64-kb/s PCM, as described in Chapter 3. The fact that 64-kb/s PCM as such generally requires much more bandwidth than does a 3 kHz analog signal is of itself a continuing incentive to seek lower bit rate forms of digitized speech.

Another impetus toward the development of low bit rate speech has come from the desire to store speech elements digitally for future synthesis and output in response to information requests. Speech that can be expressed in the smallest number of bits requires the least amount of digital memory, and thus is the least expensive in terms of storage costs.

What information rate actually corresponds to the spoken word? One could take a cavalier approach and point out that a skilled typist can convert the spoken word to typewritten form, which then can be conveyed by teletype at 75 b/s to a receiving terminal where the result can be read. That demonstration might be felt to prove that the information content of the spoken word is only 75 b/s. In fact, based upon the relative frequencies of occurrence in the English language, each set of distinctive sounds (phoneme) conveys 4.9 bits. In conversational speech about ten phonemes per second are produced. On this basis the bit rate for speech is less than 50 b/s.

What is missing in the foregoing arguments is the fact that the human auditory apparatus can scan the multidimensional attributes of the acoustic image and concentrate on minute details which may have emotional content or information about the physical environment of the source. The ability of a listener to recognize a known speaker generally is denied by these very low bit rate implementations of speech. The true information rate of the spoken word thus must be higher than 50–75 b/s.

Nevertheless, enormous economies in bandwidth might be realizable if some deterioration in quality were acceptable and if the resulting reduction in transmission costs justified the expenses of the pre- and post-processing equipment, or if the available bandwidth constituted a restriction in meeting further demands for service.

As noted at the beginning of Chapter 2, speech can be digitized on a waveform, parametric, or hybrid basis. The waveform coder operates on a basis independent of what the waveform is, i.e., the waveform may be male or female speech, a signaling tone, dial tone, or busy tone. The parametric coder, however, is designed specifically for one type of input, usually speech. Therefore, for example, it may try to determine at any given moment whether its input is voiced or unvoiced and, if voiced, its pitch, its amplitude, and the nature of its frequency content. If the input is speech, the parametric coder does what it is intended to do, perhaps well, perhaps poorly. However, the device is only built to respond to speech. It will attempt to treat a dial tone as if it were speech and may produce something that sounds totally different. Accordingly, the parametric coder can be used for speech only. Nonspeech signals on the system must be routed in other ways to achieve proper system operation.

The hybrid coder often is a waveform coder whose operation has been tailored or optimized for speech in some way. Examples are devices that are built to adapt to the nature of their inputs at a rate that corresponds to the syllabic rate in human speech. Other hybrid coders may contain spectral gaps in their frequency responses, and thus may respond well to tones of some frequencies but not others.

This chapter discusses a variety of coding techniques that have been devised to digitize a speech channel at a rate less than the 64 kb/s commonly used in the public switched network.

4.2. SPECIAL PCM TECHNIQUES

Several techniques have been applied to PCM to enable the transmission of a voice channel at less than 64 kb/s. These techniques include coarsely quantized PCM, often aided by adaptive quantization, as discussed next in Section 4.2.1, and nearly instantaneous companding, discussed in Section 4.2.2. Another technique, digital speech interpolation (DSI), is applied to large numbers of channels (often 40 or more). The discussion of DSI is reserved for Chapter 5.

4.2.1. Coarsely Quantized PCM

An examination of the factors that cause standard PCM to require its 64-kb/s rate shows that there are only two ways of reducing the rate. Either the sampling rate or the fineness of quantization or both must be reduced. Reductions in the sampling rate from 8000 to about 6000 can be achieved with only moderate degradation in quality. Below that the result starts to sound very muffled or distorted because its bandwidth is being narrowed appreciably by a low pass filter or, if not, distortion is occurring due to a problem called *aliasing,* in which unwanted frequency components are produced in the voice band. Reduction of the fineness of quantization can be done to a greater extent without such a serious lowering of the quality. Moreover, *adaptive quantization* can be introduced to help maintain the quality.

When quantization levels are present in abundance, as is the case for $B \geqslant 7$, there is no real need to be concerned about adapting the levels to the input. However, when B is smaller, especially for $B \leqslant 4$, adaptive quantization produces a significant quality improvement. By definition, adaptive quantization is step-size adaptation to the signal variance based on quantizer memory. The step size generally is modified for every new input sample, based on a knowledge of which quantizer slots were occupied by the previous samples. The result is called *adaptive PCM* (APCM). Alternatively, a fixed quantizer can be preceded by a time varying gain which tends to keep the speech level constant.

For example, in the case of a one-word (one sample) memory, let the quantizer output be

$$Y_r = H_r \Delta_r / 2 \tag{4–1}$$

where $\pm H_r = 1, 3, 5, \ldots, 2^B - 1$, $\Delta_r > 0$, $B \geqslant 2$. Thus the various output levels Y_r are governed by the step size Δ_r. The next step, Δ_{r+1}, is chosen according to the equation

$$\Delta_{r+1} = \Delta_r M(|H_r|), \tag{4–2}$$

i.e., a multiplying factor is applied to Δ_r, depending on the magnitude of H_r for the previous step. Thus the adaptation logic matches the step size, at every sample, to an updated estimate of the signal variance.[1]

Figure 4–1 shows a 3-bit quantizer characteristic. Notice that the multiplier M depends upon the step being used by the signal at any given moment, i.e., on the magnitude $|H_r|$. The assumption, valid for speech, is made that the input probability density function $p(x)$ is symmetrical about a mean value of zero.

The general advantage of adaptive quantization is an increased dynamic range for a given number of bits per sample.

Table 4–1 lists step size multipliers for APCM coding of low-pass filtered speech. Note that the negative steps are treated the same as the positive

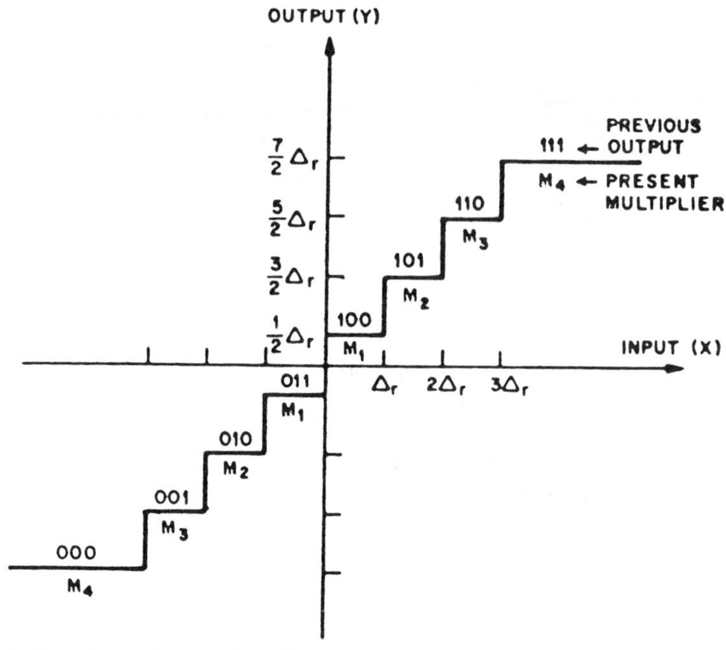

Fig. 4–1. Three-bit adaptive quantizer characteristic: one-word memory. (Courtesy N. S. Jayant, Ref. 1, © IEEE, 1974.)

ones. Consequently, there are two M values for $B = 2$, four for $B = 3$ and eight for $B = 4$. The values are not critical. For example, for $B = 3$, adequate choices are $M_1 < 1$, $M_2 = M_3 = 1$, $M_4 > 1$. However, step size increases should be more rapid than step size decreases, because overload errors tend to harm the SNR more than do granular errors. The general shape of the optimal multiplier function for $B > 2$ is shown in Fig. 4–2.

Table 4–1. APCM Step-Size Multipliers.*

B	2	3	4
M_1	0.60	0.85	0.80
M_2	2.20	1.00	0.80
M_3		1.00	0.85
M_4		1.50	0.90
M_5			1.20
M_6			1.60
M_7			2.00
M_8			2.40

* Courtesy N. S. Jayant, Ref. 1., © IEEE, 1974.

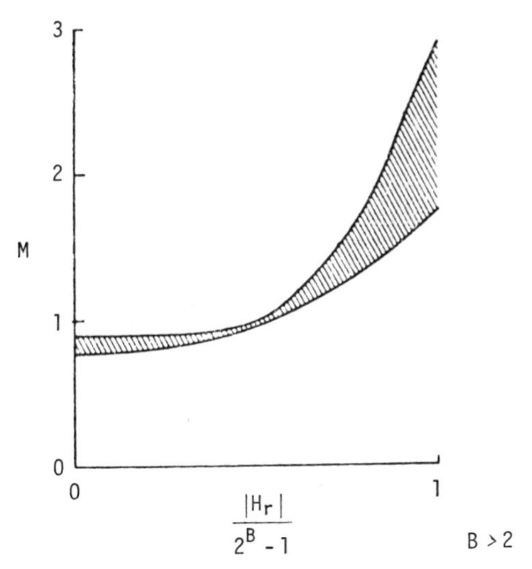

Fig. 4–2. General shape of optimal multiplier function for adaptive quantization of speech. (Courtesy N. S. Jayant, Ref. 1, © IEEE, 1974.)

The shaded region shows the extent to which M may vary for changes in B as well as in input signal characteristics. The curve of Fig. 4–2 does not include the case $B = 2$ because in this case the distinction between the expected magnitudes of granular and overload errors is small.

Other approaches to achieving APCM have been devised. The adaptations may be achieved by switching between two invariant quantizers, with the switching being based on the use of quantizer memory. Alternatively, instead of instantaneous adaptation, as discussed thus far, the adaptation may be done at a syllabic rate. In this case, the step size Δ_r is adapted with a time constant of 5–10 ms, rather than every sample. The disadvantage of the use of syllabic techniques is that they are somewhat more complex to implement than is instantaneous adaptation.

4.2.2. Nearly Instantaneous Companding

Section 3.3 discussed the use of companding in PCM transmission. As described there, the companding is instantaneous in that each speech sample is companded as such. Later, Section 4.4.2 will discuss the use of syllabic adaptation in connection with delta modulation. There, changes in the coding occur at a syllabic rate.

In nearly instantaneous companding (NIC), also known as block companding, the companding is done based upon the amplitude of a small block (e.g., 10) of speech samples.[2] In general, a block of N speech samples is searched to determine which sample has the largest magnitude. Each sample then is encoded (e.g., using six instead of eight bits) with the top of the "chord" of the maximum sample being the overload point. The chord consists of the second, third and fourth bits of a PCM sample, with the first bit designating the sign of the sample. The encoding of the chord is sent to the receiver together with the sample encoding. Thus the scaling information is sent only once for each block of N samples rather than with each sample. The name variable quantizing level (VQL) is used for this technique.

In general, for a block of N samples and n-bit encoding the resulting bit rate is $2W(n - 2 + 3/N)$ b/s. Thus for $W = 4$ kHz, $n = 8$ and $N = 10$, the bit rate is 50.4 kb/s. Some NIC systems (VQL) operate[3] at 32 kb/s.

NIC can be achieved by processing 15-segment μ-law or 13-segment A-law PCM for a reduced bit rate. As such, it is a form of APCM, or digital automatic gain control, with the quantization being nearly uniform and the step size being varied adaptively based upon the short-term signal level. Because speech power varies at a syllabic rate, the transmission of the power level once per block is adequate. The quantization of all samples within a given block is done with similar accuracy, thus avoiding the fine quantization of a small sample when a neighboring sample is much larger and thus more coarsely quantized, as is done in μ-law and A-law PCM.

The performance of NIC is found to be somewhat better than that of $(n - 1)$-bit companded PCM, and is largely insensitive to the statistics of the input signal, unlike many other low-bit-rate speech coding techniques. At 6.3 bits/sample, NIC speech exhibits as much as a 6 dB SNR improvement over μ255 PCM. An overall SNR $\geqslant$ 30 dB is achieved. The group delay is less than 50 μs over the 200 to 3000 Hz band.

Modem transmission at 9600 b/s can be achieved at a 10^{-6} ber on conditioned voice channels when NIC is used.

In the presence of channel noise, the sign bits are the most vulnerable to noise, as is the case with standard PCM. The scaling information is affected less often with NIC, but an error affects an entire block of samples rather than one sample only. As a result, NIC produces fewer "pops" with smaller amplitude. Listening tests at a channel bit error rate of 10^{-3} show that NIC sounds better than standard PCM.

The achievement of NIC encoding requires an inherent delay of one block at the transmitter. For $N = 10$ and $2W = 8$ kHz, this delay is 1.2 ms. Such delay may result in echo problems, since echo suppressors generally are installed at lengths of 3000 km, based upon a median incremental round-trip delay of 8 μs per kilometer of trunk length. Thus, a 2.4 ms round-trip processing delay corresponds to a 300 km foreshortening of the nominal 3000 km trunk length.

4.3. DIFFERENTIAL PCM

The formation of a differential waveform was discussed briefly in Chapter 2. It is produced by subtracting from the present value a recent past value or values, or some weighted combination of them. The basis for the quantization of differential waveforms is that once the initial level of a waveform has been established, the information content is expressed by the changes in value of that waveform. Accordingly, only the changes need to be transmitted. As a result, differential PCM (DPCM) can be used at a lower bit rate than PCM with a comparable quality of reproduction. The sampling rate, of course, is presumed to be at the Nyquist rate.

Figure 4–3 shows a typical DPCM system in block diagram form. As can be seen there, instead of direct sampling of the input, as was done in PCM, the difference between the input and a prediction signal (based on past samples) is sampled and coded. For this reason, DPCM also is known as *predictive coding.* The "prediction signal" is derived in the same way as the receiver detects the signal, and is performed in the box labeled "Integrator."

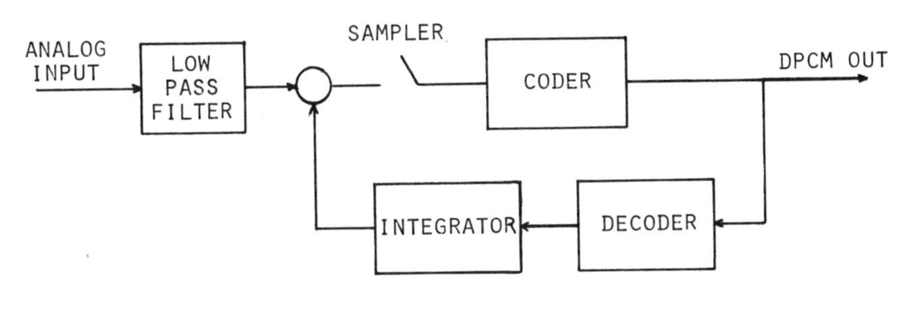

TRANSMITTING SECTION

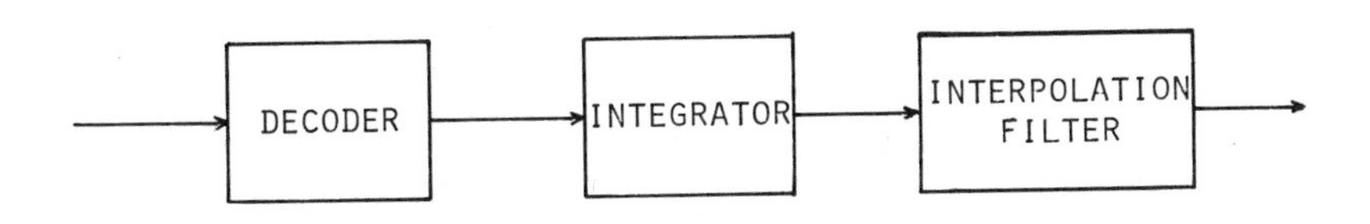

RECEIVING SECTION

Fig. 4–3. Block diagram of DPCM system using analog differencing.

Since the quantizer's input is analog but its output digital, the formation of the differential waveform requires some discussion. Figure 4–3 shows that the quantizer consists of a sampler and a coder. The coder converts sampled amplitude values to specific bit combinations. Thus an analog-to-digital conversion is performed. The decoder followed by the integrator performs the conversion back to analog. The result is formation of the differential on an analog basis, i.e., analog differencing and integration. Note that the coder and decoder need handle only the dynamic range of the difference signal.

An alternative approach is illustrated in Fig. 4–4. Here the digital form of the difference signal is summed and stored in a register, thus producing a digital representation of the previous input sample. Following this step, a decoder and integrator produce an analog version of the signal for subtraction from the analog input. The decoders of Fig. 4–4 thus must be able to handle conversions over the entire dynamic range of the input.

A third approach is illustrated in Fig. 4–5. In this approach the coder must handle the full dynamic range of the input. From the coder's output

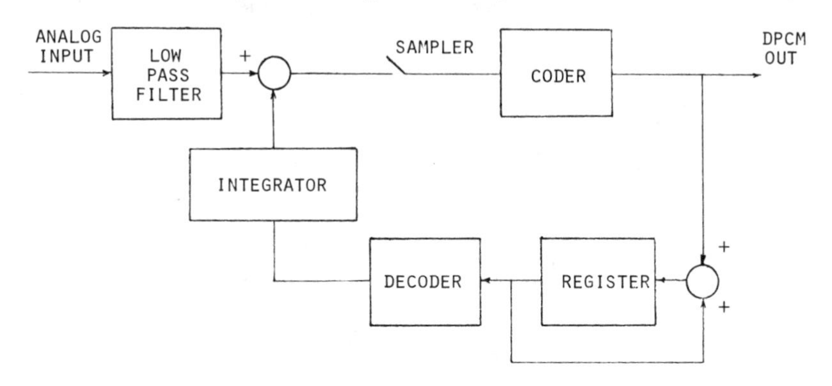

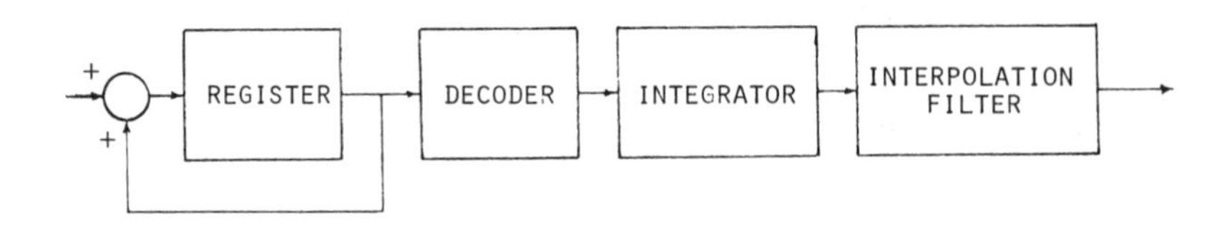

Fig. 4–4. Block diagram of DPCM system using digital integration.

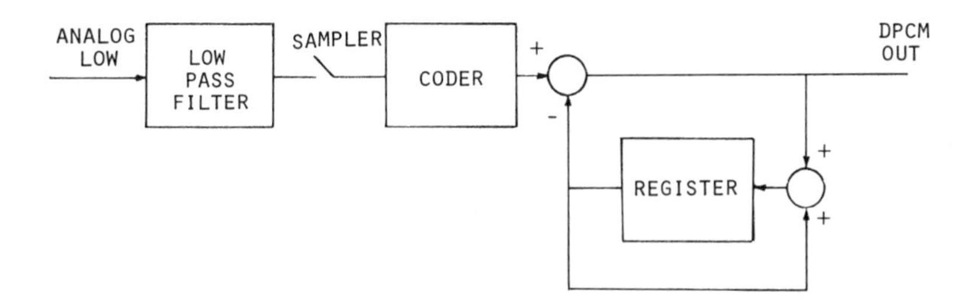

TRANSMITTING SECTION

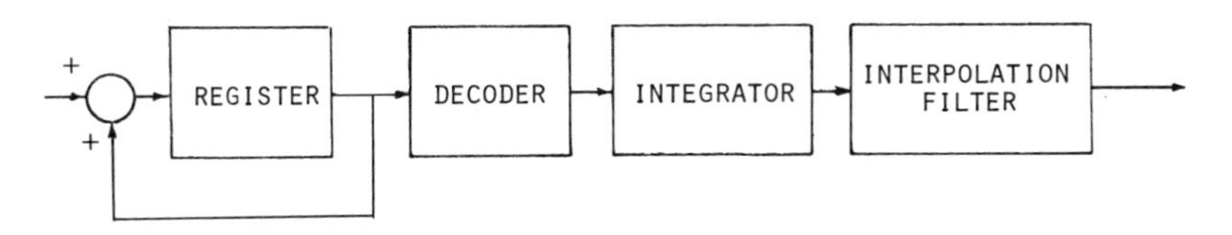

RECEIVING SECTION

Fig. 4–5. Block diagram of DPCM system using digital differencing.

is subtracted a digitally produced approximation of the code representing the previous input amplitude.

The implementations of Figs. 4–4 and 4–5, being digital, are insensitive to minor changes in circuit constants, and are easily reproduced on a large quantity basis. The advantage of the analog approach of Fig. 4–3 is its simplicity when built using lumped components, as well as its modest cost because of its relatively small dynamic range.

In Figs. 4–3, 4–4, and 4–5, the receiving section consists of components identical to those in the feedback path of the transmitting section. This results from the fact that the feedback path built into the transmitter is intended to predict the receiver's output and will, in fact, in the absence of channel errors.

4.3.1. Basic DPCM Concepts

In treating DPCM, the concept of correlation is useful. The *autocorrelation function* $\phi(\tau)$, usually called the *correlation function,* is a measure of the extent to which two values of a function $f(t)$, separated by a time τ, are related to one another. It is defined as

$$\phi(\tau) = \int_{-\infty}^{\infty} f(t)f(t-\tau)\,dt, \qquad f(t) \to 0 \text{ as } t \to \infty \tag{4–3}$$

For zero separation in the time domain, $\tau = 0$, $\phi(0)$ expresses the energy of the function. If $f(t)$ does not go to zero for large values of t, then the following alternative expression must be used:

$$\phi(\tau) = \lim_{t \to \infty} \frac{1}{2T} \int_{-T}^{T} f(t) f(t - \tau)\, dt \tag{4–4}$$

Speech sampled at the Nyquist rate exhibits good correlation between successive samples. Let τ_0 be the time interval between successive samples. For 8000 samples/second, $\tau_0 = 125\ \mu s$. Then the variance of the first difference of the samples, $D_r(1)$, is

$$D_r(1) = X_r - X_{r-1} \tag{4–5}$$

where X_r is the present input and X_{r-1} is the immediately preceding input, i.e., τ_0 seconds ago. Letting $\langle\ \rangle$ denote the mean value,

$$\begin{aligned} D_r^2(1) &= \langle (X_r - X_{r-1})^2 \rangle \\ &= \langle X_r^2 \rangle + \langle X_{r-1}^2 \rangle - 2\langle X_r X_{r-1} \rangle \\ &= \langle X_r^2 \rangle [2(1 - C_1)] \end{aligned} \tag{4–6}$$

where C_1 is the value of the correlation function at $\tau = \tau_0$. Thus if $C_1 > 0.5$, the variance of the first difference $D_r(1)$ clearly is smaller than the variance of the speech signal itself. Actually, a variance reduction can be demonstrated[1] for all values of C_1. The quantization error power is proportional to the variance of the signal present at the quantizer input. Therefore, by quantizing $D(1)$ instead of X and using an integrator to reconstruct X from the quantized values of $D(1)$, the result is a smaller error variance and thus a better SNR. Alternatively, for a given SNR, DPCM permits a smaller B and thus operates at a lower bit rate.

Note that in the three approaches to DPCM (Figs. 4–3, 4–4, and 4–5) the differential is obtained by feeding back the estimate of the receiver's output to the transmitter's input. As a result, quantization errors do not accumulate. If quantization errors should cause the feedback signal to drift away from the input, the next encoding of the difference signal corrects automatically for the drift. Channel errors, of course, will cause receiver output deviations, but these deviations do not persist because of the finite time constants used in the integrator.

How many past samples of a speech waveform should be used in obtaining a predicted value? Certainly some improvements are to be expected through the use of more than one correlation coefficient. The answer depends upon the extent to which significant correlation exists among successive samples. Fig. 4–6 illustrates the autocorrelation functions of speech signals from two male and two female speakers. Sampling was done at 8 kHz and the correlations were done over a 55-second duration. Thus the correlation index i = 1 corresponds to 125 μs. The upper curves show the results with a 0–3400 Hz filter in use, while the lower curves show the results with a 300–3400-Hz filter. As expected, the presence of lower frequency components causes high correlation up to greater values of the correlation parameter, where $C_1 = \varphi(\tau_0)$, $C_2 = \varphi(2\tau_0)$, ... $C_i = \varphi(i\tau_0)$.

The gain G of a DPCM system is the number of dB by which the SNR is improved over a corresponding PCM system. Figure 4–7 shows values of G obtained as a function of the number of predictor coefficients n used in the system. In each case, the maximum, average, and minimum for four speakers are plotted. The low-pass filtered speech (0–3400 Hz) provides greater gain than the bandpass filtered speech (300–3400 Hz) because of the greater low-frequency energy of the low-pass filtered speech and hence the greater adjacent sample correlations.

The gain advantage achieved by differential encoding is directly related to its redundancy removal. The disadvantage of DPCM is that changing input speech material generally degrades the performance of a DPCM system operating with a predictor that has been designed on the basis of average speech statistics. One rather significant problem is that the voiced segments of speech usually have $C_1 > 0.5$ while noiselike fricatives usually have $C_1 \simeq 0$. Voiced segments occur more frequently than unvoiced segments, so

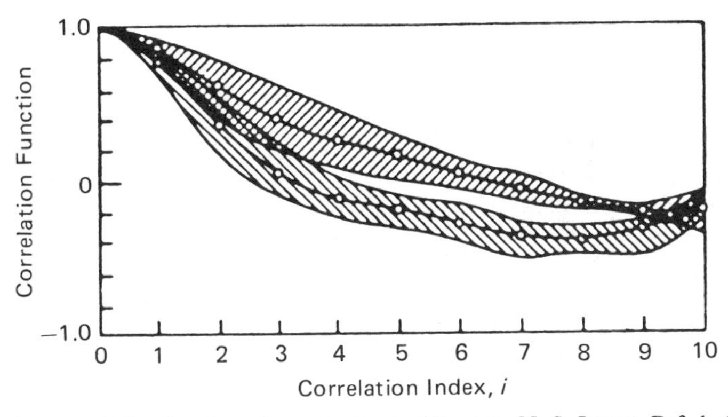

Fig. 4–6. Autocorrelation function of speech signals. (Courtesy N. S. Jayant, Ref. 1, © IEEE, 1974.)

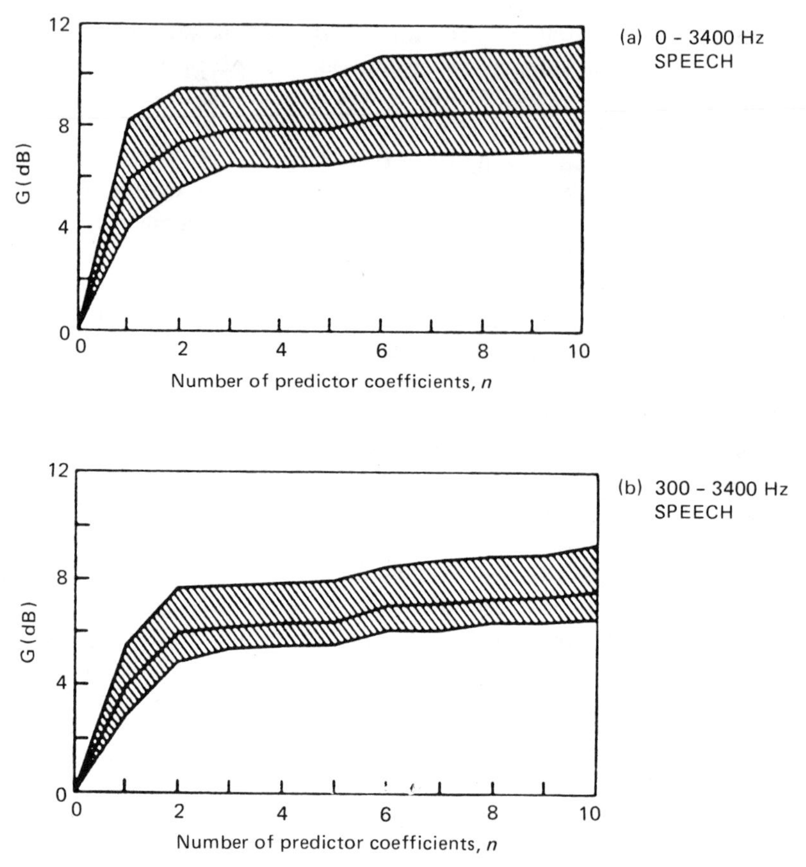

Fig. 4–7. Gain of DPCM system versus number of predictor coefficients. (Courtesy N. S. Jayant, Ref. 1, © IEEE, 1974.)

the way they are coded tends to control overall speech quality. Fricatives (unvoiced sounds) are relatively tolerant to granular quantization noise.

The foregoing problems make the use of adaptive quantizers and predictors highly desirable on DPCM systems.

4.3.2. Adaptive DPCM (ADPCM)

The adaptive quantizer follows changes in signal power. In fact, the one-word adaptive quantizer, as described for APCM, is applicable also for ADPCM. The step size multipliers, however, are slightly different from those for APCM, and are shown in Table 4–2. These ADPCM step size multipliers are very close to the theoretical optimum for white gaussian signals, for which $C_1 = 0$. The reason is that the input samples of a DPCM quantizer are less correlated than those of a PCM quantizer because of the differentiating process.

In addition to an adaptive quantizer, an adaptive predictor (part of the integrator function) is used in ADPCM. It responds to changes in the short-term spectrum of speech. There are two methods for adapting the predictor coefficients to changes in the short-term spectrum of speech. One, called *backward-acting prediction,* measures the power level of a section of the speech waveform, often a syllable or less in length. The resulting information is used to determine what gain should be applied to the following speech sections. This method is satisfactory if the power levels change relatively slowly.

The other method for adapting the predictor coefficients is called *forward-acting prediction.* This method uses the following steps:

1. Store finite sections of the speech waveform (typically a syllable in length).
2. Calculate the correlation function for each section.
3. Determine the optimum predictor coefficients. (The set of coefficients is sometimes called the *vector*).
4. Update the predictor at intervals based on the length of the stored section.

The predictor coefficients are then sent to the receiver as low bit rate "housekeeping" information. Because these coefficients can be coarsely quantized and are updated only at a syllabic rate, the amount of channel capacity they require is a very small percentage of a full voice channel.

Table 4–2. ADPCM Step-Size Multipliers.*

B	2	3	4
M_1	0.80	0.90	0.90
M_2	1.60	0.90	0.90
M_3		1.25	0.90
M_4		1.75	0.90
M_5			1.20
M_6			1.60
M_7			2.00
M_8			2.40

* Courtesy N. S. Jayant, Ref. 1, © IEEE, 1974.

The advantage of forward-acting prediction is that the encoder and decoder use gain values that are directly related to the speech sections from which they are determined. Such devices as shift registers and charge-coupled devices allow forward-acting prediction to be done in a straightforward manner.

The use of the adaptive predictor means that variable predictor coefficients can be used to achieve a greater SNR gain over standard PCM than would be effective with a fixed predictor. These coefficients are derived from the correlation coefficients. Table 4–3 shows that for only one coefficient ADPCM has nearly the same gain as DPCM, whereas an increased number of coefficients results in an appreciable gain of ADPCM over DPCM.

A gain of about 20 dB can be achieved by means of the Atal-Schroeder adaptive predictor,[4] which exploits the quasi-periodic nature of the speech signal to obtain a more complete signal prediction than is provided by those predictors using ten or fewer coefficients. This requires an increased amount of computation in order to determine the pitch period. In the Atal-Schroeder method, the signal redundancy is removed in two stages. One predictor removes the quasi-periodic nature of the signal while another removes the formant information from the spectral envelope. (See Section 4.8, "Vocoders," for a definition of formant.) The first predictor consists of a gain and a delay adjustment. The second one provides a linear combination of the past values of the first predictor's output. The term *adaptive predictive coding* (APC) often is used to describe the Atal-Schroeder technique. (See Section 4.6).

Other approaches to exploiting the quasi-periodic nature of speech provide useful speech reproduction for the low bit rate applications discussed in Section 4.8. Some involve the replication of an entire pitch-period long segment of speech. The limitation of this approach is the existence and perceptual significance of very small variations of the pitch period in voiced speech.

Because ADPCM is capable of providing toll-quality speech at 32 kb/s, it is being seriously considered as a potential replacement for 64-kb/s PCM for interoffice transmission, not only in the U.S. but internationally by the CCITT. For this purpose, it must be able to handle both speech and data signals without their being previously identified. In addition, it must be able

Table 4–3. Comparison of Nonadaptive and Adaptive Predictors (SNR Gain over PCM, dB).*

Order of predictor	1	3	5	10
Nonadaptive DPCM	5.4	8.4	8.6	9.0
ADPCM, 4 ms update	5.6	10.0	11.5	13.0
ADPCM, 32 ms update	5.6	9.8	11.1	12.6

* Courtesy N. S. Jayant, Ref. 1, © IEEE, 1974.

to handle many tandem encodings. Other requirements include the ability to perform well in the presence of analog impairments, and with minimum coding delay to minimize potential echo problems. Moreover, moderate levels of channel noise must not cause adverse effects. Simplicity is important in keeping costs commensurate with the savings in bandwidth.

To enable the handling of data as well as speech, a device called a dynamic locking quantizer (DLQ) has been devised with two "speeds" of adaptation.[5] The rapid power variations of speech cause it to operate in its "unlocked" mode, while the relatively constant power of data causes it to operate in its "locked" mode. In addition, an adaptive transversal predictor with adaptive tap coefficients tracks the nonstationary sample correlations of speech waveforms, providing adaptation to speech, data, or tone signals.

Performance tests at 32 kb/s of ADPCM/DLQ show it to have an SNR about 2 dB lower than that of 64-kb/s PCM for tones. However, 32-kb/s ADPCM/DLQ fully meets CCITT Recommendation G.712, Total Distortion, Method I (Narrowband Noise). Listener tests also show good to excellent opinions even after eight encodings for synchronously operated 32 kb/s ADPCM/DLQ. In the presence of channel errors, 32-kb/s ADPCM/DLQ outperforms 64-kb/s PCM for error rates higher than 1.5×10^{-4}.

4.4. DELTA MODULATION (DM)

Delta modulation (DM), more correctly known as delta coding, is a very simple coding technique to implement. It is a one-bit version of DPCM, which means that it uses only a two-level quantizer. The output bits convey only the polarity of the difference signal. A delta modulator uses a simple first-order predictor (an integrator). To compensate for the fact that the quantizer has only two levels, the adjacent sample correlation is increased by oversampling, i.e., sampling at a rate well beyond 2W.

4.4.1. Linear Delta Modulation (LDM)

The principle of linear delta modulation is illustrated in Fig. 4–8. The delta modulator approximates the input time function by a series of linear segments of constant slope, hence, the name linear delta modulation (LDM). If the Y_r denote successive output bits and the X_r denote the corresponding sampled input levels, then

$$Y_r - Y_{r-1} = \Delta b_r \tag{4–7}$$

where $b_r = \text{sgn}\,(X_r - Y_{r-1})$, i.e., b_r, the transmitted channel symbol (either a 0 or a 1), is determined by whether the input is greater than, or less

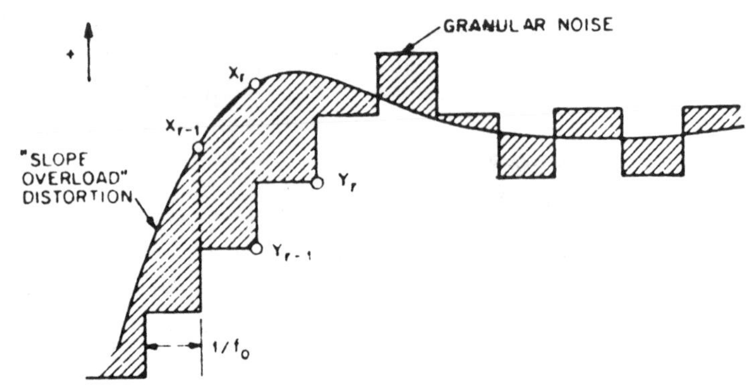

Fig. 4–8. Principle of linear delta modulation. (Courtesy N. S. Jayant, Ref. 1, © IEEE, 1974.)

than, the accumulated (integrated) value of the output. Δ is the size of the step.

Two types of problems can arise in the production of delta modulation. First, the input can change so rapidly that the output cannot keep pace with it. This is called *slope overload* distortion. The solution might seem to be that of designing in a larger step size Δ. However, this would aggravate the second problem, *granular noise.* During relatively flat inputs, the quantizer tends to jump between its two levels. The larger Δ is, the greater is the granular noise.

Figure 4–9 is a block diagram of a linear delta modulator. The input signal is sampled at a rate $f_0 \gg 2W$. Then a staircase approximation to the input, as in Fig. 4–8, is constructed. This is done in the accumulator, which feeds back to the quantizer input the previously received value of Y_r, i.e., Y_{r-1}. Note that the feedback circuit of the transmitter constitutes a replica of the receiver, except for the output filter. This is a characteristic of all of the differential-type digital coding techniques. At the receiver, the value Y_r is incremented by a step Δ in the direction b_r at each sampling instant. Then Y_r is low-pass filtered to the original signal bandwidth. The low-pass filter rejects out-of-band quantization noise in the staircase function Y_r.

Thus, for each signal sample, the transmitted channel symbol is the single bit b_r. The bit rate equals the sampling rate f_0.

If ΔV is the height of the unit step in volts, the quantized noise power using a single integration[6] is

$$N_0 = 2W(\Delta V)^2/3f_0 \qquad \text{watts} \tag{4–8}$$

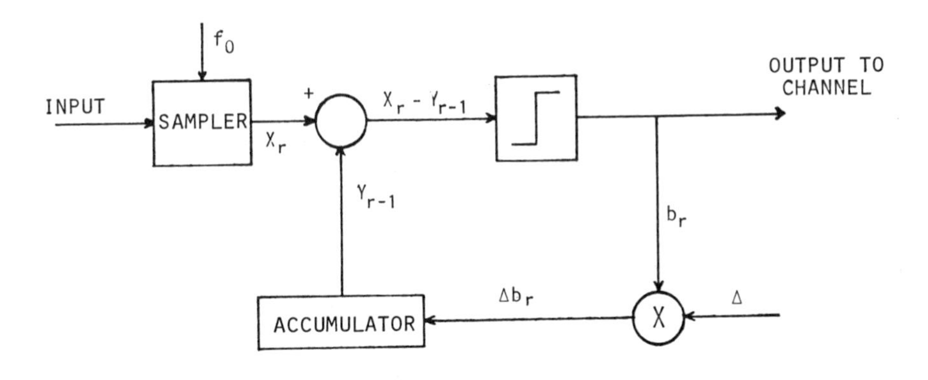

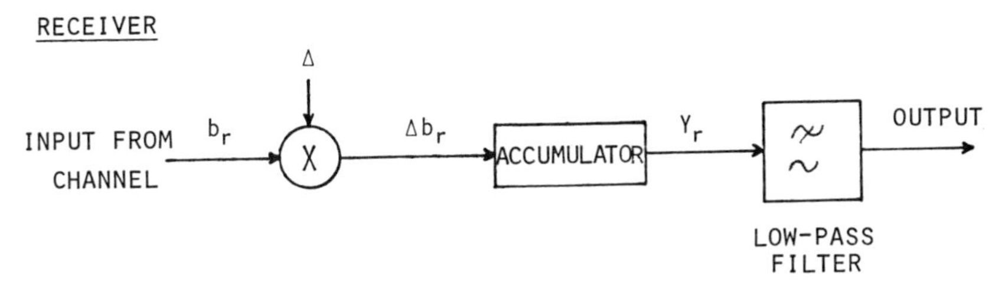

Fig. 4–9. Linear delta modulator block diagram.

The maximum amplitude at frequency f that can be transmitted without overloading in a single-integration coder[7] is

$$A = f_0 \Delta V / 2\pi f \qquad \text{volts} \tag{4–9}$$

The average signal power[8] is

$$S_0 = f_0^2 (\Delta V)^2 / 4\pi^2 f^2 \qquad \text{watts} \tag{4–10}$$

If T expresses the sample time, $1/f_0$, then the performance of linear delta modulation compared with that of 2-bit DPCM can be displayed as shown in Fig. 4–10.

The SNR of linear delta modulation depends significantly on the normalized step size, as indicated in Fig. 4–11. Shown there are the SNR results for an input signal that was gaussian band-limited flat to $1/2T$ Hz with unit power. The values of T are in time units. This figure shows that by doubling

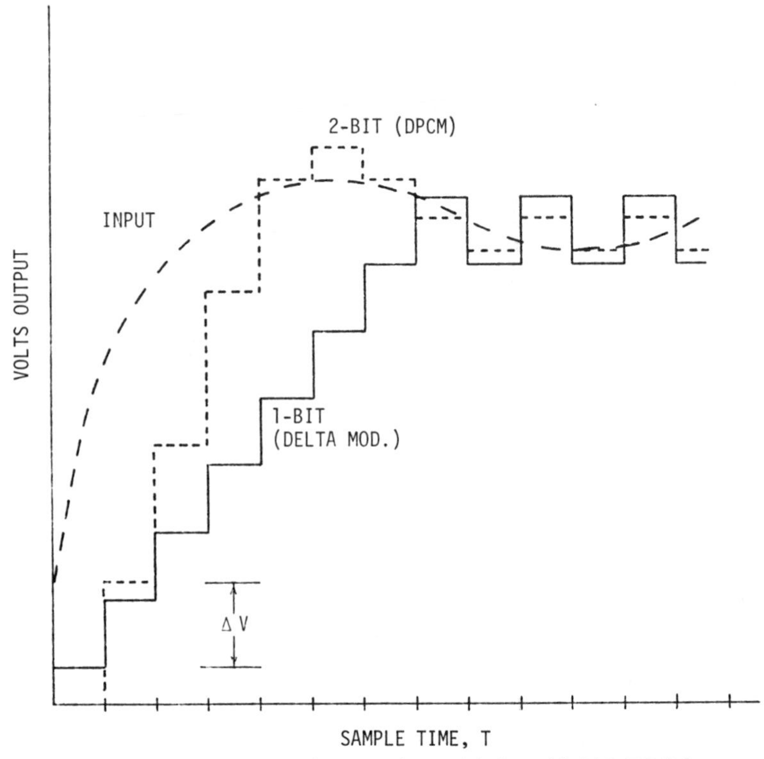

Fig. 4–10. Comparison of linear delta modulation with 2-bit DPCM.

the output digital rate, the SNR is improved by nearly 9 dB (theoretically 9 dB). This compares with an improvement of only 6 dB in the case of PCM.

For a sine wave of frequency f_s Hz, delta modulation with a first-order predictor (single integrator) in the feedback loop[9] provides a SNR of

$$\text{SNR} = 10 \log_{10} (f_0^3/f_s^2 W) - 14 \qquad \text{dB} \qquad (4\text{–}11)$$

For a sine wave of frequency f_s Hz, delta modulation with a second-order predictor (double integrator) in the feedback loop[10] provides an SNR of

$$\text{SNR} = 10 \log_{10} (f_0^5/f_s^2 f_m^2 W) - 32 \qquad \text{dB} \qquad (4\text{–}12)$$

where f_m is the frequency above which second-order prediction takes place in the feedback loop.

The advantage of double integration is that it provides a nearly constant

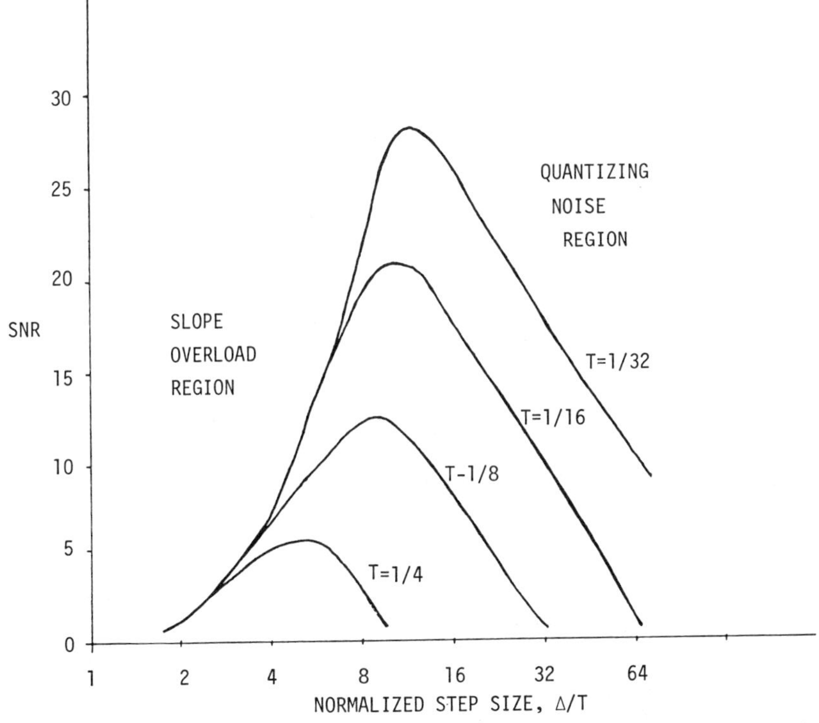

Fig. 4–11. Performance of linear delta modulation. (Copyright 1982, Bell Telephone Laboratories. Reprinted by permission.)

overload probability for all components of the average speech spectrum. In addition, its gain is 15 dB/octave. The disadvantage is possible instability when it is used with a very strong instantaneously adaptive logic such as the one-bit memory scheme, as will be discussed under adaptive delta modulation. The solution to this problem is to increase the length of the quantizer memory or to use delayed encoding in which the magnitudes of the higher order predictor coefficients are increased.

The optimum step size Δ_{opt} is given by Abate's Rule:[11]

$$\Delta_{opt} = \langle (X_r - X_{r-1})^2 \rangle^{1/2} \ln 2F \tag{4–13}$$

The results of an experimental verification of Abate's Rule are shown in Fig. 4–12. The curves were obtained from a simulation using Gaussian signals with a uniform power spectrum. To the left of the curve peaks the SNR is controlled by slope-overload distortion, whereas to the right of the peaks

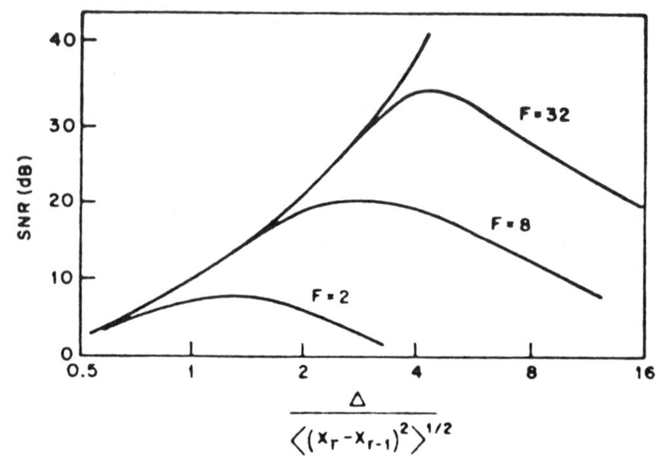

Fig. 4–12. Experimental verification of Abate's Rule. (Courtesy N. S. Jayant, Ref. 1, © IEEE, 1974.)

the SNR is limited by granular noise. The SNR can be seen to increase as F^3, or about 9 dB per octave of bit rate increase.

Companding can be done by varying the step size of a delta modulator. Other variations also can be done to advantage, as will be seen next in the discussion of adaptive delta modulation.

4.4.2. Adaptive Delta Modulation (ADM)

A variable step size can be used to improve the dynamic range of a DM coder. The concept is illustrated in Fig. 4–13. The variable step size increases during a steep segment of the input but decreases while slowly varying portions of the input are occurring.

Observing a single sample of quantizer output does not provide an indication of overload or underload because there are only two quantizer levels. However, step size variation *is* done based on the quantizer output, and thus no separate housekeeping information needs to be transmitted. This is accomplished as follows: Sequences of quantizer outputs of length two or more are inspected to determine how to adjust the step size. If successive bits b_r and b_{r-1} are alike, the assumption is made that slope overload may be occurring. On the other hand, if $b_r \neq b_{r-1}$ the assumption is made that granular noise is being produced. A formula[12] has been devised for adaptation. It is

$$\Delta_r = \Delta_{r-1} P^{b_r b_{r-1}}, \qquad P \geqslant 1 \tag{4–14}$$

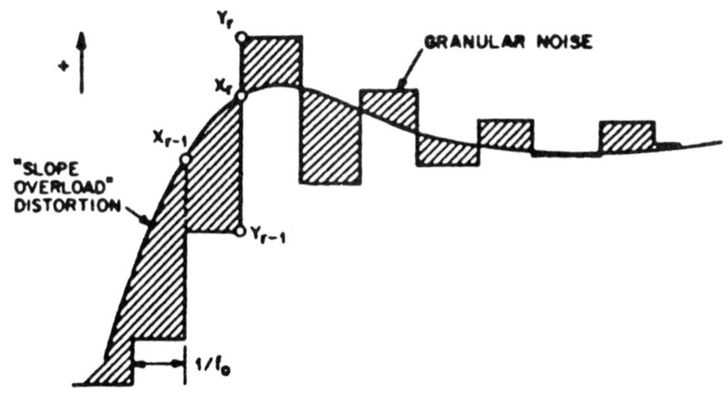

Fig. 4–13. Principle of adaptive delta modulation. (Courtesy N. S. Jayant, Ref. 1, © IEEE, 1974.)

where P is a factor giving the rate of step size increase or decrease. For linear (nonadaptive) delta modulation, $P = 1$. To minimize the quantization error power for speech encoding, the optimum value is $P_{opt} \simeq 1.5$.

Adaptive DM has about a 10-dB SNR advantage[13] over linear DM for $W = 3.3$ kHz, $f_0 = 60$ kHz, and $P = 1.5$. Typical attainable dynamic ranges for ADM are 30–40 dB.

Figure 4–14 is a block diagram of an adaptive delta modulator. Input sampling is done at a rate $f_0 \gg 2W$. Then a staircase approximation to the input is constructed as in Fig. 4–13. This is done in the accumulator, which derives the previously received value of Y_r, i.e., Y_{r-1}. Unlike linear DM, however, the step size Δ (here called Δ_r) is not fixed. Instead, it is computed by adaptation logic based upon a formula such as Eq. (4–14).

One form of adaptation used for ADM is *syllabic adaptation.* The time constant is on the order of 5–10 ms, providing for smooth changes in step size. The result has been called *continuous delta modulation* (CDM). A block diagram is shown in Fig. 4–15. The companding is syllabic because of the use of LPF 1, typically a 100 Hz filter. Step size control is derived from the number of ones in the bit stream. This number reflects the input signal level through the feedforward control provided by the low-frequency envelope signal S_{en}.

Syllabic adaptation decreases the granular noise in the output and thus produces a clean sound, but results in increased slope overload distortion and thus loss of crispness. Since changes are produced at a syllabic rate, the system is more tolerant to channel errors than is ordinary PCM. A significant disadvantage of syllabic adaptation is that it is more complex to

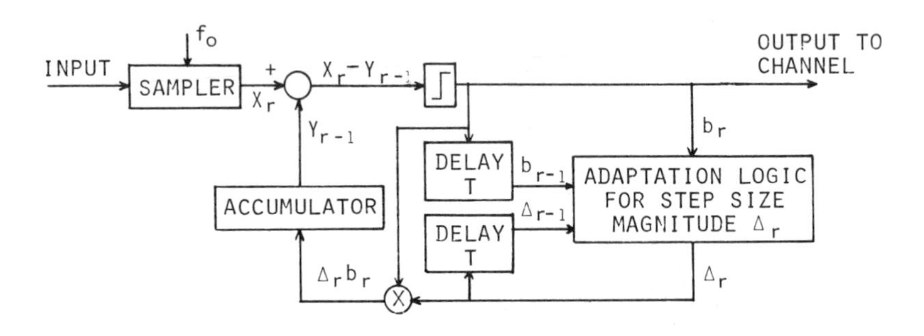

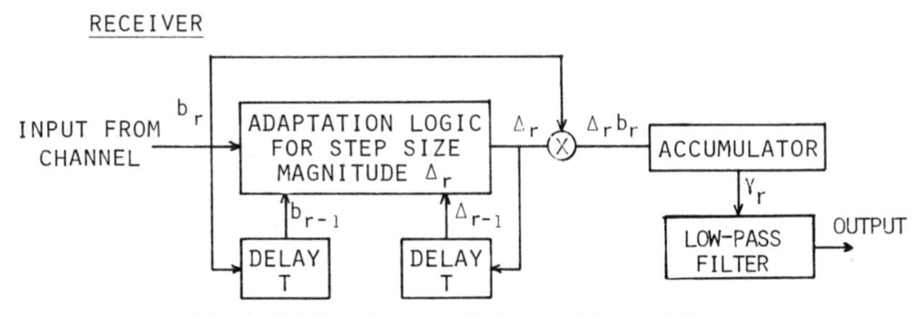

Fig. 4–14. Block diagram of adaptive delta modulator.

implement than is instantaneous adaptation, especially for the time-division multiplexing of several speech channels.

4.4.2.1. Continuously Variable Slope Delta Modulation (CVSD). Variable slope delta modulation is achieved by encoding differences in slope rather than differences in amplitude. The result is a system that is responsive to a wider range of signal voltages than would be the case otherwise.

Variable slope delta modulation is implemented by establishing a set of discrete values of slope variations. For example, in the case of 3-bit error encoding, the slopes are determined by the last three transmitted bits. Each of a set of standard slopes (e.g., 1, 2, 4, 6, 8, 16) is available for selection. As usual, the direction (+ or −) is controlled by the current bit, with one bit being transmitted at each sampling interval. Table 4–4 shows the rules used for changing the slope magnitude.

The most commonly used form of variable slope DM is called *continuously variable slope delta modulation* (CVSD). The slope changes usually are done at a syllabic rate, and may result from either (1) a sensing of an overload when a train of ones or zeros appears in the output, or (2) an exponential

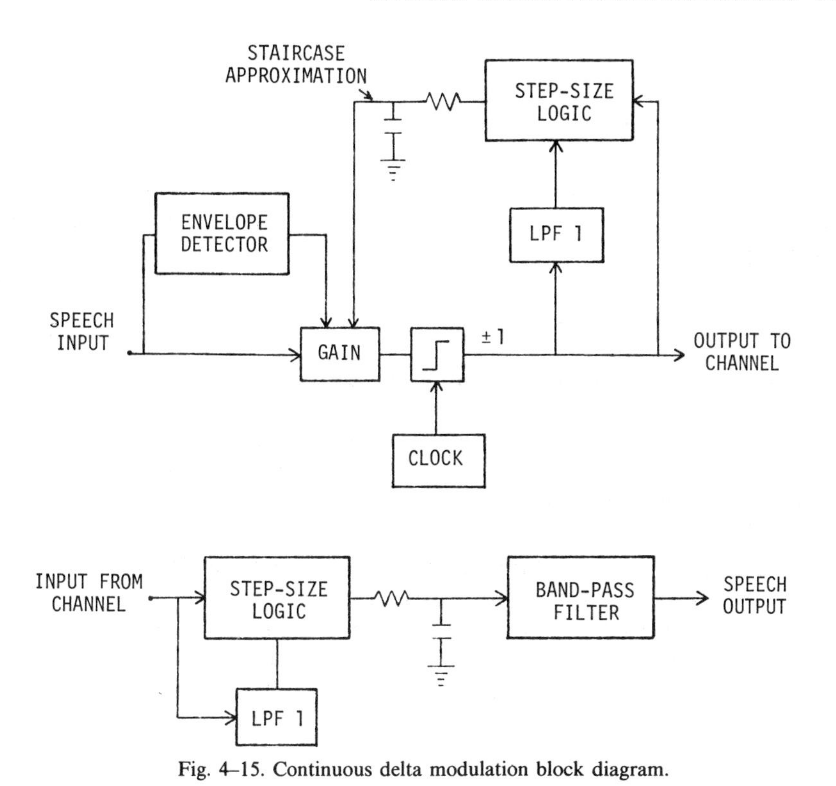

Fig. 4–15. Continuous delta modulation block diagram.

Table 4–4. Variable Slope Delta Modulation Rules.

CURRENT BIT	PREVIOUS BIT	PRE-PREVIOUS BIT	SLOPE MAGNITUDE RELATIVE TO PREVIOUS SLOPE MAGNITUDE
0	0	0	Take next larger magnitude*
0	0	1	Slope magnitude unchanged
0	1	0	Take next smaller magnitude†
0	1	1	Take next smaller magnitude†
1	0	0	Take next smaller magnitude†
1	0	1	Take next smaller magnitude†
1	1	0	Slope magnitude unchanged
1	1	1	Take next larger magnitude*

* If the slope magnitude is already at the maximum, this instruction is not obeyed.
† If the slope magnitude is at the minimum, this instruction is not obeyed.

increase or decrease of the feedback error voltage until the overload pattern ceases to exist, as indicated by a change in bit sense (one to zero or zero to one). A block diagram of a CVSD coder-decoder (codec) is provided in Fig. 4–16.

The CVSD codec senses two or three bit values at a time. The pulse height control function changes the quantization step size of the receiver according to the following rules:

1. If the group (two or three) of successive output bits are of the same sign, the step size is increased.
2. In all other cases the step size is decreased.

Thus the increases and decreases do not occur equally. A syllabic time constant (several ms) is used for the step size decreases. The resulting performance shows lower quantization noise than that of variable slope DM.

4.4.2.2. Digitally Controlled Delta Modulation (DCDM). Digitally controlled delta modulation (DCDM) uses syllabic companding but derives the step-size information directly from the bit sequence,[14] thereby avoiding the need for speech envelope detection as in CDM. For example, a step-size increase may follow the detection of a specified number of consecutive

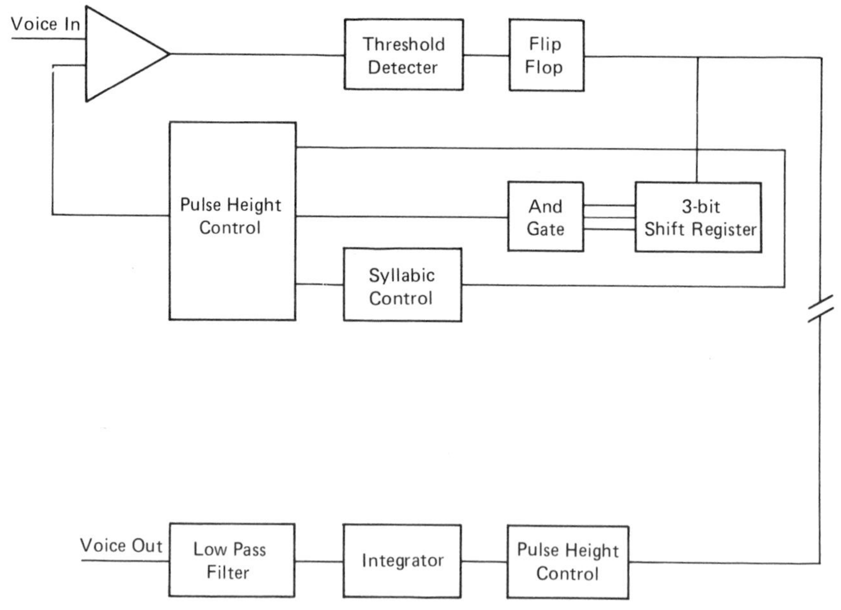

Fig. 4–16. CVSD codec block diagram.

bits of the same polarity. DCDM works best above 16 kb/s. At lower rates, the correlation between adjacent signal samples becomes too low to allow observations on the bit stream to be useful for step-size control. At these lower rates, direct monitoring of the speech envelope provides better control results than observations of the bit stream, but toll quality voice is not achieved.

4.5. SUBBAND CODING (SBC)

The various forms of PCM, DPCM, DM, etc., described in Sections 4.2–4.4 are time-domain coding techniques in that the speech is coded on a time-domain basis. A frequency-domain coder, on the other hand, does its coding based upon the spectral content of speech. The subband coder is a prime example of frequency-domain coding.[15] The speech signal is divided into a number of separate frequency components, each of which is encoded separately. The frequency-domain approach has two distinct advantages. One is that the number of bits used to encode each frequency component can be selected so that the encoding accuracy is placed where it is needed in the frequency domain. The second advantage is that bands with little or no energy may not be encoded at all. The main differences among the frequency-domain techniques usually lie in the degree of prediction used in each.

In subband coding (SBC), as the name implies, the speech spectrum is divided into subbands. This division is based on the fact that the quantizing distortion is not uniformly detectable at all speech frequencies. Thus coding based on subbands offers a possibility for controlling the distribution of the quantization noise. The overall speech band actually is partitioned so that each subband contributes equally to the articulation index (AI), as shown in Table 4–5. The AI indicates the average contribution of each part of the spectrum to the overall perception of the spoken sound.[16] Figure 4–17 shows

Table 4–5. Subbands for Equal Articulation Index.*

SUBBAND	FREQUENCY RANGE (HZ)	CONTRIBUTION TO AI
1	200–700	20%
2	700–1310	20%
3	1310–2020	20%
4	2020–3200	20%
		80%

NOTE: 80% AI corresponds to 93% intelligibility. The remaining 20% would come from frequencies below 200 Hz and above 3200 Hz.

* Reprinted with permission from *The Bell System Technical Journal.* Copyright 1979 and 1976, AT&T.

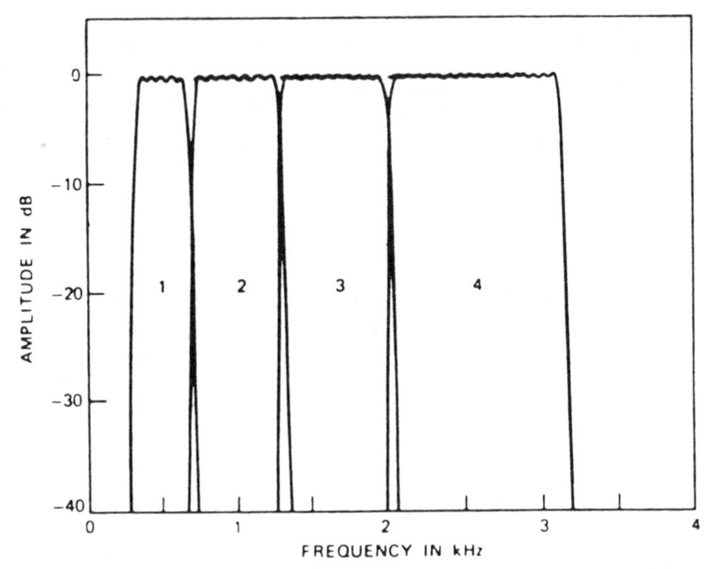

Fig. 4–17. Subband coding-partitioning in the frequency domain. (Reprinted with permission from *The Bell System Technical Journal,* Copyright 1983 as a book, AT&T.)

the partitioning in the frequency domain. SBC is implemented by translating each of the subbands downward in frequency so that the low end of each is essentially dc. This facilitates sampling rate reduction. Figure 4–18 shows the sequence of operations in SBC transmission and reception. The term *decimate* refers to a reduction of the sampling rate consistent with the width of the band to be transmitted. At the receiver, the term *interpolate* refers to the smoothing of the decoder output.

The process shown in Fig. 4–18 takes place simultaneously in each of the four bands, as portrayed in Fig. 4–19. APCM is used in each band because it provides low sample-to-sample correlation of the low-pass translated, Nyquist sampled signals.

The transmission bit rate can be reduced by limiting the subband widths, thus allowing spectral gaps, as shown in Fig. 4–20. If these spectral gaps can be placed in frequency ranges where there is little contribution to intelligibility, their effect will be minimal. Moreover, quantizing distortion is not equally detectable at all frequencies. As the spectral gaps are made larger, however, to reduce the bit rate, the intelligibility begins to suffer. A variable-band SBC coding scheme has been devised[17] which attempts to identify and encode those spectral components (formants) of the signal that are most significant to listeners. The two upper bands are allowed to vary in accordance with the dynamic movement of the vocal tract resonances, as measured by

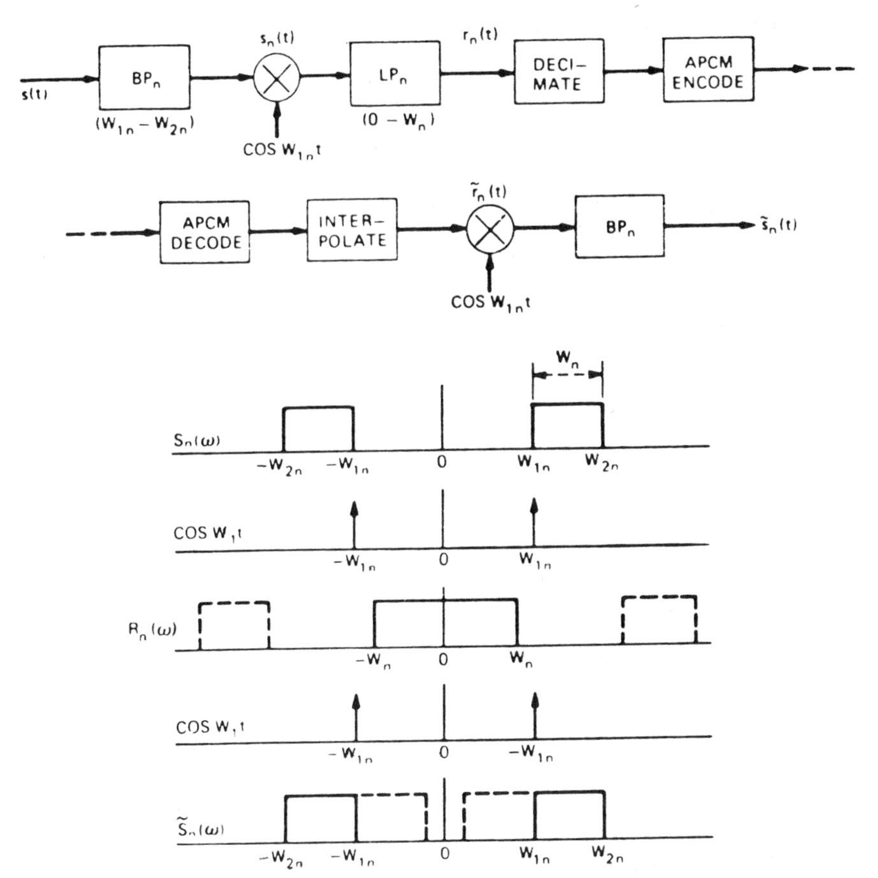

Fig. 4–18. Subband coding-transmission and reception. (Reprinted with permission from *The Bell System Technical Journal,* Copyright 1976, AT&T.)

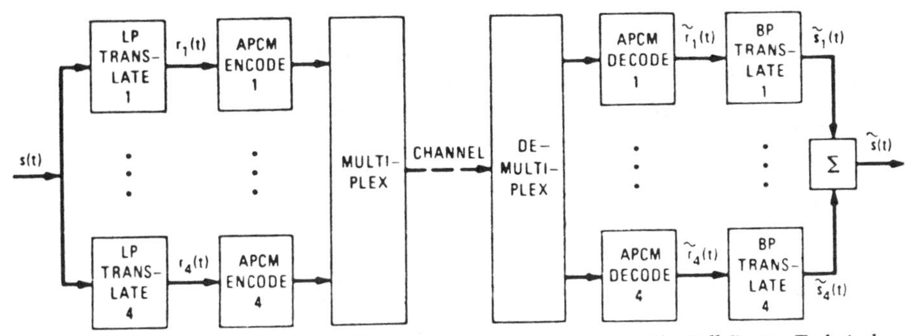

Fig. 4–19. Four-band SBC system. (Reprinted with permission from *The Bell System Technical Journal,* Copyright 1976, AT&T.)

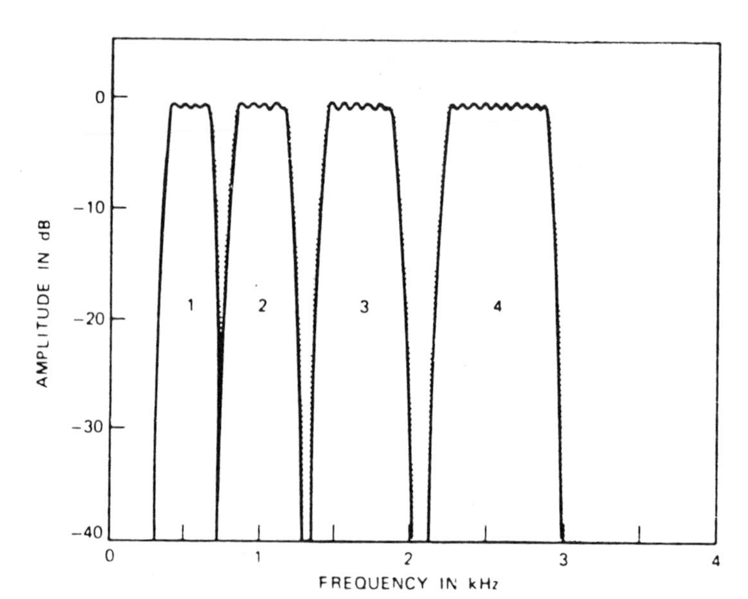

Fig. 4–20. SBC with limited subband widths. (Reprinted with permission from *The Bell System Technical Journal,* Copyright 1979 and 1976, AT&T.)

a zero crossing technique. The result is a 4800 b/s coder which provides moderate quality intelligible speech.

A technique that eliminates the use of the modulators and achieves a theoretical maximum efficiency in sampling is the *integer band sampling technique,* illustrated in Fig. 4–21. These advantages are somewhat offset, however, by the fact that the bands no longer can be chosen only on the basis of their contribution to the articulation index.

In integer band sampling the signal subbands are chosen to have lower cut-off frequencies mf_n and upper cut-off frequencies $(m + 1)f_n$, where m is an integer and f_n is the bandwidth of the nth band. A sampling rate of $2f_n$ is used on each incoming subband. The received signal is recovered by decoding it and then bandpassing it to its original signal band. Table 4–6 shows the frequencies and sampling rates for 16-kb/s SBC, and Table 4–7 shows the same parameters for 9.6-kb/s SBC.

The sampling rate reduction needed to keep the overall bit rate low results in aliasing in each of the subbands. This aliasing is introduced on both sides of each band edge and is called interband *leakage.* The amount of leakage depends upon how sharply the subband filters attenuate out-of-band energy. The solution to this problem[18] is the use of the quadrature mirror filter bank (QMFB). In the QMFB the quadrature relationship of the subband

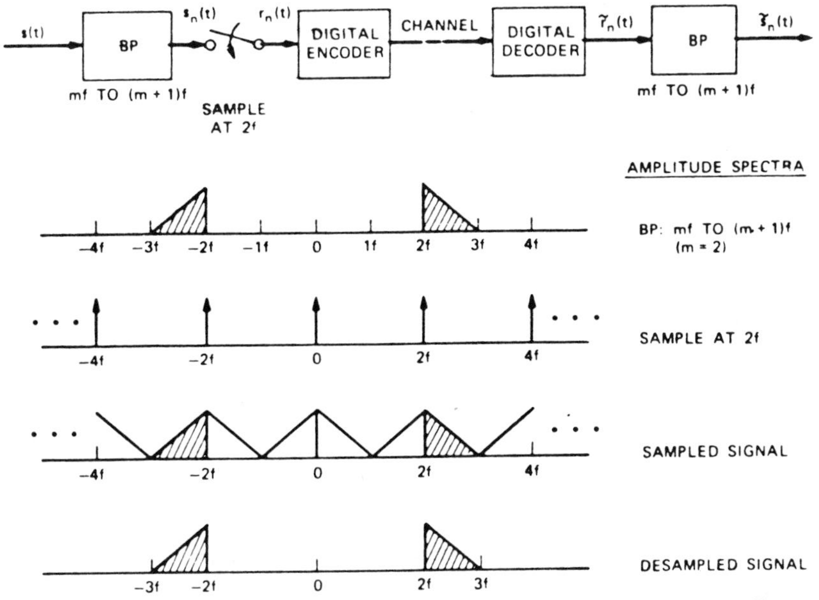

Fig. 4–21. Integer-band sampling for SBC. (Reprinted with permission from *The Bell System Technical Journal,* Copyright 1979 and 1976, AT&T.)

signals results in a cancellation of the unwanted images. This cancellation is achieved down to the quantization noise level of the coders.

SBC compares favorably with both ADPCM and ADM, as shown by the curves of Figure 4–21. These curves are based on the preferences of 12 listeners, each making 16 comparisons. Two ADPCM systems were used, one (3 bits/sample) at 24 kb/s, and the other (2 bits/sample) at 16 kb/s. An ADM coder was operated at 10.3 kb/s, 12.9 kb/s, and 17.2 kb/s to

Table 4–6. Frequencies and Sampling Rates for a 16-kb/s SBC.*

SUBBAND	CENTER FREQUENCY (HZ)	SAMPLING RATE (S^{-1})	DECIMATION† FROM 10 KHZ	QUANTIZATION (BITS)
1	448	1250	16	3
2	967	1429	14	3
3	1591	1667	12	2
4	2482	2500	8	2

NOTE: Bandwidth = 1/(2 × sampling rate).

† Decimation is the sampling rate reduction by an integer factor *m* (retain only one out of every *m* samples of filter output).

* Reprinted with permission from *The Bell System Technical Journal.* Copyright 1976, AT&T.

Table 4–7. Frequencies and Sampling Rates for a 9.6-kb/s SBC.*

SUBBAND	CENTER FREQUENCY (HZ)	SAMPLING RATE (s^{-1})	DECIMATION† FROM 10 KHZ	QUANTIZATION (BITS)
1	448	800	25	3
2	967	952	21	2
3	1591	1111	18	2
4	2482	1538	13	2

NOTE: Bandwidth = 1/(2 × sampling rate).

† Decimation is the sampling rate reduction by an integer factor *m* (retain only one out of every *m* samples of filter output).

* Reprinted with permission from *The Bell System Technical Journal.* Copyright 1979 and 1976, AT&T.

obtain the lower pair of curves of Fig. 4–22. The curves show SBC to be preferable to ADPCM below about 22 kb/s and to be preferable to ADM below 18 to 20 kb/s. Tests of SBC at 7.2 kb/s have shown its quality to be only slightly poorer than at 9.6 kb/s. However, at 4.8 kb/s, SBC's quality was considerably poorer than at 9.6 kb/s because of increased band limiting and the gaps between the bands needed to achieve this lower rate.

SBC may be preceded by *time domain harmonic scaling* (TDHS), which is a method for frequency scaling the short-term spectrum of speech to reduce the bandwidth and thus to reduce the sampling rate and therefore the bit rate.[19] At the receiving end the spectrum is expanded again to its original form. TDHS also can be used independently of other techniques.

One TDHS/SBC coder has been developed using 17 kb/s SBC with 2:1 TDHS. With the addition of side information for pitch and framing, the resulting bit rate is 9600 b/s.

4.6. ADAPTIVE PREDICTIVE CODING (APC)

The principle of predictive coding (see Section 4.31) is to remove redundancy by subtracting from the signal that part which can be predicted from its past. The reason for the adaptive approach is that a fixed predictor cannot predict speech values efficiently at all times because of the nonstationary nature of speech, i.e., the periodicity changes with time.

Predictive coders have been built in which the processed speech has fair voice naturalness and intelligibility even when acoustic background noises are present. In addition, the coding can be done efficiently without the need for large codebook memories. The design principle on which an adaptive predictive coder (APC) is built is that the parameters of a linear predictor are optimized to obtain the minimum mean-square error between the predicted value and the true value of the signal, as is done by the Atal-Schroeder predictor described in Section 4.3.2. Figure 4–23 is a block diagram of an

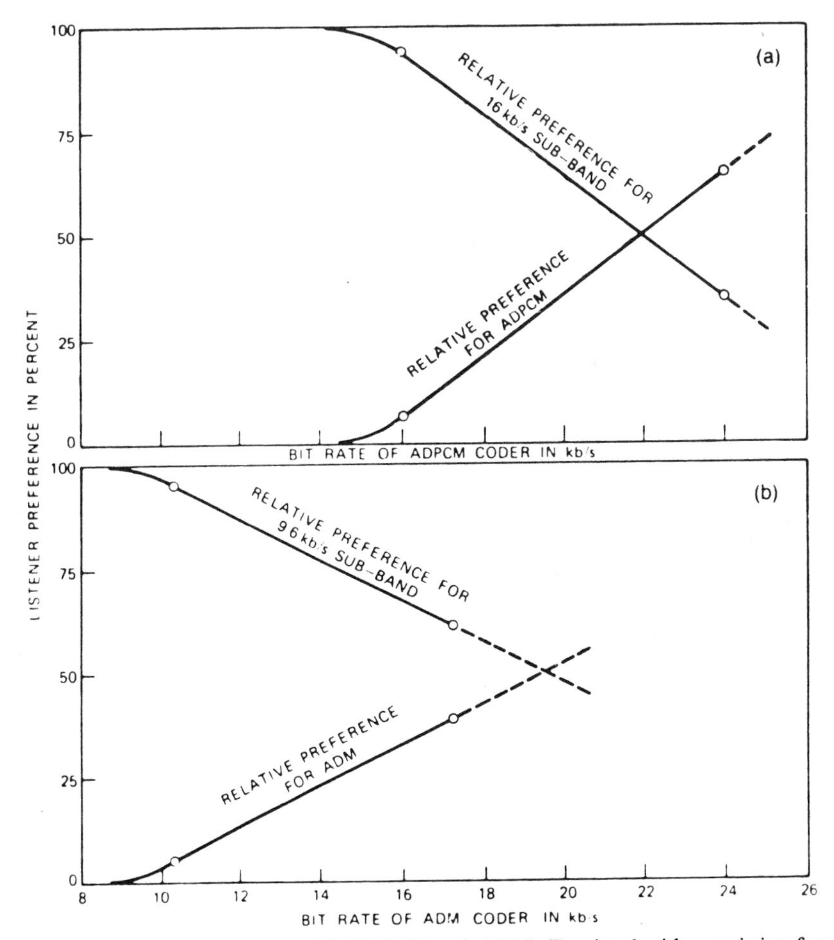

Fig. 4-22. Comparison of SBC with ADPCM and ADM. (Reprinted with permission from *The Bell System Technical Journal*, Copyright 1979 and 1976, AT&T.)

adaptive predictive coder.[20] The sampler operates at the Nyquist rate. The system provides 13 to 20-dB improvement in speech signal-to-quantizing noise (depending on the speaker) over PCM using identical quantizing levels.

As can be seen from the foregoing discussion, APC is primarily a waveform coding technique but one in which prediction parameters are sent separately. Actual digitized speech samples are transmitted. It thus is not to be confused with the parametric coding technique known as linear predictive coding (LPC), described under Section 4.8.1, which does not send digitized speech samples as such, but instead sends parameters which can be used to construct a simulation of the original speech.

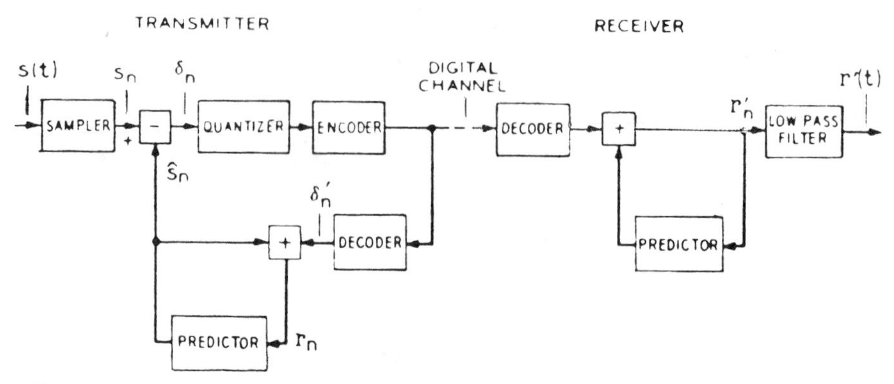

Fig. 4–23. Adaptive predictive coder block diagram. (Reprinted with permission from *The Bell System Technical Journal,* Copyright 1976, AT&T.)

Transmission errors $\leqslant 10^{-3}$ do not degrade the speech perceptibly. The prediction process, however, may produce errors in the predicted pitch resulting in a reverberant sound, although occasional pitch errors do not severely affect voice naturalness. Granular noise results from 1-bit quantization of the error signal. The signal-to-quantizing-noise ratio can be improved through pre-emphasis at 6–12 dB/octave above 500 Hz before sampling and A/D conversion and de-emphasis at the output of the synthesizer (receiver). The reason for the choice of 500 Hz is that the spectrum of the human voice tends to fall off beyond 500 Hz whereas the spectrum of quantization noise tends to be flat.

4.7. ADAPTIVE TRANSFORM CODING (ATC)

Adaptive transform coding (ATC) is similar to SBC in that the speech band is divided into a number of subbands. A fast transform algorithm, such as a 128-point discrete cosine transform, then translates the input to the frequency domain. The transformed outputs then are smoothed and decimated to 16 values. Finally, the transformed coefficients are encoded using APCM. Figure 4–24 is a block diagram[21] of an ATC system. The complexity of the ATC system is comparable to that of ADPCM with a variable predictor. Based upon talker-listener tests, ATC is claimed to be superior to SBC and ADPCM in the 9.6–24-kb/s range.[22]

4.8. VOCODERS

The actual information rate needed to convey conversational speech has been the subject of much study and discussion. Intelligibility is not the only factor

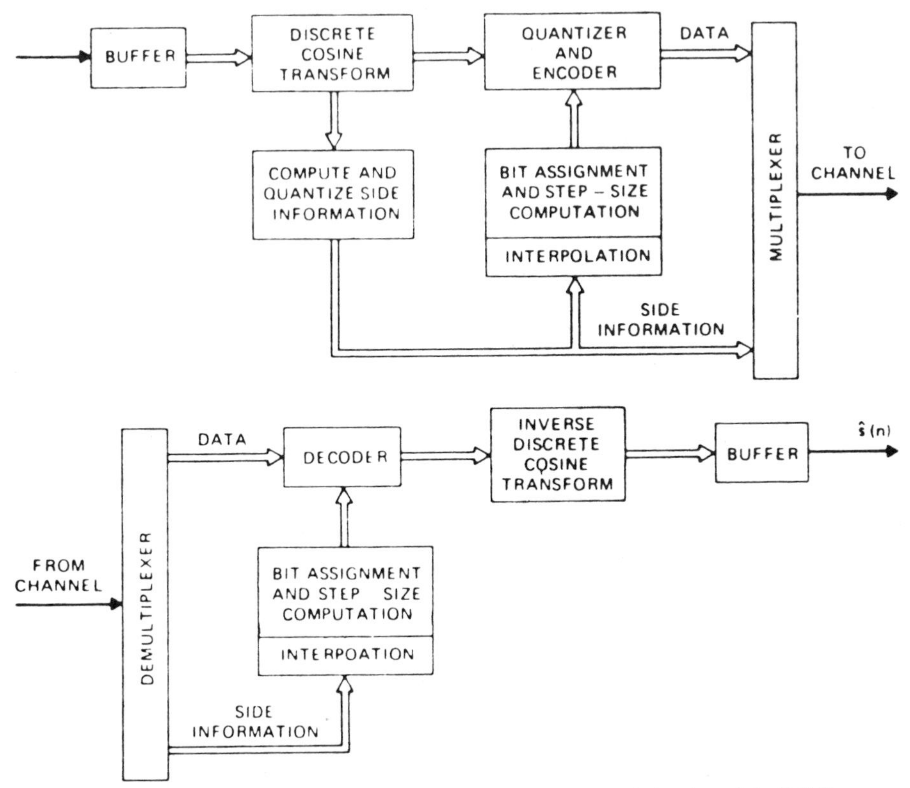

Fig. 4–24. Adaptive transform coder system. (Reprinted with permission from *The Bell System Technical Journal,* Copyright 1979 and 1976, AT&T.)

of significance. Also important are speaker recognition and naturalness. Based upon the relative frequencies of occurrence of different sound elements (phonemes) in the English language, each phoneme conveys about 4.9 bits. In conversational speech about 10 phonemes/second are produced. Thus the bit rate for speech might be said to be less than 50 b/s. However, the human ear can sense minute details that may indicate emotional content or information about the physical environment of the source. These facts indicate that enormous economies in bit rate, and thus in bandwidth, might be realizable in telephony systems if some deterioration in quality were acceptable and if the expenses of the pre- and post-processing equipment justified the reduction in transmission costs. Alternatively, if further demands for service were restricted because of limited spectrum availability, then the very low bit rate techniques would look attractive.

The foregoing considerations have been instrumental in the development of devices called *vocoders* (voice coders), whose operation depends upon a parametric description of the characteristics of speech rather than actual coding of its waveform. Vocoders often exhibit a synthetic sound because of difficulties in accurately measuring and coding the pitch or because the spectral energy measurements made do not preserve all of the perceptually significant attributes of the speech spectrum.

Vocoder operation generally depends upon a parametric description of the transfer function of the human vocal tract. The voice provides an acoustic carrier, the pitch, for the speech intelligence, which appears largely in the form of time variations in the envelope of the radiated signal. The three types of sounds with which a vocoder must deal are voiced sounds, fricatives, and stops.

The *voiced sound* is a nearly periodic sequence of pulses having a spectrum that falls at 12 dB/octave. Its network equivalent is a current source exciting a linear, passive, slowly time-varying network.

A *fricative* is a sustained voiceless (unvoiced) sound produced from the random sound pressure that results from turbulent air flow at a constricted point in the vocal system. It covers a broad band with gentle attenuation at the band edges. Its network equivalent is a series voltage source whose internal impedance essentially is that of the constriction and typically is of large value.

A *stop* is produced by an abrupt release of air pressure built up behind a complete occlusion. Its spectrum falls as $1/f$. Its electrical equivalent is a step function.

In the present state of the art, the synthetic quality of vocoder speech is not appropriate for commercial telephony. However, numerous special applications exist already, and quality is continuing to improve.

4.8.1. Linear Predictive Coding (LPC)

Since the design of a vocoder depends on speech characteristics, many attempts have been made to model such characteristics. One model that is widely used is Fant's linear speech production model.[23] Its parameters are easily obtained using linear mathematics. The result is a *linear prediction* model, which forms the basis for the LPC vocoder.

Figure 4–25 shows a digital model of speech production in which, at any moment, the speech is either voiced and simulated by an impulse train having the proper pitch period, or unvoiced and simulated by a random function source. The impulse train or random function source then is multiplied by a gain factor, following which it is filtered by a time-varying device whose parameters are reflection coefficients of an acoustical tube which simulates the human vocal tract.

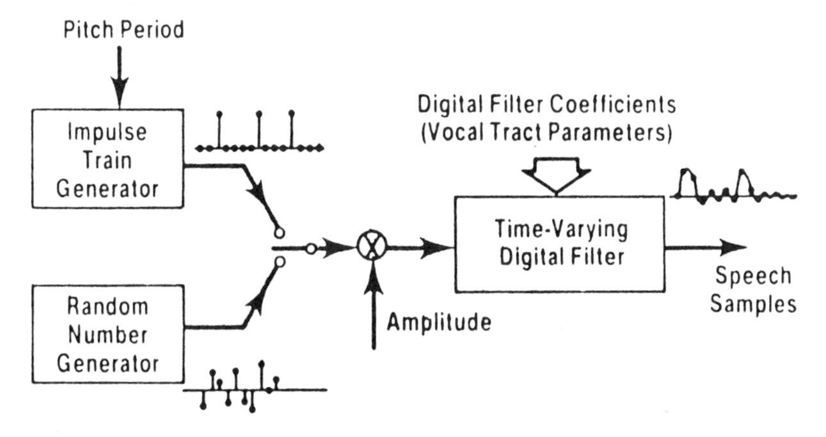

Fig. 4–25. Digital model of speech production. (Courtesy R. W. Schafer.)

The foregoing functions are implemented as shown in Fig. 4–26. Here the transmitter function is called "analysis," while the receiver function is called "synthesis." With reference to the discussion of APC in Section 4.6, an LPC system has been said[24] to be "an APC system in which the prediction residual has been replaced by the pulse and noise sources," as in Fig. 4–25. Figure 4–27 shows the way in which LPC approximates the speech spectrum envelope as a function of the number of predictor coefficients p. Implementations using $p = 8$ and $p = 10$ provide reasonably good quality speech

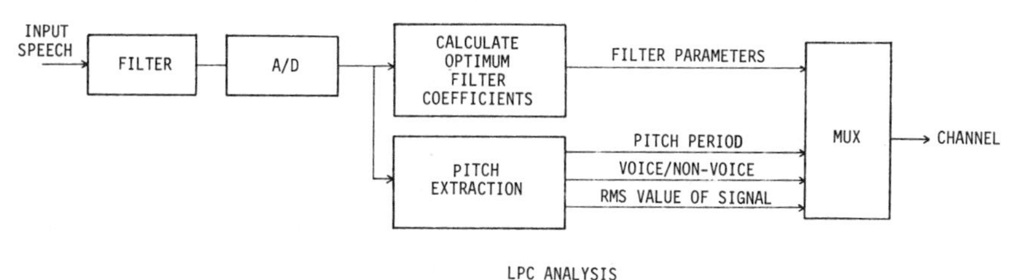

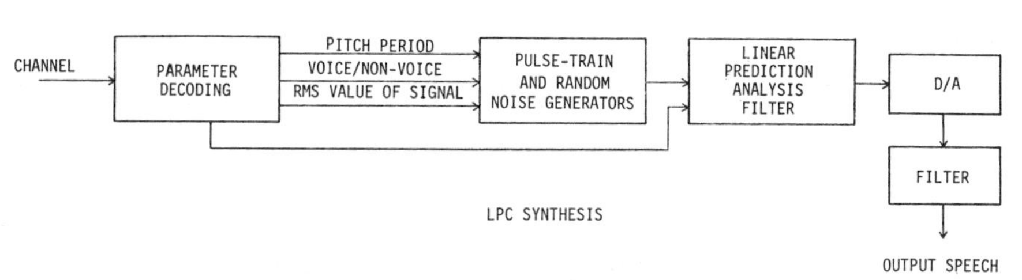

Fig. 4–26. Implementation of an LPC system.

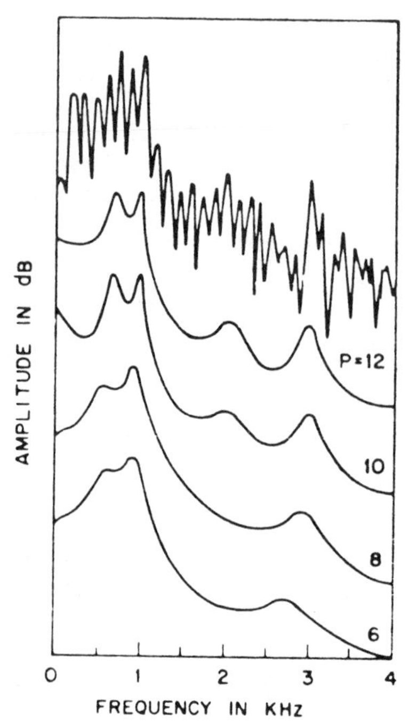

Fig. 4–27. Actual speech spectrum compared with LPC approximations. (Courtesy J. L. Flanagan et al., Ref. 24, © IEEE, 1979.)

at 2.4 kb/s, a rate at which LPC vocoders often operate. This rate is based on sending 48 bits every 20 ms or 54 bits every 22.5 ms. The bit allocation used for the LPC-10 algorithm established by the U.S. Defense Department as a standard for LPC voice transmission is shown in Table 4–8.

The predictor coefficients define an "all pole" spectrum, which is the same as that of a vowel. They can also handle narrow-bandwidth formants and continuously follow the movements of formant frequencies. (A formant is a resonance frequency of the vocal tract tube.)

A problem arises in the detection of pitch in telephony. It results from the fact that the fundamental frequency (50–200 Hz for males; 100–400 Hz for females) may be physically suppressed by the low-frequency roll off of the voice frequency circuits. Even though the fundamental is suppressed, it is still fully perceived by the listener because of what is called the *residue pitch* effect of hearing. The listener hears the harmonics of the pitch and, as a result, perceives the presence of the fundamental.

Whether a speech signal is from a telephone transmitter or not, it is forced into a simplistic model of voiced and unvoiced sounds. However, the excitation

Table 4-8. Bit Allocation for the LPC-10 Algorithm.

	VOICED	UNVOICED
Pitch/voicing	7	7
RMS	5	5
Sync	1	1
$K(1)$	5	5
$K(2)$	5	5
$K(3)$	5	5
$K(4)$	5	5
$K(5)$	4	0
$K(6)$	4	0
$K(7)$	4	0
$K(8)$	4	0
$K(9)$	3	0
$K(10)$	2	0
	54	33
Error correction	0	20

for voiced sounds is not always well represented by a periodic pulse train, just as the excitation for unvoiced sounds is not always well represented by a noise generator. These differences between real and simulated voice account for the artificial sound often exhibited by LPC and other vocoder techniques.

LPC systems using the LPC-10 algorithm and operating on 2400 b/s channels have been shown to achieve a 92% intelligibility score on the diagnostic rhyme test using male speakers. Although the tests were based on an error-free channel, an error rate as high as 10^{-3} has a negligible effect on LPC performance.

LPC offers a speech quality that, while not suitable for telephony in the present state of the art, nevertheless is better than other vocoder techniques requiring a comparable bit rate, usually 2400 b/s. Accordingly, research efforts are aimed at improvements in LPC. Improvements to the basic "simplistic" model that may prove useful include those which provide better duplication of the short-time amplitude spectrum and better duplication of the natural acoustic interactions between the source and the system. An improved model may be one in which the sound source and resonant system are allowed to load one another acoustically, i.e., to interact. Such a model would account not only for voiced and unvoiced sounds, but also for sound radiation from the yielding side walls of the vocal tract, as well as for detailed loss factors due to viscosity, heat conduction, and radiation resistance. Preliminary results indicate that such an improved model requires no increase in information rate over a simplistic one, but does make more effective use of the information available to it.

4.8.2. Other Vocoders

A vocoder can operate on either a time domain or a frequency domain basis. (The LPC vocoder described in Section 4.8.1 is a time-domain vocoder.) The parametric description of the vocal-tract transfer function, on which a vocoder operates, can take a variety of forms, either in the frequency or the time domain. Examples of frequency domain vocoders include the *channel vocoder,* in which values of the short-time amplitude spectrum of the speech signal are evaluated at specific frequencies. Another frequency-domain vocoder is the *formant vocoder,* which uses the frequency values of major spectral resonances. Time-domain vocoders include the *autocorrelation vocoder,* which operates on specified samples of the short-time autocorrelation function of the speech signal. Another is the *orthogonal function vocoder,* which sends the coefficients of a set of orthonormal functions that approximate the speech waveform. The *cepstrum vocoder* uses the Fourier transform of the logarithm of the power spectrum.

4.9. HYBRID (WAVEFORM-PARAMETRIC) TECHNIQUES

The difficulties encountered in pitch detection have resulted in a number of hybrid techniques which avoid the problems and complexities of pitch detection by coding the lower frequencies on a waveform basis but the higher frequencies on a parametric basis. Hybrid arrangements of SBC, APC, and LPC have been devised for this purpose. These combinations generally operate at 4.8–9.6 kb/s.

4.10. PERFORMANCE

Research on efficient coding techniques over the years has produced some reasonably good-sounding results at bit rates well below that of 64-kb/s PCM. Why, then, except for ADPCM/DLQ, is there no strong move at the present time toward the use of such techniques in the public switched network? Although such reasons as standardization and compatibility with existing plant in place certainly may be cited, together with questions about the acceptability of slight quality degradations by the general public, probably the strongest impediment to the widespread utilization of efficient coding techniques is their lack of transparency in the case of nonspeech signals. A telephone network, in order to transmit speech, also must handle a variety of in-band signaling and supervision tones. These are handled very well by 64-kb/s

PCM, even when many A/D conversions have to be placed in tandem. The various efficient coding techniques, however, do reasonably well if only one or two A/D conversions are involved in a connection, but beyond that their performance degrades appreciably.

In the future, with an expected extensive use of signaling and supervision outside the voice channel, as in common-channel signaling (CCS), e.g., CCITT No. 7 signaling, many aspects of the transparency problem will be removed, thus creating an environment in which the benefits of efficient coding techniques may be usable. In such an environment, data would be transmitted on a direct digital basis without the need for modems. This would be done by bringing the digital stream directly to the subscriber (see Section 5.12) and using digital format converters such as data service units (DSU) or channel service units (CSU) to convert the digital format of the computer to the communications line format.

Numerous evaluations of efficient coding techniques have been performed using both human speakers and listeners and computerized simulations. As explained in Sections 2.6 and 3.4, performance may be compared on either an SNR or a listener preference basis.

Computer simulations have been used to compare ADPCM and ADM with logarithmically companded PCM (log PCM). Figure 4–28 shows the results of these comparisons. SNR was measured in response to the input sentence, "A lathe is a big tool." For log PCM, the sampling frequency was 8 kHz, while for ADPCM it was set at 6.6 kHz. The ADPCM system had an adaptive quantizer and a first-order predictor. The bit rate, Bf_0, was varied by varying B, the number of bits per sample. For ADM, a one-bit memory was used. Since $B = 1$ for ADM, the bit rate was varied by changing the sampling rate.

The curves of Fig. 4–28 show that ADPCM has a constant gain over PCM because of the differential coding advantage. It also is superior to ADM at all bit rates but has the added complexity of a multibit quantizer. The SNR for ADM increases as the cube of the bit rate, whereas the PCM SNR increases exponentially. ADM is simpler to implement than PCM and provides a better SNR below 48 kb/s. APCM is not shown but is better than log PCM for $B \leqslant 5$. At $B = 3$, the advantage of APCM over PCM is 6 dB.

The adaptive techniques all exhibit distinct advantages over their nonadaptive counterparts. For example, at a sampling rate of 56 kHz, ADM is 10 dB better than LDM. At $B = 3$, ADPCM is 3 dB better than DPCM.

Log companding also produces considerable performance improvements over uncompanded PCM. Measured improvements are 24 dB to 30 dB, the equivalent of four to five quantization bits.

Perceptual and subjective effects are determined from the responses of

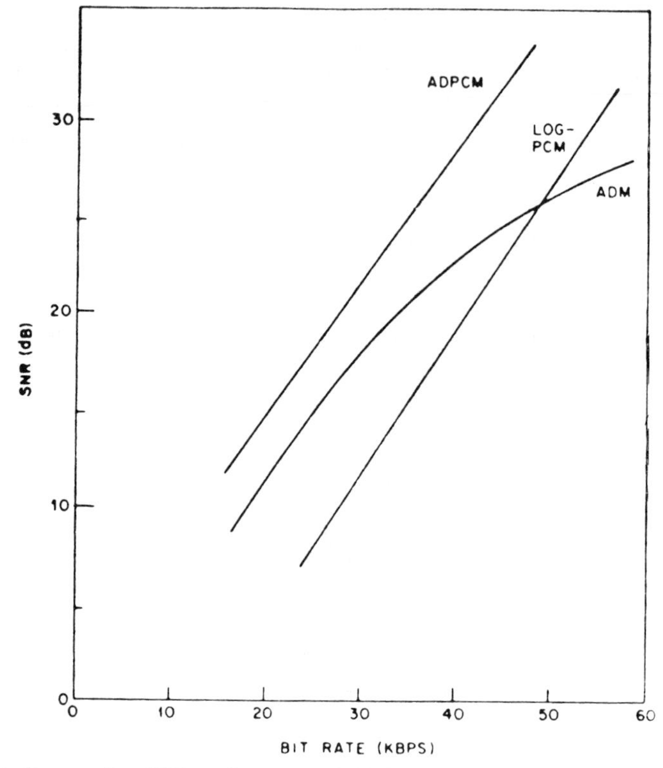

Fig. 4–28. Comparative SNR performance of ADPCM, ADM and Log PCM. (Courtesy N. S. Jayant, Ref. 1, © IEEE, 1974.)

listeners. Such results are more meaningful than the analytically determined SNR values for several reasons. First, the quantization error sequence has signal-dependent (or signal-correlated) components. In addition, signal-dependent noise (distortion) does not have the same annoyance value as does independent additive noise of equal variance.

The SNR may overstate PCM performance for $B < 6$, where dithering is used to mask the noise. With low B operation, amplitude correlated noise is more objectionable than additive noise of the same variance. On the other hand, at high B with nonuniform quantization, the quantization error is correlated with the input amplitude. Such amplitude correlated noise tends to be less annoying than additive noise of equal variance.

A comparison of the objective (SNR) and subjective (preference) ratings of ADPCM and log PCM is shown in Table 4–9. Perceptual effects to which the differences can be attributed[1] are the following:

1. ADPCM errors have more low frequency (slope overload) distortion, so are more correlated with speech than is PCM noise, which is relatively white.
2. ADPCM noise contains more pitch-related buzziness than does PCM.
3. ADPCM suppresses quantizing noise better during silent intervals than does PCM.

Thus, for the same SNR, ADPCM errors have a more signal-correlated distribution in time, making the error variance less objectionable.

In the case of delta modulation, signal-correlated errors take the form of temporal bursts of slope-overload distortion. The perceptual quality of ADM is determined largely by granular errors. For example, in the case of a 20-kb/s system with a 1-bit memory, referring to Eq. (4–14), a value of $P = 1.5$ maximizes the SNR, but $P = 1.2$ produces some interesting results. They are: (1) a 40% increase in overload noise power, but (2) a 1 dB decrease in SNR, as a result of (3) a 30% decrease in granular noise, which is only 2% of the total noise power. Above 20 kb/s, SNR results are believed to provide a conservative estimate of DM performance.

Performance characteristics of logarithmically quantized DPCM at various numbers of quantization bits and various sampling frequencies[25] are shown in Fig. 4–29. The results consist of preference judgments by a pool of 17 listeners. The comparisons were based on listener preferences relative to a white-noise degraded speech signal having the indicated SNR. The isopreference contours connect points of equal subjective quality.

The performance of CVSD in noisy channels makes it useful in military environments. As an introduction to this matter, Fig. 4–30 shows how the word intelligibility of VSD varies with bit rate as well as with channel error

Table 4–9. Comparison of Objective and Subjective Performance of ADPCM and log PCM.*

OBJECTIVE RATING (SNR)	SUBJECTIVE RATING (PREFERENCE)
7-bit PCM	7-bit PCM
6-bit PCM	4-bit ADPCM
4-bit ADPCM	6-bit PCM
5-bit PCM	3-bit ADPCM
3-bit ADPCM	5-bit PCM
4-bit PCM	4-bit PCM

* Courtesy N. S. Jayant, Ref. 1, © IEEE, 1974.

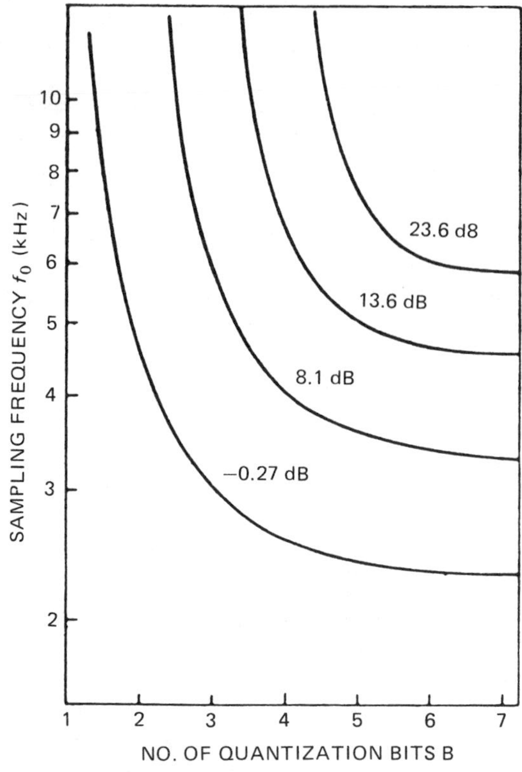

Fig. 4–29. Performance of logarithmically quantized DPCM: isopreference curves. (Courtesy Chan and Donaldson, Ref. 25, © IEEE, 1971.)

rate, designated "bit error rate, percent" in the figure.[26] Note that VSD is essentially unaffected by error rates as high as 5%. By comparison, PCM is significantly degraded by only a 0.1% error rate.

The intelligibility values shown in Fig. 4–30 are word intelligibilities. Sentence intelligibility may be higher. Comparable performance is claimed for "two-channel DM," which uses an auxiliary delta modulator to transmit a slope-envelope signal as a basis for step size control.[27]

The performance degradation caused by channel errors is quantizer dependent. DPCM is more tolerant of random bit errors than is PCM because the error spikes are related to waveform differences rather than to the total waveform. Delta modulation also can be made more resistant to errors than PCM for equal bit rates and equal channel error rates. However, instantaneously adapting DM is more vulnerable to channel noise than is slowly companded DM or LDM. Syllabically companded DM, designed for use over noisy channels, performs well in low bit rate applications (<20 kb/s).

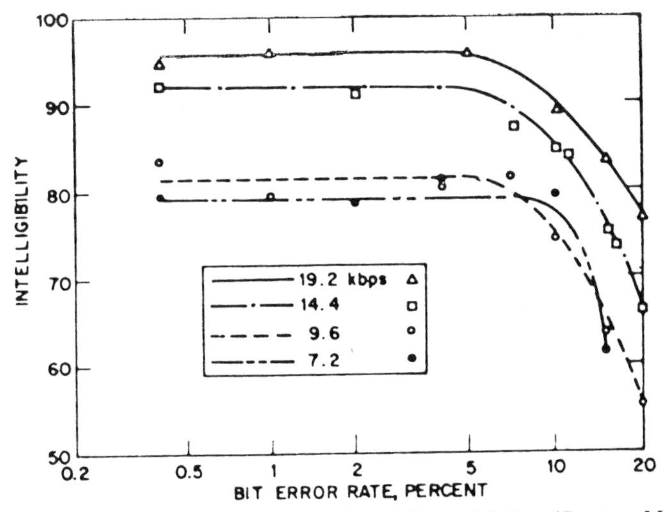

Fig. 4–30. Intelligibility performance of variable slope delta modulation. (Courtesy M. Melnick, Ref. 26, © IEEE, 1973.)

Clustered bit errors may occur in land mobile radio links. Here the slow signal fading causes bit-error patterns in which temporal correlations are usually very obvious. Differential systems such as ADM can be designed to provide a high degree of noise suppression. Such designs usually involve operation of the delta modulator in a slope-overloading mode.

Figure 4–31 shows the SNR for CVSD at 38.4 kb/s for tones of various frequencies and levels. These curves are useful in evaluating the performance

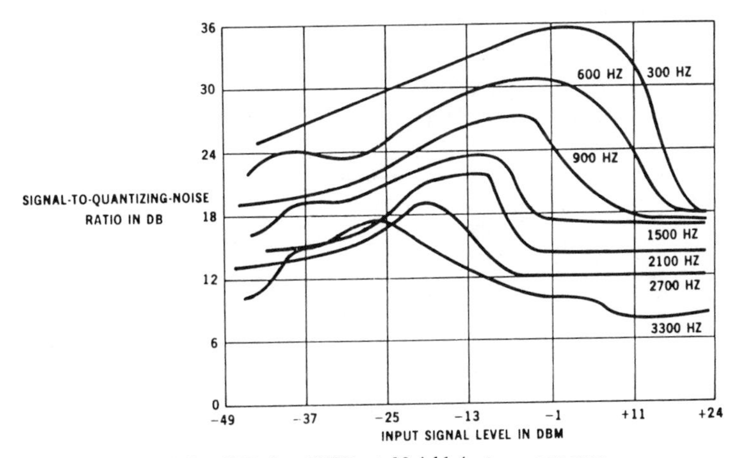

Fig. 4–31. SNR for CVSD at 38.4 kb/s: tone response.

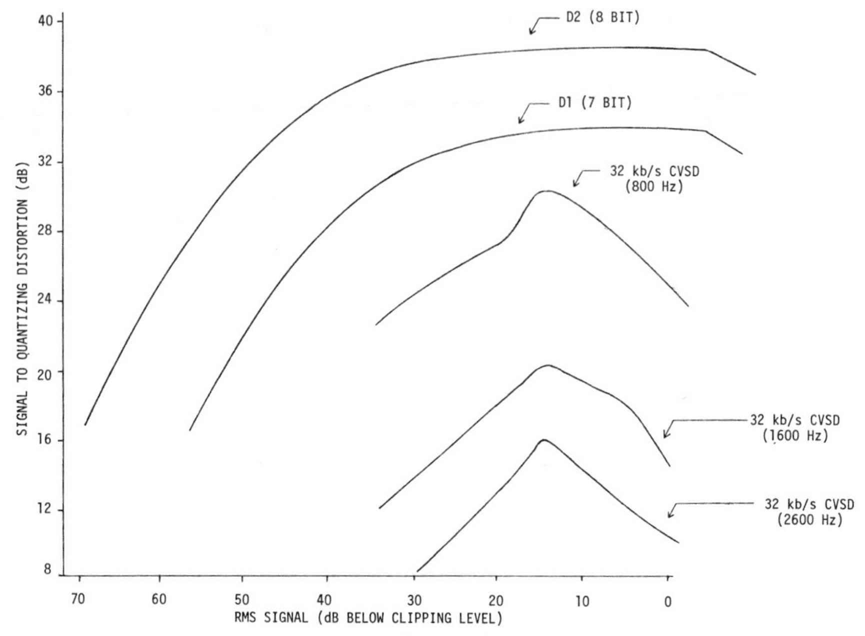

Fig. 4–32. Comparison of CVSD with PCM for signaling and supervision purposes.

of a CVSD system with respect to signaling and supervision as well as data modem tones. Figure 4–32 compares CVSD at 32 kb/s with 7-bit and 8-bit PCM as used in the *D*1 and *D*2 channel banks, respectively. Figure 4–33 displays the overall frequency response of a CVSD encoder/decoder to voice frequency transmission for a fixed amplitude.

Repeated encoding of speech may occur by identical or nonindentical quantizers in tandem. Moreover, direct digital conversions among adaptive and

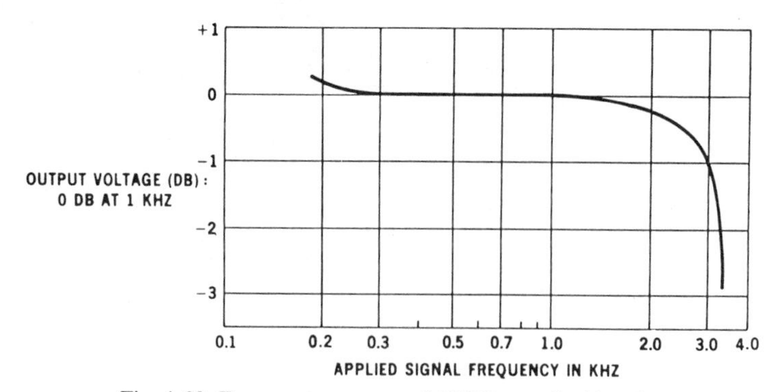

Fig. 4–33. Frequency response of CVSD encoder/decoder.

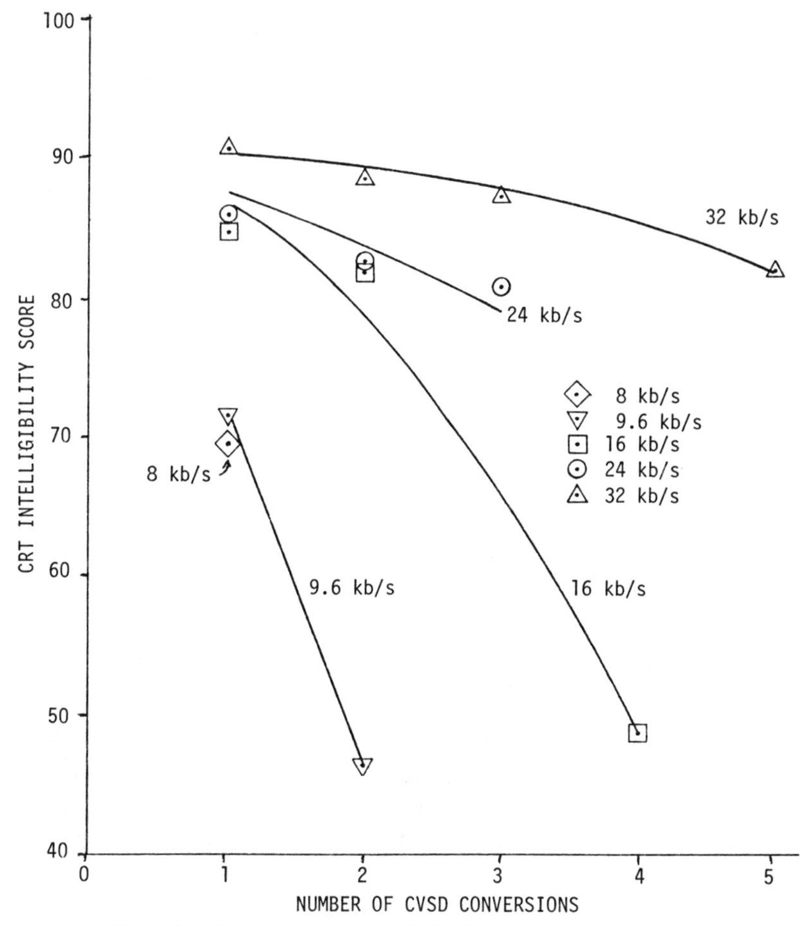

Fig. 4–34. Intelligibility versus number of CVSD conversions at various bit rates.

nonadaptive formats may be required. While CVSD performs fairly well through a single analog-digital conversion, additional conversions rapidly result in performance degradation which is increasingly severe at the lower bit rates, as shown in Fig. 4–34. These curves are based on the use of the consonant recognition test, which measures how well listeners distinguish one consonant from another. Figure 4–35 displays the way in which CVSD intelligibility varies with channel error rate. Note that a 10^{-3} bit rate, which definitely degrades 7-bit and 8-bit PCM, has an almost negligible effect on CVSD.

In summary, PCM provides toll quality voice transmission with transparency to other analog signals, i.e., a "universal" voice channel. CVSD provides

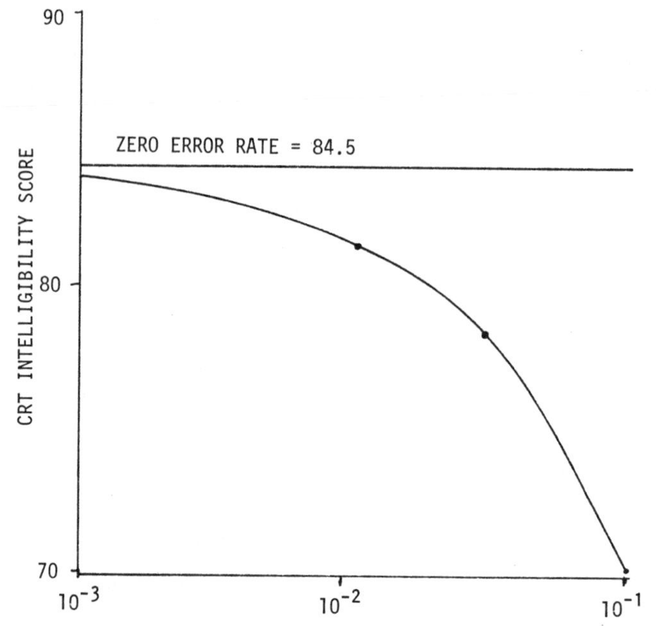

Fig. 4–35. Intelligibility versus channel error rate for one CVSD conversion at 16 kb/s.

a voice quality only modestly degraded from that of PCM, but at half the bit rate or less. However, it shows marginal to unsatisfactory performance with nonvoice analog signals. Vocoders not only perform very poorly on nonvoice analog signals, but often exhibit a synthetic sound because of difficulties in pitch extraction. In addition, the spectral energy measurements made in the analysis (transmit) process do not preserve all of the perceptually significant attributes of the spectrum.[28]

Table 4–10 is a summary of some of the more popular speech digitization and compression methods with respect to considerations that are important in voice transmission.

PROBLEMS

4.1 Explain why a true information rate for speech cannot be calculated readily.

4.2 If the same individuals always use an adaptively quantized system, is the adaptive feature of any use to them, or could they achieve the same results with a properly adjusted fixed quantizer?

4.3 Explain why instantaneously adaptive systems are more sensitive to background (room) noise than are syllabically adaptive ones.

4.4 Explain why differentially coded systems are not as susceptible to channel noise as are nondifferential systems.

Table 4–10. Summary of Speech Digitization and Compression Methods.

VOICE DIGITIZATION METHODS	IMPORTANT CONSIDERATIONS				
	REQUIRED BIT RATE (KBPS)	QUALITY		ESTIMATED ERROR RATE REQUIREMENT	DATA TRANSMISSION PERFORMANCE (TO 4.8 KB/S)
		NATURALNESS	INTELLIGIBILITY		
Waveform Digitization Methods					
PCM	48–64	satis.	satis.	10^{-4}	satis.
DPCM	32–56	satis.	satis.	10^{-3}	marg.
ADPCM	24–48	satis.	satis.	10^{-3}	satis.
ADM (CVSD)	16–24	satis.	satis.	10^{-2}	unsatis.
Parametric Methods					
Time-domain analysis					
LPC	2.4–4.8	satis.	satis.	10^{-3}	imposs.
Spectral Analysis					
Voice excited vocoder	7–16	marg.	satis.	10^{-2}	imposs.
Channel vocoder	2.4–4.8	marg.	marg.	10^{-2}	imposs.
Formant vocoder	0.6–2.4	unsatis.	unsatis.	10^{-2}	imposs.

4.5 A speech waveform sampled at 8000 times per second is found to have a correlation coefficient C_1 of 0.95. The sampling rate then is doubled to 16 000. Would you expect C_1 to increase or decrease? Why?

4.6 Two codecs are each designed for subband coding at 16 kb/s. Will the two work properly with one another without further specification? In what ways might the two differ?

4.7 List and discuss the reasons why 64-kb/s PCM continues to be installed as the standard digital voice technique by the common carriers. What conditions might change this situation in the future?

4.8 Explain why sentence intelligibility generally is higher than word intelligibility. Why is word intelligibility higher than phoneme intelligibility?

REFERENCES

1. Jayant, N. S., "Digital Coding of Speech Waveforms: PCM, DPCM, and PM Quantizers," *Proceedings IEEE,* Vol. 62, pp. 611–632 (May 1974).
2. Duttweiler, D. L., and Messerschmitt, D. G., "Nearly Instantaneous Companding for Nonuniformly Quantized PCM," *IEEE Transactions on Communications,* Vol. COM-24, No. 8, pp. 864–873 (August, 1976).
3. The VQL Codec," Aydin Monitor Systems Document 901–0022, Ft. Washington, PA.
4. Atal, B. S. and Schroeder, M. R., "Adaptive Predictive Coding of Speech Signals," *Bell Syst. Tech. J.,* Vol. 49, pp. 1973–1986 (Oct. 1970).
5. Petr, D. W., "32 kb/s ADPCM/DLQ Coding for Network Applications," *Globecom '82,* (Miami, FL, Nov. 29–Dec. 2, 1982), pp. A8.3.1–A8.3.5.
6. *Reference Data for Radio Engineers,* Howard W. Sams & Co., Inc. Indianapolis, IN, Sixth Edition, 1975.
7. See note 6.
8. See note 6.
9. Schindler, H. R., "Delta Modulation," *IEEE Spectrum,* Vol. 7, pp. 69–78, (Oct. 1970).
10. See note 9.
11. Abate, J. E., "Linear and Adaptive Delta Modulation," *Proceedings IEEE,* Vol. 55, pp. 298–308 (Mar. 1967).
12. Jayant, N. S., "Adaptive Delta Modulation with a One-Bit Memory," *Bell Syst. Tech. J.,* pp. 321–342 (March 1970).
13. See note 12.
14. Greefkes, J. A., "A Digitally Companded Delta Modulation Modem for Speech Transmission," *ICC '70 Conference Record,* pp. 7–33 to 7–48 (June, 1970).
15. Crochiere, R. E., Webber, S. A., and Flanagan, J. L., "Digital Coding of Speech in Subbands," *Bell Syst. Tech. J.,* Vol. 55, No. 8, pp. 1069–1085 (Oct. 1976).
16. Crochiere, R. E., and Sambur, M. R., "A Variable-Band Coding Scheme for Speech Encoding at 4.8 kb/s," *Bell System Technical Journal,* Vol. 56, No. 5, pp. 771–779, (May–June, 1977).
17. See note 16.
18. Crochiere, R. E., "Sub-Band Coding," *Bell System Technical Journal,* Vol. 60, No. 7, pp. 1633–1653, (Sept. 1981).
19. Daumer, W. R., "Subjective Evaluation of Several Efficient Speech Coders," *IEEE Transactions on Communications,* Vol. COM-30, No. 4, pp. 655–662, (April, 1982).
20. See note 2.

21. Tribolet, J. M., Noll, P., McDermott, B. J. and Crochiere, R. E., "A Comparison of the Performance of Four Low-Bit-Rate Speech Waveform Coders," *Bell Syst. Tech. J.,* Vol. 58, No. 3, pp. 699–712 (March, 1979).
22. Tribolet, J. M., Noll, P., McDermott, B. J., and Crochiere, R. E., "A Comparison of the Performance of Four Low-Bit-Rate Speech Waveform Coders," *Bell System Technical Journal,* Vol. 58, No. 3, pp. 699–712, (March, 1979).
23. Markel, J. D. and Gray, A. H., Jr., *Linear Prediction of Speech,* Springer-Verlag, Berlin, Heidelberg, 1976.
24. Flanagan, J. L., et al., "Speech Coding," *IEEE Trans. Comm.,* Vol. COM-27, No. 4, pp. 710–737 (April 1979).
25. Chan, D. and Donaldson, R. W., "Subjective Evaluation and Pre- and Postfiltering in PAM, PCM and DPCM Voice Communication Systems," *IEEE Trans. Comm. Tech.,* Vol. COM-19, pp. 601–612 (Oct. 1971).
26. Melnick, M., "Intelligibility Performance of a Variable Slope Delta Modulator," *Proceedings International Conference on Communications,* (Seattle, WA, June, 1973), pp. 46–5 to 46–7.
27. Greefkes, J. A., "Code Modulation Systems for Voice Signals Using Bit Rates Below 8 kb/s," *Proceedings, International Conference on Communications* (Seattle, WA, June, 1973), pp. 46–8 to 46–11 (June, 1973).
28. Rabiner, L. R., and Schafer, R. W., *Digital Processing of Speech Signals,* Prentice-Hall, Englewood Cliffs, NJ, 1978.

5
Digital Techniques in the Telephone Network

5.1. INTRODUCTION

This chapter deals with the various functions that are needed to make a digital telephone system operate. It begins with the subjects of synchronization and time-division multiplexing and then covers the basic functions of error coding, scrambling and channel coding. Figure 5–1 relates these functions to one another as well as to source coding as they might be found in a generalized digital transmission system. Signaling and supervision are covered next, followed by monitoring and digital speech interpolation.

The part of the telephone system that remains analog in most areas is the subscriber set and the local loop. However, progress is being made in the digitization of the local loop, and a section of the chapter is devoted to this topic. This is followed by digital processing techniques at the switch, which serves to introduce the more extensive coverage of this subject in Chapters 10–14. The final portions of the chapter deal with speech recognition, and with computer voice input.

5.2. SYNCHRONIZATION

Synchronization is the process whereby the receiver is made to sample the incoming bit stream so that each bit is properly identified and so that the bits constituting a particular channel are combined to provide the proper output. The process is called *frame synchronization or alignment.* The timing signal is made available to the receiver from the received bit stream itself through the process of timing recovery.

5.2.1. Frame Synchronization

Framing is the determination of which groups of bits constitute quantized levels (characters) and which quantized levels belong to which channels. Framing allows the separation of one signal from another by counting pulse positions relative to the beginning of a frame. Thus code words representing

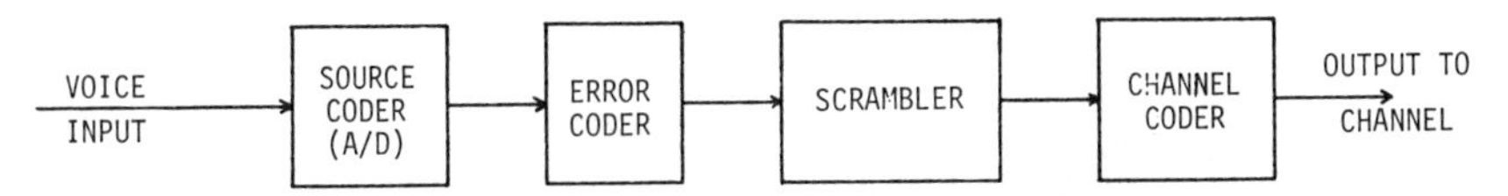

Fig. 5–1. Digital transmission system functions. *Note:* The source coder is 7- or 8-bit PCM in standard telephony systems.

the elements of each signal can be extracted from the combined bit stream and reassembled into a single stream of pulses associated with a particular channel's signal.

The framing of each character is accomplished by defining a synchronization character, commonly called *sync.* The bit arrangement of the sync character is significantly different from that of any of the regular bit combinations being transmitted. In fact, the sync character has an irregular pattern, i.e., no periodicities. The result is that when the sync character is preceded and followed by regular characters, there is no likely successive pattern of bits that equals the bit pattern of the sync character.

In the formation of a DS1 digital stream (1.544 Mb/s), a framing bit is added after each sequence of 24 8-bit words to supply synchronizing information for the receiver system. This is known as the *added digit* framing strategy, i.e., a dedicated digit position in the frame is used. It represents a negligible reduction in the information rate.

A frame thus contains the bits from one sample of each channel being transmitted, plus signaling and frame alignment bits. Figure 5–2 illustrates the frame structure for the CCITT 24-channel PCM system providing two signaling channels, as commonly used in the U.S. The corresponding signal is designated DS1. Note that 12 frames constitute a multiframe, lasting 1.5 ms. Since each frame contains one sample from each channel, the 12 frames at 1.5 ms provide an 8000/s sampling rate on each of the 24 channels. Frame alignment assures that Channel 1 from the transmit terminal is interpreted correctly as Channel 1 at the receive terminal, etc. The frame alignment signal, as shown in Fig. 5–2, is 101010. It is conveyed by Bit 1 in the odd-numbered frames. This bit is separate from those used to convey speech, and thus does not detract from speech quality. It is generated in the transmit terminal and used as a reference for all the following digits up to the next frame alignment signal.

The multiframe alignment signal, 111000, identifies the sixth and twelfth frames, which carry signaling states in the least significant bit position of each time slot. Thus the composite framing sequence is 110111001000.

At the receive terminal the frame alignment sequence is recognized by the receive terminal logic. This sequence becomes the fixed reference from which the logic determines the position of the character signal for Channel

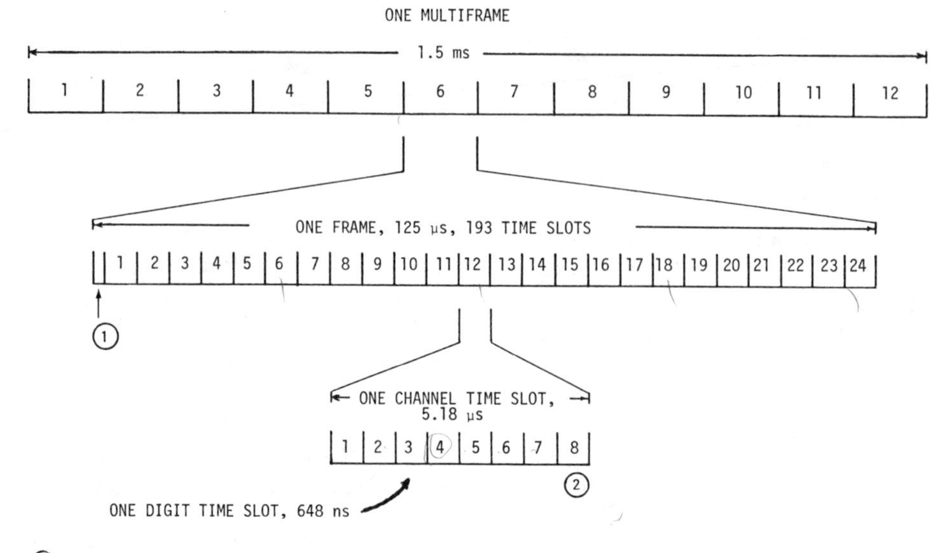

Fig. 5–2. Frame and multiframe structure for 24-channel PCM system (DS1).

1 and all the following channels. An extended framing format[1] is being implemented in new designs of DS1 level equipment which frame on a pattern contained in the framing bit position of the DS1 signal. It is used where both transmitter and receiver have the hardware and software capabilities to handle it. The extended format extends the multiframe structure of Fig. 5–2 from 12 to 24 frames and redefines the 8 kb/s-framing bit pattern into 2 kb/s for mainframe and robbed bit signaling synchronization (instead of 8 kb/s as before). As a result, it provides 2 kb/s for a cyclic redundancy check code, called CRC-6 because it contains six check bits, and 4 kb/s for a terminal-to-terminal data link.

The CRC-6 code can detect most errors in the DS1 signal, and can be used for such functions as false framing protection, protection switching, end-to-end performance monitoring, automatic restoral after alarms, line verification, and the determination of error free seconds. Thus if false framing should occur, parity checks at the receiver would reveal it. Protection switching refers to the replacement of a malfunctioning chain of repeaters with a properly operating spare chain. Line verification refers to the determination that the line is operating properly. These functions are discussed further in Section 5.9.

The 4-kb/s data link may be used for protection switching, alarms, loop-

Table 5–1. Extended Superframe *F*-Bit Assignments.*

FRAME NUMBER	BIT NUMBER	*F* BIT ASSIGNMENTS			ROBBED BIT SIGNALING
		FPS	FDL	CRC	
1	0		*m*		
2	193			CB_1	
3	386		*m*		
4	579	0			
5	772		*m*		
6	965			CB_2	*A*
7	1158		*m*		
8	1351	0			
9	1544		*m*		
10	1737			CB_3	
11	1930		*m*		
12	2123	1			*B*
13	2316		*m*		
14	2509			CB_4	
15	2702		*m*		
16	2895	0			
17	3088		*m*		
18	3281			CB_5	*C*
19	3474		*m*		
20	3667	1			
21	3860		*m*		
22	4053			CB_6	
23	4246		*m*		
24	4439	1			*D*

FPS = Framing Pattern Sequence (2 kb/s)
FDL = Facility Data Link Message Bits, *m* (4 kb/s)
CRC = Cyclic Redundancy Check (Check Bits $CB_1, \ldots, CB_6$) (2 kb/s)
* Courtesy American Telephone and Telegraph Company

back, received line performance monitoring, supervisory signaling, network configuration information and general maintenance information.

Table 5–1 shows the extended multiframe (superframe) structure. The F-bit is the first bit of the 193-bit frame, as shown in Fig. 5–2.

Figure 5–3 shows the frame structure for another 24 channel PCM system approved by CCITT and providing common channel signaling. All bits are usable for speech or information, i.e., full 8-bit quantization occurs at all times. It is based on CCITT Recommendation G. 733.

5.2.2. Timing

In addition to the maintenance of frame alignment, the timing signals within both the transmit and the receive terminals must occur at the same average rate. The transmit clock governs the rate at which the overall digital signal

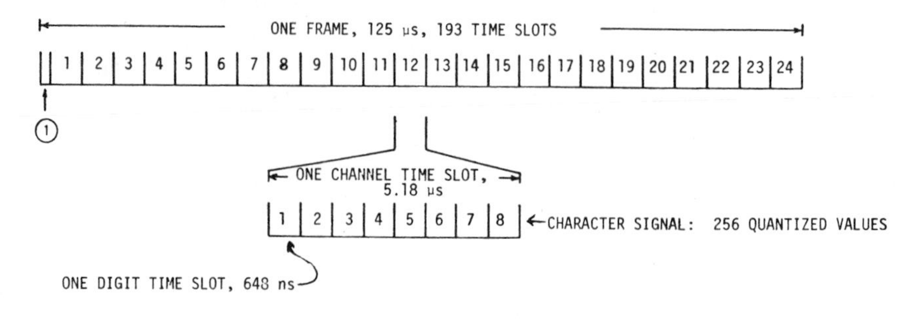

Fig. 5–3. Frame structure for 24-channel PCM system (DS1) with common channel signaling.

is produced. If the receive terminal were to use its own internal clock, any slight difference in rate would cause a loss of frame alignment. Accordingly, the receive terminal must derive its timing from the received digital signal. Thus the receive terminal is slaved or synchronized to the transmit terminal.

For conversational voice transmission, two directions are involved, so each terminal must receive as well as transmit. While each direction of transmission could have independent timing, the need to time-division multiplex numerous channels together makes it imperative that there be only one timing source in an all-digital network. Section 13.3.2 discusses digital synchronization in the public switched network.

5.2.3. Time Recovery

Timing information can be recovered from a digital signal provided a sufficient number of transitions exists in the received signal. The timing extractor is a circuit tuned to the timing frequency. This circuit has a sufficiently high Q to provide an adequate output during a sequence of zeros in the received signal. The desired output is amplified and limited to produce a square wave at the signaling rate. This square wave then controls a clock-pulse generator that generates narrow pulses which are alternatively positive and negative at the zero crossings of the square wave. Then a delay circuit in the timing path adjusts the timing pulses so they occur at the middle of each signal pulse interval. This technique, called *forward-acting timing,* is used in self-timed digital repeaters.[2]

Note that ordinarily a string of zeros is produced as the binary number corresponding to the smallest quantized value. Thus this code would be produced by many of the channels when traffic is light. In order that this condition not have an adverse effect on timing recovery, special measures must be taken. One approach is to have the coder generate a binary output in which alternate digits are inverted (ADI). In other words, a string of all zeros becomes 01010101. Another approach, called *zero code suppression,* is to place a one in the next-to-least significant bit. Thus 00000000 becomes 00000010.

Two conditions may cause the receiver to go into its "sync search" mode: (1) when transmission begins, and (2) when a bit stream drop-out occurs. The "sync search" function operates as follows: With reference to Fig. 5–2, beginning with an arbitrary bit position, every 193rd bit is sent into a shift register. When twelve such bits have filled the register, its contents are compared with the frame alignment character stored in another register. Actually, the frame alignment signal appears only in the odd frames (see Figs. 5–2 and 5–3) with the multiframe alignment signal appearing in the even numbered frames when common-channel signaling is not being used. The foregoing process is repeated by examining each of the 193 bit positions within the frame until the match is found. When a match is detected on two successive characters, the receiver raises a "character available" flag every eight bits, thus delineating the eight-bit characters of the 24-channel T1 system. Each "character" corresponds to a sample of the speech waveform of one of the 24 channels being transmitted.

The occurrence of a bit stream drop-out, as mentioned in the preceding paragraph, indicates either system failure or a very high error rate. Usually three consecutive frame alignment signals must be received incorrectly before the system is assumed to have lost alignment.

5.3. TIME-DIVISION MULTIPLEXING

The concept of time-sharing a channel is a very old one in communications technology. In the early days of telephony, the party-line concept was commonly used. Now, scarce radio telephone channels are shared among many users. Another sharing concept, digital speech interpolation (DSI), uses the fact that one party to a conversation usually is listening at least half the time, and that hesitations, pauses and other silent intervals also occur. Accordingly, for example, 100 two-way channels can be used to carry over 200 conversations by assigning channels in either direction only to those users who are talking at the moment.

Time-division multiplexing differs from the foregoing time-sharing concepts in that it involves time sharing among continuous users. Continuous band-limited signals are sampled at discrete times and the full channel is assigned to each input for a short interval, Fig. 5–4 illustrates the concept. Here

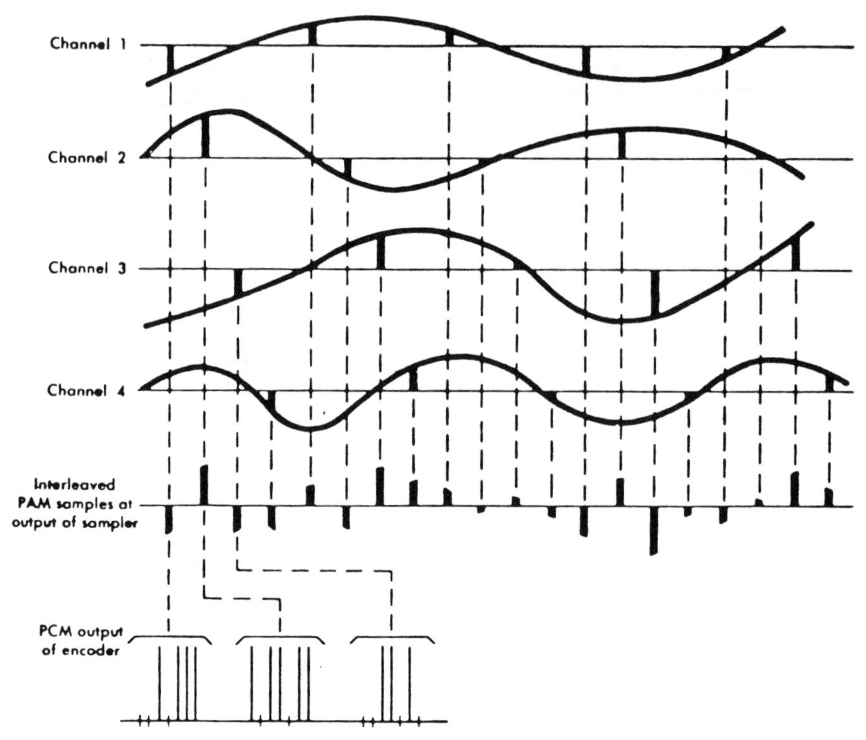

Fig. 5–4. Time-division multiplexing. (Copyright 1982, Bell Telephone Laboratories. Reprinted by permission.)

each of four analog signals are sampled in accordance with Nyquist's theorem. The sampling is timed so that the samples can be combined on an interleaved basis at the output of the sampler. This type of arrangement, with 24 channels instead of four, forms the basis for the frame structure illustrated in Fig. 5–2.

The channel bank time-division multiplexes a number of voice channels (e.g., 24) together into a digital stream (e.g., at 1.544 Mb/s) based on sampling each 3.4 kHz voice channel at 8000 times per second. The samples are time-interleaved, allowing them to be combined on a TDM basis. Companding using $\mu = 255$ is done to achieve a 40 dB dynamic range. Thus the functions of these banks include filtering, sampling, compressing, coding, multiplexing, synchronizing, and framing at the transmitter, with the inverse of these processes at the receiver.

The various channel banks differ to some extent in detail, but all perform the same basic functions. Each of one or more 24-channel digroups (North American standards) is processed into one or more DS1 pulse streams. The resulting signal is binary (unipolar) within the channel bank itself, but is converted to a bipolar format for digital transmission as will be defined in Section 5.6.

Within the channel banks, plug-in channel units provide the interface between analog signal transmission circuits and digital transmission circuits. Separate interfaces are used for voice and for signaling. Various types of channel units provide for the various types of signaling arrangements such as common channel and channel-associated, as well as for the many types of trunks and special service circuits, and for 2-wire or 4-wire subscriber loop operation.

5.4. ERROR CODERS

Figure 5–1 showed that an error coder follows the speech coder in an overall digital speech transmission system. The error coder is used in those systems in which the detection and/or correction of channel errors is important. In telephony, the primary requirement for error coding is in mobile radio applications, especially land mobile, where the problem of rapid fades justifies the complexity of error coding, as well as the resultant increase in transmitted bit rate. Radiotelephone systems planned for cellular service[3] in the future may use ADPCM at 32 kb/s or ADM at 40 kb/s. Table 4–9 shows that such systems can tolerate bit error rates of 10^{-3} and 10^{-2}, respectively. Higher error rates, however, may result from the rapid fading that accompanies unfavorable propagation conditions. Accordingly, such techniques as forward error correction may be needed to maintain voice quality.

Forward error correction involves the addition of redundant digits to a bit stream according to pre-established rules. The receiver then examines all received bits, using the redundant ones to determine what errors may have occurred and to correct these errors. In performing such coding, the speech signal may be handled in blocks or groups of blocks whose length does not exceed the duration of a syllable. *Block coding* is the term used to describe an arrangement in which the redundant bits in a given block relate only to actual speech or information bits of the same block. However, if the redundant bits also check speech or information bits in preceding blocks, the code is called *convolutional.* While forward error correction can provide appreciable improvements in minimizing speech distortion during rapid fades or drop-outs less than a syllable in length, it cannot aid in the case of longer-term signal drop-outs, such as are experienced in fringe areas relative to a mobile base station.

In addition to its use in improving speech quality under difficult transmission conditions, as are encountered in mobile radiotelephony, error coding has obvious applications in maintaining the extremely low bit error rate ($<10^{-7}$) essential in signaling and supervision.

5.5. SCRAMBLERS

In the digital encoding of speech, the most probable speech amplitudes are the lowest ones, because of the large number of pauses in conversational speech. In addition, the use of waveform encoding means that the frequent zero crossings of any waveform will produce zero amplitudes to be encoded. The result may then consist of appreciable trains of zeros in the transmitted stream. Since the receiver must derive its clock from the received data stream, it must receive enough ones on a regular basis to allow its timing circuits to function as intended, as discussed in Section 5.2.3. Figure 3–2 illustrated one method of minimizing the long train of zeros caused by speech pauses. This method is called zero code suppression. Scramblers also may be used to prevent the occurrence of long runs of zeros.

Another reason for the use of scramblers is that a data stream, such as one produced by signaling and supervision functions, may have a repetitive pattern with high discrete frequency components. If such a pattern were transmitted, it could constitute a serious source of interference to other DS1 streams using the same facilities because of the line spectrum it produces. Scrambling can be accomplished by adding a pseudo-random digital sequence to the transmitted stream, and then subtracting the same sequence from the received stream. This can be accomplished by the use of pseudo-random generators that are built alike, started in synchronism, and clocked at the same rate. The clocking rate, of course, is derived directly from the digital stream itself. Figure 5–5 illustrates the scrambling and subsequent descrambling of a digital stream.

With reference to Fig. 5–5, a pseudo-random sequence is added to the stream to be scrambled. This is done using modulo-2 addition, in which the rules are: $0 + 0 = 0$, $0 + 1 = 1$, $1 + 0 = 1$, and $1 + 1 = 0$. The sum then constitutes the scrambled stream which is transmitted. The received stream is assumed to be identical to the transmitted stream. Thus, for purposes of illustration, the channel has been assumed to be error free. At the receiver a locally generated pseudo-random stream is added, again modulo-2, to the received stream. The resultant descrambled stream is found to be identical to the original stream to be scrambled. Note that the received stream, on which the receiver's timing circuits function, bears no resemblance to the descrambled stream.

Figure 5–6 illustrates the way a pseudo-noise generator is built. The period of such a generator is $T = (2^n - 1)/F_0$, where n is the number of shift register stages and F_0 is the clock frequency.

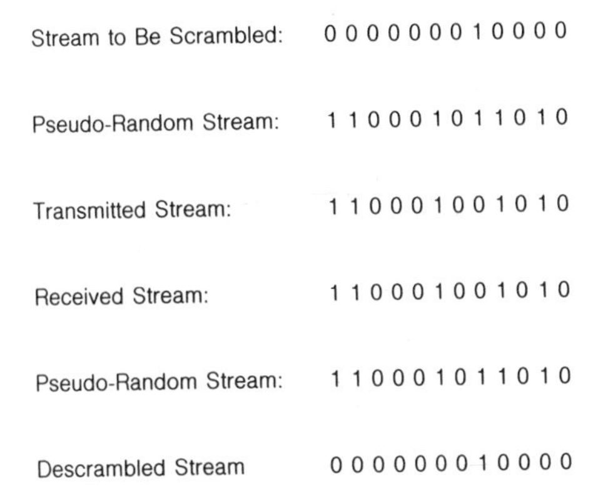

Stream to Be Scrambled:	0 0 0 0 0 0 0 1 0 0 0 0
Pseudo-Random Stream:	1 1 0 0 0 1 0 1 1 0 1 0
Transmitted Stream:	1 1 0 0 0 1 0 0 1 0 1 0
Received Stream:	1 1 0 0 0 1 0 0 1 0 1 0
Pseudo-Random Stream:	1 1 0 0 0 1 0 1 1 0 1 0
Descrambled Stream	0 0 0 0 0 0 0 1 0 0 0 0

Fig. 5–5. Scrambling and descrambling of a digital stream.

A simpler alternative that is commonly used derives the added digital sequence by inverting the original message sequence (as in zero-code suppression) and then using modulo-2 addition of the bits of the original and inverted sequences.

By eliminating periodicities, the scrambler reduces the amplitude of any discrete spectral lines (that otherwise would exist) to negligible levels. Because the scrambling process removes spectral lines, the interchannel interference between digital systems can be kept to a low level.

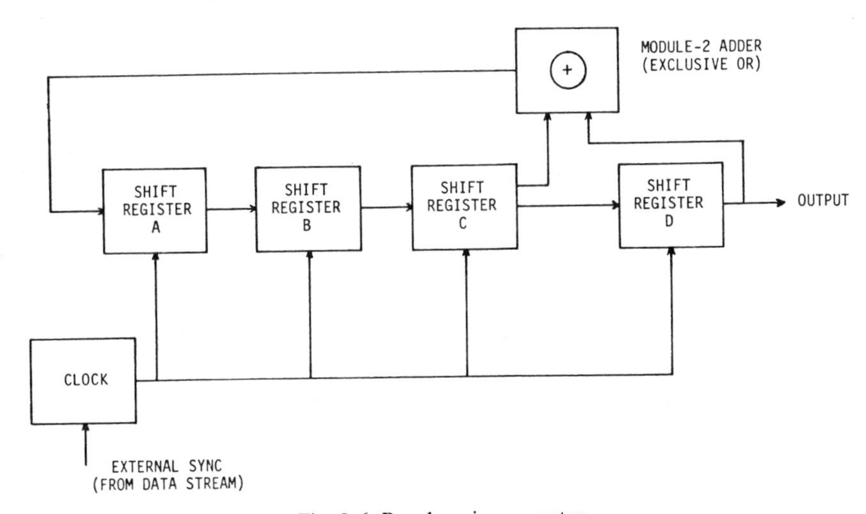

Fig. 5–6. Pseudo-noise generator.

5.6. CHANNEL CODERS

The channel coder is the device that processes the binary input into a multilevel or modified binary signal. This processing may alter the spectral shape of the signal since, in general, converting a two-level signal to three or more levels tends to decrease the amount of spectrum it occupies.

The formats which have been selected for use on the public switched network provide reliability together with efficiency of operation in terms of spectrum utilization and power consumption.

Local loop digital transmission at subrate speeds (2.4, 4.8 and 9.6 kb/s) is accomplished using a bipolar return-to-zero (BRZ) signal with a 50% duty cycle, as illustrated in Figure 5–7. Conversion from the customer's format to BRZ occurs at the customer's interface with the local loop. Note that zero is sent as 0 volts whereas the first one is sent as a positive voltage during 50% of the bit period. The second one is sent as a negative voltage of equal magnitude to the positive voltage. Subsequent ones continue to alternate in polarity. Accordingly, the transmitted signal has no dc component.

A departure from the alternating positive and negative pulse rule for ones is called a *bipolar violation* (BPV). Noise on the channel thus can cause a BPV to be produced. However, they are produced intentionally for several purposes. They are used when a trouble condition exists on the facility, when an idle code or more than six zeros (seven for 56 kb/s) are being sent, or when equipment tests are being done.

If a digital sequence from a customer contains more than six (seven for 56 kb/s) sequential zeros, the transmitting channel coder inserts BPVs, thus breaking the sequence. These BPVs are detected by the receiving equipment, which replaces them by the proper number of zeros before forwarding the sequence to the customer's receiving terminal equipment. As a result, no restriction needs to be placed on customer digital sequences.[4]

At the central office various subrate signals may be combined together to produce a 64 kb/s stream known as the DS0 signal. It is a nonreturn-to-zero (NRZ) signal, i.e., has a 100% duty cycle, and is used for intraoffice

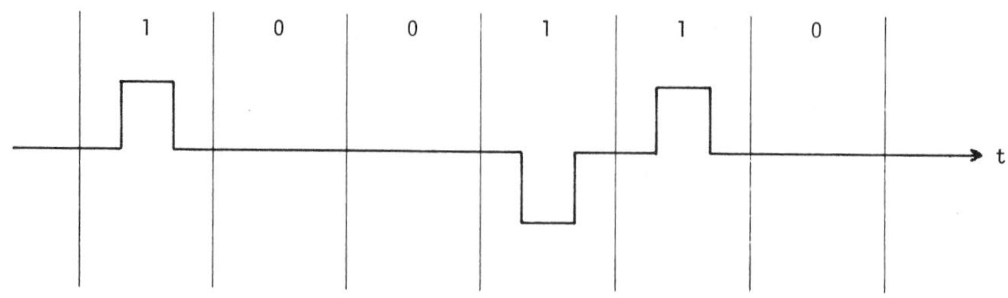

Fig. 5–7. Bipolar return-to-zero signal with 50% duty cycle.

transmission. To the extent that the subrate signals add to less than 64 kb/s, *pulse stuffing* is used in combining them. Pulse stuffing is a process in which enough time slots are added to an incoming signal to make this signal operate at the specific rate controlled by the clock circuit of the transmitter.[5] Pulses inserted into these new time slots carry no information. The signal must be coded so that these noninformation bits can be recognized and removed at the receiving terminal. The European term for this process is *justification.*

The use of pulse stuffing requires that the output channel rate be higher than the aggregate input rate to allow for the addition of the *stuff* (also called *null*) *bits.* The stuff bits are added in a prescribed manner so that they are identifiable at the receiver, which must remove them.

For interoffice transmission, 24 DS0 signals are combined to produce a DS1 signal at a 1.544-Mb/s rate. Because of possible minor differences in the DS0 signals, pulse stuffing is used in this combining process. The DS1 signal has the BRZ format with a 50% duty cycle. Its bit allocation was shown in Figs. 5–2 and 5–3.

The various factors that affect the design of these signals include (1) interconnection and transmission requirements and (2) compatibility with terminal equipment. Under interconnection and transmission requirements are included transmission rate, signal format (polar, bipolar, or multilevel), pulse amplitude, number of consecutive zeros allowable, and the location of parity bits in the bit stream. Compatibility requirements include the multiplexing method, framing, and signaling formats.

Two DS1 signals can be combined using an M1C multiplexer to form a DS1C signal at 3.152 Mb/s. The DS1C uses a BRZ 50% duty cycle format, with an additional 64 kb/s for synchronization and framing. In this process, time slots are added to the received signal (pulse stuffing) so that the signal produced operates at a rate controlled by the transmitter's clock.

The DS2 signal (6.312 Mb/s) is produced by multiplexing four DS1 signals together in an M12 multiplexer. Synchronization of the four incoming signals is obtained through the addition of stuff bits since the DS1 signals may originate from different sources. Control and framing bits also are added. The DS2 format is called *bipolar with six-zero substition* (B6ZS). For sequences of five or fewer zeros the stream is bipolar. If six or more zeros occur in a row, the output depends on the polarity of the one pulse that preceded them. If the polarity was positive, the output produced is $0 + - 0 - +$, where $+$ and $-$ denote the polarity of the digit one. Thus two bipolar violations have been caused. However, if the polarity was negative, the output produced is $0 - + 0 + -$. Again, two bipolar violations have been produced. These violations occur in the second and fifth bit positions of the sequence. The receiver recognizes them and substitutes zeros instead.

The DS3 signal (44.736 Mb/s) is produced by an M13 multiplexer. Within this multiplexer four DS1 streams are combined to produce the DS2, and seven DS2s then are combined to produce the DS3. The polar[6] format is maintained internally, which means there are no bipolar violations. The output stream is bipolar with a 50% duty cycle, and is modified with three-zero substitution (B3ZS). Each group of three consecutive zeros is replaced by *B0V* or by *00V*, where B is a one pulse that adheres to the bipolar rule, whereas *V* is a one pulse that violates the bipolar rule. The selection between *B0V* and *00V* is made so that the number of *B* pulses between consecutive *V* pulses is odd.

The DS4 signal (274.176 Mb/s) is the result of multiplexing six DS3 signals together in an M34 multiplexer. Pulse stuffing is used, as at the lower levels. The DS4 is polar binary, which means that one is positive (100% duty cycle) and zero is negative (100% duty cycle), as measured from the center conductor to the outer conductor of the coaxial cable used for transmission.

Table 5–2 summarizes the foregoing description, and shows the timing tolerances in parts per million. Figure 5–8 illustrates how the various levels of the digital hierarchy are related. A CSU is a channel service unit; a DSU is a data service unit. These are devices that convert the customer's digital data signals to the format used by the transmission facilities. Local loop arrangements for digital voice are discussed in Section 5.10.

Figure 5–9 compares the unipolar, bipolar, and polar binary waveforms.

The foregoing discussion indicated that codes may require frequent format conversions in digital transmission. Noteworthy are the signal format differences indicated in Fig. 5–9. Another type of conversion is one involving a

Table 5–2. Summary of Repetition Rates and Codes in the Digital Hierarchy.*

SIGNAL	REPETITION RATE (MB/S)	TOLERANCE (PPM)	FORMAT	DUTY CYCLE (%)
Subrate	0.0024	—	bipolar	50
	0.0048			
	0.0096			
	0.056			
DS0	0.064	†	bipolar	100
DS1	1.544	±130	bipolar	50
DS1C	3.152	± 30	bipolar	50
DS2	6.312	± 30	B6ZS	50
DS3	44.736	± 20	B3ZS	50
DS4	274.176	± 10	polar	100

† Expressed in terms of slip rate.

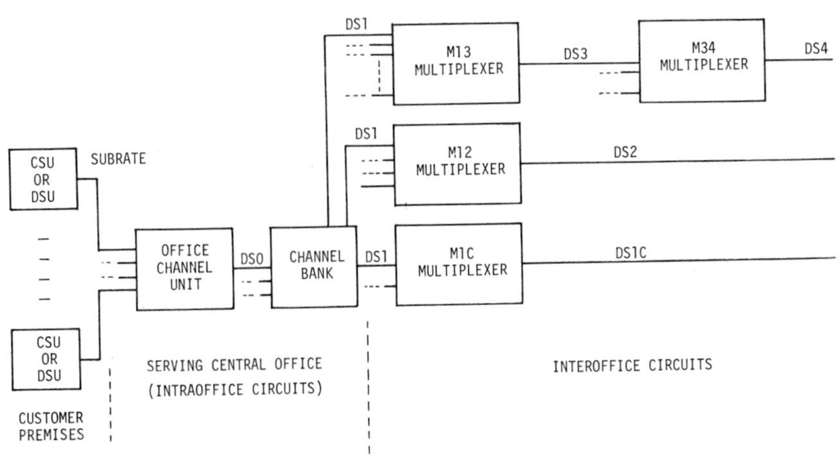

Fig. 5–8. Relationships among levels of the digital hierarchy.

format associated with transistor-transistor logic (TTL), in which zero is nominally 0 volts and one is nominally 5 volts; this format must be converted to the BRZ line format. In the case of TTL levels, tolerances of ±2 volts may be placed around the nominal values.

A conversion from a 2-level to a 3-level format is required in one method

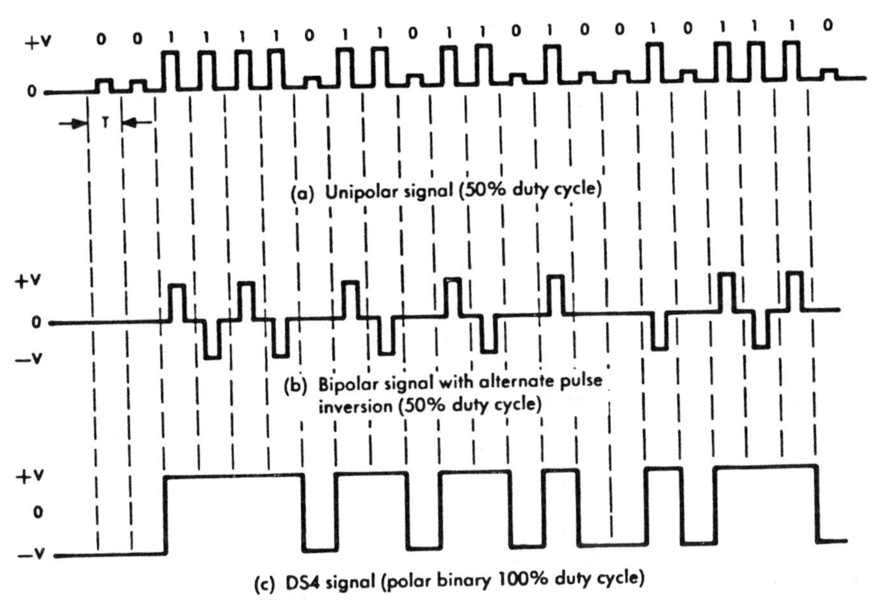

Fig. 5–9. Signal waveform comparison. (© American Telephone and Telegraph Company, 1977.)

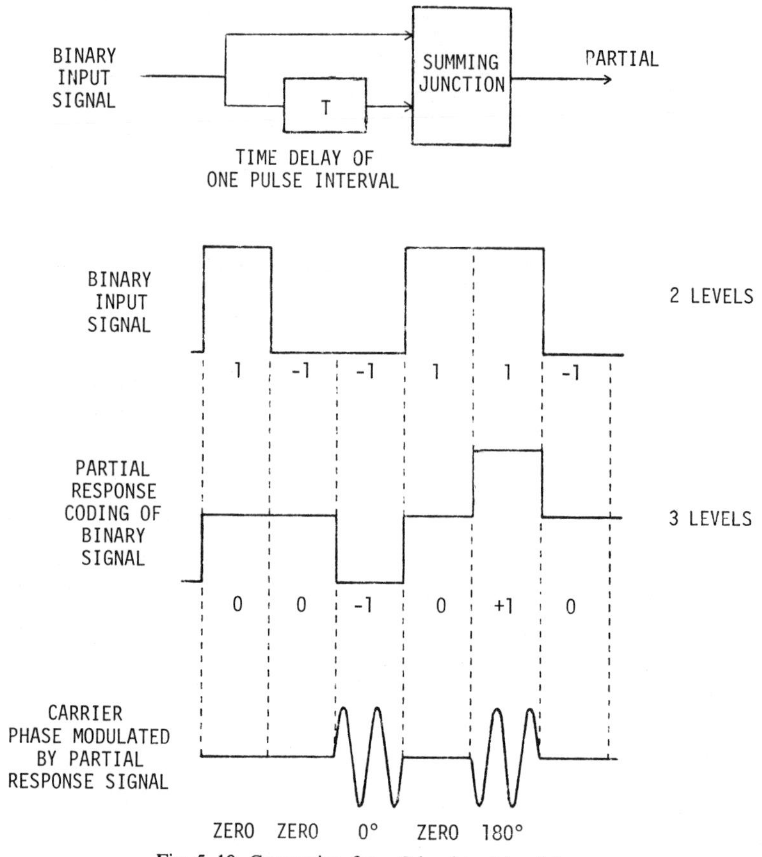

Fig. 5–10. Conversion from 2-level to 3-level format.

of implementing quadrature partial response signaling (QPRS), (see Section 6.2.5) to achieve bandwidth compression, as shown in Fig. 5–10. In this case, the input is a binary signal while the output is obtained by combining the present and the immediately preceding input symbols.

5.7. SIGNAL IMPAIRMENTS IN TRANSMISSION

5.7.1. Types of Impairment

Transmission impairments can be categorized in a variety of ways. One of these ways is shown in Fig. 5–11. Note that the vertical dimension on the figure denotes level or amplitude, while the horizontal dimension denotes

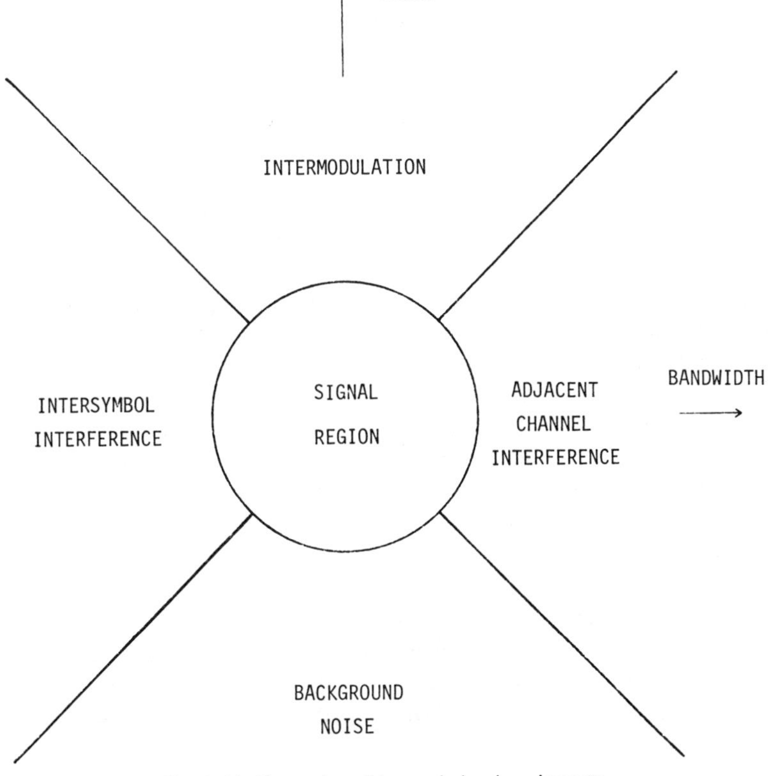

Fig. 5–11. Categories of transmission impairments.

bandwidth. Thus too large a signal level results in intermodulation as repeater characteristics are pressed to their limits. Intermodulation can produce interference between digital and analog transmissions that use different portions of the spectrum on a given facility. Background noise causes impairments because it alters the zero crossing times of the signals. Background noise includes random and impulse noise as well as crosstalk. Adjacent channel interference and intersymbol interference occur on facilities whose bandwidth is shared on a frequency division basis. Intersymbol interference is a response in a given time slot produced by a symbol in another slot, usually the preceding one. Attempts to place transmissions too close to one another in the frequency domain result in adjacent channel interference if channel passbands are large enough to encompass some of the adjacent channel. Conversely, if relatively narrow filtering is done, intersymbol interference is the result.

All forms of impairment result in bit errors. Bit error rates as low as 10^{-7} can disturb the accuracy with which signaling and supervision are accomplished. A bit error rate in excess of 10^4 can degrade PCM voice, while bit error rates in excess of 10^3 may degrade other forms of digital voice, as shown in Table 4–9.

5.7.2. Echo Cancellers

Echoes arise from impedance mismatches in terminating equipment. Often echoes cause as much subjective annoyance on long-distance telephone calls as do noise and low speech level. The subjective annoyance of echoes increases with both delay and level. Echo problems have been controlled through the judicious use of circuit loss as well as by the use of the echo suppressor, which opens the transmission path from the listener to the talker.[7]

For circuit lengths in excess of 3000 km, the use of loss (known as "via net loss") results in unacceptable received levels.[8] Echo suppressors, on the other hand, cannot provide satisfactory service under conditions in which both parties attempt to speak simultaneously (double-talk), and also may chop or clip low speech levels. In addition, while echo suppressors often are used on one-way single-hop satellite circuits, their performance has been found to be unacceptable[3] on two-way (full-hop) satellite circuits, where the echo delay is a full 540 ms.

The foregoing problems of excess loss, chopping, and clipping are best solved through the use of the echo canceller. Echo cancellation was proposed in the early 1960s, but could not be implemented on a widespread basis at that time because of its high cost. However, the use of very-large-scale integra-

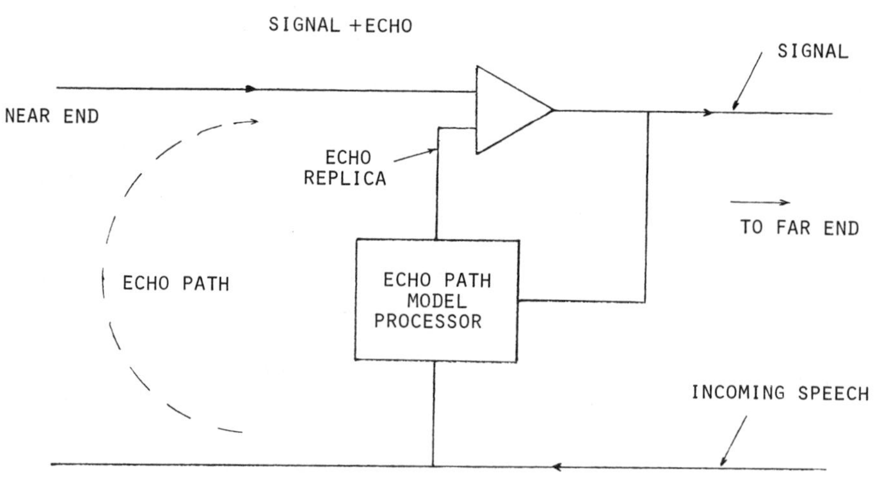

Fig. 5–12. Principle of echo cancellation. (Courtesy Suyderhoud et al, Ref. 7, © IEEE, 1976.)

tion (VLSI) has allowed an entire echo canceller to be placed on a single chip,[10,11] thus rendering the concept practical economically.

Figure 5–12 illustrates the basis on which echo cancellers are implemented. An adaptive model of the echo signal is built using tapped delay-line principles. The modelled echo then is subtracted from the actual echo, thereby leaving the speech without echo. Accordingly, the failings of the echo suppressor technique do not arise. Adaptive feedback processing is used to obtain a model of the echo path response. This model then is stored in memory. Following this, the incoming signal is processed by convolution with the stored impulse response, thus providing a close replica of the actual echo signal. This replica then is subtracted from the actual echo signal on the sending side. As a result, the echo is removed, leaving the speech unaffected.

5.8. SIGNALING AND SUPERVISION

Signaling and supervision functions including addressing, ringing, and on- or off-hook supervision. Other functions are information signals and test signals.

Signaling on loops currently is done on an analog basis, but this will change with the future digitization of the local loop plant, as described later in this chapter. Signaling on trunks is done digitally where the trunks have been implemented to carry a digital stream. Address and supervisory bits are assigned as indicated in Fig. 5–2 for Signaling Channel A and Signaling Channel B. Thus, the least significant of the eight quantization bits is robbed every sixth frame. This means that each of the 24 channels is quantized on an 8-bit basis $5/6$ of the time and on a 7-bit basis $1/6$ of the time. (In some older systems still in service only 7-bit quantization is used, in which case the eighth bit is totally devoted to signaling and supervision.) The receiving portion of the system performs the inverse operations. The incoming signal from the digital line is converted to binary form. A framing circuit then searches for, and synchronizes to, the framing bit pattern. This provides assurance that the locally generated timing pulses are in synchronism with the incoming pulse train. The signaling digits then are sorted out and directed to the individual channels.

5.9. MONITORING AND MAINTENANCE

To maintain service quality, a terminal is removed from service if a failure occurs or if the quantizing noise exceeds the maintenance limit. The maintenance limit and the time allowed for restoration of service depend on the required grade of service as well as the number of message channels affected. Terminals for 672 channels (T3) or more must be restored rapidly, thus requiring automatic switching to spares, whereas smaller terminals may be restored by manual connection to spares or by repair.

Received framing information shows how the clock circuits are performing at both ends, as well as the performance of the digital facility. Frequent misframing initiates alarms at the terminal so that proper action can be taken. Thus some maintenance features and functions are incorporated within the transmission systems whereas others are provided by external arrangements that may include record keeping as well as operational features.

A *red alarm* (red light and audible alarm) is initiated when a loss of signal or framing occurs in one direction. This alarm is triggered by a circuit in the receiving channel bank that is affected. Along with the initiation of this alarm, the associated transmitting channel bank is alerted. The result is the lighting of the *yellow alarm* at the transmitting bank. The existence of such alarms may initiate certain trunk processing functions. For example, those busy network trunks that are involved in the failure are removed from service until repairs can be completed. A resupply (*blue*) signal is substituted for a failed signal. It satisfies the line format at the bit rate at which it is inserted, but does not carry message or framing information for lower levels in the digital hierarchy. Its purpose is to prevent or minimize protection switching or the sounding of alarms, especially on downstream multiplex equipment and channel banks.

The monitoring of repeatered lines is done from offices at the ends or along the routes of the lines. The signal is examined for code format violations, such as bipolar violations. In addition, violations of successive zeros restrictions and loss of signal also are monitored. Defective repeaters are identified by a built-in fault location system.

5.10. DIGITAL SPEECH INTERPOLATION

During ordinary telephone conversations, the transmission path in each direction is used only about 40% of the time.[12] The usage is this low because only one person usually talks at a time and because there are natural pauses in speech. This fact has been used to advantage in a technique called *time assignment speech interpolation* (TASI) for over twenty years on long-haul analog channels such as those of the transatlantic cable system. It allows an approximate doubling of channel capacity because each talker uses the circuit less than half the time. A digitized version of TASI is one type of digital speech interpolation (DSI). In the case of PCM voice, a special TASI technique known as *channel augmentation by bit reduction* can be used to absorb overloads that would otherwise result in clipping.

In operation a TASI-type system serves a number of trunks N via a smaller number of channels n by connecting, at any moment, only those trunks on which speech activity (a speech "spurt") is present. Special TASI common signaling channels, whose propagation delay is the minimum possible for a given transmission medium, are used to notify the far end which trunk is connected to which channel. A voice-path delay of 25–50 ms is used to allow time for the connection information to be acted upon.[13] A delay of 50 ms eliminates processing clip caused by the switching not being complete in time for the speech spurt.

Another form of clipping, called *competitive clip,* occurs when all channels are active and an additional trunk has a speech spurt to be connected. Such a new spurt must await the availability of the next channel and thus becomes clipped, i.e., its start is cut off.

Generally a clip of 50 ms or less is not detectable by the user.[14]

Because data tones normally cannot tolerate clipping, data transmission is provided service on which TASI is not being used.

Under conditions such that the number of trunks being used actively for voice exceeds the available number of channels, the least significant bits of the PCM channels can be robbed to form additional voice channels. In such a system used by INTELSAT,[15] the probability of a clip longer than 50

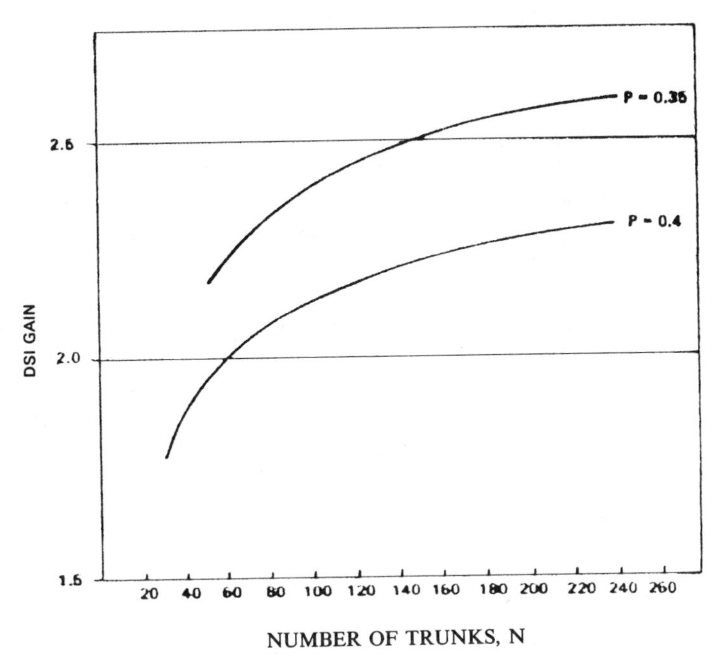

Fig. 5–13. DSI gain versus number of trunks, N (Courtesy Verma, Ramasastry and Monsees, from NTC '78 Conf. Record © IEEE, 1978)

ms is limited to less than 2% by serving 240 trunks with 127 normal channels and 16 overload channels derived by robbing least significant bits. To assure the continuous achievement of <2% clips of >50 ms duration, a 1000 Hz square wave is sent one second out of 10 on one of the trunks. At the receiving end, its on-off durations are measured to detect the extent of any clipping. If clipping beyond pre-established limits occurs, an alarm message is generated and sent back to the originating encoder.

DSI gain is defined as (Number of Trunks)/(Number of Channels, including assignment channel).

Figure 5–13 shows how DSI gain varies with the number of incoming trunks.[16] The parameter *P* is the "freeze-out" percentage, which is defined as 100 × (Average of clip time)/(Average of talk spurt time), for a given channel.

To date, DSI has been used largely on satellite circuits. However, its use on a significant number of terrestrial circuits now is beginning. The use of DSI offers the possibility of appreciable increases in the number of voice channels that can be carried by digital radio systems, which are discussed in Chapter 7. For example, long-haul systems at 6 GHz can provide 2016 digital voice channels in a 30 MHz radio channel bandwidth through the use of 64-QAM. The use of DSI could double this number of voice channels.

5.11. DIGITAL REPEATERS

A major advantage of digital transmission is its ability to perform well at much higher channel noise levels than are feasible for analog transmission. This advantage results directly from the use of regenerative repeaters in digital transmission. The regenerative repeater senses the state of the incoming noisy and distorted signal and reconstructs a clean output signal.

The functions performed within a regenerative repeater are illustrated in Fig. 5–14. The amplifier compensates for signal level losses prior to the repeater and provides a uniform output level through the use of an automatic line buildout circuit,[17] which corrects for the effects of temperature changes as well as for variations in repeater spacing. The equalizer's function is to optimize the attenuation versus frequency characteristic for pulse transmission by keeping pulse distortion low while suppressing out-of-band noise.

Timing information is obtained from the signal after the amplification and equalization have been done. A fixed delay is used in the timing path to cause the timing pulses to occur at the middle of each signal pulse interval. The regenerator thus looks at the incoming signal at the middle of each pulse interval. Positive clock pulses gate the incoming pulse stream into the regenerator. Negative clock pulses turn the regenerator off. In this manner, the width of the output pulses is controlled.

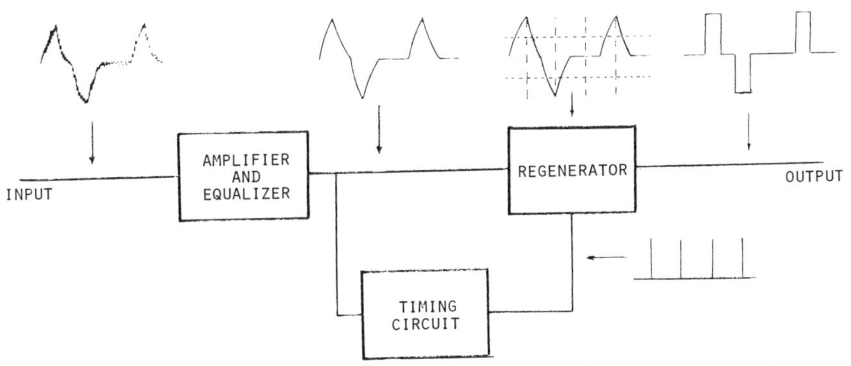

Fig. 5–14. Regenerative repeater functions

Threshold circuits are used to determine whether a positive, zero, or negative signal state exists at such times. When either the positive or negative threshold is exceeded, a pulse of the proper polarity, duration and amplitude is produced as the output. Otherwise, the output remains zero.

The tolerable channel error rate places a limit on repeater performance. Errors are produced both by near-end crosstalk and by receiver front end noise, which has a gaussian characteristic.

The channel error rate produced by receiver front-end noise places a limit on the maximum distance between repeaters. Thus if transmission is to be done at an error rate not exceeding a specified level, the signal-to-noise ratio must be at least a minimum value based on the equation[18]

$$\text{b.e.r.} = 0.5\ \text{erfc}(V_p/\sqrt{2}\sigma_n) \tag{5–1a}$$

where V_p = peak signal level
σ_n = rms level of gaussian noise.

Eq. (5–1a) is based on simple two-level transmission. In the event of multilevel transmission, e.g., $m > 2$, a more general equation is used:

$$\text{b.e.r.} = [(m-1)/m]\text{erfc}(V_p/(m-1)\sqrt{2}\sigma_n) \tag{5–1b}$$

Figure 5–15 is a plot of Eq. (5–1b). Even for bit error rate values on the order of 10^{-10}, the signal-to-noise ratio requirements are very modest compared with those of analog transmission.

At each digital repeater the signal is received and detected, whereupon a new output waveform is produced. This process is called *waveform regeneration*. It is a primary advantage of digital transmission. To achieve an end-to-end bit error rate of 10^{-7} in transmission through a large number of repeaters, an adequate signal-to-noise ratio must be present at the input to each

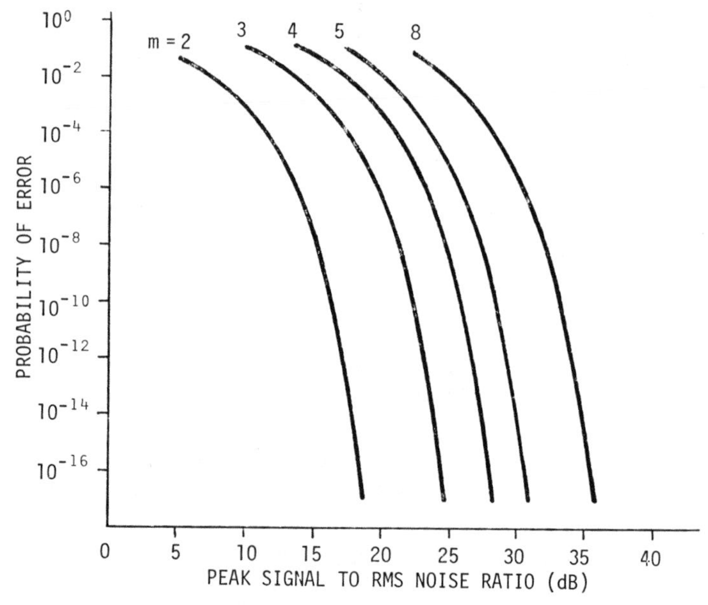

Fig. 5–15. Bit error rate for multiple decision thresholds (© 1982, Bell Telephone Laboratories. Reprinted by permission)

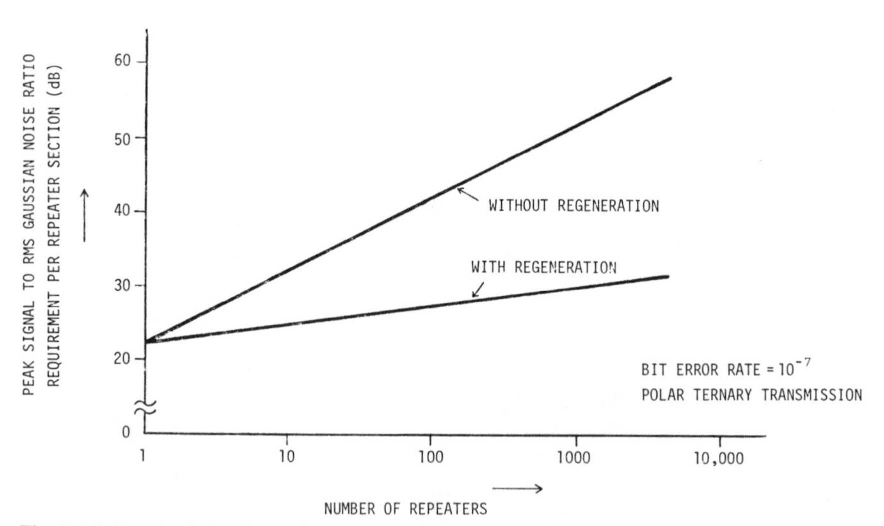

Fig. 5–16. Required signal-to-noise ratio versus number of repeaters (© 1982, Bell Telephone Laboratories. Reprinted by permission)

repeater. In the case of three-level (polar ternary) transmission, Fig. 5–16 shows the input signal-to-noise ratio requirement both with and without waveform regeneration. The advantage of regeneration is quite clear from this figure.

5.12. DIGITIZATION OF THE LOOP PLANT

The transition to digital technology within the telephone industry started in the exchange plant, and is being extended to long-haul transmission. The final part of the transition, within the loop plant, now is beginning to materialize. The description of the digital loop plant (subscriber lines) contained in this section is based on work being done by the common carriers as well as by manufacturers to develop the systems to be used.[19,20]

Because the investment in subscriber lines by wireline common carriers everywhere is so vast, the utilization of that investment to a maximum extent is important. Accordingly, the use of two-wire transmission between the subscriber's instrument and the serving central office is a significant factor. Important design considerations are compatibility with related systems, crosstalk, and impulse noise from the central office. System parameters include transmission rate, transmitted power, and the choice of line codes. Design uniformity is important with respect to component interchangeability, but the design must be flexible enough to accommodate variations in the loop plant and local implementation details, as well as manufacturing variations in components.

5.12.1. Transmission Modes

Although many end instruments operate on a totally wired basis, the use of radio transmission for part or all of the local loop is gaining in importance. In addition, the use of fiber optics (see Chapter 9) is beginning to be seen in the Integrated Services Digital Network concept (see Chapter 14). This section is divided into discussions of wired systems (5.12.1.1) and radio systems (5.12.1.2).

5.12.1.1. Wired Systems. The three basic transmission modes envisioned for two-wire operation are *hybrid, frequency-division multiplexing* (FDM), and *time compression multiplexing* (TCM). The hybrid system achieves isolation through the use of echo cancellers, with transmission in both directions occurring at the same time. The principle of echo cancellation is described in Section 5.7.2. The FDM system uses separate frequency bands for the two directions of transmission, thus accomplishing them simultaneously. TCM separates the two directions in the time domain. Accordingly,

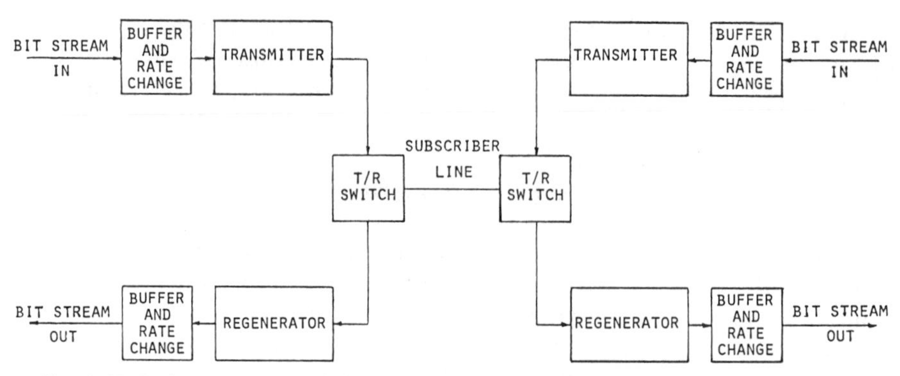

Fig. 5–17. Basic components of time compression multiplex (TCM) local loop transmission system. (Courtesy Ahamed et al., Ref. 19, © IEEE, 1981.)

the line rate must be more than twice the bit rate. Thus for standard 64 kb/s PCM, a line rate greater than 128 kb/s must be used. Design parameters for TCM are the block length transmitted, the line length (and thus pulse transit time), and the amount of idle time needed for line transients to decay. A dependable hybrid system requires high quality echo cancellers. The TCM system, however, is readily implemented with digital components, and is less sensitive to gauge discontinuities and bridged taps. Figure 5–17 illustrates the components of a TCM system.

The maximum line length is limited by both propagation time and crosstalk. The propagation time limitation has been shown[21] to be related to the system parameters as follows. Assume that the line is to be used for 64-kb/s PCM plus 8-kb/s data in each direction, and that one additional bit for every nine is to be used for housekeeping purposes. This results in 80 kb/s in each direction. Let each burst in each direction have a length of $k \cdot 10$ bits. A burst then will have a duration of $k \cdot 125$ μs.

Let τ_g = group delay per unit length of line in μs/km

τ_l = length of a $k \cdot 10$ bit burst in μs.

Allowing a total time of $k \cdot 125$ μs for two-way transmissions, the maximum distance that can be served by the line is

$$d_{max} = [k \cdot 125 - 2\tau_l]/2\tau_g \text{ km} \tag{5–2a}$$

Let ζ = line bit rate/input bit rate

= compression factor.

Then $\tau_l = k \cdot 125/\zeta$ μs. Since $\tau_g \simeq 5$ μs/km for typical local line conductors,

$$d_{max} = k \cdot 12.5(\zeta - 2)/\zeta \text{ km} \tag{5–2b}$$

For example, at a line rate of 256 kb/s, $\zeta = 3.2$. For 10 bit bursts, $k = 1$, so $d_{max} = 4.7$ km, generally adequate for more than 95% of the subscribers. This limit only takes pulse travel time into account. Crosstalk will be considered in Section 5.12.2.

5.12.1.2. Radio Systems. Numerous factors are contributing to the rapid increase in the use of radio to replace the wired local loop in given situations either partially or entirely. Present systems operate on an analog basis, but digital implementations are under consideration. Partial replacement of the wired local loop is seen in the popularity of the cordless analog end instrument. Cordless systems to date have been wired to the user's location, usually on a 2-wire local loop basis. The end instrument base unit, however, contains a low power FM transmitter and receiver, as does the handset. Full duplex operation is achieved through the use of frequency pairs in the 46- and 49-MHz bands, or the 49- and 1.7-MHz bands, with separate frequency pairs being available to minimize interference between nearby cordless systems. The power levels used allow operation to distances of about 250 m between the handset and the cradle. When placed in the cradle, the handset's self-contained battery is recharged. Extension telephones are implemented as part of the wired portion of the system.

Radio common-carrier systems use wired local loops from the common carrier's switching exchange to a mobile base station which covers an entire community, often on a competitive basis, and serves numerous mobile units, most of which are installed in automobiles, but some of which are hand held. Existing systems use analog (FM) modulation, but work is being done toward the use of digital modulation in such systems.[22] One approach, using 64 kb/s PCM, takes alternate data words (8-bit speech samples) and forms "odd" and "even" streams.[23] One stream is delayed by 3–7 ms and the streams are interleaved prior to transmission using BPSK. At the receiver the previously undelayed stream is given the same 3–7 ms delay and interleaved with the other stream. As a result, an error burst will place words (speech samples) with bit errors next to words that are correctly received. Error detection can identify words in error, and interpolation then is used to eliminate the effects of the errors. Since the duration of deep fades often is 5 ms or less, this system showed itself capable of providing toll quality speech in the presence of a ber of 10^{-3}.

Cellular radio systems, like the radio common carrier systems, currently use analog (FM) modulation in providing service to both mobile and handheld units. Because cellular systems make extensive use of space diversity reception, optimum received signal combining techniques using PSK signals have been studied looking toward the implementation of digital cellular radio systems.[24] Optimum combining is important not only in handling fading problems, but also those of cochannel interference as it occurs in cellular radio systems.

Other uses of radio transmission to the end instrument are found in the single-channel-per-carrier (SCPC) systems used in satellite transmission, as discussed in Section 8.3.5.1, and the digital termination (bypass) systems discussed in Section 7.5.2.1.

The remaining discussion of this section deals with two-wire transmission to digital end instruments.

5.12.2. Crosstalk

While both near-end crosstalk (NEXT) and far-end crosstalk (FEXT) can limit transmission range, TCM allows the self-NEXT problem to be avoided by sending bursts from the central office to all subscriber lines simultaneously. (Other subscriber lines still can be affected by NEXT, however.) In general, two crosstalk determinations must be made: (1) when the line is the source of interference and (2) when the line is disturbed from the outside. Situation (1) calls for reducing the transmitted power, while situation (2) calls for increasing it. The combined problem tends to limit transmission range.

FEXT may result from signals coming from subscribers near the exchange. In fact, with TCM, FEXT is the main crosstalk problem. Subscribers far from the exchange can suffer from this problem. Accordingly, FEXT places a limit upon the maximum subscriber line length. Based upon a 10^{-8} bit error rate, a signal to FEXT ratio of 21 dB, and an estimated 20 interferers, the maximum line lengths for various line bit rates have been determined.[25] Thus for a 200-kb/s line rate the maximum length is 7 km, while for a 400-kb/s line rate the maximum length is only about 2.6 km.

5.12.3. Impulse Noise

Another impairment that may limit range is impulse noise caused by transients from the central office and transmitted via analog subscriber lines. Such transients can raise the bit error rate (BER) where the received power level is low. In addition, it can introduce synchronization errors in the period between adjacent bursts. The best protection against this form of impairment is to enable the receiver only a short time interval before a received burst is expected. Measurements[26] have shown that the line bit rate should not exceed 200 kb/s on 7 km of 0.5 mm paired cable if the bit-error rate is to be kept below 10^{-7}.

5.12.4. Line Codes

Three candidate line codes are the bipolar, duobinary (partial response) and nonreturn to zero codes. They will be discussed in that order.

Considerations in the selection of a line code include the code's ability to function with minimal degradation in the presence of NEXT and impulse noise, the required receiver bandwidth, how readily clock can be recovered, and whether a dc component is present.

Bipolar code is simple to implement. It has no dc component, and its spectral peak is at half the keying rate. It is not very sensitive to line distortion and thus does not require critical cable adjustments.[27] Detection is kept simple if the code is implemented on a biphase basis, i.e., so that only two phases need be transmitted.

Partial response codes such as duobinary and modified duobinary consist of three levels and thus confine their transmitted energy to a frequency band between zero and half the keying rate. The result is increased transmitted power in this band and thus better immunity to crosstalk and impulse noise compared with the bipolar codes. The disadvantage of partial response codes is that they require sophisticated equalizers and regenerators. Modified duobinary code also requires very precise timing recovery.

Nonreturn to zero codes, like duobinary codes, may require precise low frequency equalization. Their advantages with respect to impulse noise and crosstalk resistance are achieved at the expense of additional design complexity.

5.12.5. Digital End Instruments

A fully digital telephone system is one that is completely digital all the way to the end instrument. This means that speech is encoded to PCM, DM, or some other form before it leaves the subscriber's set and, likewise, arrives at the subscriber's set in digital form, to be decoded there. The technical feasibility of such a set is not new, but its economic feasibility has come into focus only with the appreciable cost reductions that have occurred with advances in integrated circuits. The ease with which other subscriber digital services can be integrated into such an all digital telephone helps to make it even more attractive. Moreover, with the use of TCM, as described in Section 5.12.1, hybrid transformers are not needed, and singing and echo problems do not exist.

Digital end instruments producing 2.4 kb/s LPC are in use for secure communication purposes. Four such instruments can be time-division-multiplexed to produce a 9.6 kb/s stream which can be handled by a modem on a single conditioned voice bandwidth line. Alternatively, one or more such end instruments can operate in locations where 2.4 kb/s digital lines are available. The widespread appearance of digital end instruments probably will occur first in business systems in which the private branch exchange (PBX) and the end instruments all are provided by a single supplier. Elsewhere, the change to digital end instruments is expected to proceed slowly because of the lack of proper interface protocols and digital subscriber lines.

5.12.5.1. Extensions. A special problem related to the digital end instrument is that of extension service.[28] Each extension must be given its time slots, and correct synchronization must be maintained among all end instruments. Moreover, any rewiring of the local plant must be kept to a minimum. This means the continued use of the existing two-wire local plant.

The foregoing requirements cause serious consideration to be given to a two-wire system using the TCM transmission mode at a sufficiently high bit rate to accommodate up to three extensions. Based upon a standard 8000/second sampling rate, a 125-μs sampling period can be assigned to each direction of transmission. The central office or PBX switch and the end instrument(s) alternately transmit, with an intervening guard period to account for round-trip delay time. In the case of extension instruments, a separate reply slot can be provided for each instrument, as illustrated in Fig. 5–18. The parameters shown there are compatible with the standard T1 format used by the common carriers. The maximum number of extensions, however, must be chosen in advance, as must the guard slot duration. The latter will constitute one limit on the maximum line length between extensions and to the switch. However, the distance to the switch can be extended

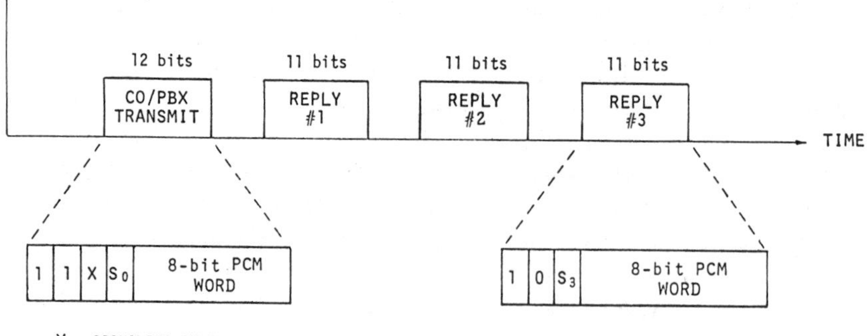

Fig. 5–18. TCM format for system with extensions. (Courtesy L. G. Abraham and D. M. Fellows, Ref. 20, © IEEE, 1981.)

through the use of a remote switch/concentration/multiplexer unit. Large extension spacings can be accommodated through the use of conference techniques at the switch. As can be seen from Fig. 5–18, special bits, designated S_n, provide a separate channel for ringing, supervision, and test functions. In addition, this channel can send display bits which can be used to inform the end instrument user which number is calling. The provision of this separate channel simplifies the design of the end instrument in return for a modest increase in line bit rate. In addition, it simplifies the transition to common channel signaling between central offices. Talking and signaling can occur simultaneously on a noninterfering basis. In fact, a rearrangement of the time slots could allow two separate simultaneous conversations to take place.

5.12.5.2. Design. Simplification and uniformity of the design of the subscriber set are paramount factors in saving overall system cost. To this end the central switch is given the processing functions, as well as the automatic testing of the end instruments attached to it.

An important feature of any end instrument is *side tone,* which gives the speaker a replica of his own voice in his receiver. Side tone can be achieved by receiving and decoding the instrument's own transmitted words at a reduced level.

Another design factor is time slot assignment. Such assignments may be either fixed or on a dynamic basis. Some form of dynamic assignment is more economical of line capacity than is fixed assignment, but this means that each instrument must search for an available slot when it goes off-hook.

Power consumption and packaging considerations also play an important role in the design of digital end instruments. Experimental transducers made of polyvinylidene fluoride (PVDF) can be used for both transmitters and receivers. PVDF is a plastic piezoelectric material. For telephony use it is aluminum-coated on both sides. The transducer functions as either a transmitter or a receiver depending on how its associated amplifier is connected. Excellent frequency response is available, well beyond telephony demands. Its impedance is on the order of 100 ohms at voice frequencies. CMOS voltage amplifiers can be used for an overall low power drain instrument. Monolithic codecs, filters and LSI logic all help to minimize the parts count of the instrument, and thus to provide high reliability.

5.13. SPEECH RECOGNITION

The remainder of this chapter is devoted to several special techniques which are facilitated by having speech in digital form. Their implementation thus can be handled readily within an overall digital telephone system. Significant application areas for these techniques are expected to develop in the integrated voice-data networks of the Integrated Services Digital Network (see Chapter 14). The special techniques are known, respectively, as speech recognition, voice input, and voice response.

5.13.1. Speech Production

Speech recognition by a piece of digital equipment requires that speech be treated in accordance with the nature of speech production. Several significant definitions follow:

diphthong A gliding monosyllabic speech item that starts at or near the articulatory position for one vowel and moves to or toward the position for another. (Examples: *bay, boat, buy, how, boy, you.*)

formant A frequency at which the vocal tract tube resonates.

phoneme One of a set of distinctive sounds (vowels, diphthongs, semivowels, consonants).

plosive sound A sound resulting from making a complete closure (usually toward the mouth end), building up pressure behind the closure, and abruptly releasing it. (Example: *tsh, j.*)

semivowel A sound characterized by a gliding transition in vocal tract area function between adjacent phonemes. (Examples: *w, l, r, y.*)

A human sound source provides an acoustic carrier, known as the *pitch,* for the speech intelligence, which appears largely as time variations in the envelope of the signal produced. Sounds can be characterized by voiced sounds, fricatives, and stops.

A voiced sound is produced by forcing air through the glottis with the tension of the vocal chords adjusted so that they vibrate in a relaxation oscillation, thereby producing quasi-periodic pulses of air which excite the vocal tract. (Examples: *u, d, w, i, e.*) A voiced sound is made up of a nearly periodic sequence of pulses. Its network equivalent is a current source exciting a linear, passive, slowly time-varying network. Its spectrum falls off with frequency at a rate of 12 dB/octave.

A *fricative* is an unvoiced sound generated by producing a constriction at some point in the vocal tract (usually toward the mouth end), and forcing air through the constriction at a high enough velocity to produce turbulence. (Example: *sh.*) The result is a sustained random sound pressure. Its network equivalent is a series voltage source whose internal impedance essentially is that of the constriction and typically is of large value. Its spectrum is a broad band with gentle attenuation at the band edges.

A *stop* is a transient sound which may be either voiced (Examples: *b, d, g*) or unvoiced (Examples: *p, t, k*). A stop results from an abrupt release of air pressure built up behind a complete occlusion. Its network equivalent is a step function. Its spectrum falls off with frequency f as $1/f$.

5.13.2. Speech Recognition Systems

Speech recognition systems are of two levels of sophistication. The simpler ones are capable of isolated word recognition (IWR), while the more complex ones can achieve continuous word recognition (CWR). Applications for isolated word recognition include data entry (via speakerphone) under conditions such that the use of a keyboard is impractical since the user's hands are otherwise occupied. Other (speakerphone) applications include the programming of numerically controlled machine tools, with the user's hands and eyes busy with blueprints, or voice commands to industrial robots where shop floor conditions make the use of buttons difficult or impossible.

Continuous word recognition refers, ideally, to the transcription of any spoken sentence from any speaker. It has the potential for eliminating many clerical tasks as well as improving man's speed of communication through programming as well as data entry. Laboratory work has resulted in the recognition of artificial, limited vocabulary languages with rigid, predictable structures for ease of recognition.[29] One commercial device, the Nippon Electric DP-100, is claimed to achieve 99% recognition on a 50 to 150 word vocabulary.

One characteristic of a language significant to CWR is the perplexity of the language. The perplexity is the average number of words which theoretically could follow a given word. For natural language, the perplexity is greater than 50 words. For the DP-100 language, it is less.

Advanced systems capable of CWR have been built by IBM and by Carnegie-Mellon[30] for research purposes. The IBM system has a 1000 word English vocabulary with a 20 word perplexity. It requires 20 minutes to recognize a six-second sentence. Improvements over existing capabilities required to make CWR a practical reality can be stated relative to the capabilities of the most advanced system, that of IBM. These improvements are a tenfold vocabulary expansion, a tenfold perplexity expansion, a two hundred-fold speed improvement (achievable with a superconducting Josephson junction computer), and a one hundred-fold cost reduction.

Numerous reasons exist for the difficulties experienced by CWR equipment developers. One of the foremost is the fact that no one speaks the same words the same way each time. This is true also of phonemes. This means that the device must recognize a similarity of patterns rather than identical sounds. The human brain, on the other hand, in ways not fully understood, works on an associative basis which has not yet been simulated by computer.

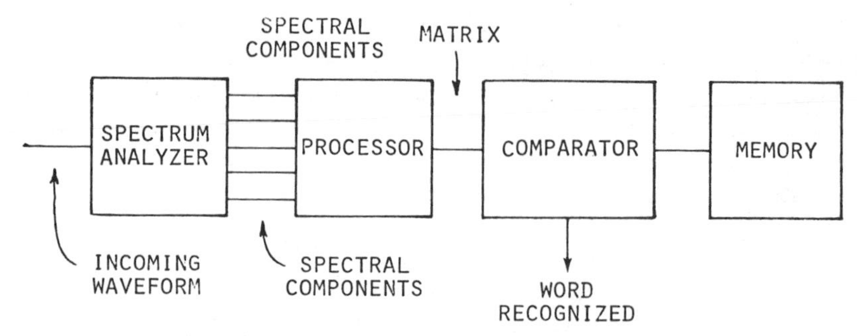

Fig. 5–19. Isolated word recognition system.

5.13.3. Speech Recognition Techniques

Isolated word recognition (IWR) systems have been devised by splitting each word into a number of equal time segments, e.g., 16. Each segment then is analyzed for a number of spectral components, e.g., 32. These components are processed to extract a set of key features. The resulting matrix (16 × 32 using the example numbers) then is matched against a number of "word templates" stored in memory. Figure 5–19 illustrates this sequence of operations in block diagram form.

Continuous word recognition (CWR) involves the use of a technique called nonlinear time warping. This technique compensates for the variations in time used by various speakers to pronounce different words. Figure 5–20 illustrates time warping and shows that it amounts to a variable delay implemented by the computer as it expands or contracts the lengths of various parts of words to improve their match with a given word template. The computer tries each of a number of path slopes and calculates the cumulative difference between the template and the signal to that point.

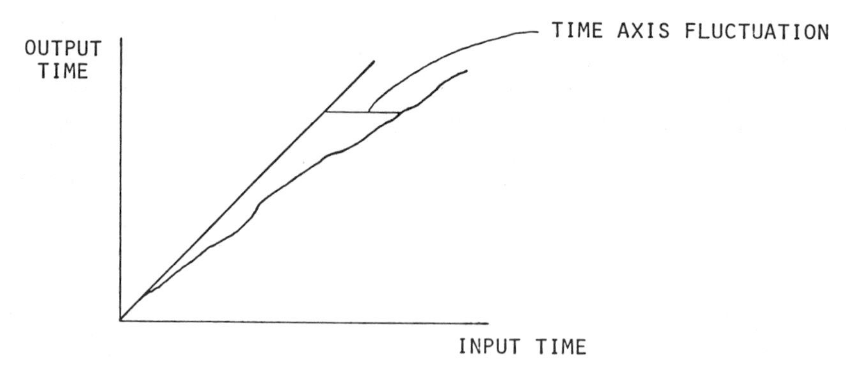

Fig. 5–20. Time warping for continuous word recognition.

5.13.4. Voice Input Systems

Word recognition systems are used in providing voice input to computers. In many cases this voice input can best be transmitted to the computer via a digital subscriber line, as described in Section 5.9. Numerous applications exist in manufacturing, and include quality control and inspection, automated material handling, and parts programming for numerically controlled machine tools. In many cases the reason for using voice input is that the inspector's hands and eyes are occupied in the inspection task. Applications include television faceplate inspection, automobile assembly line inspection, and various types of receiving inspections in manufacturing plants and service facilities. In the case of parts programming for numerically controlled machine tools, the programmer speaks each machine command into a speakerphone or headset. The system eliminates the need for a separate conversion of the information into computer-compatible format. The programmer's hands remain free to handle prints or perform calculations.

5.14. COMPUTER VOICE RESPONSE

The general principle underlying computer voice response is illustrated in Fig. 5-21. Words or phrases are stored in a computer memory called the *vocabulary*. Messages then are created by retrieving the required words and phrases and reproducing them in the proper sequence.

The choice of the digital coding method has a major impact on the amount and type of digital memory required, and therefore on its cost. Thus LPC at 2.4 kb/s requires less than 4% of the memory required by 64-kb/s PCM.

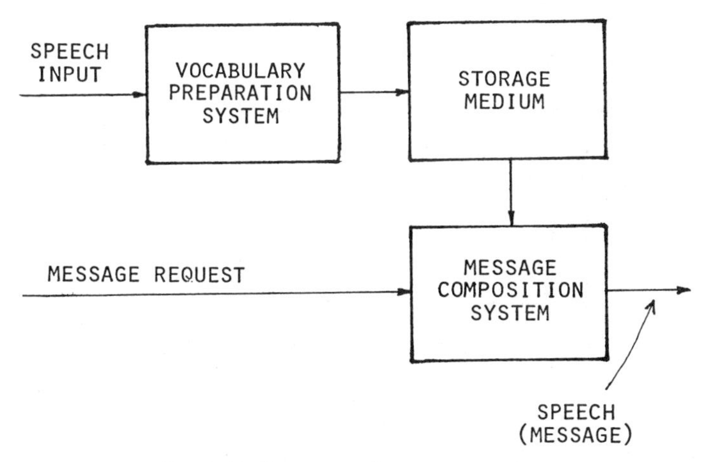

Fig. 5-21. Computer voice response system.

While LPC may not be satisfactory from a quality viewpoint, SBC at 9.6 kb/s or CVSD at 16 kb/s may be quite cost effective while allowing satisfactory voice quality as well.

Computer voice response requires the preparation of a suitable vocabulary by a trained speaker, its digitization and insertion into memory, and the establishment of a directory of word addresses. The absence of an input is used to establish the start and end of each word. Message composition by the computer then involves logical operations followed by the proper data transfers. A single computer, with different programs, can be used for both vocabulary preparation and message composition.

Figure 5–22 shows a voice response system capable of handling ten lines simultaneously.[31] The vocabulary preparation portion is used only initially,

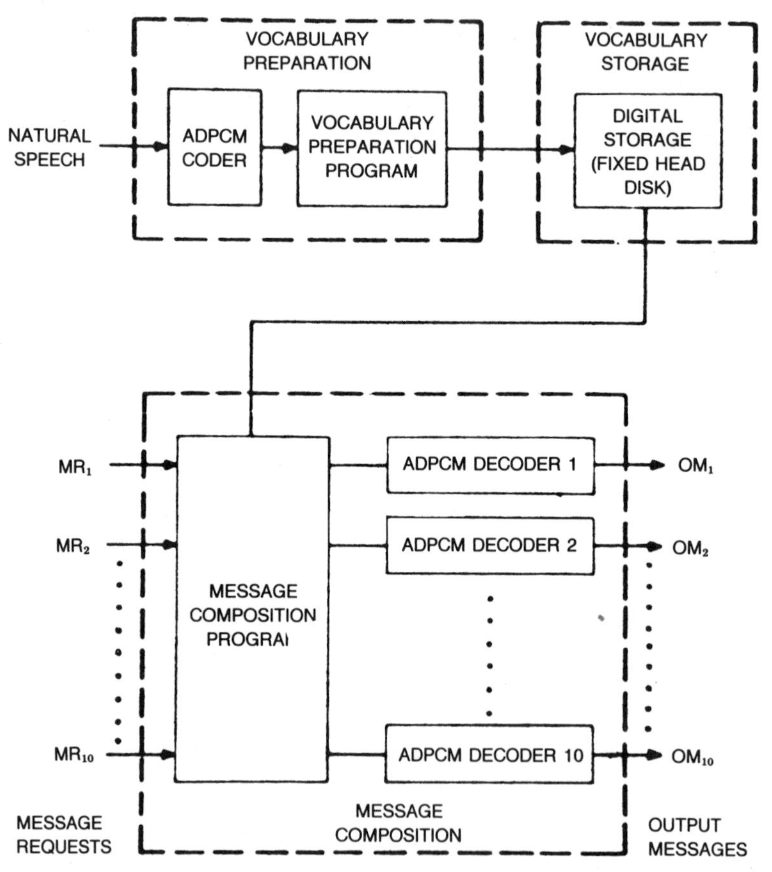

Fig. 5–22. Ten line digital voice response system using ADPCM. (Courtesy L. H. Rosenthal et al., Ref. 22, © IEEE, 1974.)

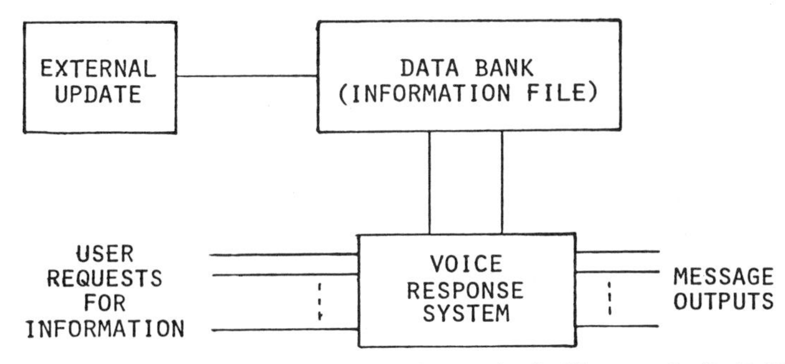

Fig. 5–23. Voice response system for information retrieval. (Courtesy L. R. Rabiner and R. W. Schafer, Ref. 23, © IEEE, 1976.)

or when changes must be made in the vocabulary. A variety of applications exists for systems such as this one. They include computer-aided voice wiring and information retrieval systems. In the case of computer-aided wiring, a wire list is recorded and the wireman actuates the system by a foot switch. He then hears the wiring instructions one at a time through a speakerphone. Usually a relatively small vocabulary and frequent modifications make computer design of the list desirable.

Information retrieval systems using voice response generally operate on the principle illustrated[32] in Fig. 5–23. A data bank is updated from an external source. This data bank then is searched by the voice response system to provide outputs as requested. Information retrieval systems include directory assistance systems in which the user can key in requests on a telephone equipped for dual-tone multifrequency (DTMF) calling. (DTMF is discussed in Section 10.5.3.2.) Other applications include a stock price quotation system and a flight information system. Enroute weather information for flyers also can be provided. A data set testing system is programmed so that, in response to an installer's request, a computer applies a custom test signal to determine if the data set, and all its options, are performing correctly. In addition, the computer advises the installer about the expected causes of any trouble.

In a speaker verification system, the user enters his claimed identity, speaks a verification phrase, and requests a transaction. A computer then performs a feature analysis on the voice, based on the verification phrase, compares the result with recorded reference data on the speaker's voice, and determines whether to accept or reject the speaker's request.

PROBLEMS

5.1 A circuit experiences a 10-μs noise burst which causes loss of synchronization. A subsequent portion of the bit stream is lost. If this loss is 1 ms long, what will be the effect on voice communication? on data communication?

5.2 The following stream is to be scrambled: 0 1 1 0 0 0 1 0. Compose three arbitrary pseudo-random streams. In each case, what is the transmitted stream? Assuming no channel errors, derive the recovered stream at the receiver's output.

5.3 Why is two-wire service important in digital subscriber loops?

5.4 Why can digital end instruments not be paralled directly for extension service as is the case with analog end instruments?

5.5 Under what conditions does channel error coding become very important?

5.6 Why do DSI systems generally require a large number of required and available channels to be effective?

REFERENCES

1. "The Extended Framing Format Interface Specifications," Technical Advisory No. 70, Issue 2, American Telephone and Telegraph Company, Basking Ridge, NJ, Sept. 29, 1981.
2. Aaron, M. R., "PCM Transmission in the Exchange Plant," *Bell System Technical Journal,* Vol. 41, pp. 99–141, (Jan. 1962).
3. Henry, P. S. and Glance, B. S., "A New Approach to High Capacity Digital Mobile Radio," *Bell Syst. Tech. J.,* Vol. 60, No. 8, pp. 1891–1904, (Oct. 1981).
4. *Telecommunications Transmission Engineering,* Vol. 2, *Facilities,* Bell System Center for Technical Education, Western Electric Company, Winston-Salem, NC, p. 168, 1977.
5. Bellamy, J. C., *Digital Telephony,* John Wiley & Sons, New York, NY, pp. 337–343, 1982.
6. *Telecommunications Transmission Engineering,* Vol. 2, *Facilities,* Bell System Center for Technical Education, Western Electric Company, Winston-Salem, NC, p. 559, 1977.
7. Suyderhoud, H. G., Onufry, M., and Campanella, S. J., "Echo Control in Telephone Communications," *1976 National Telecommunications Conference Record,* pp. 8.1–1 to 8.1–5.
8. Hatch, R. W. and Ruppel, A. E., "New Rules for Echo Suppressors in the DDD Network," *Bell Laboratories Record,* Vol. 52, pp. 351–357, (December, 1974).
9. Helder, G. K., and Lopiparo, P. C., "Improving Transmission in Domestic Satellite Circuits," *Bell Laboratories Record,* Vol. 46, pp. 202–207, (September, 1977).
10. Duttweiler, D. L., and Chen, Y. S., "A Single-Chip VLSI Echo Canceler," *Bell System Technical Journal,* Vol. 59, No. 2, pp. 149–160, (February, 1980).
11. Messerschmitt, D. G., "Echo Cancellation in Speech and Data Transmission," *IEEE Journal on Selected Areas in Communications,* Vol. SAC-2, No. 2, pp. 283–297, (March, 1984).
12. Brady, P. T., "A Statistical Analysis of On-Off Patterns in 16 Conversations," *Bell Syst. Tech. J.,* Vol. 47, No. 1, pp. 73–91, (Jan. 1968).
13. Easton, R. L., et al., "TASI-E Communications System," *ICC '81 Conference Record,* IEEE Document No. 81 CH1648–5, pp. 49.3.1–49.3.5.
14. See note 13.
15. Reiser, J. H., Suyderhoud, H. G. and Yatsuzuka, Y., "Design Considerations for Digital Speech Interpolation," *ICC '81 Conference Record,* IEEE Document No. 81 CH1648–5, pp. 49.4.1–49.4.7.
16. Verma, S. N., Ramasastry, J., and Monsees, W. R., "Digital Speech Interpolation Applications for Domestic Satellite Communications," *NTC '78 Conference Record,* IEEE Document No. 78 CH1354–0 CSCB, pp. 14.4.1–14.4.5.
17. *Telecommunications Transmission Engineering,* Vol. 2, *Facilities,* Bell System Center for Technical Education, Western Electric Company, Inc., Winston-Salem, NC, pp. 542–546 and 605, 1977.
18. *Transmission Systems for Communications,* Bell Telephone Laboratories, Inc., prepared for

publication by Western Electric Company, Inc., Technical Publications, Winston-Salem, NC, pp. 627–631, 1971.

19. Ahamed, S. V., Bohn, P. P. and Gottfried, N. L., "A Tutorial on Two-Wire Digital Transmission in the Loop Plant," *IEEE Trans. Comm.,* Vol. COM-29, No. 11, pp. 1554–1564, (Nov. 1981).
20. Abraham, L. G. and Fellows, D. M., "A Digital Telephone with Extensions," *IEEE Trans. Comm.,* Vol. COM-29, No. 11, pp. 1602–1608, (Nov. 1981).
21. Brosio, A., et al., "A Comparison of Digital Subscriber Line Transmission Systems Employing Different Line Codes," *IEEE Trans. Comm.,* Vol. COM-29, No. 11, pp. 1581–1588, (Nov. 1981).
22. Lee, W. C. Y., *Mobile Communications Engineering,* McGraw-Hill, Inc., 1982.
23. Wong, W. C. et al., "Time Diversity with Adaptive Error Detection to Combat Rayleigh Fading in Digital Mobile Radio," *ICC '83 Conference Record,* IEEE Document No. 83 CH1874–7, pp. B8.5.1–B8.5.5.
24. Winters, J., "Optimum Combining in Digital Mobile Radio with Co-Channel Interference," *ICC '83 Conference Record,* IEEE Document No. 83 CH1874–7, pp. B8.4.1–B8.4.5.
25. Hayashi, I., Mano, S. and Komlya, R., "Transmission System Architecture for Digital Subscriber Loops," *ICC '81 Conference Record,* IEEE Document No. 81 CH1648–5, pp. 48.6.1 to 48.6.6.
26. Ibid.
27. See note 17.
28. See note 14.
29. Reddy, R., and Zue, V., "Artificial Intelligence: Recognizing Continuous Speech Remains an Elusive Goal," *IEEE Spectrum,* Vol. 20, No. 11, pp. 84–88, (Nov. 1983).
30. Bisiani, R., Mauersberg, H., and Reddy, R., "Task-Oriented Architectures," *Proc. IEEE,* Vol. 71, No. 7, pp. 885–898, (July, 1983).
31. Rosenthal, L. H., et al., "A Multiline Computer Voice Response System Utilizing ADPCM Coded Speech," *IEEE Trans. Acoustics, Speech, and Signal Proc.,* Vol. ASSP-22, No. 5, pp. 339–352, (Oct., 1974).
32. Rabiner, L. R., and Schafer, R. W., "Digital Techniques for Computer Voice Response: Implementations and Applications," *Proc. IEEE,* Vol. 64, No. 4, pp. 416–433, (April, 1976).

6
Digital Transmission

6.1. INTRODUCTION

The objective of digital transmission, as applied to telephony, is to convey digital voice and nonvoice signals at sufficiently low bit error rates (b.e.r.) to assure satisfactory performance of the system from the users' viewpoints. This means, for voice, a bit error rate that generally does not exceed 10^{-4}, since 64-kb/s PCM performance is quite good if the b.e.r. is kept to this value or lower. Other voice coding techniques, such as ADPCM, can tolerate higher b.e.r. values, as was described in Chapter 4. The transmission of signaling and supervision data, however, generally requires lower b.e.r. values, achievable with error correction techniques to prevent wrong numbers, false alarms, and other undesirable occurrences.

An important aspect of digital transmission is Shannon's theorem, which relates the capacity C of a channel to its bandwidth W, and signal-to-noise ratio S/N as follows:

$$C = W \log_2 (1 + [S/N]) \text{ b/s} \tag{6–1}$$

According to Shannon up to C bits/second can be sent error free through such a channel. However, equipment becomes very complex as the Shannon limit is approached. Shannon's theorem also quantifies some well-known trends which are found in practice. The data rate that can be sent over a channel increases in proportion to the bandwidth of the channel. In addition, an increase in S/N allows an increase in data rate for a given bandwidth, although an increase in equipment complexity may be required to realize the higher data rate.

6.2. DIGITAL MODULATION TECHNIQUES

Modulation is defined as the alteration of a carrier in order to cause it to convey information. The characteristics of a carrier can be expressed in the form $A \sin (\omega t + \phi)$, where A = amplitude, ω = radian frequency, and ϕ = phase. Thus a carrier can be modulated in amplitude, frequency, or phase. Digital modulation refers to the use of a limited set of discrete values of A,

ω, and ϕ. The type of digital modulation used affects the relationship between b.e.r. and S/N. In addition, the type of modulation affects the amount of spectrum occupied by a given transmission. Moreover, the equipment complexity and cost depend upon the modulation technique selected.

Modulation of the amplitude A and the phase ϕ of a carrier can be shown on a signal-state space diagram, as illustrated in Fig. 6–1. This is also called a constellation. The detector must determine which of the four allowable A, ϕ combinations have been sent at any given time. Since four states are allowed, each state can convey a unique combination of bits. For example, the following assignment might be made:

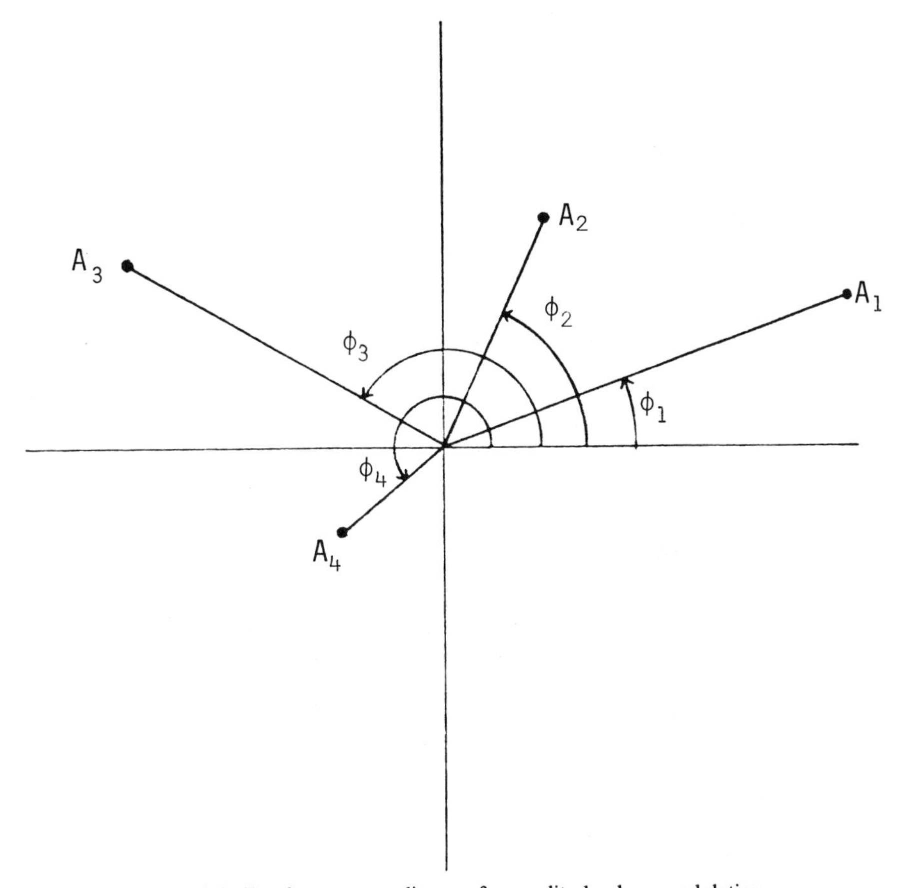

Fig. 6–1. Signal-state space diagram for amplitude-phase modulation.

Signal State	*Bits Conveyed*
$A_1\phi_1$	00
$A_2\phi_2$	01
$A_3\phi_3$	10
$A_4\phi_4$	11

For pure phase modulation, $A_1 = A_2 = A_3 = A_4$, while for pure amplitude modulation, $\phi_1 = \phi_2 = \phi_3 = \phi_4$, but the A's may take discrete negative as well as discrete positive values.

Modulation of the frequency of a carrier produces frequency shift keying, which is discussed in Section 6.1.4.

The amount of spectrum occupied by a digitally modulated signal can be shown to depend on the type of coding used to produce that signal. For example, Fig. 6–2 shows a digital sequence which is coded on a two-level basis and which thus undergoes changes every bit period. Alternatively, with

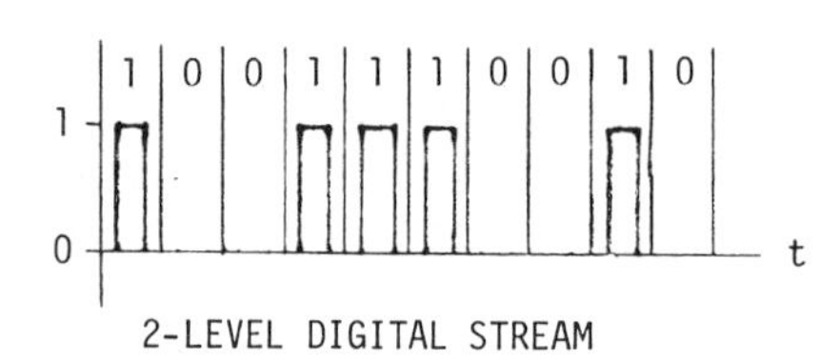

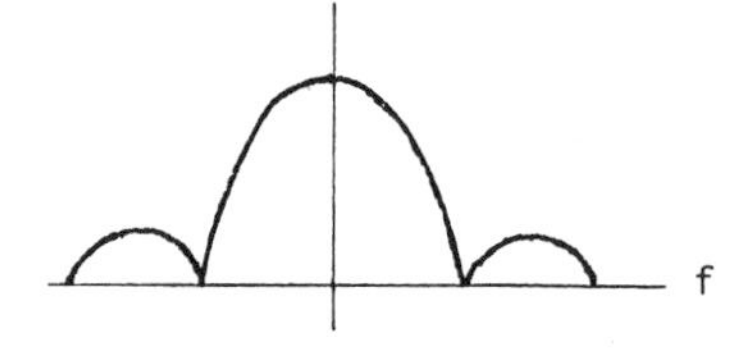

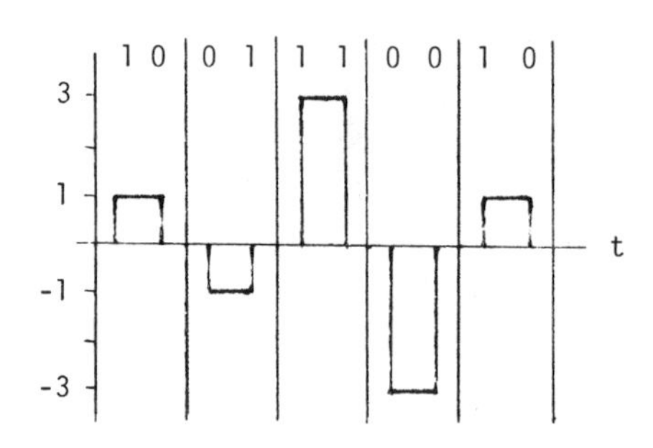

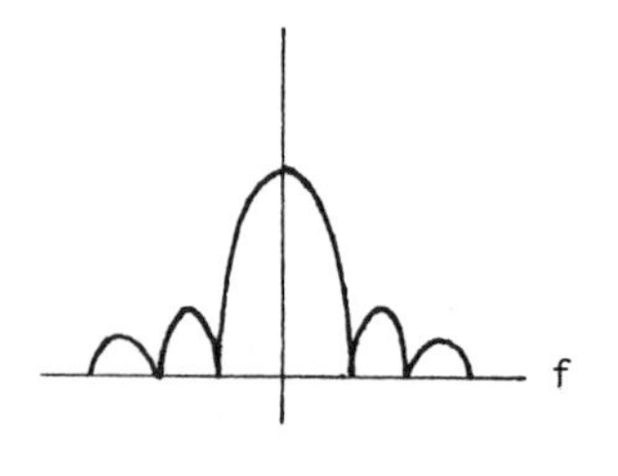

Fig. 6–2. Use of coding in reducing spectrum occupancy.

Table 6–1. Minimum Number of Voice Channels in Microwave Bands Below 12 GHz.

FREQUENCY RANGE (GHZ)	MINIMUM NO. OF VOICE CHANNELS	RADIO CHANNEL BANDWIDTH (MHZ)
2.11–2.13	96	3.5
2.16–2.18	96	3.5
3.70–4.20	1152	20
5.925–6.425	1152	30
10.70–11.70	1152	40

coding on a four-level basis, a change occurs only every other bit period. Four-level coding thus requires less frequent level changes and, correspondingly, occupies less spectrum.

Spectrum crowding has become a serious problem in the United States and other countries with large volumes of telecommunications traffic. As a result, the FCC[1] has established a rule regarding the minimum number of voice channels that a microwave radio system in the domestic public radio service (common carrier) must be capable of transmitting. Table 6–1 shows these requirements.

These requirements have made the use of multilevel modulation techniques, as well as cross-polarization, important, especially in the lower frequency bands.

In the modulation technique descriptions which follow, use is made of the ratio of the energy per bit, E_b, to the noise power spectral density ratio, N_0, which is the noise power in a 1-Hz bandwidth within the receiver's passband. This ratio can be related to the carrier-to-noise ratio, C/N, at the receiver's input in the following way.

Let C = average carrier power (watts) and let N = noise power (watts) in the receiver's bandwidth. If T_b = duration of a bit (seconds), then $E_b = CT_b$ (joules). The bit rate (b/s) then is $f_b = 1/T_b$. Letting BW = receiver noise bandwidth (Hz),

$$\frac{E_b}{N_0} = \left[\frac{C}{N}\right]\left[\frac{\mathrm{BW}}{f_0}\right] \tag{6–2}$$

Equation (6–2) expresses a numerical ratio. Most commonly, E_b/N_0 is expressed in decibel terms by taking $10 \log_{10}[C/N][\mathrm{BW}/f_b]$.

6.2.1. Phase-Shift Keying

Phase-shift keying involves the shifting of the carrier's phase among several discrete values. If only two values of phase, e.g., 0° and 180°, are used, the result may be called *biphase-shift keying* (BPSK). Each phase represents one bit and the coding is done at 1 b/s per baud. If four values of phase are used, e.g., 45°, 135°, −135°, and −45°, the result is called *quadrature* (or quaternary) *phase-shift keying* (QPSK). Each phase can represent two bits, so the coding is done at 2 b/s per baud. Systems using 8-PSK (3 b/s per baud) also are in use.

To detect phase changes, the receiver must have a phase reference. There are two ways of obtaining this reference, known, respectively, as the differential and the coherent methods. The detection efficiency of differential phase-shift keying (DPSK) is known to be about 1 dB below that of coherent phase-shift keying (CPSK) for two-level modulation, and approaches 3 dB for multi-level modulation. Accordingly, CPSK is preferred in those applications in which small losses in signal-to-noise ratio are significant, as is the case in satellite transmission, especially the downlink (see Chapter 8). However, CPSK requires the production and extraction of a local carrier phase reference at the receiver. Usually a coherent phase estimate is obtained through the use of phase-locked loop (PLL) techniques.[2] Because both the transmitter and receiver have inherent frequency instabilities and phase jitter, the band-width of the carrier recovery loop cannot be made arbitrarily small. As a result, a somewhat noisy phase estimate is obtained, and only partially coher-ent reception actually can be claimed.

DPSK avoids the carrier phase recovery problem and thus is immune from slow carrier phase fluctuations. Under noisy phase estimation conditions and significant intersymbol interference the detection efficiency of DPSK may approach that of CPSK. Excellent comparative treatments of DPSK and CPSK have been provided by Prabhu and Salz,[3] and by Spilker.[4]

In the case of differential phase shift keying, the previous phase is used as the reference. Thus for differential QPSK, the various bit pairs might be conveyed in accordance with the list and diagram of Fig. 6–3. While receivers using differential phase detection are relatively simple to build, better perfor-mance can be achieved through the use of coherent receivers with built-in carrier recovery circuits, as illustrated in Fig. 6–4. The improvement is equiva-lent to up to 2.5 dB in E_b/N_0, or to as much as two orders of magnitude in bit error rate. In Fig. 6–4, showing a coherent QPSK receiver, the bandpass filter minimizes interference and noise not actually in the band being received. The multipliers (balanced mixers) provide the function of coherent linear demodulation. The carrier recovery circuit may be any of several types of circuits[5] designed with internal feedback loops to track carrier frequency or phase. A phase splitter then provides outputs at the carrier frequency that are separated by 90°. Harmonics of the carrier wave produced by the multipliers are attenuated by the low-pass filters.

Some modems, such as the General Telephone and Electronics (GTE)

TO SEND	CHANGE PHASE BY
00	+45°
01	+135°
10	-135°
11	-45°

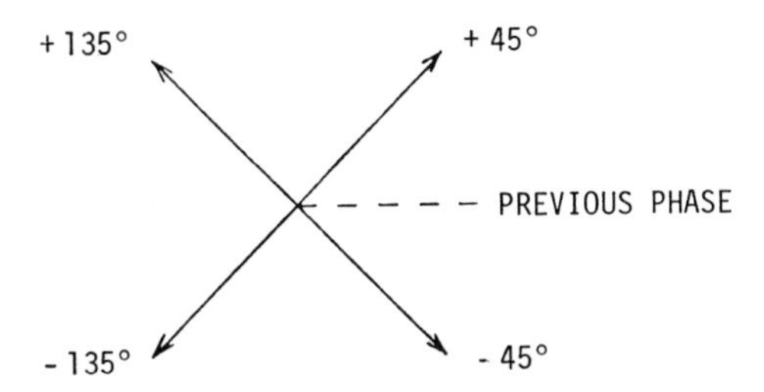

Fig. 6–3. Differential QPSK. (From D. R. Doll, *Data Communications,* © John Wiley & Sons, Inc., 1978. Reprinted by permission of John Wiley & Sons, Inc.)

Multi-rate Microprocessor Modems, can be programmed to provide either differential or coherent detection.

The symbol timing recovery circuits[6] perform nonlinear signal processing to derive the symbol rate clock, and thus to determine the symbol sampling times, for the detectors. The detectors are threshold comparators whose outputs are logic one or zero states. The output summation circuit performs a

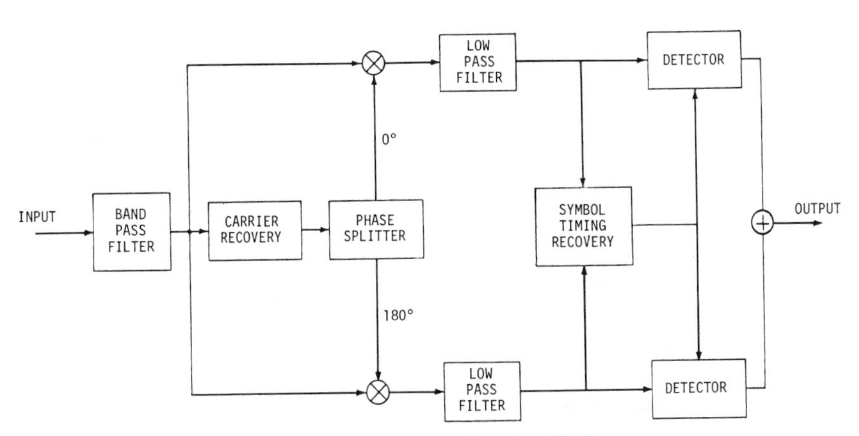

Fig. 6–4. Coherent receiver for QPSK.

parallel to serial conversion, thus providing the demodulated output bit stream.

Individual descriptions of M-PSK with $M = 2, 4, 8$ and 16 follow next. The b.e.r. performance of an M-PSK system for any value of M for additive white Gaussian noise is given[7] by the expression:

$$\text{b.e.r.} = (1/n)\ \text{erfc}\ [\sqrt{(nE_b/N_0)}\ \sin(\pi/M)] \qquad n > 1 \qquad (6\text{–}3)$$

where n = number of bits per keying interval
M = number of phase states. (Thus $2^n = M$.)

Equations (6–5), (6–6), and (6–7), which follow later, are derivable directly from Eq. (6–2) by substituting $n = 2, 3$, and 4, respectively, into Eq. (6–3).

All of these equations for bit error rate are based upon thermal noise, i.e., noise having a Gaussian amplitude distribution. This is the most common type of noise encountered in telecommunication systems.

6.2.1.1. Bi-Phase Shift Keying (BPSK). In bi-phase shift keying only two phase states, 180° apart, are used. One state thus represents the digit zero, and the other represents the digit one. Since the transmitter must be keyed once for each digit, the system provides 1 b/s per baud. The Nyquist theorem states that one baud can be transmitted in a bandwidth of 0.5 Hz, but BPSK signals are sent using two sidebands, so the result is, at most, 1-b/s per Hz. This is a theoretical limit. Its practical implications are discussed in Section 6.2.6.1.

The bit error rate (b.e.r.) performance of BPSK in thermal noise using coherent detection is

$$\text{b.e.r.} = 0.5\ \text{erfc}\ \sqrt{E_b/N_0} \qquad (6\text{–}4)$$

where

$$\text{erfc}\ x = (2/\sqrt{\pi}) \int_x^{\infty} \exp(-y^2)\, dy$$

E_b = energy per bit
N_0 = noise power spectral density

Note that Eq. (6–4) has a value only 0.5 as great as might be expected from Eq. (6–3). This is because BPSK requires only one phase detector and it produces errors for only one polarity of noise.

6.2.1.2. Quadrature-Phase Shift Keying (QPSK). In quadrature-phase shift keying, also known as quaternary phase-shift keying, four phase states are used, the adjacent ones being separated by 90°. Each phase state is made to represent a pair of bits or a symbol, i.e., 00, 01, 10, 11. Thus a pair of bits are sent each time the transmitter is keyed; accordingly, its theoretical limit is 2.0-b/s per Hz. (See Section 6.2.6.1.)

The b.e.r. performance of QPSK using coherent detection is

$$\text{b.e.r.} = 0.5 \text{ erfc} \sqrt{E_b/N_0} \tag{6–5}$$

In this case, the energy per symbol $E_s = E_b \log_2 M$, where M is the number of levels, or bits/symbol. Thus

$$E_s = E_b \log_2 4 = 2\mathrm{E}_b$$

Its performance in the presence of thermal noise thus is the same as that of BPSK with respect to E_b/N_0. However, in the presence of such impairments as continuous wave (cw) interference and linear delay distortion, Eq. (6–5) no longer is valid, and QPSK is found to degrade more rapidly than does BPSK.

6.2.1.3. Eight-Phase Shift Keying (8-PSK). In eight-phase shift keying, eight phase states are used, the adjacent ones being separated by 45°. Each phase state is made to represent a symbol consisting of a sequence of three bits, i.e., 000, 001, 010, etc. Thus three bits are sent each time the transmitter is keyed; accordingly, the technique provides a theoretical limit of 3-b/s per Hz. (See Section 6.2.6.1.) The b.e.r. performance of 8-PSK using coherent detection is

$$\text{b.e.r.} = \tfrac{1}{3} \text{ erfc } [\sqrt{(3E_b/N_0)} \sin (\pi/8)] \tag{6–6}$$

The 8-PSK technique is used in a number of digital microwave radio systems operating in the 6- and 11-GHz bands. A total of 1344 PCM voice channels (90 Mb/s) can be transmitted within a standard 30-MHz channel allocation in the 6-GHz band. Figure 6–5 is a block diagram of the transmitter (providing a 70-MHz IF output) and Fig. 6–6 shows the corresponding receiver.

For transmission the 90-Mb/s stream is split into three 30-Mb/s streams. The transmit logic circuits produce two 4-level (amplitude) streams, which are used to modulate two quadrature carriers using double sideband suppressed-carrier amplitude modulation, as shown in Fig. 6–7. The power combiner, accordingly, produces the 8-PSK output centered at 70 MHz, and occupying the 55–85-MHz spectrum. The inverse functions are accomplished

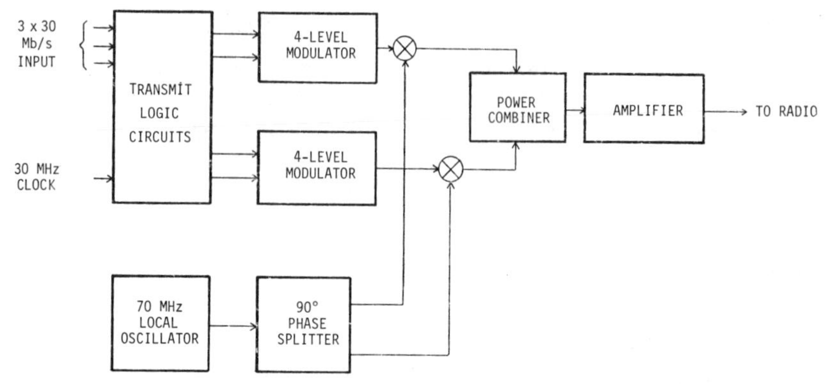

Fig. 6–5. Block diagram of 8-PSK transmitter.

in the receiver (Fig. 6–6), which shows the carrier (70-MHz VCXO) and symbol timing recovery (30 MHz VCXO) chains, with the VCXOs being voltage controlled crystal oscillators.

6.2.1.4. Sixteen-Phase Shift Keying (16-PSK). In sixteen-phase shift keying, sixteen phase states are used, the adjacent ones being separated by 22.5°. Each phase state is made to represent a symbol consisting of a sequence of four bits, i.e., 0000, 0001, 0010, 0011, etc. Thus four bits are sent each time the transmitter is keyed; accordingly, 16-PSK provides a theoretical limit of 4-b/s per Hz. (See Section 6.2.6.1.)

The b.e.r. performance of a 16-PSK system is given by the expression

$$\text{b.e.r.} = \tfrac{1}{4}\,\text{erfc}\,[\sqrt{(4E_b/N_0)}\,\sin(\pi/16)] \tag{6–7}$$

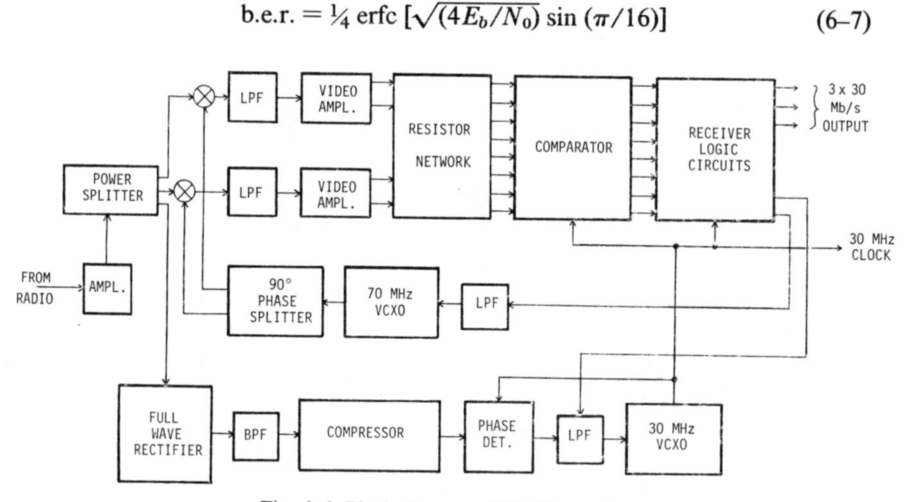

Fig. 6–6. Block diagram of 8-PSK receiver.

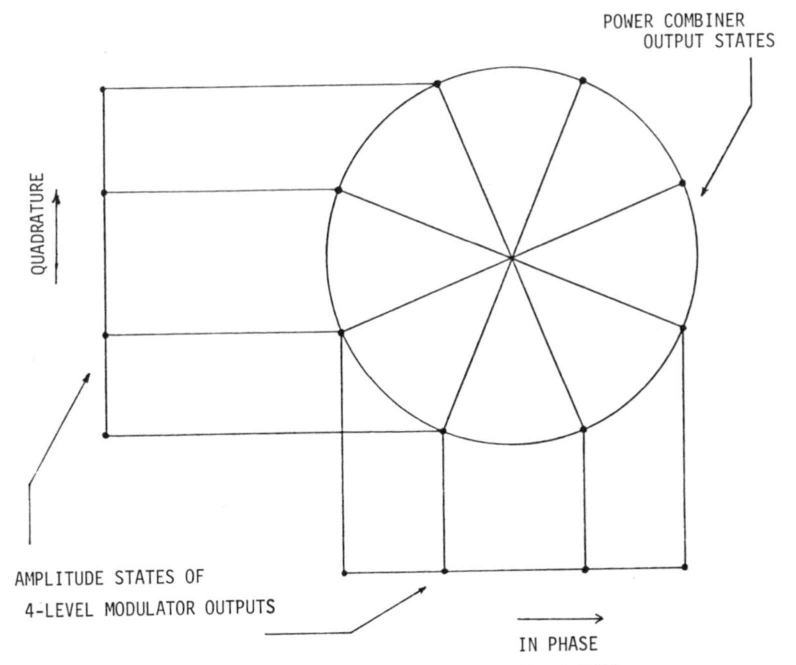

Fig. 6–7. Amplitude levels used to produce 8-PSK.

No 16-PSK systems are in commercial use at present because of the superior performance of 16-APK, as described in Section 6.2.3.

6.2.2. Amplitude Shift Keying and Quadrature Amplitude Modulation

The phase-shift keyed systems provide good performance (b.e.r. versus E_b/N_0) provided phase jitter can be kept adequately low. They are insensitive to amplitude jitter. However, channel noise actually consists of both amplitude and phase fluctuations. Accordingly, phase jitter begins to cause severe performance limits when attempts are made to implement more than eight phases, as may be desired for efficient use of the spectrum. While many practical 8-PSK systems are being implemented, the achievement of 4 bits per symbol (16 signaling states) is done most effectively using amplitude, as well as phase shift. In fact, comparisons of 16-PSK with 16-state techniques involving amplitude as well as phase shift show that 16-PSK may require up to 3.5 dB more carrier-to-noise ratio; moreover, the intersymbol interference produced by filtering causes greater degradation to 16-PSK and higher order PSK systems than to systems that also involve amplitude shift, as will be seen later from Figs. 6–17, 6–18, and 6–19.

The concept of amplitude shift keying was introduced already in the discussion of 8-PSK, where it is used as part of the process of generating the 8-PSK signal. In fact, all of the M-PSK outputs can be generated by quadrature combinations of amplitude shift signals. The term *quadrature amplitude modulation* (QAM) is used to describe the combining of two amplitude shifted streams in quadrature. In its simplest form it can be produced by the quadrature combination of two two-level signals, as illustrated in Fig. 6–8. The resulting amplitude-phase diagram will be recognized as being the same as that of QPSK. The difference[8] is that the QAM system uses premodulation and post-detection low-pass filters while the QPSK system has post-modulation and predetection bandpass filters. Accordingly, QAM and QPSK systems have identical transmitted spectra and identical b.e.r. performance, as expressed by Eq. (6–5).

The diagrams of Figs. 6–7 and 6–8 illustrate the positions taken by the transmitted amplitude and phase as the transmitter is keyed by various symbols (bit combinations). Each symbol corresponds to a specific point on such a diagram. Sometimes the diagram is called the *signal state space diagram,* or the *constellation.* On a normalized basis, the distance between the points may be referred to as the *Euclidean distance.* The greater this distance, the greater will be the extent to which a given modulation technique is noise resistant. This is why 16-PSK is not commonly used. A constellation with greater Euclidean distances can be devised, for example, using 16-QAM. For comparison, Fig. 6–9 shows 16-PSK (a) and 16-QAM (6), both of which

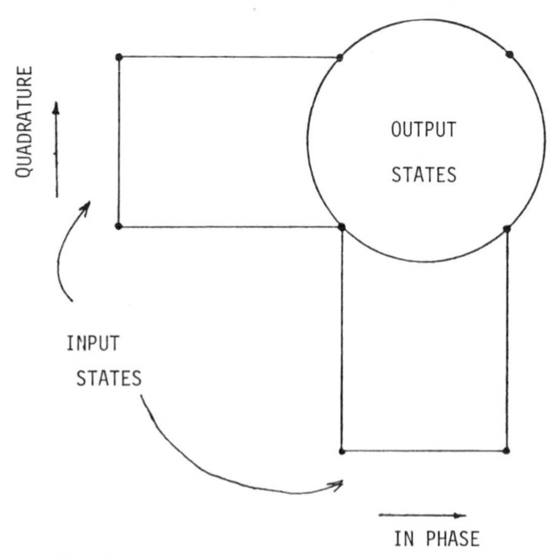

Fig. 6–8. Quadrature amplitude modulation states.

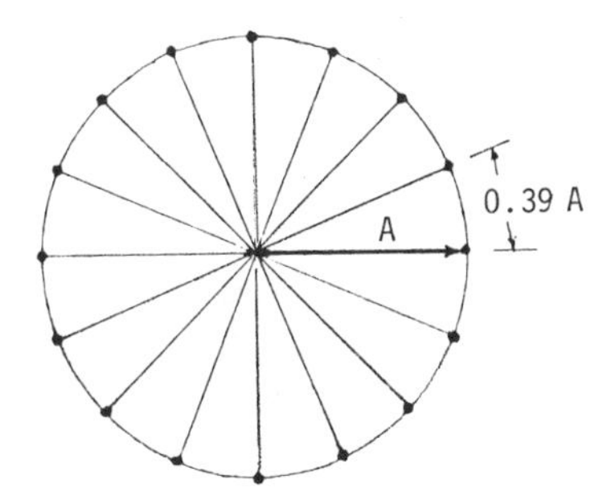

(a) Constellation for 16-PSK

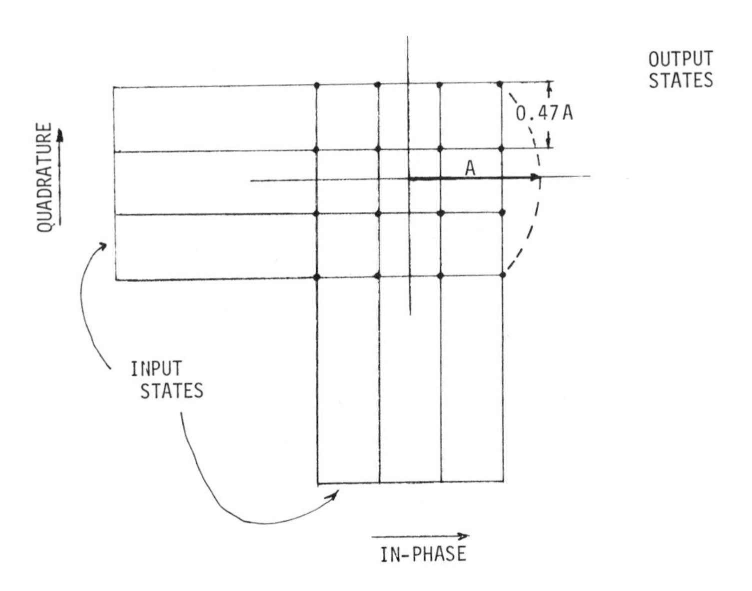

(b) Constellation for 16-QAM and its derivation

Fig. 6–9. Constellations for 16-PSK and 16-QAM.

are of equal peak amplitude. Note the much closer spacing of the signal states in 16-PSK compared with those of 16-QAM. For the 16-QAM, the spacing is 0.47A, whereas for the 16-PSK, the spacing is only 0.39 A, where A is the unmodulated carrier amplitude. As shown in Fig. 6–9 (b) for QAM, two quadrature 4-amplitude signals are combined, but their timing and magnitudes produce sixteen well-spaced constellation points using various amplitudes as well as various phases of the carrier.

6.2.3. Amplitude-Phase Keying

Amplitude-phase keying (APK) is a general term used for any digital modulation system in which both the amplitude and the phase of the carrier are altered to produce the various symbol states. Thus QAM is one form of APK. Figure 6–10 is a generalized block diagram of an APK modulator using suppressed-carrier QAM, while Fig. 6–11 shows the corresponding demodulator. In these figures, f_b refers to the bit rate of the two-level digital stream being transmitted. In the modulator, A amplitude levels are produced and combined in phase as in Figs. 6–8 and 6–9. The digital stream rate thus decreases by a factor $\log_2 A$ at this point. The receiver derives its carrier using a carrier-recovery circuit similar to the type used for the coherent

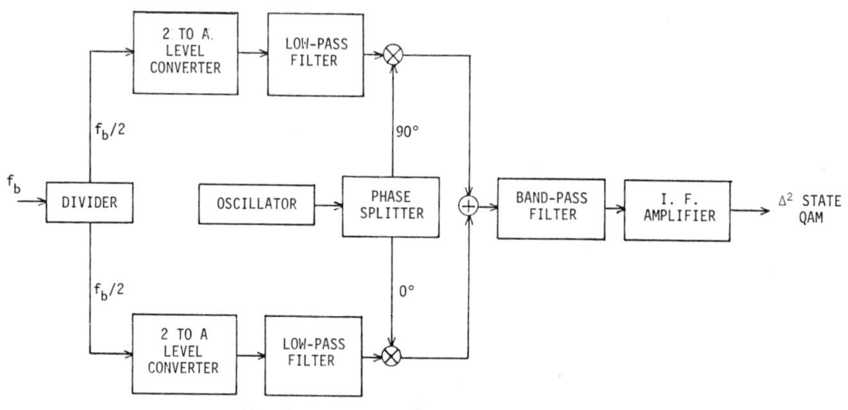

Fig. 6–10. Generalized QAM modulator.

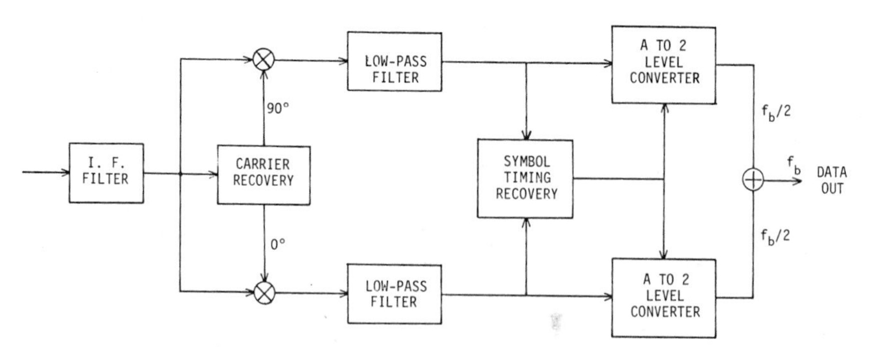

Fig. 6–11. Generalized QAM coherent demodulator.

demodulation of phase-shift keying. Conversion from A to two levels then is achieved by individual threshold comparators, each with a separate preset threshold level. The outputs of these threshold comparators are then combined in digital logic to produce a stream at a rate of $f_b/2$. Finally the $f_b/2$ streams from the in-phase and quadrature parts of the demodulator are combined to reproduce the output bit stream f_b.

The b.e.r. performance of a 16-QAM system is given[9] by the expression

$$\text{b.e.r.} \simeq \frac{3}{8} \operatorname{erfc} \sqrt{\frac{2}{3} \frac{E_b}{N_0}} \tag{6–8}$$

while that of a 64-QAM system is given[10] by the expression

$$\text{b.e.r.} \simeq \frac{7}{24} \operatorname{erfc} \sqrt{\frac{1}{7} \frac{E_b}{N_0}} \tag{6–9}$$

One popular APK system uses 16-QAM to provide up to six DS1 streams (144 voice channels) in a 3.5-MHz bandwidth in the 2 GHz microwave band. Another example is the 400-Mb/s system developed by the Nippon Telegraph and Telephone Public Corporation (NTTPC), and occupying a 120-MHz bandwidth centered at an IF frequency of 1.7 GHz. These systems have been developed in spite of the nonlinearity problems that exist with respect to the amplitude modulation of microwave facilities.

Systems using 64-QAM now allow 2016 voice channels to be transmitted in the 6 GHz microwave band. Such systems are manufactured by Western Electric, Rockwell-Collins and Fujitsu.

The transmission of digital bit streams through relatively linear analog voiceband facilities can be done using a large number of signal states, if necessary. For example, several manufacturers now offer a 14.4-kb/s modem which is keyed at 2400 baud.* The signal structure uses hexagonal packing, as illustrated in Fig. 6–12, which is a photograph of the constellation used by the Codex SP14.4 Data Modem. A data rate of 14 400 b/s keyed at 2400 baud requires 6 bits per symbol and thus $2^6 = 64$ signal states. Accordingly, the carrier-to-noise ratio required for a given bit error rate is significantly higher (see equations (6–8) and (6–9)) than that of a system using 16 or fewer states. On the conditioned telephone facility,† this modem operates with a b.e.r. $\leqslant 10^{-5}$.

A large number of APK constellations have been devised and numerous studies have been done relative to their performance.[11]

---

* 16.0 kb/s modems keyed at 2667 baud also are available.

† Four-wire 3002 D1 conditioned or CCITT M 1020 lines.

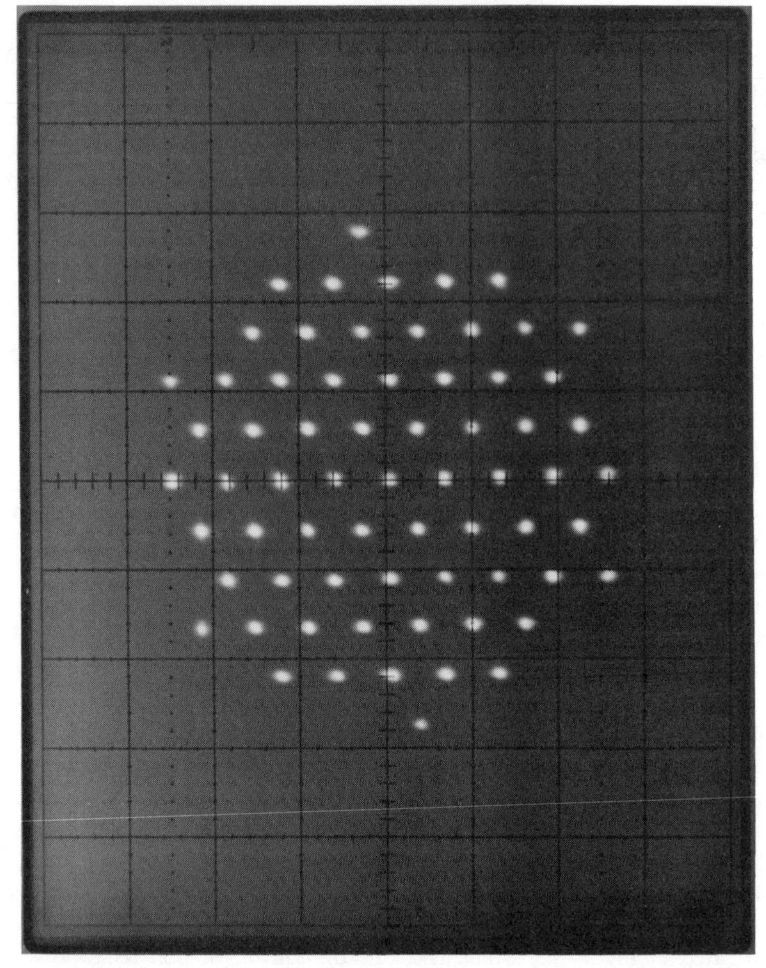

Fig. 6–12. Signal state space array for 64-state modulation used in 14.4 kb/s modem. (Courtesy Codex Corp.)

6.2.4. Frequency-Shift Keying

Frequency-shift keying (FSK) generally uses two frequencies to represent two binary states, e.g., zero and one. Noncoherent detection is accomplished with circuits tuned to each of these frequencies. The implementation of FSK thus can be simple, and limiters can be used to reduce the effects of selective fading, which is common on the high-frequency (HF) transmission circuits on which FSK often is used. FSK is a commonly used technique for transmission at and below 1200 b/s.

The deviation ratio d of an FSK signal is defined as $d = \Delta f/b$ where Δf = peak-to-peak frequency deviation and b = bit rate. With noncoherent detection, d must be at least 1.0 to prevent significant overlap of the filter passbands. This constraint on d can be eliminated by using a discriminator to convert the frequency variations to amplitude variations, whereupon envelope detection is used.

Continuous phase FSK (CP-FSK) avoids abrupt phase changes at the bit transition instants by using observation intervals greater than one bit period in length.[12] The result is rapid spectral roll-off and thus narrower filter bandwidths than would be possible otherwise. With coherent detection, $d = 0.715$ is the optimum deviation ratio with respect to transmitted power and occupied bandwidth in the presence of thermal noise if decisions are limited to one bit interval.

A useful special case of CP-FSK is called *minimum shift keying* (MSK). For MSK the peak frequency deviation is equal to ± 0.25 b, and coherent detection is used. Thus $d = 0.5$ for MSK. MSK achieves performance identical to coherent PSK with the efficient spectral characteristics of CP-FSK by extending the decision interval to two bit periods for binary detection.[13] An additional advantage of MSK over coherent CP-FSK is the possibility of a relatively simple self-synchronizing implementation.[14]

In the absence of bandwidth or amplitude limiting between the modulator and the demodulator the bit error rate performance of MSK is identical to that of BPSK,[15] as expressed by Eq. (6–4), i.e., for a given ratio of signal to thermal noise, the two techniques yield the same bit error rate. However, in the presence of the amplitude and bandwidth limitations that often occur in a transmission system, MSK suffers less degradation than BPSK because MSK exhibits less out-of-band power.

6.2.5. Correlative Techniques

Correlative techniques use a finite memory to change the baseband digital stream to a new form.[16,17] This new form provides improvements in coding efficiency from a spectrum occupancy viewpoint. Correlative techniques produce intersymbol interference on a planned and controlled basis and are sometimes called *partial response techniques*[18] because the resulting filtered pulses no longer exhibit zero outputs at those sampling instants reserved for adjacent pulses. Figure 6–13 illustrates the response of a bandwidth-limited system to a rectangular pulse. The result ideally is zero intersymbol interference. However, if the pulse tails have finite amplitude at the times

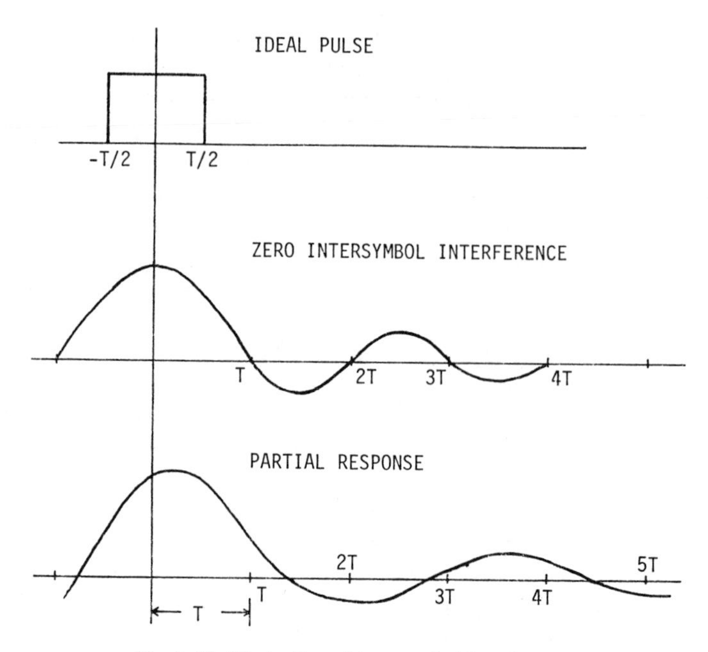

Fig. 6–13. Illustration of intersymbol interference.

T, 2*T*, . . . , *nT* reserved for adjacent pulses, the result is intersymbol interference.

Correlative techniques are called duobinary, modified duobinary, and polybinary. The name *duobinary* means doubling the speed of binary. This is done by converting an ordinary two-level binary sequence into one that uses three levels. This conversion involves intersymbol interference extending over one bit interval, as will be seen soon. If the intersymbol interference extends over two bit intervals, the technique is called *modified duobinary. Polybinary* is the term used to denote the fact that the correlation span includes more than three bits. The greater the correlation span, the more extensively the energy of the sequence is concentrated at the lowest frequencies. As a result, the number of bits/second per Hertz of bandwidth can be increased significantly over that of systems not using such techniques. In addition, certain unique patterns are produced among the correlative pulse trains. These patterns can be monitored at the receiver, thus allowing error detection without the addition of redundant bits at the transmitter.

Figure 5–10 showed the introduction of controlled interference between input bits. The "partial response" is produced by adding the binary input signal to itself delayed by one bit interval using the following rules: $1 - 1 = 0$; $-1 - 1 = -1$; $1 + 1 = 1$. The resulting three-level signal is found to occupy significantly less bandwidth than the original two-level signal.

Correlative techniques have been developed in which the processes of coding and modulation are combined in a single step. An example is duobinary AM-PSK modulation, as illustrated in Fig. 5–11. Here the −1 level corresponds to a 0° carrier shift, the 0 level corresponds to zero output, and the +1 level corresponds to a 180° carrier shift. Because the carrier reverses phase according to duobinary rules, the bandwidth is less than it would be for simple on-off keying by a factor of two.

A decision feedback receiver is used to remove the intersymbol interference. Shown in Fig. 6–14 is the concept for such a receiver for a polybinary signal in which the intersymbol interference extends over four bit intervals.

Quadrature partial response signaling (QPRS) is achieved with two 3-level duobinary signals phase modulated in quadrature, as illustrated in Fig. 6–15. This technique is being used in several microwave radio systems that operate in the 2-GHz band. The advantage of using correlative techniques in such systems is that they are relatively simple to implement, being comparable to a four-phase modulation system, while being less complex than the eight-phase systems. With this simplicity, they can still achieve the spectral efficiency (96 voice channels, or 6.312 Mb/s, in a 3.5-MHz bandwidth) required for operation at 2 GHz. This is a 2-b/s per Hertz efficiency.

The Western Electric 1A-RDS "data under voice" system uses modified duobinary coding with four-level inputs to provide a 1.544 Mb/s stream for data transmission in the 500 kHz of baseband spectrum available below the FDM mastergroups in many microwave radio systems (see Section 6.8 and Fig. 6–26).

A seven-level modified duobinary signal processor is also used in some 2-GHz radio systems. Three positive, three negative and a zero amplitude level are obtained digitally. The binary input at 6.312 Mb/s (96 PCM voice channels) is converted to 3.156 M baud (2-b/s per baud), which occupies a baseband of 1.578 MHz. As applied to a standard FM radio terminal,

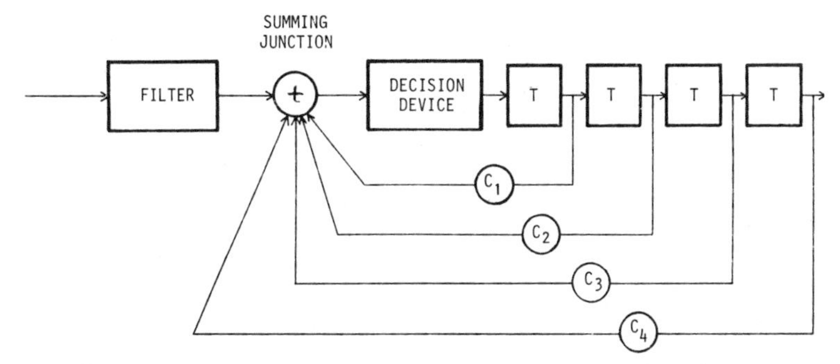

Fig. 6–14. Decision feedback receiver for polybinary signal with correlation span extending over four bit intervals.

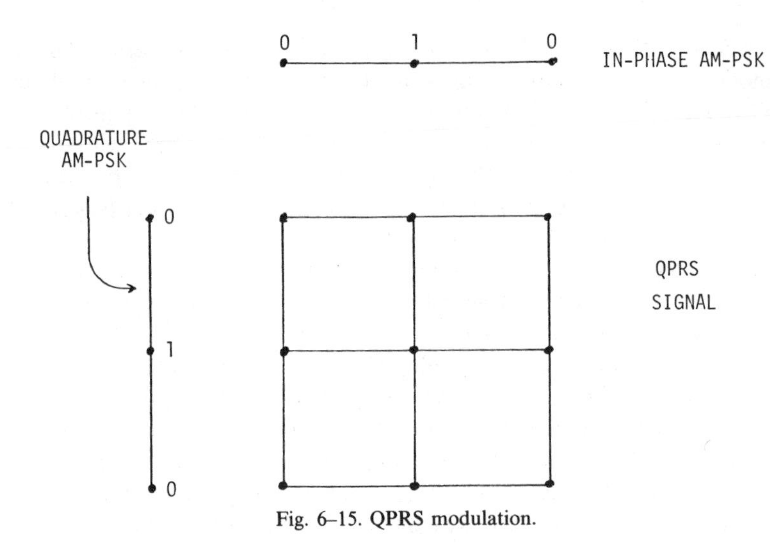

Fig. 6–15. QPRS modulation.

the resulting double sideband signal occupies 3.156 MHz, for an overall 2-b/s per Hertz. It thus meets the FCC requirements for T2 transmission at 2 GHz. A block diagram of this radio system is shown in Fig. 6–16. There is no dc component in the baseband spectrum, and a negligible amount of energy at low frequencies. Accordingly, a voice frequency order wire channel is placed into the first 8 kHz of the baseband spectrum. Note the presence of the error detector at the output. This detector monitors the bit patterns from the binary threshold detector and thus achieves error detection without the need for redundant bits in the input bit stream.

Error detection is accomplished using the following rules, based upon observation of the behavior of the "extreme," or top and bottom, levels of the

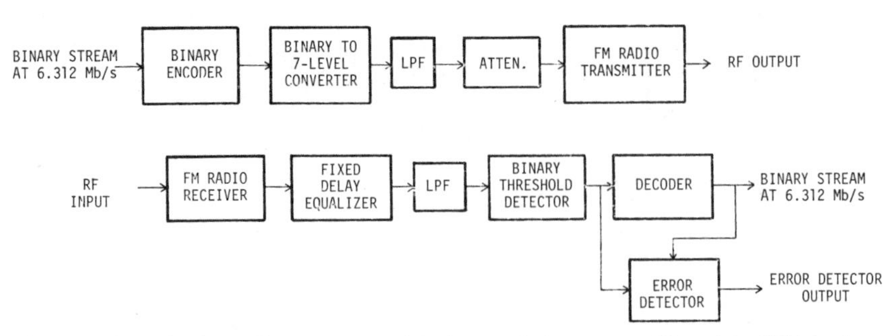

Fig. 6–16. Seven level modified duobinary digital transmission system. (Copyright GTE Lenkurt, Incorporated, San Carlos, CA.)

data stream. Thus for a (+, 0, −) system, the (+) and (−) levels are the extreme levels. Violation of the rules below indicates the presence of an error.

Rule 1. For a duobinary (+, 0, −) system, the polarities of two successive bits at the extreme levels are opposite if the number of intervening bits at the zero level is odd; otherwise, the extreme bits have the same polarity.

Rule 2. For a modified duobinary (+, 0, −) system, divide the pulse train into *even* and *odd* trains. Both trains follow the same pattern. For either train, two successive bits at the extreme levels must have opposite polarity.

Note that only the extreme levels are usable for error detection. This also is true for systems with larger numbers of levels, i.e., (−3, −1, +1, +3), etc. Intermediate levels may be formed in more than one way and thus are not suitable for error detection.

Example. Given the following modified duobinary stream:

$$0 - - + 0 0 0 - 0 + 0 - 0 + + 0$$

The "even" and "odd" streams are

$$- + 0 - + - + 0 \quad \text{(even)}$$
$$0 - 0\ 0\ 0\ 0\ 0 + \quad \text{(odd)}$$

Since any two successive bits at the extreme levels have opposite polarity, no bit errors are apparent.

The use of single-sideband techniques with 7-level correlative coding also has been described.[20] Such an arrangement provides 4-b/s per Hertz, thus allowing 192 PCM channels (12.63 Mb/s) to be transmitted in the 3.156 MHz bandwidth available in a 2-GHz radio system.

6.2.6. Comparison of Modulation Techniques

Three basic questions arise in the choice of a digital modulation technique:

1. What is its bandwidth efficiency (b/s per Hertz)?
2. How well does it perform in the presence of:
   - Channel noise?
   - Filter limitations?
   - Other impairments?
3. How complex is it to implement?

This section provides quantitative answers to the first two categories of questions. The third, complexity, can be answered only in a qualitative fashion. General comments are provided below where appropriate.

6.2.6.1. Bandwidth Efficiency. Bandwidth efficiency is expressed in bits per second (b/s) per Hertz of bandwidth. To determine it for a given modulation technique, the number of bits/symbol, or b/s per baud, *n*, must be established first. The number of logic levels is $M = 2^n$. For a zero-memory (noncorrelative) coding technique, the maximum symbol rate per unit bandwidth is 2 baud/Hertz. This is valid in baseband, or at RF for single-sideband modulation techniques. Most modulation, however, involves either frequency or phase shift, and thus double sidebands. The result is a maximum symbol rate of 1 baud/Hertz. This implies the use of filters with infinitely steep cut-off characteristics, sometimes designated $\alpha = 0$ filters, where α denotes the filter's roll-off factor. In general, a filter with a roll-off factor α will have an amplitude versus frequency characteristic for impulse transmission[21] that begins to roll off at a frequency of $(1-\alpha)f_c$, where f_c is the cut-off or -3 dB frequency. The response is very low (e.g., -80 dB or lower) at and beyond a frequency of $(1+\alpha)f_c$. Accordingly, the maximum double-sideband transmission rate is $1/(1+\alpha)$ baud/Hz. Thus if $\alpha = 0.25$, the maximum symbol rate is 0.8 baud/Hz.

Table 6–2 compares various digital modulation techniques on the basis of their bandwidth efficiency, as established by the foregoing principles. The filter is assumed to have an ideal roll-off factor, $\alpha = 0$, since some systems in the field approximate $\alpha = 0$ very closely. For example, some of the operational digital radio systems in the 6-GHz microwave band (see Chapter 7) manufactured by Raytheon, Collins, and Nippon Electric use 8-PSK and achieve a 90-Mb/s rate in a 30-MHz radio channel bandwidth by using very sharp cut-off filters and suffering a corresponding penalty of 4 dB or more as a result. However, this design enables them to meet the regulatory requirements for spectral efficiency in that band.

Figure 6–17 shows that increased bandwidth efficiency carries with it a requirement for increased carrier-to-noise ratio *C/N*, but that the techniques that use only phase shift require more *C/N* than do those that involve both amplitude and phase changes. Fig. 6–17 is based upon a b.e.r. of 10^{-8}. The numerals refer to the number of signaling states required in each case. Equipment complexity is not displayed. The Shannon limit is based upon a zero b.e.r. However, equipment complexity increases substantially as one attempts to get closer to this limit.

6.2.6.2. Modulation Technique Performance. Modulation techniques can be evaluated based upon their performance in the presence of various impairments. A detailed discussion of this subject is beyond the scope

Table 6–2. Bandwidth Efficiency of Digital Modulation Techniques.*

TYPE OF MODULATION	NUMBER OF LOGIC LEVELS M	B/S PER BAUD	MAXIMUM ACHIEVABLE EFFICIENCY	
			BASEBAND (B/S PER HERTZ)	MODULATED CARRIER (B/S PER HERTZ)
Amplitude (AM)	2	1	2	1
Frequency shift keying (FSK)	2	1	2	1
7-Level modified duobinary†	7	2	4	2
SSB 7-level modified duobinary	7	2	4	4
2-Phase shift keying (BPSK)	2	1	2	1
4-Phase shift keying (QPSK)	4	2	4	2
8-Phase shift keying (8-PSK)	8	3	6	3
16-Phase shift keying (16-PSK)	16	4	8	4
16 Quadrature AM (16-QAM)	16	4	8	4
64 Quadrature AM (64-QAM)	64	6	12	6

† Three of the four bit combinations can each be represented by either of two logic levels. The fourth combination is represented by only one level.

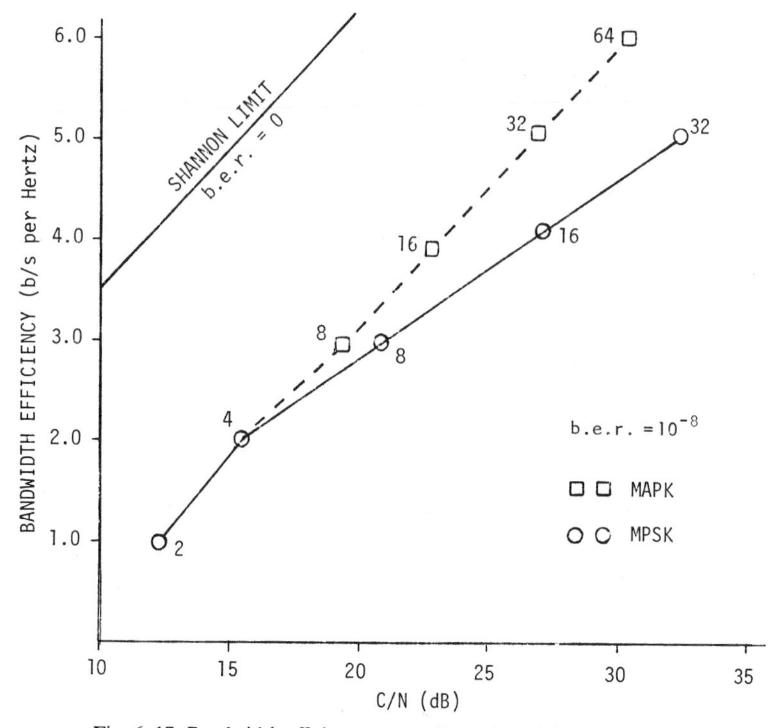

Fig. 6–17. Bandwidth efficiency comparison of modulation methods.

of this text, but two major types of impairments will be considered: channel noise and filter bandwidth limitations.

Figure 6–18 shows how several of the systems of this chapter perform in the presence of channel noise, with performance[22] being expressed in terms of b.e.r., designated as $P(e)$. The rms C/N is specified in the double-sided Nyquist bandwidth in each case.

Figure 6–19 shows the effect of filter bandwidth on several of the systems in terms of C/N degradation.[23] In this figure, B is the 3-dB double-sided bandwidth of a Gaussian filter and T_s is the duration of a symbol.

6.3. T-CARRIER SYSTEMS

The T1 Digital Transmission System is used for the transmission of PCM voice between the central offices of major metropolitan areas and between cities for distances up to about 80 km. Multipair exchange cables of a variety of standard gauges (19, 22, 24, and 26 gauge copper pair and 17- and 20-gauge aluminum pair) are used. Regenerative repeaters (capable of wave-

C/N(dB)	BPSK	QPSK	8-APK	8-PSK	16-APK	16-PSK
7	7.7×10^{-4}					
8	1.9×10^{-4}					
9	4.0×10^{-5}					
10	4.6×10^{-6}	$>10^{-3}$				
11	2.9×10^{-7}	2.3×10^{-4}				
12	7.2×10^{-9}	4.6×10^{-5}				
13	$<10^{-10}$	4.6×10^{-6}	$>10^{-3}$			
14		3.1×10^{-7}	5.9×10^{-4}	$>10^{-3}$		
15		1.0×10^{-8}	1.6×10^{-4}	7.8×10^{-4}		
16		$<10^{-10}$	3.1×10^{-5}	2.6×10^{-4}		
17			4.1×10^{-6}	5.5×10^{-5}	$>10^{-3}$	
18			3.2×10^{-7}	8.6×10^{-6}	4.0×10^{-4}	
19			4.6×10^{-9}	8.0×10^{-7}	8.9×10^{-5}	
20			$<10^{-10}$	2.2×10^{-8}	1.3×10^{-5}	
21				1.0×10^{-10}	1.7×10^{-6}	$>10^{-3}$
22				$<10^{-10}$	9.0×10^{-8}	4.5×10^{-4}
23					1.7×10^{-9}	8.3×10^{-5}
24					$<10^{-10}$	1.0×10^{-5}
25						5.1×10^{-7}
26						3.7×10^{-8}
27						1.0×10^{-10}
28						$<10^{-10}$

Fig. 6–18. Performance of *M*-ary PSK and *M*-ary APK coherent systems.

form reshaping) are placed at nominal spacings of 1.6 km with closer spacings (e.g., 0.8 km) near offices. Digroups consisting of 24 voice channels, each at 64 kb/s, are arranged in the frame structure of Fig. 6–20. As shown there, a framing bit is added to each digroup. The T1 system thus transmits a digital stream which is made up as follows:

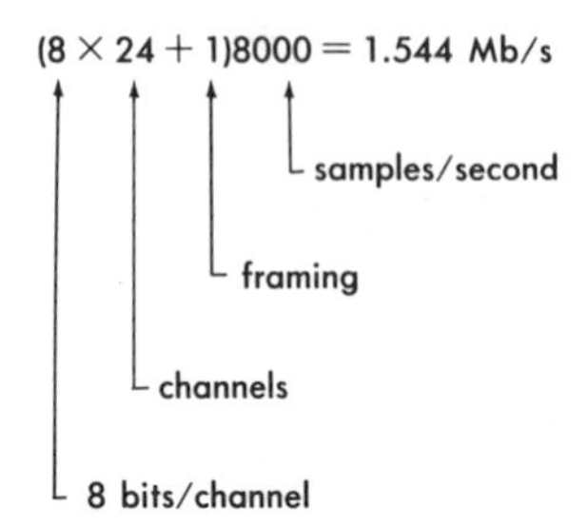

In heavily populated areas use is made of metropolitan area trunk (MAT) cable,[24] designed to minimize crosstalk where large numbers of circuits are required between central offices. For the much longer systems needed between major cities, the T1 Outstate (T1/OS) System has been developed.[25] It can function with as many as 200 repeaters in tandem. Repeater spacings are

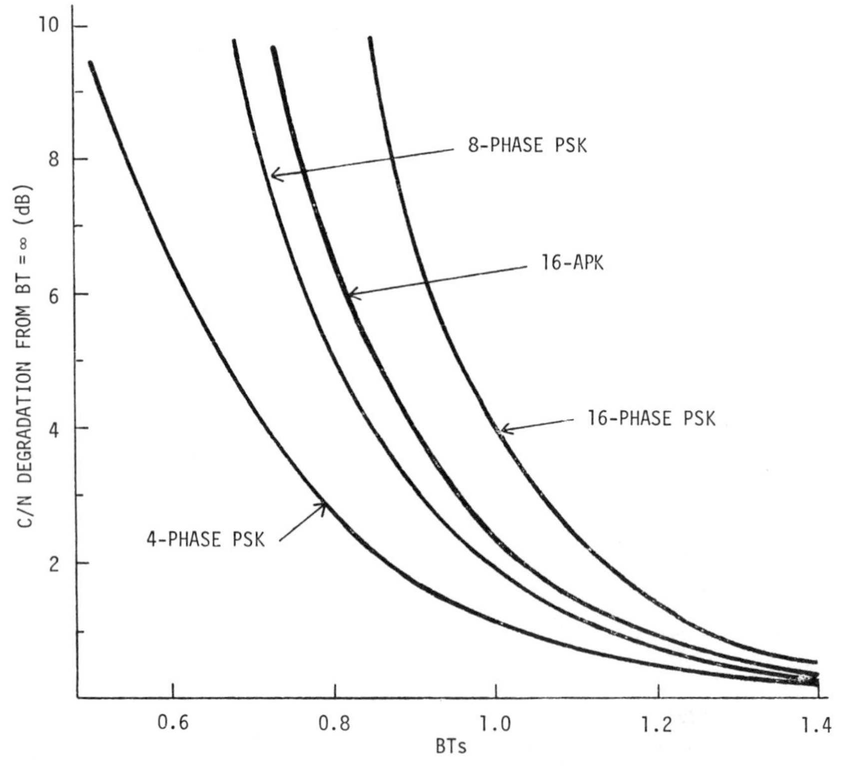

Fig. 6–19. *C/N* degradation caused by gaussian filter band limitation. (Courtesy Ishio, et al., Ref. 16, © IEEE, 1976.)

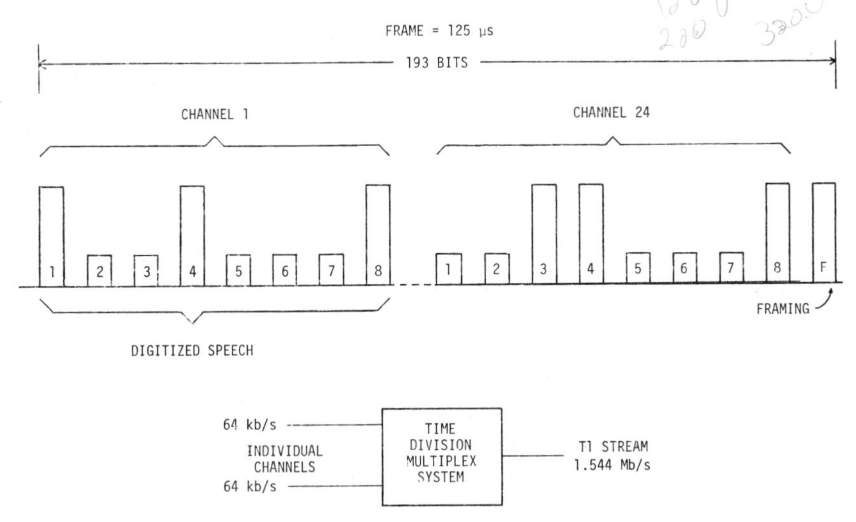

Fig. 6–20. T1 carrier frame structure. (From D. R. Doll, *Data Communications,* © John Wiley & Sons, Inc., 1978. Reprinted by permission of John Wiley & Sons, Inc.)

determined based upon the need to meet well-established performance objectives. A *failure time* is said to occur whenever the b.e.r. $\geqslant 10^{-6}$. Failure time is not to exceed 0.01% per year (52 minutes) over a two-way 400-km path based upon propagation outages. In Canada, only 0.02% per year (104 minutes) is allowed for all causes of unavailability (propagation plus hardware causes).

Of critical importance to users is the number of error-free seconds (EFS) or the error-free interval (EFI) which the transmission facility achieves. This is the time duration, or the number of bits, between error bursts. An error burst, in turn, is a series of bits which starts and ends with an erroneous bit and which is separated from other error bursts by a number of error-free bits. An error burst longer than 300 ms is called an *outage.*

The derivation of T-carrier transmission channels through the retrofit of existing analog transmission systems is discussed in Section 6.8.

T-carrier systems for rates in excess of 1.544 Mb/s are discussed in Section 6.4. North American digital stream rates are designated by the letters DS. Thus DS1 is 1.544 Mb/s, DS2 is 6.312 Mb/s, DS3 is 44.736 Mb/s, and DS4 is 274.176 Mb/s.

Other rates, especially multiples, are in use. They include DS1C (3.152 Mb/s), twice the DS3 rate ($\simeq$90 Mb/s), and three times the DS1 rate ($\simeq$135 Mb/s). Some military systems operate at six times DS1, or 9.696 Mb/s, and at eight times DS1, or 12.928 Mb/s.

6.4. THE DIGITAL HIERARCHY

Section 6.3 described the T1 Digital Transmission System for the transmission of 24 PCM voice channels at an overall 1.544-Mb/s rate. This is the most commonly used system in the U.S., Canada and Japan, and is the basic building block in the digital hierarchy. Both this system and a 30-channel system have been included in the CCITT Recommendations. The 30-channel system (actually 32 channels if the two channel time slots for frame alignment and signaling are counted) operates at a 2.048-Mb/s rate and is used extensively in Europe as well as in the Latin American countries. Higher level digital multiplex rates are listed in Table 6.3.

The types of multiplexing equipment used to translate the signals of the North American system between levels are shown in Fig. 6–21. The multiplexers must be capable of providing the proper interface parameters in terms of transmitted signal format as well as bit stream organization. These parameters include transmission rate, signal format (polar, bipolar or multilevel), pulse amplitude, parity bit locations and allowable number of consecutive zeros. In addition, signal compatibility with the terminal equipment used at the ends of the facility must be maintained. Thus the message signal, framing and signaling formats must be maintained.

At the T1 rate, 1.544 Mb/s, the most common application is the channel bank handling 24 PCM voice channels. Other applications include data terminals, often with a composite of lower bit rate data streams, and some limited-motion video terminals. Some high-speed facsimile machines also operate at the T1 rate. T1 lines are generally a maximum length of about 80 km with repeater spacings not exceeding 1.6 km. An exception is the T1/OS

Table 6–3. Digital Multiplex Hierarchies

LEVEL	NORTH AMERICAN	JAPANESE	EUROPEAN (CEPT)†
First	1.544 Mb/s (24 ch)	1.544 Mb/s (24 ch)	2.048 Mb/s (30 ch)
Second	6.312 Mb/s (96 ch)	6.312 Mb/s* (96 ch)	8.448 Mb/s (120 ch)
Third	44.736 Mb/s (672 ch)	32.064 Mb/s (480 ch)	34.368 Mb/s (480 ch)
Fourth	274.176 Mb/s (4032 ch)	97.728 Mb/s (1440 ch)	139.268 Mb/s (1920 ch)
Fifth		400.352 Mb/s (5760 ch)	565.148 Mb/s (7680 ch)

NOTE: As stated in Section 6.3, many systems are being implemented using multiples of these basic levels. Many fiber optic transmission systems are being designed to operate at 90, 405, or 432 Mb/s.
* Alternative: 7.876 Mb/s (120 ch).
† Consortium of European Postal and Telegraph Organizations.

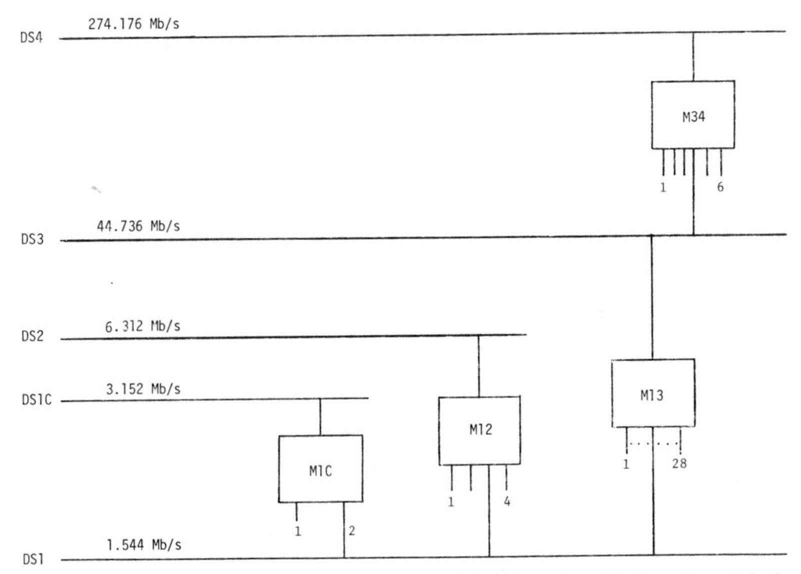

Fig. 6–21. North American digital hierarchy and multiplexing plan. (© American Telephone and Telegraph Company, 1977.)

system[26] described in Section 6.3. Increases in system length well beyond 80 km have been achieved through improved equipment reliability and maintenance procedures, as well as increased cable shielding to reduce NEXT. Improved fault location procedures as well as protection switching also are used to maintain service standards.

The T1C rate provides for the transmission of two T1 signals which generally are not synchronized with one another. A 64-kb/s stream is added for synchronization and framing by a process called *pulse stuffing* to produce a 3.152-Mb/s rate. Pulse stuffing involves the addition of enough time slots to each signal so that the T1C signal rate is that of the clock circuit in the transmitter. The stuffed pulses carry no information, but the signal is coded so that they can be recognized and removed at the receiving terminal. Details of the multiplexing process are discussed in Ref. 24. Generally the T1C system uses the same transmission media and maximum length as the T1.

The T2 transmission rate, 6.312 Mb/s, can handle 96 PCM voice channels or one motion-compensated conference video channel. T2 lines may extend to distances as great as 800 km. Stuff pulses are used in combining four T1 lines together to provide the input to a T2 line, since the T1 lines may have been timed by independent (unsynchronized) clocks. In addition, the necessary control and framing bits must be added. Details of the multiplexing process and the framing are contained in Ref. 11.

As shown in Fig. 6–21, an M13 multiplexer combines 28 T1 lines into a T3 line. Since the T3 line can handle 672 PCM voice channels, it can readily transmit a 600-channel mastergroup. A system capable of twice the DS3 rate can handle full 80- to 85-Mb/s color video with 9-bit coding.

The highest line rate in the North American digital hierarchy, the T4, operates at 274.176 Mb/s, and can handle six T3 lines, which are input to it through an M34 multiplexer. Systems designed for higher rates use multiples of the lower rates.

6.5. MULTIPLEXING

Multiplexing is accomplished by two classes of devices, muldems and transmultiplexors. The muldem, a contraction for multiplexor-demultiplexor, is used to combine message channels at one level of the digital hierarchy to produce a higher level and, at the receiving end, to reproduce the channels at the original level. A transmultiplexor is used to convert TDM signals to FDM, and vice versa, and thus serves as an interface device between digital and analog networks.

6.5.1. Muldems

Figure 6–21 showed what muldems are used in the digital hierarchy. In the M12 multiplexor two DS1 signals in bipolar format are converted to unipolar signals. The second input signal is inverted (all zeros are changed to ones and all ones are changed to zeros). Pulse stuffing then is done and the two bit streams next are interleaved bit by bit. The multiplexed bit stream then is scrambled such that each output bit is the modulo-2 sum of the corresponding input bit and the preceding output bit.[27] A control bit sequence then is multiplexed into the stream to permit the correct demultiplexing of the two DS1 streams, as well as the deletion of the stuff bits at the receiving end.

The M12 multiplexor combines four DS1 streams[28] together with the needed control, framing and stuff bits, since the original streams may have come from sources with separate clocks. Prior to multiplexing, the second and fourth input streams are inverted. At the M12 output the combined stream is unipolar and is converted to bipolar with a 50% duty cycle. The format, known as B6ZS, was described in Section 5.6.

The M13 multiplexor (see Ref. 24) contains two multiplexing steps. First, up to four DS1 streams are combined to produce a DS2 stream. Then up to seven DS2 streams are combined to produce the DS3 stream. Internally, all streams are in the polar format, so there are no bipolar violations. The output DS3 stream uses the B3ZS format with a 50% duty cycle. Within the DS3 stream, a frame length of 4760 bits is established. Each frame consists of seven 680 bit subframes, corresponding to the fact that a DS3 stream contains the information from seven DS2 streams. Each subframe is divided into eight blocks of 85 bits, with the first bit used for control purposes and the remaining 84 bits containing message information.

The M34 multiplexor (see Ref. 11) combines six DS3 streams into a single DS4 stream. The resulting DS4 stream contains polar binary signals in which the ones are 100% duty cycle positive voltage pulses, measured from the center conductor to the outer conductor of the coaxial cable and the zeros are correspondingly negative voltage pulses. A DS4 superframe is 4704 time slots. Each superframe is divided into 24 frames having 196 time slots each. Each frame, in turn, contains two subframes of 98 time slots. The first two of these 98 slots are used for control bits, while the remaining are for information bits.

At the various levels of the digital hierarchy the bit stream rates must conform to the tolerances shown in Table 5–2. In addition, the timing of one bit stream relative to another in the multiplexing process must be controlled carefully. This timing is achieved by pulse stuffing. Upon demultiplexing, the stuffed pulses are removed. To prevent gaps, and consequent jitter, upon removal of the stuffed pulses, elastic stores are used. These are specially designed buffers whose output repetition rate can be controlled precisely using phase-locked loop techniques. When pulse stuffing is performed at the correct rate, the elastic stores never become completely filled or completely empty.

Figure 6–22 illustrates the basis upon which jitter is removed in a regenerative repeater using an elastic store. The clock rate is recovered from the received stream and also is used to write bits from the receiver into the elastic store. The bit level in the elastic store is monitored continuously. Short duration changes in the level are smoothed via the low-pass filter. An increase in level beyond the medium range causes the frequency of the

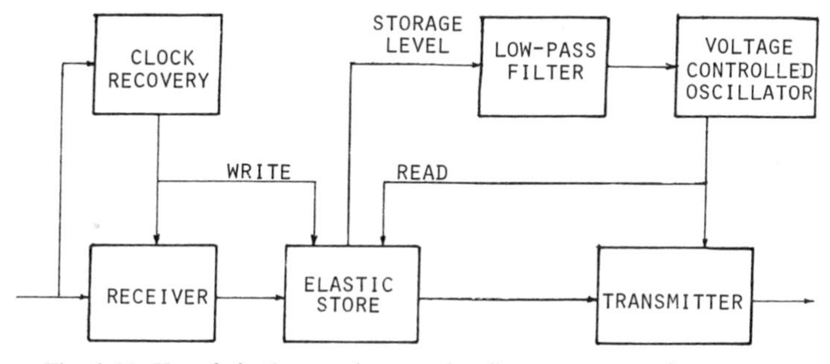

Fig. 6–22. Use of elastic store in removing jitter at regenerative repeater.

voltage-controlled oscillator to increase. This, in turn, means that bits will be read out of the elastic store at an increased rate, thus reducing the level. The voltage-controlled oscillator also serves as the transmit clock. A decrease in the elastic store level correspondingly causes a decrease in the frequency of the voltage-controlled oscillator. This decrease allows the level in the elastic store to increase, since bits then are read out of it more slowly.

6.5.2. Transmultiplexors

Both digital and analog transmission facilities now exist together in telephone networks. A given connection may involve a tandem combination of both facility types, such as may occur at a junction or at a transmission node where digital facilities connect to analog facilities. Accordingly, the need exists for the conversion of TDM signals to and from FDM signals. The device which performs this function is called the transmultiplexor. It is equivalent to a back-to-back connection of a digital demultiplexor and an analog multiplexor, or to an analog demultiplexor and a digital multiplexor. An example is the interfacing of five DS1 streams with two analog supergroups (120 voice channels), or the interfacing of two European 30-channel systems

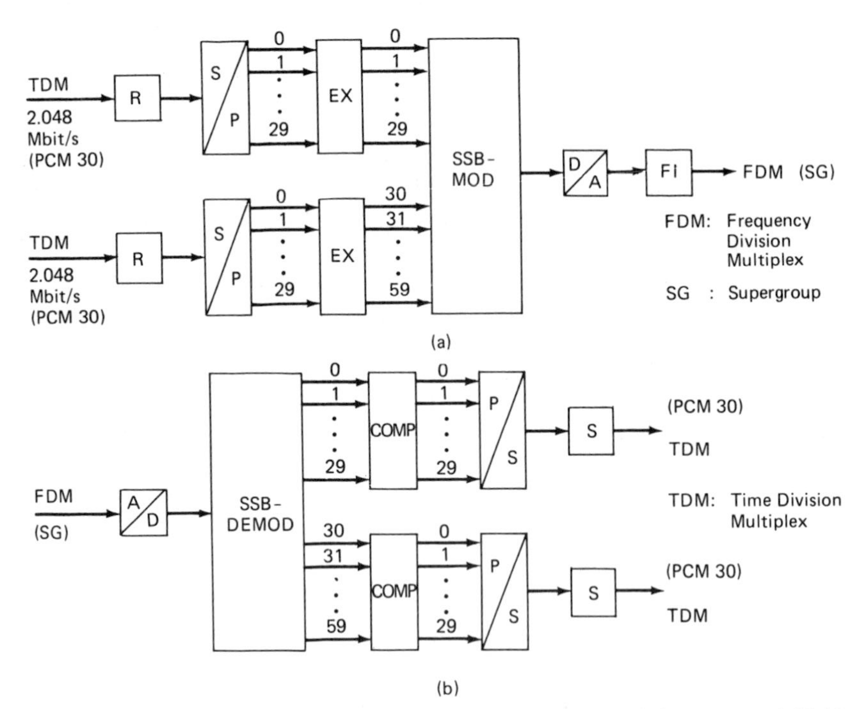

Fig. 6–23. Block diagram of a 60-channel transmultiplexor. (Courtesy Scheuerman and Göckler (29). © IEEE).

with a single analog supergroup (See Fig. 6–23). The latter is a type of application that is significant in providing international connections between systems using North American and European standards.

In Fig. 6–23(a), two European primary (30 channel) bit streams, each at 2.048 Mb/s and designated PCM 30, are applied to receivers R, which provide for their synchronization and signal level adjustment. Serial-parallel (S/P) converters then convert the TDM signals from serial to parallel form. The expandor (EX) then converts the signals from their compressed (A-law) form to linearly encoded samples. Following this step, a bank of single-sideband modulators (SSB-MOD) produces the FDM signal using one of several digital signal processing methods. Finally, a digital-to-analog (D/A) converter is used, followed by an analog filter (FI) to smooth the output, consisting of an FDM supergroup (SG).

Figure 6–23(b) illustrates the reverse process of conversion of an FDM SG to two European primary streams. Here the box labelled COMP is a compressor. Requirements for a 60-channel transmultiplexor based upon CCITT Recommendations G.792 and G.793 are shown in Table 6–4.

Table 6–4. Requirements for 60-Channel Transmultiplexors.

Amplitude response, A	
600 Hz to 2400 Hz	-0.5 dB $\leq$ A $\leq$ 0.6 dB
Maximum allowance at band edges	
300 Hz	-0.6 dB $\leq$ A $\leq$ 1.7 dB
3400 Hz	-0.6 dB $\leq$ A $\leq$ 2.4 dB
Group delay (absolute value)	$\tau \leq 3$ ms
Distortion	
1000 Hz to 2600 Hz	$\Delta\tau \leq 0.5$ ms
600 Hz to 1000 Hz	$\Delta\tau \leq 1.5$ ms
500 Hz to 600 Hz, 2600 Hz to 2800 Hz	$\Delta\tau \leq 2.0$ ms
Minimum crosstalk attenuation between any two channels	
Intelligible crosstalk	65 dB
Unintelligible crosstalk	58 dB
Maximum idle channel noise with all channels loaded except the one measured, relative to peak signal level	-80 dB
Maximum in-band rms nonlinear distortion, relative to peak signal level	-40 dB
Out-of-band signaling frequency	3825 Hz
Pilot frequency	3920 Hz

NOTE: Levels and delays are those of the multiplexed analog signal with the digital parts looped.
BASIS: CCITT Recommendations G.792 and G.793.
SOURCE: Scheuermann and Göckler [29]

Digital signal processing techniques can be used to implement transmultiplexors. Four major approaches[29] of this type are called (1) the bandpass filter bank, (2) the low-pass filter bank, (3) the Weaver structure method, and (4) the multistage modulation method. Two additional techniques[30] are the polyphase method and the Classen-Mecklenbräuker method. Numerous papers have been published on transmultiplexors[31] in Europe, Japan, and North America.

Because the FDM hierarchy uses single sideband for each voice channel, a major factor in transmultiplexor implementation is single sideband modulation and demodulation. These operations can be viewed as forms of digital sample rate interpolation and decimation. Digital interpolation consists of taking one sequence $a(n)$ and producing another sequence $b(m)$ whose samples occur r times as fast. Interpolation can be used to extract any portion of the spectrum of Fig. 6–24(a), including single sidebands. Decimation is the process of sample rate reduction, i.e., starting with the sequence $b(m)$ and producing a sequence $a(n)$ from it. Thus let $a(n)$ be a sequence whose sample period is T seconds. The spectrum of this signal is shown in Fig. 6–24(a). It is periodic with a period of $1/T$ Hz. Let $H(f)$ be a low-pass

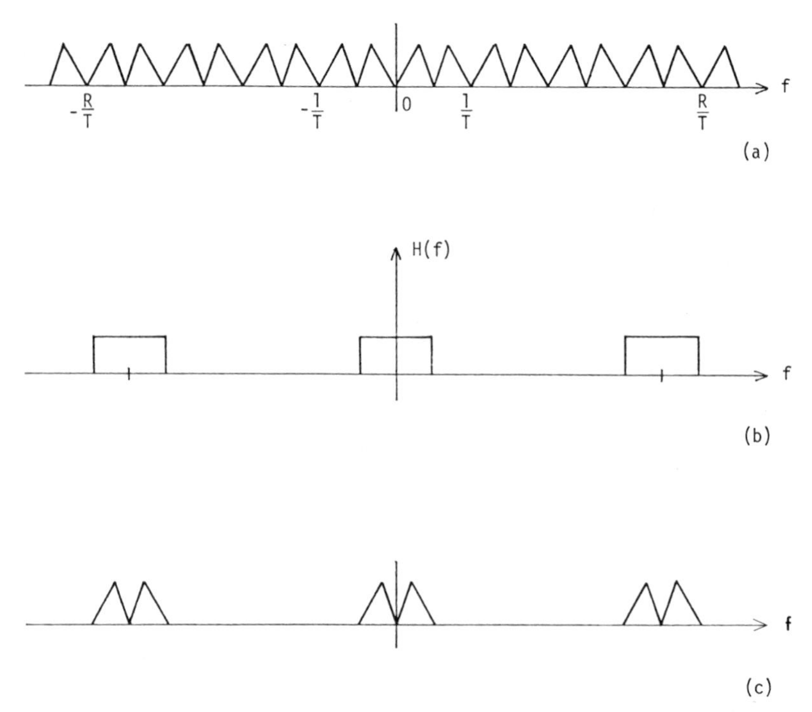

Fig. 6–24. Interpolation spectra. (Courtesy S. L. Freeny, Ref. 21, © IEEE, 1980.)

digital filter with a cutoff frequency of $1/2T$ Hz and operating at a sampling rate of r/T Hz [see Fig. 6–24(b)]. If the low-rate signal is applied to this filter (considering intervening input samples to be zero), the filter will produce an r/T rate output whose samples are the same as those that would be obtained in the interpolation process. For group band operations, a suitable r value is 14. Since $1/T = 8$ kHz, $r/T = 112$ kHz. As indicated in Fig. 6–24(c), the filter removes the unwanted signal images. The result is a spectrum that is the same as would have resulted from sampling an analog signal at the high rate initially, to the extent that $H(f)$ approximates an ideal low-pass filter.

The result of using interpolation is known as the bandpass method of digital single sideband generation. Rather than multiplying by a sine wave, as in the analog case, the "carrier" is built in because of the periodic nature of the digital spectrum. For the bandpass demodulation process, the composite FDM stream is applied simultaneously to twelve bandpass filters (one for each channel in the group spectrum) whose outputs then are decimated by a factor of $r = 14$. The per-channel computation rate for demodulation thus is the same as for modulation.

To prevent the resulting FDM signal in the modulation process from having a mixture of upper and lower sidebands, the sign of the input sequences of every other channel must be alternated. This shifts the spectra of each of the inputs by half the sampling rate, or 4 kHz. Thus in any given frequency slot, an upper sideband is exchanged for a lower one or vice versa. For the bandpass demodulator, a similar sign alteration must be done on half of the output channels after decimation.

The Weaver method is described in Refs. 29 and 30, the multistage modulation method in Ref. 29, and the polyphase and Classen-Mecklenbräuker methods in Ref. 30.

6.6. ERROR CONTROL

Chapter 3 illustrated the fact that PCM encoded speech does not demand unusually low error rates for good performance. A bit error rate of 10^{-4} or less is quite satisfactory. The other voice coding techniques, as discussed in Chapter 5, perform quite well at even higher bit error rates, such as 10^{-3} or even 10^{-2}. Why, then, should concerns arise about detecting and correcting errors? First, error detection is important as a means of monitoring facility performance. An increased bit error rate may indicate line degradation which may get worse with time. Second, although digital voice does not demand ultra low bit error rates, such functions as signaling, supervision, and voice-band data need to be accomplished accurately. For example, if wrong numbers caused by channel and equipment errors are to be minimized, a means of error correction may be desirable to assure a bit error rate of 10^{-7} or less. A further need for error control arises in the case of mobile and portable radiotelephone systems, including cordless telephones, operating under conditions close to link margin limits, i.e., near the signal-to-noise threshold.

6.6.1. Error Detection

Many error detection schemes are based on a parity check code. A simple parity check code for detecting a single error involves the addition of a computed bit to a sequence of information bits. For example, suppose the information bits are 1 0 1 1 0 1 1. To these seven bits an eighth bit is added so that the modulo-2 sum of the bits equals zero (or one, if the designer so chooses). Thus the sequence would become 1 0 1 1 0 1 1 1. This is also called a row check. At the end of a block of bits, a row of parity check bits, each checking its own column, also could be added.

The information transmission rate, when using the simple row check on seven information bits, will be only ⅞ of the value it would have if all eight bits were information bits. Thus this is called a *rate ⅞ code,* or the code is said to have an *efficiency* of ⅞. However, the lower the efficiency, the more effective the error detection can be in terms of the number of errors that can be detected. Note that a single parity bit detects an odd number of errors, but that an even number of errors is undetected by this technique.

6.6.2. Error Correction

In land mobile radiotelephone as well as cordless or portable telephone applications, the radio channel may exhibit severe phase distortions for various positions of the end instrument. In addition, momentary signal drops below the threshold may occur. Such circumstances call for error correction, illustrated by the following example[32] of what is known as a Hamming code for single error correction.

Consider information to be grouped into blocks of four binary digits. Let three parity digits be added to each block for a total block length of seven digits. Let the information digits be labeled I_1, I_2, I_3, and I_4, and let the parity digits be labeled P_1, P_2, and P_3. Each parity digit will be selected for even parity (an even number of ones) among itself and a selected subset of the information digits. The rules for determining each parity digit are given in the encoding-decoding table of Fig. 6–25. The top row of the table is labeled P_1. The P_1 value is selected so that there will be an even number of ones among I_1, I_2, I_3, and P_1. The P_2 value is selected so there will be an even number of ones among I_1, I_3, I_4, and P_2. The P_3 value is selected

	I_1	I_2	I_3	I_4	P_1	P_2	P_3
P_1	X	X		X	X		
P_2	X		X	X		X	
P_3		X	X	X			X

Fig. 6–25. Encoding-decoding table for a simple error correcting code. (From L. Lewin, ed., *Telecommunications: An Interdisciplinary Survey,* Artech House, Inc., Dedham, MA, 1979.)

so there will be an even number of ones among I_2, I_3, I_4, and P_3. For example, assume the information digits are $I_1 = 1$, $I_2 = 0$, $I_3 = 0$, and $I_4 = 1$. Select P_1 so there will be an even number of ones among I_1, I_2, I_4, and P_1, that is, among 1, 0, 1, and P_1. Therefore, P_1 must be 0. Similarly, the rules indicate that $P_2 = 0$ and $P_3 = 1$. Thus the code word, complete with its parity digits, is 1 0 0 1 0 0 1. Correspondingly the rate is $4/7$.

To illustrate the use of this code in error correction, assume that the above code word, 1 0 0 1 0 0 1, is sent over a communication system. At the receiving end of the system the parity checks indicated by the encoding-decoding table are made. More specifically, the number of ones among I_1, I_2, I_4, and P_1 is determined. This is *parity check* number 1, and if the number of ones is even, the parity check is said to be correct; otherwise, it is said to have failed. Similarly, parity checks numbers 2 and 3 are made according to the definitions given by the second and third rows of the encoding-decoding table. For example, if an error occurs, the block may be received as 1 0 1 1 0 0 1. At the receiver the first parity check is correct, but the second and third both fail. The assumption then is made that only one error has occurred, and reference to the encoding-decoding table shows the location of the error.

If the error were in I_1, the table shows that I_1 enters into parity checks 1 and 2 but not 3. Thus the first two parity checks would fail while the third one would be correct. This is not the result of the parity checks, so the error cannot be in I_1. Each of the other possible positions is considered; the only position in which an error could cause the first parity check to be correct and the other two to fail is I_3. Accordingly, the error must be in digit I_3. I_3 then is changed, producing 1 0 0 1 0 0 1, which was the block transmitted. This procedure is called *error correction.*

For situations in which the probability of double errors is significant, this code is unsuitable, but other codes are available.

6.6.3. Block and Convolutional Codes

A block (or frame) is a grouping of bits established by a sending station. The example in the previous section illustrated a block code, in which the redundant bits relate only to the information bits of the same block. If, on the other hand, the redundant bits also check the information bits in previous blocks, the code is called *convolutional.* With reference to the propagation problems of mobile and portable radiotelephone systems, note that nothing can handle a complete signal drop-out. However, a very short duration signal impairment (e.g., less than a syllable in length) may be bridged by a suitable convolutional code.

6.7. PAIR-GAIN SYSTEMS

Pair-gain systems, sometimes called subscriber loop systems, provide savings in wire pairs by combining a number of voice channels using time-division multiplex. Their use has been fostered not only because of wire pair savings, but also because they enable service additions in areas of rapid growth as well as in areas where emergency conditions have caused system outages. Pair-gain systems also are useful in serving small communities remote from a central office.

The first digital pair-gain system[33] was the *subscriber loop multiplex* (SLM). These systems are still in use and serve up to 80 subscribers via a 24 channel T1 line using concentrators. The channel codecs use 57.2 kb/s delta modulation. The system uses two pairs of 22-gauge wire, and can use remote powering to achieve distances up to 80 km. As many as six remote terminals can be placed along the line to interconnect with individual station lines along the route, with up to 40 lines being served by a remote terminal. Signaling and supervision are handled within the digital bit stream. The SLM system interfaces with both the end-office switch and the subscriber on an analog basis, thus providing system transparency.

The *Subscriber Loop Carrier-40* (SLC-40) system serves 40 subscribers via a T1 line using 38 kb/s ADM without concentration. As does the SLM, the SLC-40 uses two pairs of 22-gauge wire and works up to 80 km using remote powering. However, all subscriber line interfaces must be at a common point. Analog interfaces are provided to the end-office switch as well as the subscriber.

The SLC-96 was placed into service in 1979.[34] It uses PCM, the same as the T-carrier systems. Accordingly, it is fully compatible with other digital switching and transmission equipment used in the public switched network.

6.8. RETROFIT

Many digital transmission systems are being added to existing analog systems on a retrofit basis. The combination thus is a hybrid system providing the analog channels as before, but adding digital channels in a portion of the baseband spectrum that previously was unused. Often this is more economical than building completely new digital systems. The most commonly used arrangement is known as *data under voice* (DUV), and is shown in Fig. 6–26. Here the baseband spectrum below 564 kHz on a microwave radio system is utilized. None of the FDM mastergroups occupy this portion of the baseband spectrum. To achieve the needed spectral efficiency of over 3

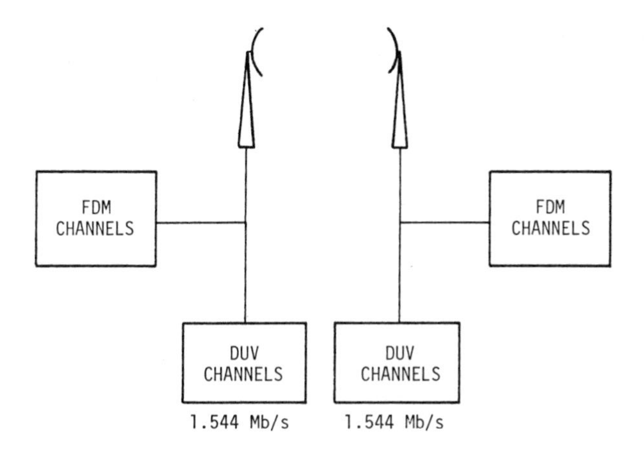

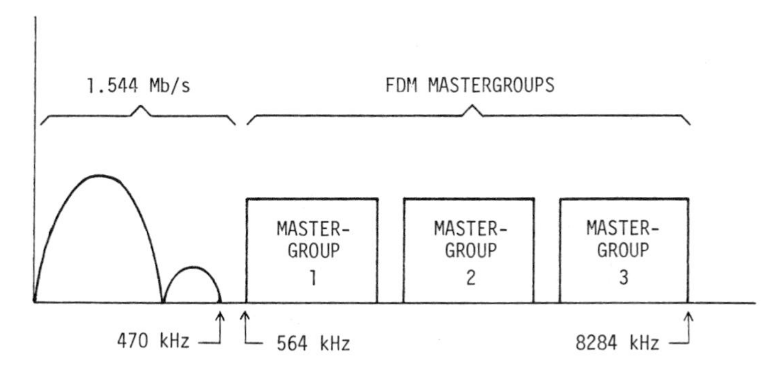

Fig. 6–26. Data under voice (DUV) system concept.

b/s per Hertz, a seven level partial response coder is used, as was described in Section 6.1.5, Correlative Techniques. This confines the T1 spectrum to the 0–470 kHz band. The resulting baseband signal is used in the 1A-RDS system. Figure 6–27 is a block diagram of the transmitting system. The elastic store performs the function of removing timing jitter. Scrambling is performed next to reduce the level of any discrete spectral components that otherwise might interfere with the FDM transmissions. The low-pass filter performs necessary spectrum shaping to suppress unwanted energy above 386 kHz. The monitoring and status indicator circuits perform in-service monitoring as well as failure indications. If the input signal should be lost, a "DS1 substitution signal," consisting of all ones, is sent instead. The 386-kHz pilot is transmitted for synchronization of the receiver circuits. Pre-emphasis is added because the resulting multilevel output becomes part of the overall baseband (see Fig. 6–26) which is frequency modulated onto the microwave carrier. In any FM system pre-emphasis is important in allowing the receiver to attenuate (de-emphasize) the background noise, which increases in level with baseband frequency.

Figure 6–28 shows the DUV receiving system, providing an output signal that is a replica of the DS1 input at the transmitting end. The various processes are the inverse of those of the transmitter. The signal monitoring and status indicators show how the receiver is performing.

Several alternative techniques for adding digital transmission to an analog

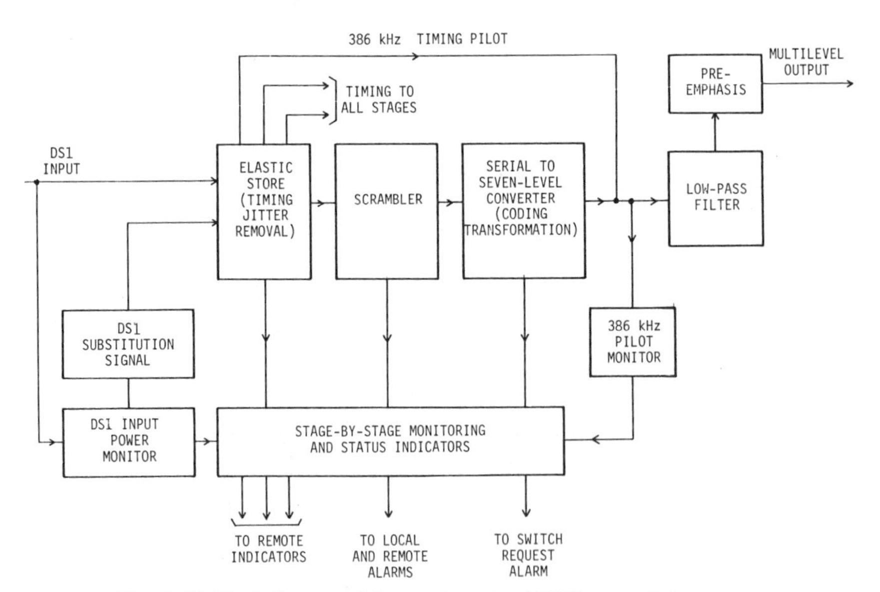

Fig. 6–27. Block diagram of data under voice (DUV) transmitting system.

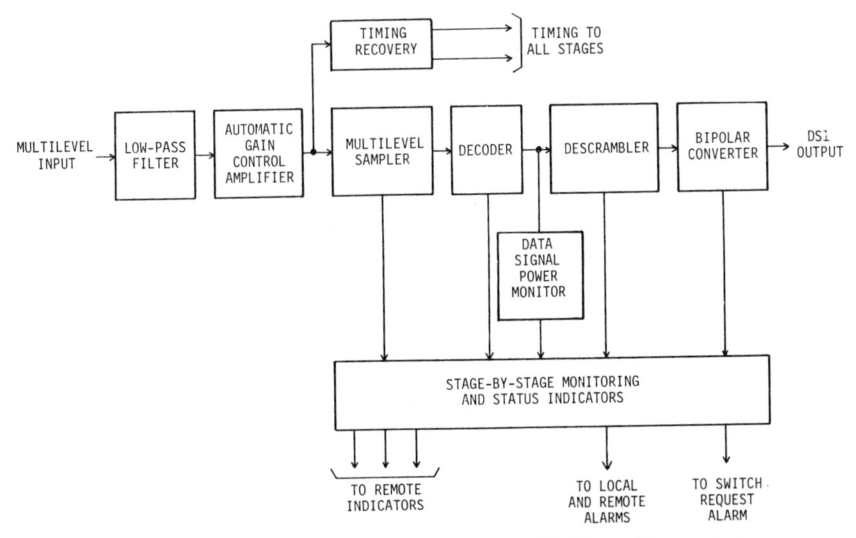

Fig. 6–28. Block diagram of data under voice (DUV) receiving system.

facility have been developed.[35] These are known as *data above voice* (DAV) and *data above video* (DAVID). They allow up to 3.152 Mb/s to be sent above the top of the baseband of an FDM message or video signal on an existing analog facility. The bands used for digital transmission normally cannot be used for analog transmission. Both DAV and DAVID are sent by converting the bipolar data to a binary stream, scrambling, and then translating to the desired frequency, at which the modulation form is QPSK. A bandpass filter keeps the QPSK separate from the lower frequency analog transmissions.

6.9. TESTING AND FAULT DETECTION

A line monitor[36] is used on T1 lines to perform pulse quality measurements, and to sense bipolar violations, as well as the presence of more than 15 consecutive zeros. The redundancies inherent in partial response coding also can be used in performance monitoring, as described in Section 6.2.5. Excessive violations cause an office alarm to sound, and may initiate protection line switching. A fault location system identifies which repeater section is causing troubles. To accomplish fault location the line under test must be removed from service. A special test signal then is sent from a fault location test set.[37] Other T1 system tests include the measurement of transmission pair losses and tests of repeater performance.

An order wire consisting of a loaded pair parallels the digital line and is used for maintenance coordination.

Lines operating at the T1C and T2 rates use fault location systems similar to those used for T1 lines. Portable battery operated monitors provide measurements at span terminating frames for T2 lines as well as at various access points along the lines, including the output of each regenerator.

PROBLEMS

6.1 What is the maximum (theoretical) data rate that can be sent error free over a voice bandwidth channel with a 30-dB ratio of signal to channel noise?

6.2 Using Table 6–1, explain why BPSK cannot be used in the 3.7–4.2-GHz band. Would the result be any different if dual polarization were used to double the number of voice channels?

6.3 Why are the higher stream rates of the digital hierarchy not integral multiples of the lower stream rates?

6.4 Explain why the tolerable channel noise level decreases as the number of modulation levels increases, assuming the bit error rate is to be held constant.

6.5 A digital transmission engineer has a 20-kHz bandwidth channel in which to transmit a 56-kb/s digital stream. Can QPSK be used to meet this requirement? If not, would 8-PSK be feasible? Would 16-QAM be preferable?

6.6 Why can a 9.6 kilobaud transmission rate not be sent on a voice bandwidth line? What is the maximum rate according to Nyquist's theorem?

REFERENCES

1. Federal Communications Commission, "Establishment of Policies and Procedures for Consideration of Applications to Provide Specialized Common Carrier Services in the Domestic Point-to-Point Microwave Radio Service and Proposed Amendments of Parts 2 and 21 of the Commission's Rules," FCC 74–657, Docket No. 18920, Washington, D.C., July, 1974.
2. D'Andrea, A. N., and Russo, F., "Noise Analysis of a PSK Carrier Recovery DPLL," *IEEE Transactions on Communications,* Vol. COM-31, No. 2, pp. 190–199 (February, 1983).
3. Prabhu, V. K., and Salz, J., "On the Performance of Phase-Shift-Keying Systems," *Bell System Technical Journal,* Vol. 60, No. 10, pp. 2307–2343 (December, 1981).
4. Spilker, J. J., *Digital Communications by Satellite,* Prentice-Hall, Englewood Cliffs, NJ, pp. 295–335, 1977.
5. Spilker, J. J., *Digital Communications by Satellite,* Prentice-Hall, Englewood Cliffs, NJ, 1977.
6. Feher, K., *Digital Communications: Microwave Applications,* Prentice-Hall, Englewood Cliffs, NJ, 1981.
7. Bellamy, J. C., *Digital Telephony,* John Wiley & Sons, New York, pp. 496–497, 1982.
8. Feher, K., *Digital Modulation Techniques in an Interference Environment, Multi-Volume EMC Encyclopedia,* Vol. 9, Don White Consultants, Inc., Gainesville, VA.
9. Bic, J.-C., Duponteil, D., and Imbeaux, J.-C., "64-QASK Sensitivity to Modem Imperfections and to Interferences," *IEEE Globecom '82 Conference Record,* Miami, FL, Nov. 29–Dec. 2, 1982, IEEE Document CH1819–2/82–0000–0322, pp. B3.3.1–B3.3.6.
10. See note 9.
11. Thomas, C. M., Weidner, M. Y. and Durrani, S. H., "Digital Amplitude-Phase Keying

with *M*-ary Alphabets," *IEEE Trans. Comm.* Vol. COM. 22, No. 2, pp. 168–180, (Feb. 1974).
12. Osborne, W. P., and Luntz, M. P., "Coherent and Noncoherent Detection of CPFSK," *IEEE Trans. Comm.*, Vol. COM-22, pp. 1023–1036 (Aug. 1974).
13. deBuda, R., "Coherent Demodulation of Frequency-Shift Keying with Low Deviation Ratio," *IEEE Transactions on Communications,* Vol. COM-20, No. 3, pp. 429–435 (June, 1972).
14. Weinberg, A., "Effects of a Hard Limiting Repeater on the Performance of a DPSK Data Transmission System," *IEEE Trans. Comm.*, Vol. COM-25, pp. 1128–1133, (Oct. 1977).
15. Mathwich, H. R., Balcewicz, J. F., and Hecht, M., "The Effect of Tandem Band and Amplitude Limiting on the E_b/N_0 Performance of Minimum (Frequency) Shift Keying (MSK)," *IEEE Trans. Comm.*, Vol. COM-22, No. 10, pp. 1525–1540 (Oct. 1974).
16. Lender, A., "The Duobinary Technique for High Speed Data Transmission," *IEEE Transactions on Communication and Electronics,* Vol. 82, pp. 214–218 (May, 1963).
17. Lender, A., "Correlative Digital Communication Techniques," *IEEE Transactions on Communication Technology,* Dec. 1964, pp. 128–135.
18. Kretzmer, E. R., "Binary Data Communication by Partial Response Transmission," *1965 IEEE Annual Communication Conference,* Conference Proceedings, pp. 451–455.
19. Feher, K., *Digital Communications: Microwave Applications,* Prentice-Hall, Englewood Cliffs, NJ, 1981.
20. Lender, A., Rogers, R., and Olszanski, H., "4 Bits/Hz Correlative Single Sideband Digital Radio at 2 GHz," *Proceedings of the IEEE International Conference on Communications,* ICC-79, Boston, MA, June, 1979, pp. 5.2.1 to 5.2.5.
21. Feher, K., *Digital Communications: Microwave Applications,* Prentice-Hall, Englewood Cliffs, NJ, pp. 47–50, 1981.
22. See note 6.
23. Ishio, H., et al., "A New Multilevel Modulation and Demodulation System for Carrier Digital Transmission," *Proceedings of the IEEE International Conference on Communications,* ICC-76, Philadelphia, PA, June, 1976, pp. 29.7 to 29.12.
24. *Telecommunications Transmission Engineering,* Vol. 2, *Facilities,* Bell System Center for Technical Education, 1977.
25. Haury, P. T., and Romeiser, M. B., "T1 Goes Rural," *Bell Laboratories Record,* Vol. 54, pp. 178–183, (July/Aug., 1976).
26. See note 25.
27. Booth, T. L., *Digital Networks and Computer Systems,* John Wiley & Sons, Inc., New York, 1971, pp. 41–52.
28. Moore, J. D., "M12 Multiplex," *1973 Conference Record,* IEEE International Conference on Communications, Vol. 1, pp. 22–20 to 22–25.
29. Scheuermann, H., and Göckler, H., "A Comprehensive Survey of Digital Transmultiplexing Methods," *Proc. IEEE,* Vol. 69, No. 11, pp. 1419–1450 (Nov. 1981).
30. Freeny, S. L., "TDM/FDM Translation as an Application of Digital Signal Processing," *IEEE Communications Magazine,* Vol. 18, No. 1, pp. 5–15 (Jan. 1980).
31. Special Issue on Transmultiplexers, *IEEE Transactions on Communications,* Vol. COM-30, No. 7, Part 1, (July, 1982).
32. Maley, S. W., "Telecommunications Systems," Chapter 10 of *Telecommunications: An Interdisciplinary Survey,* ed. by L. Lewin, Artech House, Inc., 1979.
33. *Telecommunications Transmission Engineering,* Vol. 2, *Facilities,* Bell System Center for Technical Education, Western Electric Company, Inc., Winston-Salem, NC, pp. 85–88, 1977.

34. Brolin, S., Cho, Y. S., Michaud, W. P., and Williamson, D. H., "Inside the New Digital Subscriber Loop System," *Bell Laboratories Record,* pp. 110–116, April 1980.
35. See note 6.
36. Blair, R. W., and Burnell, R. S., "Monitors Take the Pulse of T1 Transmission Lines," *Bell Laboratories Record,* Vol. 51, pp. 55–60 (Feb. 1973).
37. See note 17.

7
Microwave Transmission*

7.1. INTRODUCTION

Microwave radio systems now provide about 70% of all the toll message circuit distance within the United States, and are responsible for an appreciable portion of the world's circuits as well. While much of this message circuit distance is on an analog basis, digital service constitutes an increasing portion of microwave transmission as well. Microwave systems not only provide feeder service to long haul coaxial backbone routes but also provide cross-country service through the use of multiple repeater sites. Thus individual circuit lengths range from less than 30 km to over 6400 km with route cross-sectional capacities from less than 60 to more than 22 000 circuits. Applications include not only telephone message channels but also wideband data and television transmission.

In microwave system terminology, a short-haul system is one covering a distance up to about 400 km between end points. Such applications include intrastate and feeder service. Long haul systems exceed 400 km and are used for interstate and backbone routes.

The modulation technique used for microwave systems is predominantly angle (frequency or phase) modulation, with PSK predominating for digital systems.[1] The use of APK is growing, however.

This chapter presents an overview of microwave radio systems with an emphasis on their use in digital telephony applications.

7.2. CHARACTERISTICS OF MICROWAVE PROPAGATION

Several factors contribute to the degradation of a microwave radio signal as it travels from one repeater to another. These are spreading loss and absorption, fading (caused by refraction and multipath effects), and polarization shifts. They are described next in that order.

* Much of the information in this chapter is summarized from Bernhard E. Keiser, "Wideband Communications Systems," Course Notes, © 1981.

7.2.1. Spreading Loss and Absorption

All forms of electromagnetic radiation involve the emission of power from a source, and the spreading of that power in many directions. Thus if p_t is the power radiated from an isotropic source and if d is the radius of an imaginary sphere centered on the source and representing the locus of all points a distance d from the source, then the power density at a distance d from the source is $p_t/4\pi d^2$ watts/m². The spreading of power from the source can be exemplified by the spreading of light flux from a flashlight bulb. The greater the distance, the less is the power received.

The actual power p_r received through a receiving antenna with an effective area A_r on the surface of the sphere is

$$p_r = p_t A_r / 4\pi d^2 \text{ watts} \tag{7-1}$$

The area A_r may be thought of as the area of a reflector associated with the receiving antenna, but diminished by the overall efficiency of the antenna. Figure 7–1 shows the geometrical relationships involved in this concept. Figure 7–1(a) illustrates the isotropic source just discussed. This source can

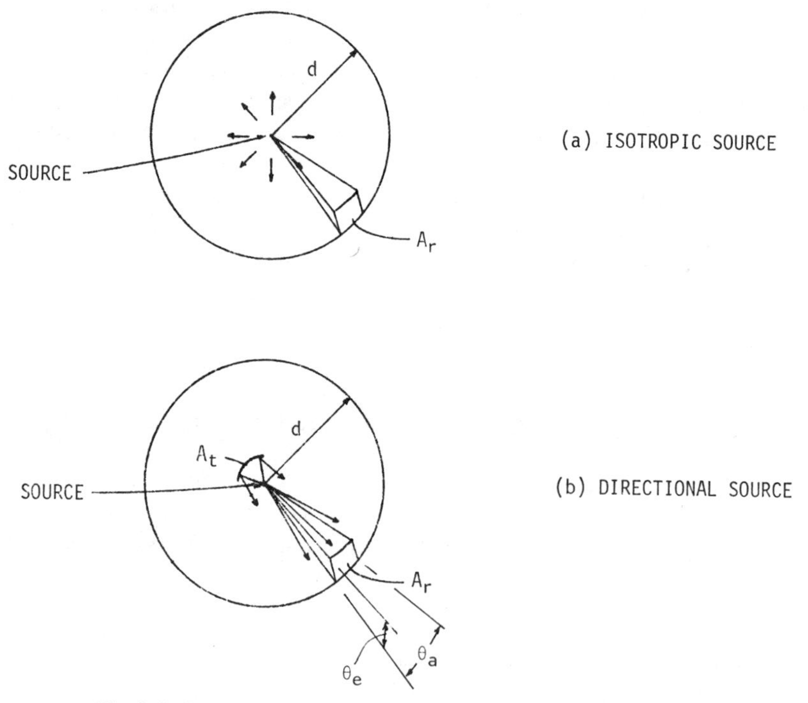

Fig. 7–1. Geometrical relationships resulting in free space path loss.

be made directional, as illustrated in Fig. 7–1(b), by placing a reflector having an effective area A_t behind it. In this case, the source exhibits an on-axis gain of

$$g_t = 4\pi A_t/\lambda^2 \tag{7–2}$$

where λ is the wavelength of radiation and A_t is the physical area of the transmitting antenna multiplied by its efficiency. The value of λ in meters can be determined from the operating frequency f in Hertz by the relationship $\lambda = c/f$, where c = speed of light = 3×10^8 m/s. Such an antenna concentrates the radiation into a solid angle Ω steradians where

$$\Omega = \lambda^2/A_t \tag{7–3}$$

The expressions for gain and solid angle can be combined to yield the result

$$g_t = 4\pi/\Omega \tag{7–4}$$

Thus the smaller the beam angle of an antenna, the greater its gain. The solid angle Ω can be expressed as the product of an azimuthal beam width θ_a and an elevation beam width θ_e.

For a directional antenna equation (7–1) becomes

$$p_r = p_t g_t (A_r/4\pi d^2) \tag{7–5}$$

Use of Eq. (7–2) yields

$$p_r = p_t (4\pi A_t/\lambda^2)(A_r/4\pi d^2) \tag{7–6}$$

Eq. (7–6) can be rearranged to express the received power in terms of the transmitting and receiving antenna gains as follows:

$$p_r = p_t (4\pi A_t/\lambda^2)(4\pi A_r/\lambda^2)(\lambda/4\pi d)^2 = p_t g_t g_r (\lambda/4\pi d)^2 \tag{7–7}$$

The reciprocal of the last term in Eq. (7–7), $(4\pi d/\lambda)^2$, is the *spreading loss* or *free-space path loss.* Generally it is expressed in decibel form as 20 log $(4\pi d/\lambda)$, or in the more convenient form

$$\text{Free-space path loss} = 32.5 + 20 \log d_{\text{km}} + 20 \log f_{\text{MHz}} \tag{7–8}$$

where d_{km} = path length in km
f_{MHz} = frequency in MHz.

The free-space path loss is the most significant loss between the microwave transmitter and receiver of a link. However, it is a fixed loss that is known in advance and can be accounted for in design.

An additional loss between the transmitter and receiver is absorption, produced primarily by water vapor in the air. Figure 7–2 shows that absorption increases with frequency and also with rain intensity. Typical microwave path lengths of 30–45 km are essentially unaffected by rain, but at frequencies of 11 GHz and higher, path lengths may have to be kept shorter, especially in areas that experience significant amounts of precipitation.

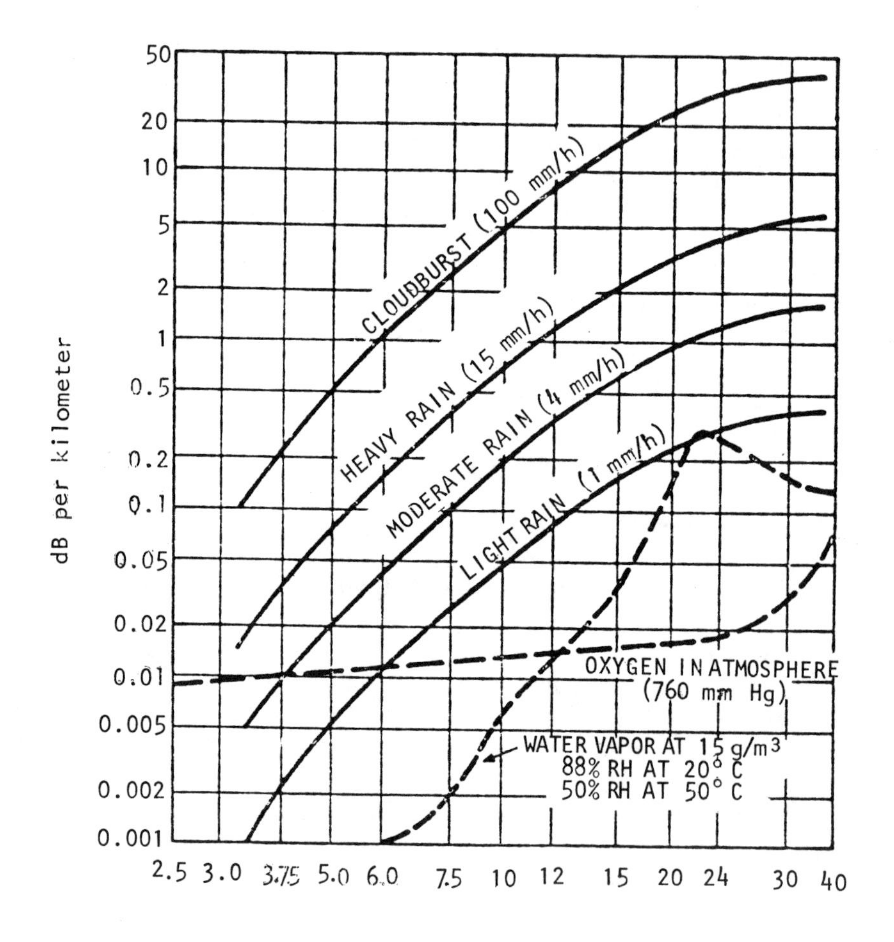

Fig. 7–2. Absorption in the atmosphere. (Copyright 1982, Bell Telephone Laboratories. Reprinted by permission.)

7.2.2. Fading

Fading on a microwave path is caused by atmospheric refraction changes and by multipath propagation. The atmosphere tends to bend microwaves back toward the earth in such a way that the earth nominally appears to have a radius that is about 4/3 of its true radius. Light rays experience a similar bending, but for them the effective earth radius is about 6/5 of the true earth radius. The ratio of the effective to the true earth radius usually is designated by the letter K.

Refraction is not constant. When heavy ground fog is present, or when there is very cold air over a warm earth, atmospheric refraction may become substandard, i.e., $K < 1$, in which case the rays tend to be bent upward away from the earth. Other conditions may exist in which K becomes infinite, i.e., the rays bend along with the curvature of the earth for distances of 100 km or more, resulting in a situation called *overreach* in which the rays appear at repeaters more distant than intended.

Ray interference is produced when a ray reaches its destination by more than one path. Often one path is the direct one and the other path is one in which a ray has been reflected from the ground. Ray interference also can result from an irregular variation in dielectric constant with height. Unlike atmospheric refraction, ray interference tends to be fast and frequency selective. Figure 7–3 illustrates the statistics of ray interference at 4 GHz. Fortunately, the deep fades are only of brief duration.

In the design of microwave facilities (see Section 7.4) the probability of fades below a given level must be ascertained and a fade margin must be incorporated into the power budget corresponding to the percent availability for which the system is being designed. Space diversity is helpful in combatting fading.

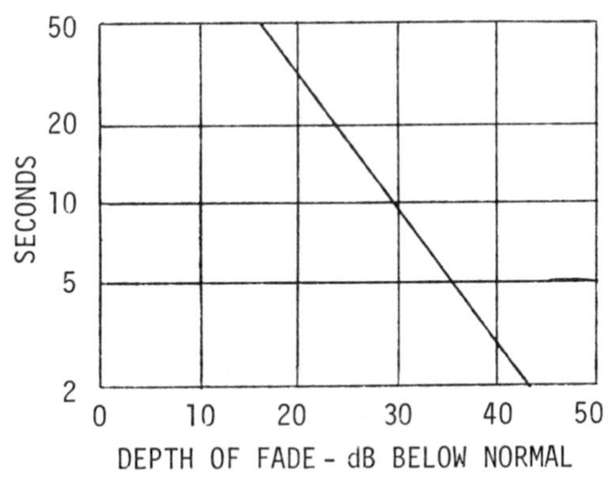

Fig. 7–3. Median duration of fast fading on 50 km paths. (Copyright 1982, Bell Telephone Laboratories. Reprinted by permission.)

7.2.3. Polarization

In order to minimize adjacent channel interference along a given microwave path, alternate polarizations, vertical and horizontal, are used. In addition, the alternation of polarizations along a path from one repeater to another can aid in minimizing the overreach interference problem. Although precipitation tends to convert linear polarization into elliptical, especially at frequencies above 10 GHz, the undesired, or cross-polarized, signal generally is 25–30 dB below the desired signal, making polarization a very useful technique in microwave radio systems. In fact, separate digital channels can be sent on the two polarizations.

7.3. MICROWAVE SYSTEM ENGINEERING

Numerous factors affect the performance of a digital radio system. Some of the key factors are discussed next.

7.3.1. Frequency Allocations

The most heavily used bands available for microwave radio relay by the common carriers within the U.S., and their usage, is as follows:

Frequency Bands (MHz)	*Usage*
2110–2130 2160–2180	Service channels for long haul administrative purposes and switching
3700–4200	Long haul
5925–6425	Long and short haul
10700–11700	Short haul

Additional common carrier bands are available in some countries. For example, in Canada the available bands are 1700–2300 MHz, 3550–4200 MHz, 5925–8275 MHz and 10 700–11 700 MHz. Higher frequencies also have been allocated in the U.S. and Canada, but are not in common use for transmission beyond metropolitan areas because of propagation problems during precipitation.

The Federal Communications Commission (FCC) has divided the common carrier bands in the U.S. into radio channels, and has specified the minimum number of voice channels per radio channel, as was shown in Table 6–1. In addition, to minimize problems of adjacent channel interference, the mean output power in any 4-kHz band must be attenuated below the mean total output power of the transmitter. This attenuation must be such that the transmitter's output spectrum fits within the "FCC mask." For operation below 15 GHz, this mask is shown in Figure 7–4 and is defined by the equations

$$A(f) = \begin{cases} 0 \text{ dB}, & \text{for } 0 < b < 50 \quad (7\text{–}9a) \\ 35 + 0.8(b - 50) + 10 \log_{10} B \text{ dB}, & \text{for } b > 50 \quad (7\text{–}9b) \end{cases}$$

with the provision that the maximum required value of $A(f)$ is 80 dB. In Eq. (7–9),

$A(f)$ = attenuation below the mean wideband output power level (dB), measured in a 4-kHz band
B = radio channel bandwidth (MHz)
f_0 = carrier frequency (MHz)
f = frequency at which attenuation is being specified (MHz)
$b = 100|f - f_0|/B$

Above 15 GHz,

$$A_M(f) = \begin{cases} 0 \text{ dB}, & \text{for } 0 < b < 50 \quad (7\text{–}10a) \\ 11 + 0.4(b - 50) + 10 \log_{10} B \text{ dB}, & \text{for } b > 50 \quad (7\text{–}10b) \end{cases}$$

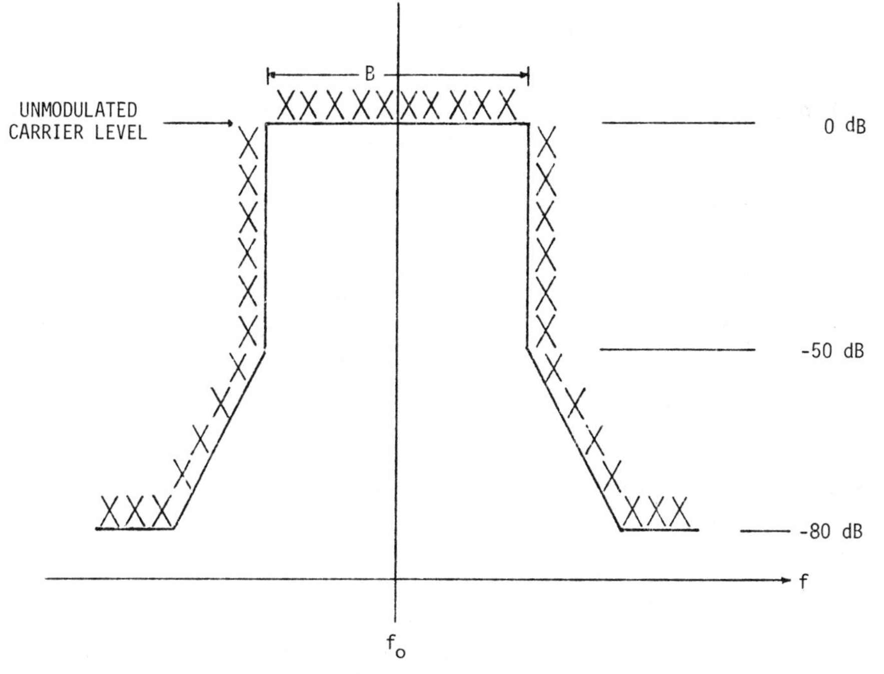

Fig. 7–4. FCC mask.

where $A_M(f)$ is the attenuation below the mean wideband power output level (dB), measured in a 1-MHz band. The maximum required value of $A_M(f)$ is 56 dB.

Filtering is a major technique used to assure the meeting of the FCC mask requirements in a given system. These requirements are especially important since otherwise the use of analog transmission in radio channels adjacent to those carrying digital transmission may be seriously affected. Other combinations, i.e., digital to adjacent digital and analog to adjacent digital, also result in degradation, but of a less serious nature.

Several system design factors are closely related to the subject of frequency allocation.[2] These include the proper choice of radio-frequency (RF) carrier, intermediate frequency (IF) and beating oscillator frequencies, as well as the likely severity of selective fading, which may alter the stability of the carrier to sideband phase relationships.

A major factor affecting the selection of the frequency band is expected system growth and thus the required number of channels. The higher frequency bands, especially 11 GHz and above, tend to be preferable from this viewpoint because more channels generally are available along any route at these higher frequencies. Protection channels also must be provided for use during deep fades or when channel equipment outages occur.

Cost factors also are significant in frequency choice. In general, equipment for the higher frequency bands (e.g., 11 GHz) tends to be more expensive, but this may be partially offset by the fact that the number of voice channels per megahertz must be larger at the lower frequencies (e.g., 4 GHz), and thus expensive filters and signal shaping techniques may be needed at these frequencies.

System costs in terms of numbers of channels consist of per site costs and variable costs. The fixed (per site) costs are independent of the number of channels carried. They include roads, buildings, towers, and antennas. Thus the fixed costs per voice channel decrease with the number of channels carried. The variable costs per channel tend to be constant for low to moderate numbers of channels, but then rise as attempts to fully load the route are made because of increasingly severe filter requirements, tight tolerances on RF and IF frequencies, and the resulting overall system complexity and cost. The optimum mix of fixed and variable costs depends upon both the present and the projected demand for route capacity.

7.3.2. Link Budget

A link budget is an essential element of any microwave radio design. Its purpose is to assure the reception of adequate signal strength through the use of adequate transmitter power and antenna sizes, while not using excessive power and thereby creating interference problems. The use of decibel expressions is convenient in structuring link budgets because the decibel values can be simply added or subtracted from one another.

An illustration of a link budget problem follows.

Problem. Determine the receiver input power for a 4-GHz radio hop given the following conditions:

Transmitter output power = 5 watts	37.0 dBm
Gain of each antenna (3 m diameter)	39.6 dB
Waveguide loss, each end	2.1 dB
Network losses, each end	1.9 dB
Path length	46 km

Solution. The free space path loss is computed using Eq. (7–8). The result is 137.5 dB. The received power then is found by adding the antenna gains and subtracting the losses. Thus

$$\text{Receiver input power} = 37.0 - 1.9 - 2.1 + 39.6 - 137.5 + 39.6 - 2.1 - 1.9$$
$$= -29.3 \text{ dBm}$$

Assuming a 20 MHz radio channel bandwidth as is standard for the 4 GHz band, the receiver noise level must be determined next. The noise figure F of a receiving system is defined as

$$F = 10 \log_{10} [([T_a + T_e]/290) + 1] \qquad (7\text{–}11)$$

where T_a = noise temperature associated with antenna in Kelvin, K.
T_e = noise temperature associated with receiver, K.

Noise figure is discussed further in Section 7.4.1.2. The system noise temperature $T = T_a + T_e$. The receiver noise level then is

$$N = kTB \text{ dBW} \qquad (7\text{–}12)$$

where k = Boltzmann's constant = −228.6 dBW/HzK.

Thus for $T_a = T_e = 290$ K, $T = 580$ K = 27.6 dBK, corresponding to $F = 6.0$ dB and in a 20 MHz bandwidth,

$$N = -128.0 \text{ dBW} = -98.0 \text{ dBm}$$

For n hops, assuming identical repeaters and identical received power levels at each receiver, the noise adds as $10 \log_{10} n$ dB. Thus for nine hops, $N = -98.0 + 9.5$, or -88.5 dBm. This obviously is well below the -29.3-dBm signal level. However, numerous factors must be accounted for between these two levels. Microwave systems are subjected to propagation fades caused by multipath transmission and by rain attenuation. Thus if a common carrier wants to keep propagation outages as low as equipment outages, a two-way annual fading allocation of 0.01% over a 400-km route (53 minutes per year) may be specified.[3] This is equivalent to seven minutes per year over a 50-km hop. The achievement of such a reliability may require a fade margin on the order of 40 dB, thus leaving 19.2 dB carrier-to-noise-plus-interference ratio in order to assure a bit error rate of 10^{-6} with the modulation technique used, including a margin for modem losses and other forms of circuit degradation. On this basis such modulation techniques as QPR-AM and 8-PSK can be utilized. Several commercial digital radio system types using 8-PSK are in operation in the 6- and 11-GHz bands.[4,5]

For operation in the 11-GHz band, where heavy rain (15 mm/h) produces an attenuation of 1.0 dB/km, a higher fade margin may be required than would be used in the lower frequency bands. Alternatively, space diversity may have to be used. Space diversity involves the use of separate receiving sites with sufficient spacing that the probability of a deep fade to both sites simultaneously is extremely low. The signals received at the two sites then are compared, and the stronger of the two is used. Some diversity systems combine the two received signals instead. Section 7.3.4 provides a further discussion of diversity.

7.3.3. Repeater Siting

Microwave repeaters must be placed so that they are within radio line of sight of one another. The term *radio line of sight* implies the fact that the waves tend somewhat to bend with the curvature of the earth during normal refraction conditions. This bending is expressed by viewing the earth's radius as being multiplied by a factor K, which often is taken to be 1.33 for radio waves. Under unusual conditions, K may drop to as low as 0.66 or increase to infinity. The distance d in km to the radio horizon over smooth earth is approximately $d = 3.55\sqrt{Kh}$, where h is the height in meters above the earth.

Repeater antenna heights must be sufficient that the microwave beam will clear all obstacles in the path. The clearance of the beam over an obstacle is expressed in terms of Fresnel zones. These zones are defined such that the boundary of the nth Fresnel zone consists of all points from which the reflected wave is delayed $n/2$ wavelengths. The distance in meters, m, from the line-of-sight path to the boundary of the nth Fresnel zone, H_n, is

$$H_n = \sqrt{N\lambda d_1 (d - d_1)/d} \tag{7–13}$$

where d_1 = path length from the reflecting point to one antenna, m
d = distance between the antennas, m.

Figure 7–5 illustrates the Fresnel zone concept by showing the first Fresnel zones[6] at 100 MHz ($\lambda = 3$ m) and at 10 GHz ($\lambda = 3$ cm). Notice that point C is beyond the first Fresnel zone for $\lambda = 3$ cm, but well within the first Fresnel zone for $\lambda = 3$ m. The indirect path d_1 will result in a received signal that may interfere with the signal received via the direct path d. The type of interference will depend on the overall loss and delay in a path d_1 compared with the loss and delay via the direct path d.

For free-space transmission, a clearance of at least 0.6 of the first Fresnel zone is required, but substantially greater clearances usually are provided because of the variability of refraction effects.

Repeater spacing thus depends upon line of sight path clearance, as well as upon fading, including rain attenuation, as will be appreciated from Section 7.3.2. Other factors affecting repeater spacing are tower costs, interference to and from other systems, and system requirements in terms of the points to be served along the route. Tower costs can be traded versus repeater spacing, with taller (costlier) towers allowing fewer repeaters overall. System

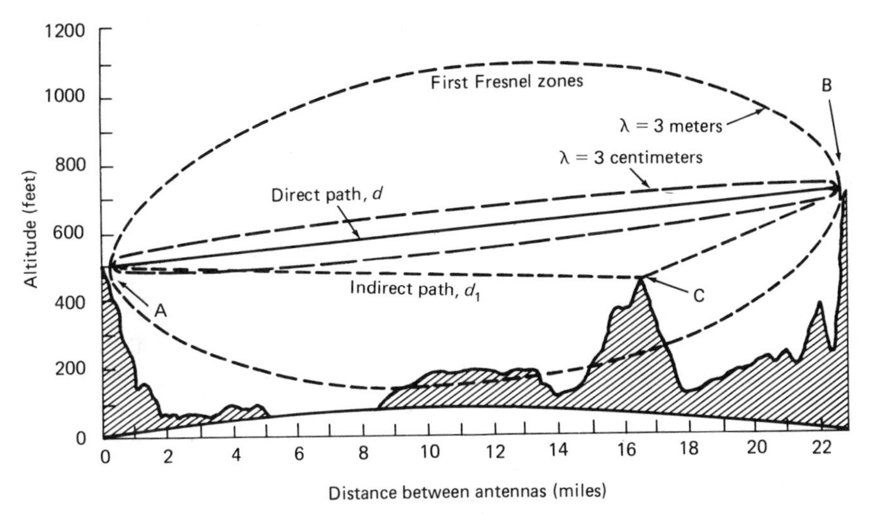

Fig. 7–5. Fresnel zones.

interference may occur when any two routes are in proximity to one another. Thus parallel paths should be avoided and, where routes must cross one another, the crossing should be done at right angles if possible.

7.3.4. Diversity

Diversity can be implemented on either a frequency or a space basis. Frequency diversity requires two separate frequencies and thus does not make efficient use of the radio spectrum. Accordingly, tight FCC regulations exist regarding frequency diversity, and industrial users are not allowed it.

Vertical space diversity is especially effective against ground- or water-reflective fading, as well as against atmospheric multipath fading. Although space diversity is more expensive than frequency diversity because of the additional antennas and waveguides it requires, it generally can provide better protection than frequency diversity, especially where the latter is limited to small frequency spacing intervals. Suitable spacings[7] are 20 m at 2 GHz, 15 m at 4 GHz, 10 m at 6 GHz, and 5–8 m at 11 GHz.

7.3.5. Reliability and Availability

The terms *reliability* and *availability* sometimes are used interchangeably. Reliability, however, often denotes the percentage of time during which the equipment performs its intended function, whereas the actual system availability is less because of propagation outages. Thus, if equipment failures disrupt service 0.01% of the time and if propagation outages also disrupt service 0.01% of the time, then the overall system availability is 99.98%.

Availability objectives for U.S. common carrier short haul systems (<400 km) specify that two-way service failure be less than 0.01% per year. This is equivalent to 53 minutes per year. Industrial microwave systems, especially electric, oil, and gas utilities, allow no more than 0.01% outage during the worst month, because of the potentially serious effects of outages on their operations.

The major factors related to equipment down time are hardware reliability, redundancy, spares availability, and power source reliability. Hardware reliability has increased appreciably with the move to all solid state components. Redundancy can be applied both to the radio equipment, in the form of duplication, and to the power source in the form of a diesel back-up generator.

Propagation outages can be minimized through the use of an adequate fade margin as well as space diversity. A useful equation[8,9,10] for determining the required fade margin M_f has been developed from the work of W. T. Barnett and A. Vigants at Bell Telephone Laboratories:

$$M_f = 30 \log_{10} d + 10 \log_{10} (6ABf) - 10 \log_{10} (1 - R) - 70 \text{ dB} \qquad (7\text{–}14)$$

where d = path length, km
R = reliability objective (one way) for a 400 km route
A = roughness factor
B = conversion factor, worst-month to annual probability
f = carrier frequency, GHz.

Values of A are:

4 for very smooth terrain, including over water
1 for average terrain with some roughness
0.25 for mountainous, very rough terrain

Values of B are:

0.5 for lake areas, especially if hot or humid
0.25 for average inland areas
0.125 for mountainous or very dry areas

To obtain the fade margin to be used for the worst month, set $B = 1$.
For example, for $R = 0.9999$ during the worst month,

$$M_f = 30 \log_{10} d + 10 \log_{10} (6Af) - 30 \text{ dB}$$

7.4. CHARACTERISTICS OF MICROWAVE EQUIPMENT

Microwave equipment has certain characteristics that distinguish it from equipment for other frequencies. This section discusses these characteristics as well as the limitations and advantages of microwave system operation.

7.4.1. Amplifiers

Microwave amplifiers are built to provide adequate transmitter power output levels, as well as to amplify weak incoming signals. Such characteristics as linearity and noise level constitute significant system design factors.

7.4.1.1. Power Amplifiers. Microwave power amplifiers exhibit nonlinearities which manifest themselves in the generation of intermodulation products, as well as errors in the reproduction of the input amplitude. In addition, a phase shift is produced that depends on the input power level. These two factors tend to alter the signal-state space diagram (see Chapter 6) in terms of both amplitude and phase. Accordingly, compensation for these nonlinearities must be achieved for the transmission of such multiamplitude digital modulation arrangements as 16-QAM and 64-QAM.

Operation in the most linear portion of a power amplifier's characteristic also is important in minimizing adjacent channel interference. Unwanted spectral sidelobes, which are inherent in the digital modulation process, may be restored by amplifier nonlinearities.

7.4.1.2. Small Signal Amplifiers. The amplifier used at the input of a microwave repeater generally is operated in the linear portion of its characteristic. Its noise figure or noise temperature, as well as gain and bandwidth, are its most important performance characteristics.

The sensitivity of a microwave receiver can be expressed in terms of the effective noise temperature T of the receiver front end, measured in Kelvin, the bandwidth, expressed in Hertz, and Boltzmann's constant, $k = 1.38 \times 10^{-23}$ W/HzK. This sensitivity, kTB watts, is the noise level at the receiver input.

The noise factor F is given by

$$F = 1 + [(T_a + T_e)/290] \tag{7–15}$$

A more commonly used term is the noise figure, which is given by $10 \log_{10} F$, and which is defined by the expression

$$NF = 10 \log_{10} \left[\frac{P_i/N_i}{P_o/N_o} \right] \tag{7–16}$$

where P_i = available input signal power
P_o = available output signal power
N_i = available input noise power
N_o = available output noise power.

The foregoing terms must all be in the same units, i.e., all in watts, milliwatts, microwatts, etc.

7.4.2. Antennas

Antennas used for microwave radio relay purposes usually are either the parabolic or the horn reflector type. Parabolic apertures normally range from 0.6 to 4.6 meters in diameter, and can be used for multi-band operation by using a separate feed for each band of operation. The gain G of such an antenna at a wavelength λ meters ($\lambda = c/f$, where $c = 3 \times 10^8$ m/s and f = frequency in Hz) is

$$G = 4\pi A\eta/\lambda^2 \qquad (7\text{–}17)$$

where A is the reflector area and η the efficiency of the antenna, typically 55–70%. The beamwidth θ of the parabolic antenna to the −3 dB points is approximately

$$\theta \simeq 70\lambda/D \text{ degrees} \qquad (7\text{–}18)$$

where D is the diameter of the reflector. Such antennas have patterns with front-to-back response ratios of 40–70 dB, although sidelobes may be only 13–20 dB down from the main lobe. The polarization is that of the feed, and is usually either horizontal or vertical. Alternate radio channels often are transmitted using alternate polarizations to help reduce adjacent channel interference. A standard parabolic antenna is pictured in Fig. 7–6.

Even though parabolic reflector antennas can be built with quite good characteristics, situations arise in which even lower sidelobe levels must be achieved than the parabola can provide. Such situations call for the use of a horn reflector antenna which uses a vertically mounted horn under a section of a parabolic surface. Figure 7–7 shows several of these antennas on a repeater tower. Such antennas not only have front-to-back ratios of 80–90 dB, but also can be built to have a beam pattern which continues to drop in amplitude off boresight rather than exhibiting the prominent sidelobes characteristic of the parabola. Figure 7–8 shows the radiation pattern envelope of a horn-parabola antenna, Andrew Type Number SHX10B, which has a 3-meter aperture and provides a 42.7 dBi gain in the 6-GHz band.[11] The envelope shown is for a horizontally polarized antenna. The actual antenna pattern, except at 0°, is equal to, or lower than, the levels described by this envelope. Minor variations may occur in the pattern itself as a result of reflections from nearby objects. The manufacturer includes a full set of digital data for each pattern in order to save time and maintain accuracy when interpreting from the analog pattern envelopes. One reason for the excellent characteristics of the horn reflector is the fact that the feed does not block the main path of radiation from the reflector, but rather is offset from this direct path. The gain and −3 dB beamwidths of the horn reflector depend on its aperture area, as do those of the parabola, so Eqs. (7–17) and (7–18) are applicable to it. Horn reflectors are suited to locations where frequency usage is heavy. Usually they allow the use of identical radio channels in the repeating and branching directions of a repeater site.

7.4.3. System Interface Arrangements

Digital radios may be built for a DS1, DS2, or DS3 interface with terminal equipment or other repeaters. In addition, some 90-Mb/s digital radios provide

Fig. 7–6. Parabolic antenna, PXL-series standard microwave antenna. (Photo courtesy of Andrew Corporation, © 1982.)

an interface consisting of three 30 Mb/s streams. These arrangements allow paths that use the 11 GHz band in the vicinity of large cities with conversion to 6 GHz for the long distance rural routes, along which the 6 GHz band is more readily available than in the large metropolitan areas.

Digital transmission via the data-under-voice (DUV) mode and similar techniques is handled by FDM repeaters which generally provide a conversion to baseband at each repeater at which trunks are dropped or provide conversion to a different frequency band along the route. Baseband repeaters are common in the 6 and 11 GHz bands.

Intermediate-frequency (heterodyne) FDM repeaters avoid the FM-to-base-

Fig. 7–7. Horn reflector antennas, SHX™ super high performance antennas. (Photo courtesy of Andrew Corporation, © 1982.)

band and baseband-to-FM steps of the baseband repeater, and thus avoid the noise associated with demodulation and remodulation. Moreover, they reduce misalignment problems because the signal deviation does not change through the repeater. Where trunks must be dropped at a heterodyne repeater, an FM terminal is used to perform the required demodulation and remodulation functions.

7.5. DIGITAL MICROWAVE RADIO SYSTEMS

A digital microwave radio is one whose instantaneous RF carrier can assume one of a discrete set of amplitude, frequency, or phase levels as a result of

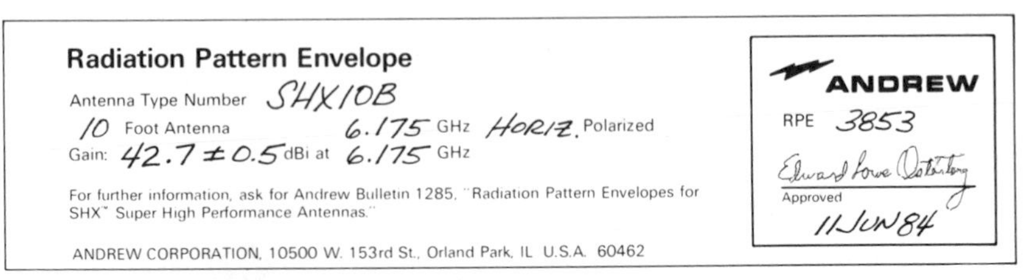
Radiation Pattern Envelope

Antenna Type Number SHX10B

10 Foot Antenna 6.175 GHz HORIZ. Polarized

Gain: 42.7 ± 0.5 dBi at 6.175 GHz

For further information, ask for Andrew Bulletin 1285, "Radiation Pattern Envelopes for SHX™ Super High Performance Antennas."

ANDREW CORPORATION, 10500 W. 153rd St., Orland Park, IL U.S.A. 60462

ANDREW

RPE 3853

Approved

11JUN84

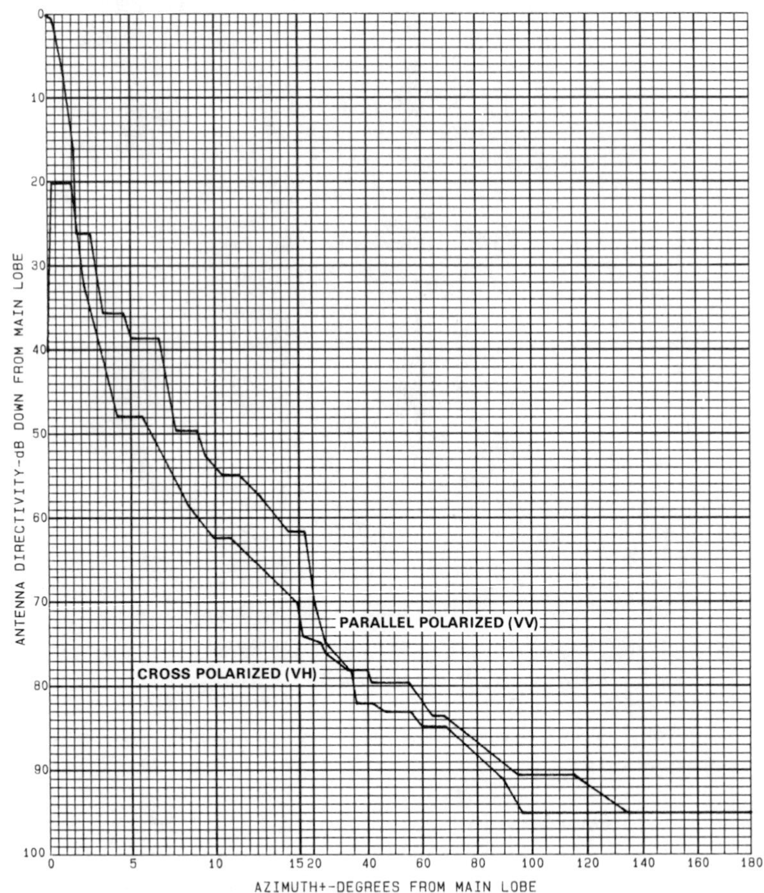

Fig. 7–8. Radiation pattern envelope of a 3 m horn-parabola antenna Type SHX10B at 6.175 GHz (Courtesy Andrew Corporation © 1984).

the modulating signal. Since digital modulation techniques produce very broadband signals, filters must be used to limit the bandwidth of the modulated signal to the assigned radio channel while still allowing transmission with a minimum amount of intersymbol interference.

A "digital radio" may include a radio that transmits a signal whose informational content is at least partly digital, i.e., the baseband may include both analog and digital signals. What, then, is a digital radio? The FCC rule on this subject with respect to an FM-digital hybrid system, states that digital modulation techniques are considered to be employed when the digital modulation contributes 50% or more to the total peak frequency deviation of a transmitted RF carrier. Thus the term "digital radio" may refer to any microwave radio that transmits PCM or other digital carrier signals, regardless of how or at what point the signals are inserted into the radio equipment.

Table 6–1 showed the minimum number of voice channels required per radio channel by the FCC. However, this number of channels does not correlate with the digital multiplex hierarchies as portrayed in Table 6–3. As a result, more voice channels per radio channel than the FCC requires may be designed into digital microwave radios in some bands.

A digital channel bank interfaces the voice channels to the digital radio. It is followed by the channel encoders, whose primary function is to encode one or more bits of the digital input signal into a symbol signal whose bit content determines the modulating effect it is to have on the amplitude, frequency or phase of the carrier. Each combination of bit values that may be encoded into a symbol signal corresponds to a logic level within the encoder. The logic level the encoder assigns to a particular bit combination defines the discrete amplitude, frequency or phase produced in the carrier at the moment of modulation.

Figures 7–9 and 7–10 show, respectively, the transmit and receive block diagrams[12] of a 135-Mb/s system for operation at 6 GHz using 64-QAM and achieving 4.5 b/s/Hz. This system is the Rockwell DST-2300. The input consists of three DS3 lines, each at 44.7 Mb/s. Input conditioners and elastic buffers are used to obtain uniform pulse quality and to allow stream synchronization. A total of 329 kb/s is added for service channel functions. Scrambling is performed to allow the data stream transitions needed for receiver clock recovery. An interface with protection channel lines is provided.

In Fig. 7–10 the system monitoring and violation monitor and removal (VMR) function is performed in the output conditioner.[13] In this unit, DS3 framing is recovered and DS3 parity is compared with parity calculated from the DS3 data. The detection of errors above a preset threshold causes switching to be initiated. The receiving equipment also contains the system DADE module, which equalizes the absolute delay between the various RF channels on a DS3 sectional basis. This helps in minimizing error bursts during switching. A data alignment capability is included to align the protection channel data to the regular channel data prior to the completion of switching. The result is found to be 10 or fewer DS3 errors per switch operation provided framing is maintained on the regular channel.

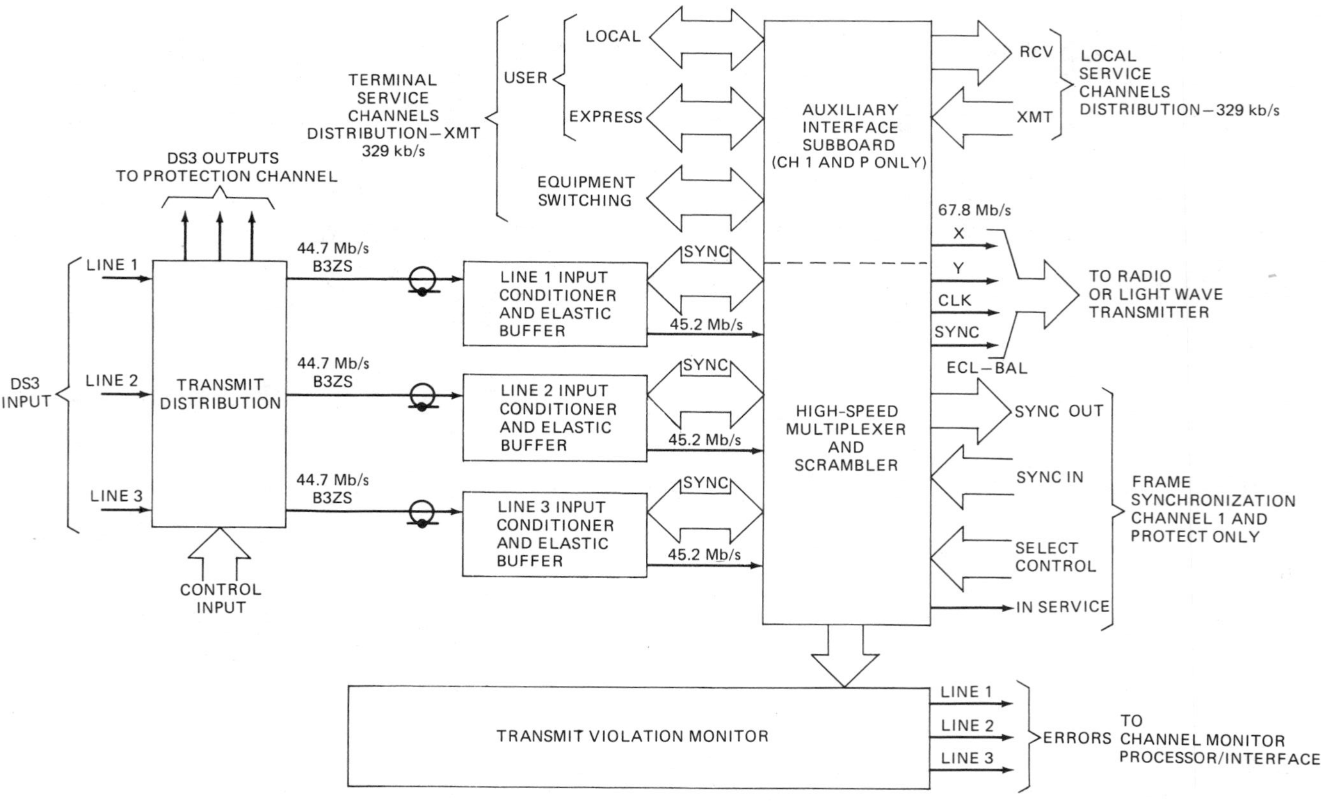

Fig. 7–9. Transmit block diagram for Rockwell DST-2300. (Courtesy Hartmann and Crossett (12). © IEEE 1983.)

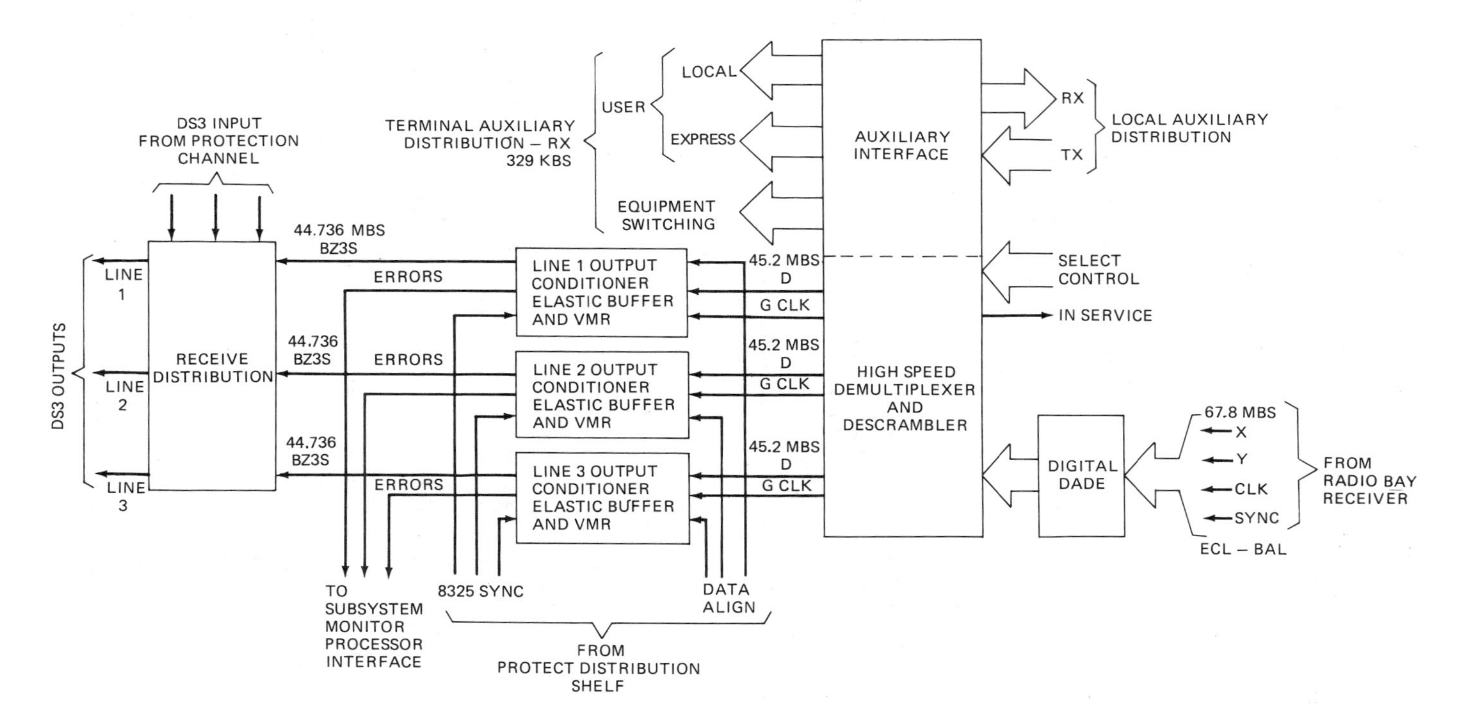

Fig. 7–10. Receive block diagram for Rockwell DST-2300 (Courtesy Hartmann and Crossett (12). © IEEE 1983.)

7.5.1. SYSTEMS FOR INTERCITY AND LONG-HAUL APPLICATIONS

One digital microwave radio is the Northern Telecom DRS-8. Its basic characteristics[4] are as follows:

Data rate	91.04 Mb/s
Bit error rate	10^{-4} at $C/N = 22$ dB (usually achieves $< 10^{-8}$ at somewhat higher C/N)
Operating frequency	7.725–8.275 GHz (or other band providing 40 MHz radio channels)
Channel bandwidth	40.74 MHz
Spectrum utilization	2.25 b/s per Hz (QPRS)

The DRS-8 has provisions for space diversity and can be overbuilt on existing structures or built as a stand-alone system. A 40-dB fade margin can be provided over 40–50-km paths. With proper repeater spacings, the availability over a 6500-km path is 99.98%.[14]

The Rockwell MDR-6 has the following characteristics:[5]

Data rate	90.258 Mb/s
Bit error rate	10^{-6} at $C/N = 29.2$ dB
Operating frequency	5.925 to 6.425 GHz
Channel bandwidth	29.65 MHz
Spectrum utilization	3.0 b/s per Hz (8-PSK)

The MDR-6 provides two asynchronous DS3 streams using the B3ZS line code. A similar radio, the MDR-11, operates in the 10.7- to 11.7-GHz band and also uses 8-PSK, but its radio channel bandwidth in 40 MHz wide, corresponding to the allocated channel widths at 11 GHz. As a result, its channel filtering is less stringent. Its spectrum utilization thus is 2.26 b/s per Hz.

While systems such as the Rockwell MDR-6 actually achieve a spectrum utilization of 3.0 b/s per Hz, they do so at a considerable carrier-to-noise ratio penalty. For example, for b.e.r. = 10^{-6}, 8-PSK theoretically requires $E_b/N_0 = 14.0$ dB. This corresponds to $C/N = 14.0 + 10 \log_{10} (3) = 18.8$ dB. The filtering required to meet the demands of the FCC mask, as outlined in Section 7.3.1, however, is largely responsible for the additional 10.4 dB of C/N needed by this system.

Accordingly, digital radio designers have turned to 16-QAM, which is capable of 4 b/s per baud. It enables the achievement of a 90.258 Mb/s rate at 22.565 M baud. With a filter $BT_s = 1.0$ in a 30 MHz wide radio channel, the roll-off factor α of the channel filters can be 0.33, which is achievable readily.

For operation in the 4-GHz band, where the radio channels are only 20 MHz wide. Fujitsu has designed a 64-QAM system[15] which achieves the 90 Mb/s rate at 15 M baud. In this case the filter roll-off factor is 0.50. This system suffers only a 2.0 dB degradation in C/N at b.e.r. $= 10^{-6}$, thus requiring 28.7 dB.

The 30 MHz radio channel bandwidth available in the 6 GHz band allows a yet higher transmission rate if 64-QAM is used, as is done in systems designed by Western Electric, Rockwell and Fujitsu. Accordingly, 3 DS-3 streams, or 135 Mb/s, corresponding to 2016 voice channels at 64 kb/s each, can be transmitted by such systems, one of which was shown in block diagram form in Figs. 7–9 and 7–10.

7.5.2. Metropolitan Area Systems

Several categories of digital radio systems have been developed for the transmission of various stream rates over relatively short distance, high traffic density routes within metropolitan areas. They include digital termination systems (DTS), private systems and common carrier systems.

7.5.2.1. Digital Termination Systems. Digital termination systems (DTS) are provided by competing common carriers using radio local loops. Each user has a small roof-top antenna, typically a 0.6 m dish, directed toward the service provider's location. Data rates are from 2400 b/s to 1.544 Mb/s at distances as great as 10 km through moderate rain. The frequency band from 10.55 to 10.68 GHz is allocated for this service both to common carriers and to private users. Additional common carrier bands are: 18.36–18.46 GHz and 18.94–19.04 GHz. A 10^{-9} to 10^{-11} bit error rate is achieved using 500-mW transmitters and 4-level FSK for distances less than 10 km.

The two available classes of service are called "extended" and "limited." Extended service carriers serve 30 or more cities and have seven 5-MHz channel pairs in each city. Limited service carriers serve fewer than 30 cities and have six 2.5-MHz channel pairs in each city. The applications include a digital electronic message service (DEMS), high speed facsimile (to 1.544 Mb/s) and teleconferencing using motion-compensated video at 1.544 Mb/s. Satellites may link the various cities.

In operation, a central control station transmits digital traffic to the remote user sites, with the data for the various users being time-division multiplexed. Each remote site monitors the signal and processes only the data addressed to it, and then responds to the central station by sending bursts of pre-assembled packets of data for a predetermined or controlled time interval on another allocated frequency.[16]

7.5.2.2. Private Systems. Digital radio systems available for private (industrial or commercial) use within metropolitan areas include the Farinon DM-18, the General Electric Gemlink and the M/A-Com MA-23DR.

The DM-18 provides T1 or T2 service[17] using FSK in the 18.36- to 19.04-GHz band. The transmitter output is 100 mW. Bit error rates of 10^{-6} are achievable over distances from 5 to 16 km, depending on the local climate. Antenna diameters are 61 cm or 122 cm.

The Gemlink[18] provides T1 or T1C service using 2-level AM (*p-i-n* diode modulator) in the 21.2 to 23.6 GHz band. The transmitter output is 20 mW. A 10^{-7} bit error rate is achievable 99.95% of the time over a 4-km distance in an average U.S. mid-latitude climate. The antenna is 38 cm × 38 cm × 23 cm deep.

The M/A-Com MA-23DR is capable of not only T1 or T1C service, but also T2(6.312 Mb/s), CEPT 1 (2.048 Mb/s), and CEPT 2 (8.448 Mb/s). The system handles the standard AMI and zero substitution line codes and provides waveform regeneration at the receiver.[19] The modulation is FSK with a ±4 MHz deviation. The transmitter uses a Gunn diode and has a 66 mW output. The antenna diameter is 0.6 m or 1.2 m. The receiver noise figure is 12 dB. At a 1.544 Mb/s rate, a 10^{-9} b.e.r. is available 99.99% of the time at distances from 4 to 13 km within the U.S. depending upon climate.

7.5.2.3. Common Carrier Systems. Digital radio facilities have been designed for T3 transmission in metropolitan areas.[20] The Western Electric 3A-RDS operates at the DS3 stream rate, using QPSK in the 40-MHz radio channels between 10.7 and 11.7 GHz. Eleven two-way streams are obtained within eleven radio channels using dual linear polarization. Regenerative repeaters are used and the spacing is 16 to 20 km in average climates and 26–32 km in dry climates. The 3.2-W transmitter output power provides better than 17.2 C/N at the receivers for a 10^{-8} bit error rate. Many 3A-RDS systems thus transmit to distances well beyond metropolitan areas.

PROBLEMS

7.1 Explain why the propagation spreading loss equation shows it increasing with frequency.

7.2 A digital radio operates in the 5.925 to 6.425 GHz band using 1.0 W transmitter powers. The antenna diameters are 3 m and the path length is 50 km. The modulation technique requires $C/N = 20$ dB to maintain an adequately low bit error rate. The receiver noise figure is 5.0 dB. Allow 3.0 dB at each end for waveguide and network losses. Assume a 55% antenna efficiency at each end. What is the fade margin?

7.3 How does the fade margin computed in Problem 7.2 compare with the level allowable during the worst month, assuming average inland terrain and a 0.999 reliability requirement?

7.4 During cloudburst conditions over a 5 km section of the path length of Problem 7.2, what additional path attenuation is encountered?

7.5 Microwave propagation in coastal regions often extends over greater distances than inland. Why does this not allow repeaters to be spaced farther apart in such areas?

7.6 The horn reflector antenna is more commonly found at terminals in large metropolitan areas, whereas the simple parabolic is more frequently found in rural locations. Why?

REFERENCES

1. Keiser, B. E., "Wideband Communications Systems," Course Notes, © 1981.
2. White, R. F., *Engineering Consideration for Microwave Communications Systems,* GTE Lenkurt, Inc., San Carlos, CA, 1970.
3. Feher, K., *Digital Communications: Microwave Applications,* Prentice-Hall, Inc., Englewood Cliffs, NJ, 1981.
4. Bell-Northern Research, *Telsis,* Vol. 5, No. 6, Special issue: DRS-8 Digital Radio System, Dec. 1977.
5. Rockwell International, Collins Technical Data Sheet, MDR-6, 6-GHz Microwave Digital Radio, Commercial Telecommunications Group, Dallas, TX.
6. *Telecommunications Transmission Engineering,* Volume 2, *Facilities,* Bell System Center for Technical Education, Western Electric Company, Inc., Winston-Salem, NC, pp. 48–51, 1977.
7. See note 2.
8. Barnett, W. T., "Occurrence of Selective Fading as a Function of Path Length, Frequency, Geography," *ICC '70 Conference Record,* IEEE.
9. Barnett, W. T., "Microwave Line of Sight Propagation With and Without Frequency Diversity," *Bell System Technical Journal,* Vol. 49, pp. 1827–1871 (Oct. 1970).
10. Vigants, A., "Number and Duration of Fades at 6 and 4 GHz," *Bell System Technical Journal,* Vol. 50, pp. 815–841 (March, 1971).
11. "Radiation Pattern Envelopes for Horn Reflector Antenna," Bulletin 1285, Andrew Corporation, 10500 W. 153rd Street, Orland Park, IL 60462, U.S.A.
12. Hartmann, P. R., and Crossett, J. A., "135 MBS-6 GHz Transmission System Using 64-QAM Modulation," *ICC '83 Conference Record,* IEEE Publication No. 83 CH1874–7, Boston, MA, June 19–22, 1983, pp. F2.6.1–F2.6.7.
13. "MDR-2000 Series Digital Radio Systems," Product Description, Collins Transmission Systems Division, Rockwell International, P. O. Box 10462, Dallas, TX 75207.
14. See note 4.
15. "Private Communication Networks," Macomnet, Inc., Rockville, MD.
16. Williams, D. S., "Local Distribution in a Digital Communications Network," *ICC '81 Conference Record,* IEEE Publication No. 81CH 1648–5, Denver, CO, June 14–18, 1981, pp. 66.3.1 to 66.3.5.

17. "Gemlink LSD-112A/122A Microwave Radio," General Electric Microwave Link Operation, Owensboro, KY, 1982.
18. "Video Microwave Systems," Bulletin 9236A, MA-23CC, M/A-COM MVS, Inc. 63 Third Avenue, Burlington, MA 01803.
19. *Telecommunications Transmission Engineering,* Vol. 2, *Facilities,* Bell System Center for Technical Education, Western Electric Company, Winston-Salem, NC, pp. 619–638, 1977.

8
Satellite Transmission*

8.1. INTRODUCTION

Communications satellites as we know them today had their beginning with *Syncom,* the first synchronous satellite, in 1963. It was soon followed by *Intelsat I* (*Early Bird*), the first commercial satellite, in 1965. Numerous military satellite systems parallel the development of the commercial systems. The first domestic commercial satellite was Telesat's *Anik* (Canadian), followed by Western Union's *Westar* (U.S.). Because of the predominance of analog voice communication during those years, however, satellite earth stations all over the world were built based upon the analog frequency division multiplex (FDM) hierarchy. Only with the advent of Satellite Business Systems (SBS) has an all-digital satellite system developed, with the initial objective, however, of data rather than voice communication.

Satellite transmission is unique in that it makes large bandwidths (i.e., hundreds of megahertz) available for intercontinental communication. Moreover, with the use of polarization and spot beaming techniques, a considerable amount of frequency reuse is feasible, which is a significant fact in view of the increasingly crowded radio spectrum.[1]

Satellite transmission is capable of providing global communication, including transmission to and from moving terminals such as ships, and considerable system development effort has been devoted to extending such capabilities to land and aeronautical vehicles.[2,3] Only the polar regions beyond 81° latitude remain beyond line of sight of the geosynchronous satellites. Since satellites that are not geosynchronous require tracking earth station antennas, coverage of the polar regions seems to be more practical by microwave or other transmission techniques.

The satellite may be regarded as a microwave repeater at a high enough elevation that most circuits require only that one repeater. Where a second such repeater is needed, the overall time delay may produce undesirable effects in attempts to carry on conversational voice transmission.

Satellite transmission developed initially on an analog (frequency division multiple access) basis. New developments in analog transmission via satellite

---

* Much of the material of this chapter is taken from B. E. Keiser, "Wideband Communications Systems," Course Notes, © 1981.

center on the use of compandored single-sideband modulation.[4,5] While the use of such techniques provides considerable spectrum economy (7200 voice channels in a 36-MHz transponder), the need exists in these systems to operate amplifiers well below (e.g., 10–15 dB) their full power output capabilities. In addition, signal-to-noise ratios on the order of 27 dB and higher are required, making such transmissions "fragile" with respect to interference. These new advances notwithstanding, the intermodulation problem resulting from transponder nonlinearities makes digital (time division multiple access) operation highly desirable, as will be explained in Section 8.3.5.2 of this chapter.

8.2. CHARACTERISTICS OF SATELLITE PROPAGATION

The propagation medium for satellite transmission tends to be a highly stable one because most of the path is through free space. The only atmospheric effects occur near the earth stations and these generally are limited by the fact that propagation usually is upward at a relatively significant angle relative to the horizon. Only at the higher latitudes, or for satellites at a considerably different longitude than the earth station, is a low angle path required.

8.2.1. The Satellite Orbit

For a satellite to remain in a fixed position in the sky relative to the earth stations it is serving, two conditions must be met. First, it must be over the equator. Second, it must be at an altitude of 35 784 km. The latter value results from the fact that the period of an orbiting satellite is given by the equation

$$T = 2\pi \sqrt{A^3/\gamma} \qquad (8\text{–}1)$$

where A = semi-major axis of ellipse (earth's radius, 6378 km, plus altitude for a circular orbit)
γ = gravitational constant = 3.99×10^5 km^3/s^2.

For a circular orbit to have a period identical with that of the earth's rotation, the period must equal one day, which is 23 h, 56 min, 4.09 s.

From a geosynchronous altitude of 35 784 km, three satellites can cover most of the earth's surface.

8.2.2. Time Delay on Satellite Paths

Although electromagnetic waves travel at the speed of light, 3×10^8 m/s, the distance to a geosynchronous satellite may range from 35 784 km for an earth station at the sub-satellite point (on the equator) to 41 677 km for an earth station at the horizon relative to the satellite. Accordingly, the one-way propagation time, neglecting electronic circuit delay, from one earth station to another may range from a minimum of 0.239 s to a maximum of 0.278 s. The total round trip delay time then ranges from 0.478 s to 0.556 s. In conversational voice, such delays begin to be noticeable, but not adversely so. In fact, each party to the conversation gets the impression that the other party is pausing a moment to collect his thoughts before replying!

The same is not true, however, if double hop satellite transmission is required, as might be the case if a domestic satellite is used in tandem with an international satellite. In such a case, the total round trip delay time is on the order of one second. When a speaker has finished talking and does not hear a reply forthcoming, he may say, "Did you hear me?" just while the reply is on its way. For such reasons international satellite transmission is being implemented increasingly on a one-way basis only, with the return path being terrestrial, i.e., submarine cable. This allows a domestic satellite to be placed in tandem with an international satellite for an overall delay of less than 0.6 second in one direction, with the delay on the return path being kept to less than 0.1 second. The combination round trip delay, on the order of 0.6 to 0.7 second, is tolerable in most cases.[6]

8.2.3. Atmospheric Attenuation

The attenuation caused by the atmosphere depends upon both the frequency and the elevation angle. This one-way attenuation is given by Fig. 8–1 for standard atmospheric conditions. It results from the presence of oxygen and water vapor in the air. Rainfall causes increased attenuation, especially at frequencies above 10 GHz.[7]

8.2.4. Rain Depolarization

As will be seen in Section 8.3, frequency reuse is an important aspect of satellite system engineering because of the increasing use of the radio spectrum. One frequency reuse technique is dual polarization. For its successful use, however, reception of a co-channel or partially co-channel signal must be such that the level of the unwanted (cross-polarized) signal is sufficiently low. In other words, the ratio of carrier to interference (C/I) must be adequate for the planned bit error rate. Typically, this calls for $C/I \geqslant 25$ dB. Two factors allow a cross-polarized signal to be present. One is the polarization purity of the hardware, mainly the antennas, and the other is the extent to which the signals become depolarized during propagation, mainly because of precipitation.

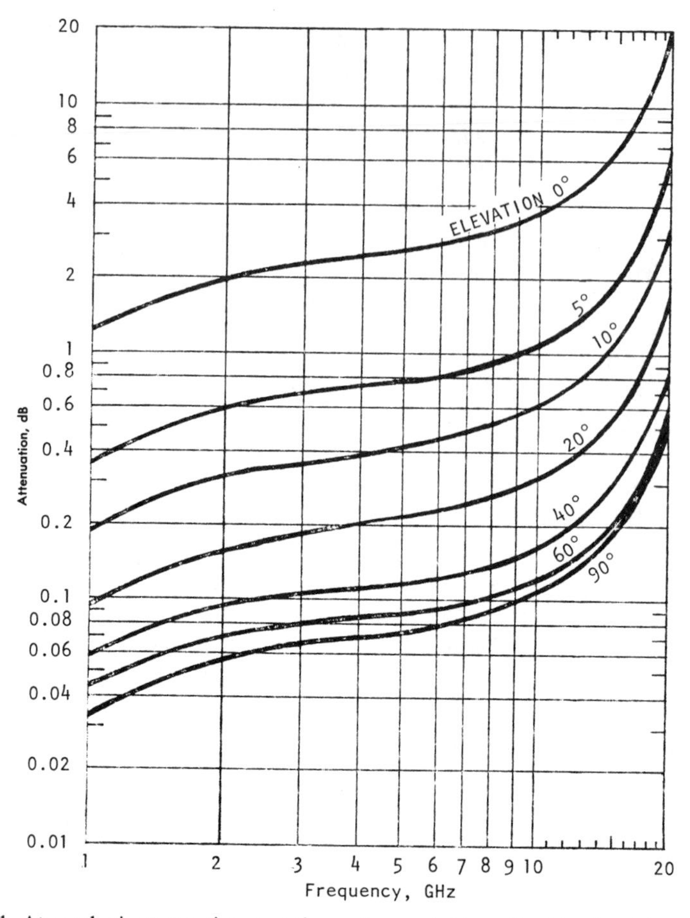

Fig. 8–1. Atmospheric attenuation over the earth to satellite path. (Reprinted by permission from *Microwave Journal.*)

Figure 8–2 illustrates the effect of rain on cross-polarization isolation for 5-km path lengths. The curves are shown for both differential phase and differential attenuation over 36-MHz transponder bandwidths for 4, 6, and 11 GHz. Differential attenuation is the difference in attenuation between the horizontal and vertical polarizations, whereas differential phase is the difference in phase shift between the horizontal and vertical polarizations.[8] As can be seen from Fig. 8–2, differential attenuation and phase are key parameters in the production of cross-polarization by precipitation.

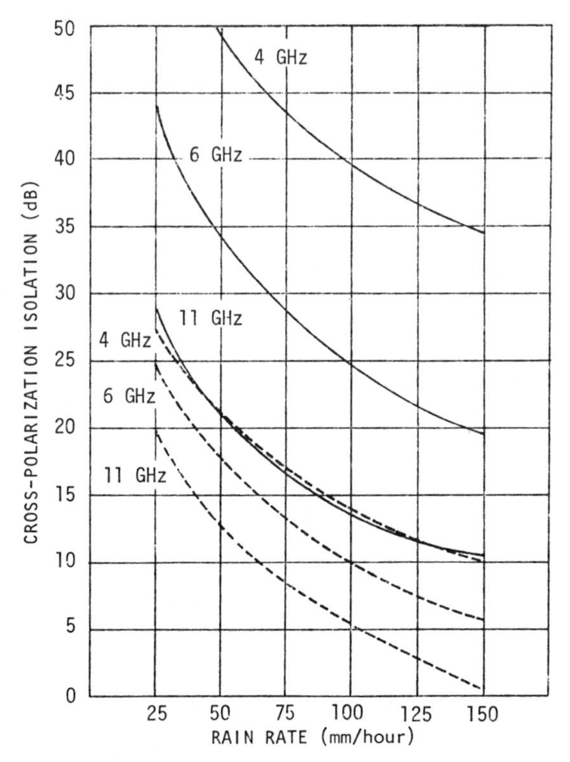

Fig. 8–2. Cross polarization isolation versus rain rate for 5 km path lengths.

8.3. SATELLITE SYSTEM DESIGN

Numerous factors, technical, regulatory and economic, affect the design of a satellite system. This section discusses the major system design factors. Section 8.4 deals with equipment characteristics and limitations. For several topics not covered here, such as echo control, diversity and reliability/availability, the reader is referred to the corresponding sections 5.7.2, 7.3.4, and 7.3.5 of earlier chapters. The echo control problem (impedance mismatches) and solution (voice-actuated half-duplex) are quite similar to those of microwave radio relay systems. Diversity is required mainly by systems operating in the 11 and 14 GHz bands in regions of heavy precipitation. Here perhaps three earth stations, each separated 10–15 km, may be required to dodge local rain cells. If the three are adequately spaced and interconnected terrestrially, outages due to heavy precipitation can be reduced to an extremely low percentage of the total annual operating time. With respect to reliability/availability, the equipment reliability problem is similar, assuming proper spacecraft design and operation. Propagation outages are confined almost entirely to low elevation angle paths which are subject to precipitation fading, as well as the other problems that affect terrestrial microwave propagation, such as multipath.

8.3.1. Frequency Allocations and Usage

Table 8–1 lists the frequency bands allocated to satellite communication and designates the type of usage in each case. In some of these bands the regulatory authorities have specified maximum power flux densities to minimize interference to existing terrestrial services, since many of these bands are not used exclusively for satellite service, but are shared with terrestrial users.

The term *allocation* means that the frequencies listed in Table 8–1 are available for the usage listed, but not all are in actual use at this time. For example, the aeronautical band is not in use as of this writing, and the bands above 17.7 GHz are currently being used only experimentally in the western hemisphere.

Table 8–1. Satellite Frequency Allocations.

BAND (MHZ)	USAGE	
225.0–400.0	military	
1535.0–1542.5	downlink	maritime
1635.0–1644.0	uplink	maritime
1543.5–1558.5	downlink	aeronautical
1645.0–1660.0	uplink	aeronautical
3700–4200	downlink	commercial
5925–6425	uplink	commercial
7250–7750	downlink	military
7900–8400	uplink	military
10950–11200 11450–11700	downlink, international	commercial
11700–12200	downlink, domestic	commercial
14000–14500	uplink	commercial
17700–20200	downlink	commercial
27500–30000	uplink	commercial
20200–21200	downlink	military
30000–31000	uplink	military
36000–38000	crosslink	

8.3.2. Link Budgets

A link budget provides a complete description of a radio transmission path from the transmitter power output to the detected signal, and often includes an expression of reception quality, expressed in bit error rate for digital systems. For a satellite system both an uplink and a downlink budget must be calculated. However, the uplink budget usually is the less critical of the two, since plentiful amounts of power normally are available to the earth-based transmitter. Uplink stations also may have significant amounts of antenna gain. Generally the satellite receiver will have a rather high noise temperature (e.g., over 1000 K) because the earth itself is noisy and no value would be gained with a low-noise receiver. Quite to the contrary, low-noise receivers cannot be made to operate as reliably as high-noise (less sensitive) ones, so are not used on spacecraft.

The criticality of the downlink budget results from several factors. The satellite is power limited by the amount of power available from the solar array and the number of transponders and pieces of control equipment among which this power must be divided. In addition, the physical size of the spacecraft antenna, and thus its gain, also is generally smaller than those on the ground. Another limit may be the available channel bandwidth, as well as the way in which a given transponder's bandwidth and power are allocated. A further limit is that in the 3700–4200-MHz band, international regulations state that the power flux density shall not exceed 12 dBW effective isotropically radiated power (EIRP) in any 4 kHz band. In a purely digital time-division-multiplexed system, each user has the entire transponder for the duration of his time slot. However, many systems allocate only a portion of the transponder to digital transmission. For example, a single channel per carrier (SCPC) system may transmit a single PCM channel using a 45-kHz wide portion of an overall 36-MHz wide transponder. As many as 800 of these SCPC systems may use the same transponder.

The basic downlink power budget may be expressed in terms of the carrier-to-noise power normalized to a 1 Hz bandwidth, C/kT, available from a transponder or a portion thereof as follows:

$$(C/kT)_A = (EIRP)_{\max} - (EIRP)_o - (TD)_d - (TPL)_d - k + (G/T)_{\text{eff}} \text{ dBHz} \quad (8\text{–}2)$$

where

$(EIRP)_{\max}$ = maximum center beam transponder effective isotropic radiated power in watts at its normal operating point, chosen to be backed off from saturation by a specified amount

$(EIRP)_0$ = power in watts devoted to other carriers in the transponder than the carrier of interest
$(TD)_d$ = tilt differential for the downlink in dB, the degradation from the maximum EIRP resulting from the position of the earth terminal relative to the boresight of the satellite antenna
$(TPD)_d$ = total path loss in dB for the downlink
= free space loss (FSL) + miscellaneous loss (m).

The expression for FSL is

$$FSL = 32.5 + 20 \log_{10} f_{\text{MHz}} + 20 \log_{10} d_{\text{km}} \tag{8–3}$$

where f_{MHz} = frequency in MHz
d_{km} = slant range in km
m = loss caused by pointing error, randome loss, atmospheric attenuation and polarization loss
k = −228.6 dBW/Hz K (Boltzmann's constant).
$(G/T)_{\text{eff}}$ = effective earth terminal figure of merit in dB/K, including the effect of thermal noise radiated from the satellite to the earth terminal.

Eqs. (8–2) and (8–3) allow a determination to be made of the available (C/kT). The next question is "what (C/kT) is required?" Let this value be designated $(C/kT)_R$. It is expressed as follows:

$$(C/kT)_R = R + (E_b/N_0) + L + M \tag{8–4}$$

where

R = 10 $\log_{10}$ (bit rate, b/s)
E_b/N_0 = ratio of energy per bit to noise power spectral density to achieve required bit error rate
L = losses resulting from use of channel filters (about 1–2 dB)
M = satellite link margin.

Typical values for M are 4 dB for the 3700–4200 MHz band, 6 dB for the 7250–7750-MHz band, and 8 dB for the 10 950–12 200-MHz frequencies.

8.3.3. Earth Station Siting

Satellite earth stations may be found in a variety of places, including building roof tops, but they are best located in valleys or other low terrain, such that their surroundings provide some shielding from interference that may reach them from terrestrial microwave relays. Before attempting to establish a new satellite earth station, a careful examination must be made of the surrounding electromagnetic environment to determine what main beam or sidelobe interference may exist between the new earth station and microwave repeaters.

An important consideration in earth station siting is the beam direction to the satellite. This can be determined using the chart of Fig. 8–3.[9] Knowledge of the earth station's latitude and of the relative longitude of the satellite allow the determination of the azimuth and elevation angles of the satellite using this chart.

Knowledge of the earth station's main beam direction is important in avoiding obstructions in that direction, as well as in determining the direction of side lobes. Side lobes may cause interference to adjacent satellites or to other satellite earth stations or nearby microwave systems.

Another type of interference that can occur is self interference outside the allocated bands. Such interference can result from nonlinearities caused

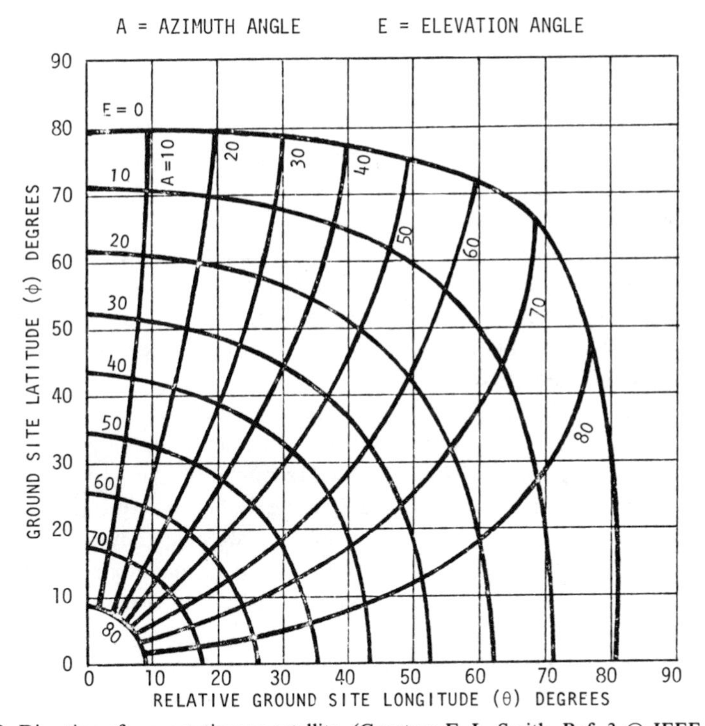

Fig. 8–3. Direction of a geostationary satellite. (Courtesy F. L. Smith, Ref. 3 © IEEE, 1972.)

by oxide layers in the waveguide feed system or by receiver or transmitter nonlinearities.

8.3.4. Frequency Reuse

The frequency allocations listed in Table 8–1 might seem to be perfectly adequate for satellite communications purposes in view of the large bandwidths available. However, a serious shortage of spectrum exists for several reasons. First, only one commercial frequency pair, the 3700–4200-MHz downlink and the 5925–6425-MHz uplink, is suitable for use under all possible weather conditions. The 11 and 14 GHz bands are suitable under most conditions, and rapidly increasing use is being made of these bands. The bands above 17.7 GHz are more severely limited by precipitation. Their use may depend upon the development of some of the special techniques discussed in Section 8.6.

Considerable use is being made of the existing allocations both domestically and internationally for video and multichannel telephony, as well as data transmission. Frequency reuse techniques include polarization diversity, usable only in the 4- and 6-GHz bands, and spot beaming, usable in all bands. Other spectrum conservation techniques include the use of multilevel modulation, as discussed in Section 6.2.

With polarization diversity the number of transponders occupying an overall band can be doubled, since the isolation between dual-polarized beams can be as great as 30 dB, compared with the C/I requirement of 25 dB noted in Section 8.2.4.

With respect to spot beaming, a satellite's antenna can be designed to cover only a portion of the U.S., or a region the size of the U.S. or western Europe. Likewise, beams can be devised to cover only hemispherical regions. In all cases, the beamwidth of the satellite antenna is significantly smaller than the 17.3° width which would cover the entire surface of the earth that is within view of the satellite.

Interest also exists in placing more satellites into the orbital arc. A 4° spacing allows receive-only earth stations with 3 m diameter antennas to operate satisfactorily in the 3700–4200-MHz downlink band. To allow more satellites to operate in this band, the spacing is being reduced to 2°. This will make the 3 m diameter antennas useless because they will be unable to receive one satellite without interference from its neighbor in the orbital arc. Low sidelobe 5 m diameter antennas, however, can operate satisfactorily with spacings significantly less than 4°.

Sidelobe interference constitutes one problem that satellite system planners must face. Both earth station and satellite antennas exhibit side lobes that are at levels on the order of −13 to −20 dB relative to the main lobe. An earth station side lobe may interfere with the signals trying to reach an adjacent satellite. Likewise, side lobes from a satellite may interfere unintentionally with an earth station.

Terrestrial microwave systems on the horizon with respect to a satellite may interfere with, or receive interference from, a satellite system. Such microwave systems might be located in the far north or far south, and have beams in a north or south direction, or they may be at medium latitudes and have beams in a direction toward a satellite at a significantly different longitude.

8.3.5. Multiple Access

Multiple access techniques allow variously located earth terminals to use portions of a satellite transponder on either a frequency- or time-division basis. In frequency-division multiple access (FDMA), each channel has some of the bandwidth and some of the power all of the time, whereas in time-division multiple access (TDMA), each channel has all of the bandwidth and all of the power, but only part of the time.

8.3.5.1. Frequency-Division Multiple Access. In FDMA the available spectrum within a satellite transponder is divided among carriers, each of which may carry a digital stream. This digital stream, in turn, may convey messages from one or more sources to one or more destinations. An arrangement often used, especially by small remote earth terminals, is single channel per carrier (SCPC). Such channels can be assigned only when actually needed, and also can be voice activated, so that each one contributes to transponder power usage only when it is actually required. This helps to minimize intermodulation interference, a problem often encountered by satellite transponders using FDMA. A detailed analysis of intermodulation in FDMA satellite transponders is provided by Spilker.[10]

Many SCPC systems use demand-assigned multiple-access (DAMA).[11] Each carrier is modulated by a bit stream that comes from the voice channel of an individual user. COMSAT has devised a form of SCPC known as the *S*ingle-channel per carrier *P*ulse-code modulation multiple-*A*ccess *D*emand-assigned *E*quipment (SPADE). The SPADE system assigns transponder capacity for use on a call-by-call basis. The overall 36 MHz transponder bandwidth is divided into 800 channels, each of which is individually accessible via an earth station. These 800 channels can serve a much larger number of telephone subscribers, the actual number depending on how often the subscribers make calls, and for what durations.

The SPADE system message channels use PCM encoded voice at 64 kb/s sent via QPSK at 32 kilobaud. The bandwidth per voice channel is

38 kHz and the channel spacing is 45 kHz. The bit error rate provided is less than 10^{-7}. Frequency stability is ±2 kHz and is held to this level using automatic frequency control (AFC). Correspondingly, oscillator phase noise and incidental FM generated in the oscillators and power amplifiers must be held to a minimum. Also a part of the SPADE system is a common signaling channel which operates at 128 kb/s with two-phase PSK. This channel allows 50 accesses (49 plus a reference), each of which is allotted a 1 ms burst within a 50-ms frame. The bit error rate is less than 10^{-7}. A central network processor controls channel assignments, passing the appropriate commands to a microprocessor at the station at each end of the link. Each microprocessor interprets the commands and sends them on to its channel units via the common channel. The remote terminals can be simplified through the use of a net control terminal, which places the more complex control equipment at a central location. Fig. 8–4 is a block diagram of a SPADE terminal.

In the SPADE system the channels are spaced uniformly. Such spacing makes efficient use of the transponder spectrum, but causes the intermodulation products to fall directly on the channels being used. However, since each carrier is voice activated, only about 40% of the carriers are on at any given time. This is equivalent to a 4 dB reduction (back-off) of the transponder's output. This reduction results in operation in a more nearly linear portion of the transponder's power amplifier characteristic, with a consequent decrease in the intermodulation products generated. In addition, the time activation of the transponder by each voice signal gates the intermodulation products in a random manner, reducing the worst intermodulation noise by 3 dB.[12]

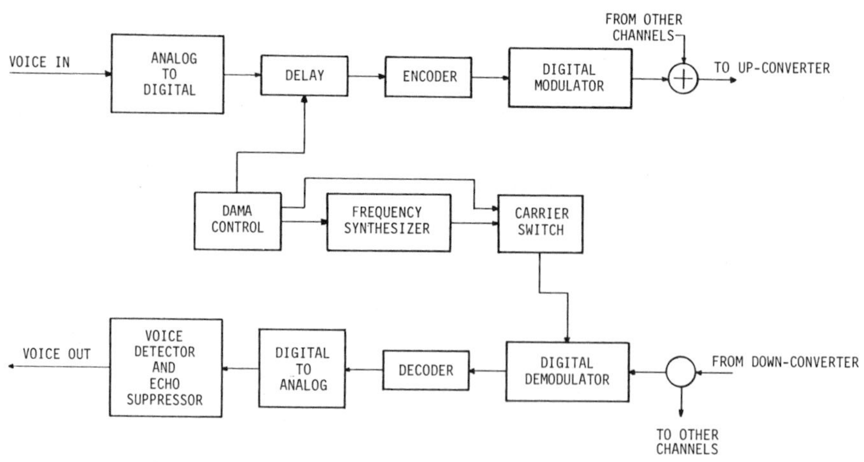

Fig. 8–4. Block diagram of SPADE terminal.

Intermodulation effects can be reduced significantly if the carriers are not uniformly spaced. One arrangement involves placing the carriers in clusters of bandwidth W.[13] The spacing between clusters is then made to exceed $3W$.

An alternative approach to the choice of carrier frequencies is one which determines the minimum bandwidth needed for complete avoidance of third- or third-and-fifth-order cross products. These results and the frequency spacing required are reported in Ref. 14.

For large numbers of available channels N, randomly spaced channels provide a better overall ratio of carrier-to-intermodulation noise power (C/IM) than the previously mentioned techniques. If W is the bandwidth of the cluster but the carriers are spread out randomly to occupy a bandwidth B, only a fraction W/B of the intermodulation power is passed through the receiver. Then, as shown in Ref. 3, the improvement in C/IM in the center channel due to the carrier spacing is

$$(C/IM)_{\text{im}} = 10 \log_{10} (B/W) + 7.7 \text{ dB} \tag{8–5}$$

For 100 38-kHz-bandwidth carriers in a total bandwidth of 36 MHz, $W =$ 3800 kHz $=$ 3.8 MHz, while $B = 36$ MHz. Then $(C/IM)_{\text{im}} = 17.5$ dB, i.e., the use of randomly spaced channels yields a 17.5 dB improvement in C/IM.

8.3.5.2. Time Division Multiple Access. Section 8.3.5.1 noted the need for transponder back-off in FDMA operation. Typical back-off values range from 3 to 6 dB and constitute a significant decrease in the power which otherwise would be available from the transponder. By comparison, TDMA operates with only one modulated carrier in the transponder at a time. The result is that efficiency in terms of power utilization can reach 90% or more since the power amplifier can be operated in saturation. Moreover, frequency guard bands are not needed in TDMA operation, and guard time slots can be kept to a very small percentage of the total on-time. Thus while frequency guard bands may occupy 10–15% of the frequency band, typical time guard slots are on the order of only 1–2% of the total on-time.

In TDMA each earth terminal transmits to the satellite in its own time slot, based on its determination of propagation time to the satellite. Guard time is used to allow for the decay of the pulses so that intersymbol interference is not caused, as well as to allow for timing inaccuracies. Pulse decay is a function of the transient responses of the filters used in both the earth stations and the satellite. These filters must have good phase linearity, i.e., low group delay distortion.

Slight motions of the satellite will cause corresponding changes in propagation time. However, satellite motion is periodic and predictable. Accordingly, elastic buffers, rather than pulse-stuffing buffers, can be used to compensate for satellite path delay variations. This presumes that highly stable clocks are used at all terminals and that the frame rate is held constant at the satellite. The sizes of the elastic buffers must be adequate to prevent overflow or underflow. This can be determined based upon the anticipated orbital characteristics of the satellite, plus a safety factor. In addition, the buffers must be reset regularly because of clock drifts at the earth terminals.

Digital transmission using TDMA is accomplished by the use of modulation during the carrier bursts. The needed phase reference must be derived during each burst since no means exists to synchronize the transmitted phases as received at the satellite. The phase reference can be obtained[15] either by (1) using a phase-locked loop or narrow-band filter which can acquire each carrier rapidly in sequence or (2) using multiple time-gated carrier recovery loops or a single time-multiplexed phase-locked loop with phase memory from frame to frame. The latter approach allows the use of narrow-band carrier recovery loops operating on each time-gated carrier.

With the development of large array antennas for use on communication satellites, narrow spot beams covering only a metropolitan area and its vicinity will become operational. The receiving and transmitting spot beams can be made independent of one another. Such antennas, together with switching on board the satellite, can allow transmission of speech packets to and from specific ground locations, thus achieving not only full transponder power utilization, but also spectrum economy through frequency reuse in various spot beam areas. Time slot coordination can be achieved by having one earth station serve as a central timing source and sending time through the satellite to all the other earth stations.[16]

Satellite-switched transmission can be accomplished using FDMA also, by using frequency bands within the transponder rather than time slots. However, the transponders then must be operated with their power levels backed off from maximum.

8.3.6. Multiplexing

The reader is referred to Section 6.5 for a general discussion of digital multiplexing. As applied to satellite transmission, PCM voice can be either on an FDM carrier or on a TDM carrier burst. The modulation usually is QPSK. The access technique may be either TDMA or FDMA. The overall designation thus might be PCM/TDM/QPSK/TDMA. The SPADE system, described in Section 8.3.5.1, is designated PCM/QPSK/FDMA. Although

FDM generally is used with FDMA and TDM with TDMA, hybrid systems also exist.[17] One is the FDM-Master Group Codec for use in the Telesat TDMA system.[18]

Because TDM is used more extensively in digital transmission than FDM, a more detailed discussion of TDM is provided in Section 5.3.

A low-loss multiplexer has been developed for satellite earth terminals to eliminate the need for a broadband high-power transmitter[19] with its reliability and efficiency limitations. Each 36 MHz channel in the 5925–6425-MHz uplink band is amplified by an individual traveling-wave tube or solid-state power amplifier. Modular units allow the addition of channels as needed. Waveguide equalizers provide for amplitude and time delay compensation. Such modular units are well suited to small, unmanned earth terminals.

8.3.7. Demand Assignment

Many earth terminals use large numbers of dedicated channels devoted to continuous or nearly continuous traffic. However, increasing recognition is being given to the fact that many channels are not really needed on a full-period basis. Demand assignment thus can achieve economies by maintaining circuits connected only for the duration of a call, or for even shorter periods of time, such as a talk spurt, as described in Section 5.9.

Demand assignment (DA) is a very old and well-established concept in telephony. A user demands the assignment of a circuit by taking the telephone instrument off hook. Receipt of dial tone then allows him to access the public switched network by dialing the desired number. A fully variable DA network pools all channels (trunks). A given channel then may be used by any station (end instrument) based upon the instantaneous traffic load. DA is most advantageous for destinations having light traffic loads. DA thus not only uses the space segment more efficiently than fixed assignment, but also allows for more efficient use of the terrestrial interconnect facilities. A result is that direct connections can be made more often, with their improved quality of service compared with the use of more tandem connections.

A semi-variable demand assignment system is one in which blocks of channels are reserved for originating and/or destination stations, but still used only on demand.

A DA system is said to be demand assigned multiple access (DAMA) when carriers (in FDMA systems) or bursts (in TDMA systems) are assigned on demand. Alternatively, when channels on existing carriers (FDMA) or time slots in existing bursts (TDMA) are assigned on demand, the system is said to be baseband demand assigned (BDA). BDA is best suited to networks with many users but only a few large earth stations, whereas fully variable

DAMA is best suited to a network with many earth stations, each of which has only low traffic requirements. Mixed approaches also are useful, depending on the requirements to be met by a system.

DAMA systems provide for access on an individual user basis. The SPADE system, described in Section 8.3.5.1, is an example of a DAMA system. Other DAMA approaches are single-channel per-burst (SCPB) TDMA and code-division multiple access (CDMA). These systems do not transmit a carrier (FDMA or CDMA) or burst (TDMA) until it is required. The carrier or burst is established only for the duration of a call. For control, a common TDM signaling channel can be used on which all stations can call each other and be aware of the available channels. These available channels can be seized on a first-come-first-served basis.

BDA systems can utilize DSI (see Section 5.9) readily since BDA involves the application of a large number of trunks (usually 40 or more) in the provision of a larger (e.g., doubled) number of two-way voice conversations.

8.3.8. Echo Cancellation

Section 5.7.2 discussed the principles of echo cancellers and their importance in long haul circuits. Echo cancellers are especially important on satellite circuits, since the round-trip delay time is on the order of 600 ms, thus making existing echoes especially noticeable. Usually an echo canceller will be used at each end of a circuit. This is called split echo control.

8.4. CHARACTERISTICS OF SATELLITE SYSTEM EQUIPMENT

A satellite system is comparable to a microwave radio relay system with a single very high altitude repeater. Accordingly, the equipment characteristics have many similarities. These include the fact that amplifiers do not have highly linear characteristics. Thus angle modulation (often QPSK) usually is preferred to amplitude modulation. Section 7.4 discusses the characteristics of microwave amplifiers and antennas in more detail.

Figure 8–5 shows the basic elements of a satellite system. They include the satellite (space segment) itself, earth stations (terrestrial segment) and a control station, whose functions are tracking, telemetry and command.

8.4.1. Space Segment

A spacecraft for telecommunications purposes consists not only of the transponders, but also of a control subsystem, solar array, batteries, monitoring systems, and a small propulsion system, as well as temperature control devices. Power for the spacecraft is furnished by a solar array which maintains a

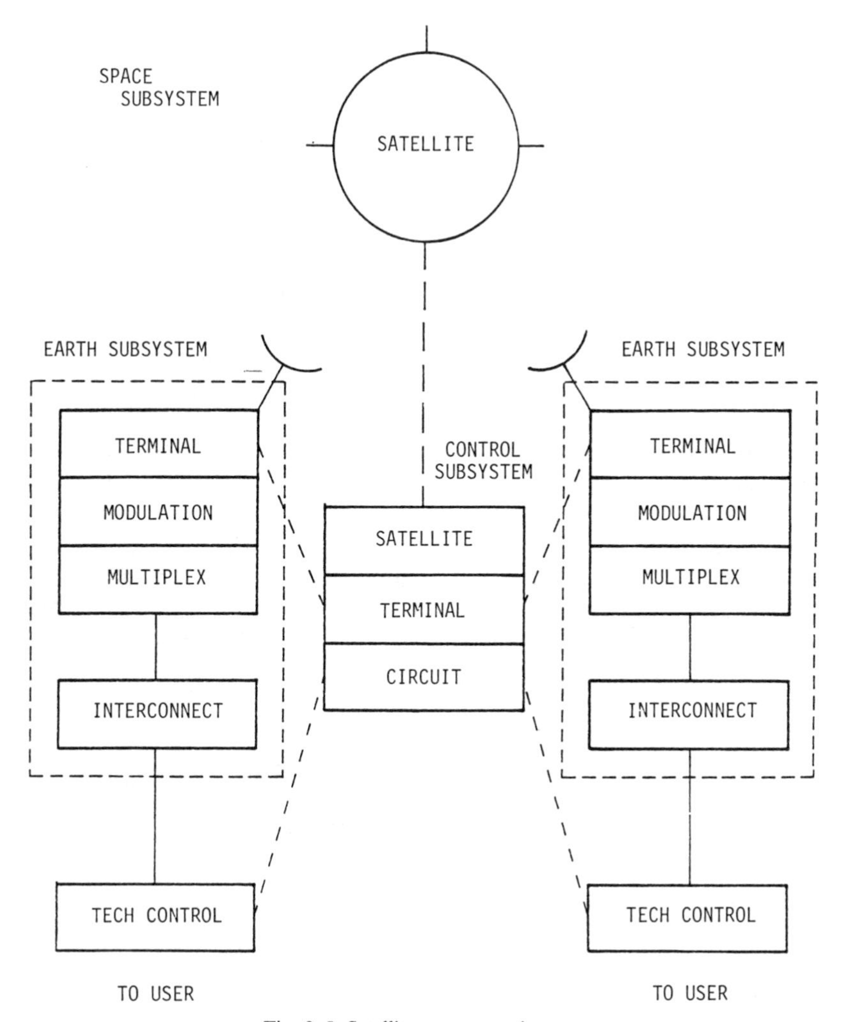

Fig. 8–5. Satellite system equipment.

charge on nickel-cadmium or nickel-hydrogen batteries. Solar cells provide efficiencies on the order of 15% or more and thus can produce at least 300 watts dc per square meter of surface area for sun rays striking the cells perpendicularly. The size of the solar array thus limits the power available on board the satellite.

For on-board power storage, nickel-cadmium batteries have been used successfully in space for many years, but their state of charge must be monitored closely. If they are allowed to discharge to less than 50% of full charge, their lifetime may be impaired. They also must be exercised, i.e., their state of charge must be varied from time to time. Nickel-hydrogen batteries do not exhibit these limitations, but are relatively new in space. However, they are being used to an increasing extent.

Twice a year, at the equinox, which occurs in the latter parts of March and September, the spacecraft enters eclipse, i.e., the sun sets on the spacecraft. This condition may last up to seventy minutes per day. During such times, the spacecraft must rely entirely upon its batteries for its power. Accordingly, satellite system designers often try to place their spacecraft about three time zones to the west of the areas they serve so that equinox will occur at about 3 A.M. local time, a time when traffic tends to be at a minimum.

The propulsion system on a spacecraft allows for the correction of its orbital position from time to time to compensate for position drift that may occur.

The remainder of this section describes the transponders and the control subsystem.

8.4.1.1. Transponders. Each satellite contains a number of transponders whose function is to receive, amplify, and retransmit the signals within a given frequency band. The retransmitted signal must be at a different frequency from the received signal to prevent the oscillation which would occur if the transmitter output were fed back to the receiver input. Because of satellite power limitations, the downlink is more critical than the uplink, so the downlink frequency band usually is lower in frequency than the uplink frequency band. A single satellite antenna with multiple feeds, however, may be used for both receiving and transmitting.

A large number of signals from various earth stations usually arrives at a given satellite. They may be of various frequencies and polarizations, and may occur in different time slots. The antenna feed arrangement is used to separate the signals into two orthogonal polarizations, often linear for domestic satellites and circular for international satellites. Following this, the different frequency bands are separated by use of channel filters so that each transponder receives the frequency band for which it is designed. Separation of signals in the time domain, if done, occurs in a processing transponder only. Dual polarization can be achieved such that interference from the oppositely (or cross) polarized signal is about 40 dB below that of the desired signal over relatively small bandwidths (e.g., 2%) and beamwidths (e.g., 2°). Over broader bandwidths and beamwidths, however, the oppositely polarized signal reaches higher levels. One *Intelsat* design achieved −27 dB over the entire 3700–6425 MHz band within a spot beam.

Precipitation degrades polarization isolation by converting linear or circular polarization to elliptical. Thus both the antennas and the propagation effects must be included in a determination of interference at a receiver from an unwanted transmitted polarization. In the 11–14-GHz bands and at higher frequencies, depolarization caused by precipitation is sufficiently severe that dually polarized transmissions are not feasible.

In general, access to a given transponder from the ground is based on the polarization and frequency of the uplink signal. Selection of a given transponder in this manner may be used to control the downlink frequency, and possibly the covered area on the ground since different transponders may be connected to different spot beams. In addition, transponders may serve different classes of users, with some transponders being dedicated to low-level mobile users while others handle large earth stations with supergroup or master group levels of traffic, or television transmissions.

A transponder frequency plan is shown in Fig. 8–6, using staggered frequencies to aid in isolating the two polarizations. Staggered plans are useful for transponders having small numbers of carriers, i.e., for television or master group applications. For SCPC applications, staggering is of only limited value.

A single conversion transponder is shown in block diagram form in Fig. 8–7. The preamplifier provides gain at the received radio frequency, but usually has an equivalent noise temperature of several thousand Kelvin. Lower noise temperatures are not needed because appreciable noise enters from the earth itself. Moreover, amplifiers with low noise figures tend to be less reliable than the noisier type typically used.

The total bandwidth of a given transponder may be as little as 36 MHz or less, as in many domestic and international satellites, or as large as 72–241 MHz, as in some of the international satellites. This transponder bandwidth may be subdivided, as is done, for example, in the SPADE system,

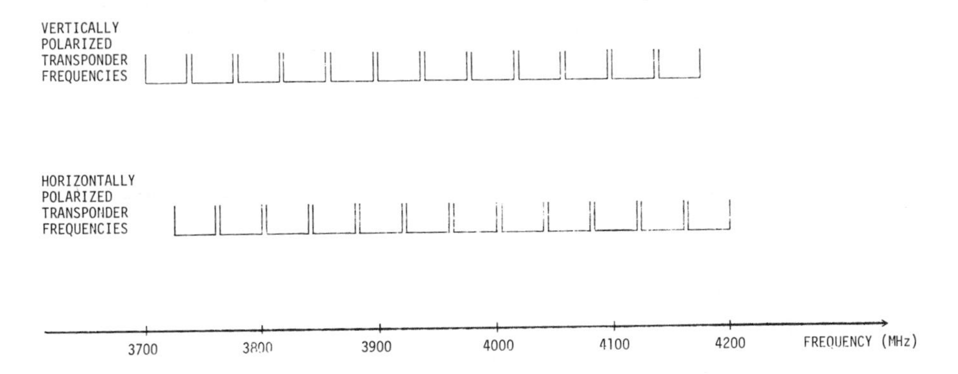

Fig. 8–6. Downlink satellite transponder frequency plan.

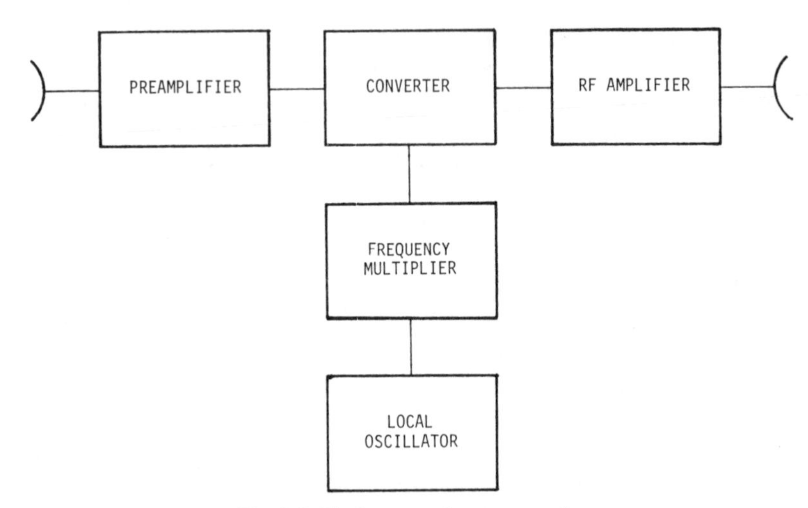

Fig. 8–7. Single conversion transponder.

which provides up to 800 carrier frequencies within a single transponder's bandwidth. (See Section 8.3.5.1.)

Most transponders provide preamplification to the uplink signal, conversion by a fixed amount to a lower frequency, and then amplification at the lower frequency which then is transmitted down to the ground. In the 4- and 6-GHz commercial bands, 2225 MHz usually is subtracted from the uplink frequency to produce the downlink frequency, whereas in the 7 and 8 GHz military bands, 725 MHz or 200 MHz is subtracted. These uplink-downlink frequency differences are the maximum ones available within the given frequency allocations[20] except for the 200 MHz difference. This relatively small difference results from the fact that the 7900–7950 MHz uplink is received on board the satellite through an earth coverage horn and is adjacent in the uplink spectrum to other channels received on the earth coverage horn. However, signals received in the 7900–7950 MHz transponder are cross-strapped to a 7700–7750 MHz downlink which is used with a narrow coverage downlink (transmit) antenna. The frequency assignment was established to place the downlink frequency band adjacent to the other downlink narrow coverage frequency bands.

Automatic gain control amplification and limiting are used to maintain uniform signal levels, but transponder gain often is commandable from the ground as well.

The downlink beam produces a coverage area on the ground known as the "footprint." An example of a footprint is shown in Fig. 8–8.

Most transponders are built on a highly redundant basis to achieve high

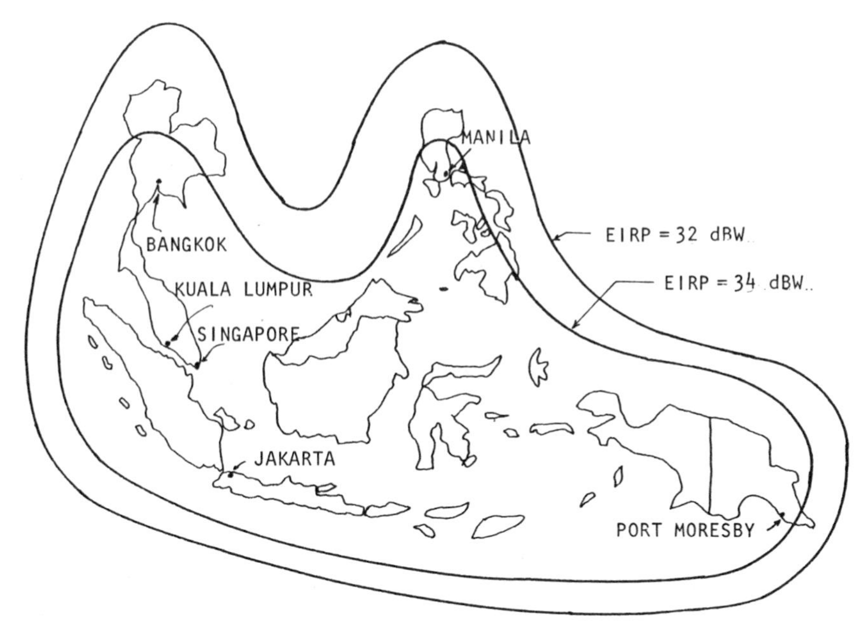

Fig. 8–8. Footprint coverage of *Palapa B* satellite (Indonesia). (Reprinted by permission of *Satellite Communications* magazine.)

reliability. The block diagrams in this chapter, for simplicity, do not illustrate these redundancies.

The transponders described thus far involved a single conversion of the uplink frequency directly to the downlink frequency, along with the needed amplification. However, various applications exist in which conversion to baseband on board the satellite is advantageous.

On-board processing of weather photograph information and earth resources data from multispectral scanners has proven feasible already. Concepts for commercial applications of on-board switching and processing include on-board demand access, also known as the "switchboard in the sky" concept, and TDMA packet transmission via spot beams. On-board demand access allows the transponder input channels to be switched by command from the ground to the correct downlink channel. TDMA packet transmission, or satellite-switched TDMA, uses a preprogrammed switching sequence in which the packet addresses are used on board the satellite to switch each packet to its proper destination via the corresponding spot beam. A packet typically is 1024 bits long, including address information.

In addition to on-board switching, signal processing can be done to regenerate uplink digital signals for downlink transmission. Such processing is useful both in commercial and military applications. Other military uses include the removal of interference accompanying the uplink signals. In this manner, the repeating of interference can be minimized, and the tendency of strong signals to "capture" the transponder AGC because of nonlinearities can be reduced.[21]

Other concepts for processing transponder applications include the use of uplink single sideband to conserve bandwidth, and its conversion on board the satellite to PCM/FM for the downlink to minimize the power required from the satellite.

Signal processing transponders are built to perform specific types of conversions. The result is a lack of flexibility with respect to changes that may be desired after the satellite has been launched. However, the advantages of processing transponders are such that their use in the future probably will increase significantly, especially with the problems of an increasingly crowded spectrum.

Figure 8–9 is a block diagram of a processing transponder. As shown there, the entire transponder input is converted to baseband, at which such functions as switching or signal processing can be performed. Then the processed baseband signals are remodulated and converted to the downlink frequency band.

8.4.1.2. Control Subsystem. The control subsystem includes a special earth station (often called the tracking, telemetry, and command station), whose function is to track the spacecraft's position, with the help of an on-board transponder, to monitor spacecraft telemetry, and to send corrective commands to the spacecraft as necessary. Telemetry from the spacecraft includes data on the solar array output, the state of charge of the batteries, the amount of propulsion fuel remaining, on-board temperatures, and the

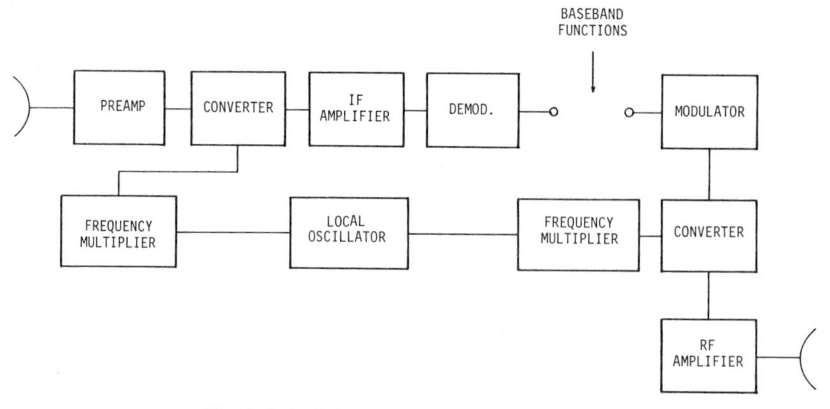

Fig. 8–9. Switching or processing transponder.

operating level and load of each transponder, including automatic gain control (AGC) level.

8.4.2. The Earth Station

Figure 8–10 is a simplified block diagram of a satellite earth terminal for digital transmission. Digital streams at the DS1 rate (i.e., T1), each conveying 24 voice channels or the equivalent, are multiplexed to form a stream at the DS3 rate (i.e., T3) or higher. A single DS3 stream can be handled using QPSK in a 36 MHz transponder. Alternatively, two DS3 streams can be transmitted using 8-PSK, and correspondingly higher rates can be sent through larger transponder bandwidths.

The frequency standard generally is of the cesium type. The tunable frequency synthesizers used in the up and down converters are phase locked to the frequency standard. The required local oscillator frequencies then are obtained by the use of phase-locked multipliers. Although Fig. 8–10 shows only a single up-conversion step, two actually may be used, with the first IF being centered at 70 MHz and the second perhaps at 700 MHz. Smaller earth terminals may use only a fixed frequency local oscillator rather than a frequency synthesizer. Earth terminals operating with satellite transponders having appreciably larger bandwidths than 36 MHz use higher IF frequencies than 70 MHz.

The portions of the block diagram described thus far are repeated for each satellite transponder through which the earth station works. A power combiner then adds the outputs of the various up converters, each of which is in a separate band of frequencies. If dual polarization is used, there will be two power combiners, one for the transponders using vertical or right-hand circular polarization and another for the transponders using horizontal or left-hand circular polarization.

Following the power combiner is a broadband intermediate-power amplifier (IPA) and a broadband high-power amplifier (HPA). Earth station power output may be on the order of 500 W to 2 kW or more, depending on the number of channels being handled. Usually the IPA and HPA are dual redundant because of the large number of channels they carry.

A diplexer is used for each of the two polarizations handled by the earth station, as well as for each frequency band pair (i.e., 4–6 GHz, 7–8 GHz, 11–14 GHz, etc.). However, dual polarization operation generally is confined to the frequency bands below 10 GHz because of the extensive depolarization caused by precipitation above 10 GHz. The diplexers and the transmit and receiver filters must provide the high degree of isolation needed to prevent the transmitter output power from feeding back into the receiver input and thus producing desensitization.

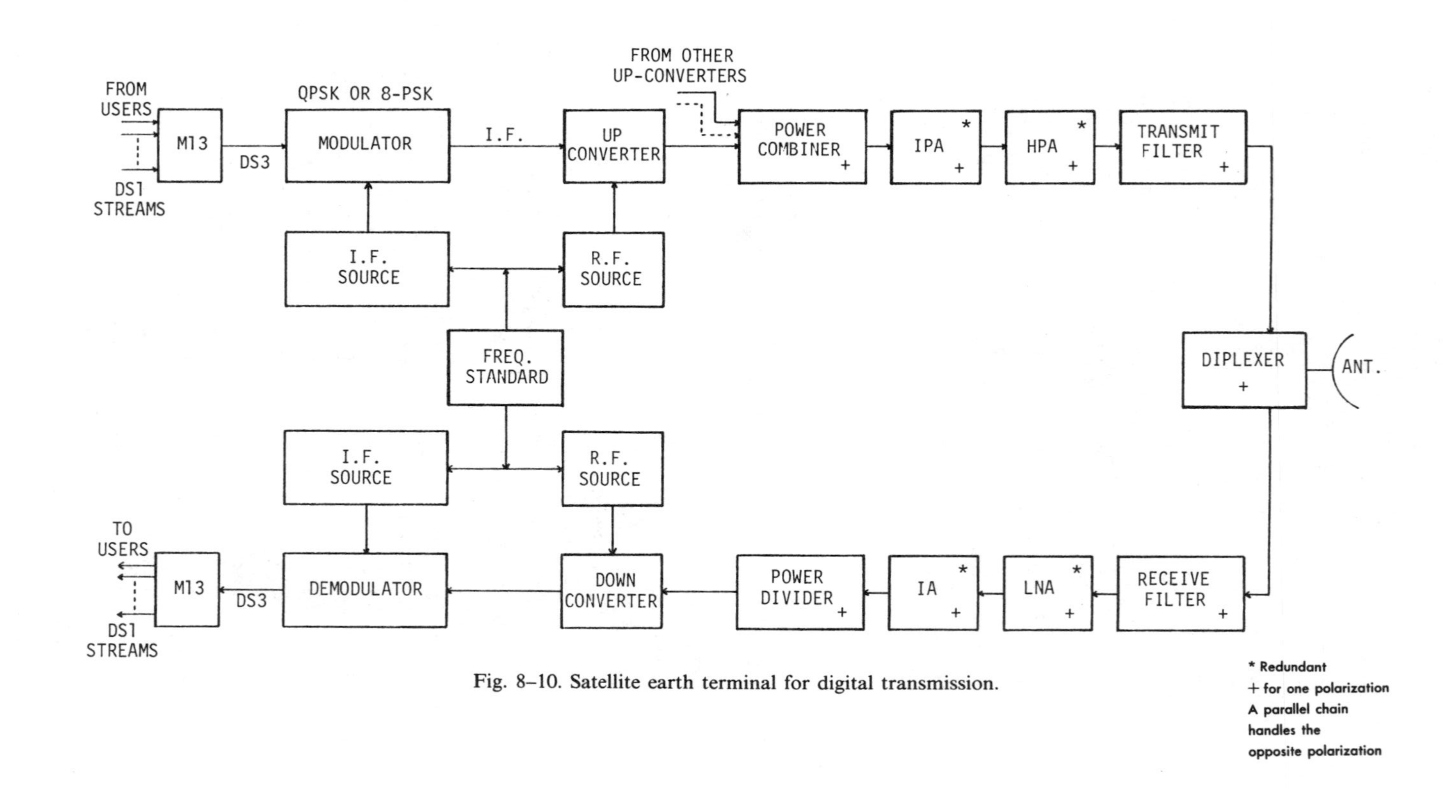

Fig. 8–10. Satellite earth terminal for digital transmission.

An important characteristic of the earth station is the ratio of the antenna gain G to the effective noise temperature T, where G is expressed in dB and T in dBK (decibels relative to one Kelvin). The ratio G/T thus is expressed in dB/K. It is a receiving system figure of merit. Large values of G/T allow lower power satellites to be used, but require more expenditure for the earth station because of the cost of larger antennas and low noise temperature amplifiers (LNAs).

A separate set of LNAs and intermediate amplifiers (IAs) is used for each received polarization and each frequency band (e.g., 4, 7, 11 GHz, etc.). In addition, the IAs and LNAs are redundant for reliability. The power divider separates the channels within the frequency domain based on the satellite's transponder frequency bands. The down conversion, demodulation, and demultiplexing steps then are the reverse of those described for the transmitting direction.

All transmitted sidebands must be attenuated sufficiently to prevent them from disturbing the receiving system. Isolation must be on the order of 140–180 dB depending on transmitted power level, bandwidth, and receiver noise level. This requires filters with good skirt selectivity but yet with well controlled delay distortion at the band edges to keep intersymbol interference low.

Earth stations used as part of the *Intelsat* system are classified as follows:

The Standard A earth station operates in the 4-GHz downlink and 6-GHz uplink bands with $G/T \geqslant 40.7$ dB/K. This can be achieved using a 32-m dish and a 55-K LNA. The standard B earth station differs from the Standard A only in that $G/T \geqslant 31.7$ dB/K. The Standard C earth station operates in the 11-GHz downlink and 14-GHz uplink bands with the following G/T values:

$$\text{Long-term } G/T \geqslant 39.0 + L_1 + \log_{10}(f/11.2) \text{ dB/K} \qquad (8\text{–}6)$$

where L_1 is the additional attenuation over clear sky all but 10% of the time and f is the carrier frequency in GHz.

$$\text{Short-time } G/T \geqslant 29.0 + L_2 + \log_{10}(f/11.2) \text{ dB/K} \qquad (8\text{–}7)$$

where L_2 is the additional attenuation over clear sky all but 0.017% of the time (1.5 hours per year). These equations are intended to keep the basic telephone channel noise power to less than 8000 pW0p for 90% of the time.

Equations (8–6) and (8–7) are applied next to illustrate the receiving antenna gain required in a particular case. The example chosen applies to the Intelsat receiving system at Etam, WV.[22] Weather statistics and propagation studies showed that clear sky attenuation, caused by mist, fog, high strato-cumulus clouds, i.e., factors not measurable by a rain gauge, is 0.4 dB. Rain conditions for all but the worst 10% of the year contribute an additional 0.55 dB attenuation. Thus $L_1 = 0.40 + 0.55 = 0.95$ dB. Rain conditions for all but the worst 1.5 hours of the year contribute an additional 9.5 dB attenuation. Thus $L_2 = 0.40 + 9.5 = 9.9$ dB. Corresponding contributors to the temperature T are 17 K for the 18° elevation angle at Etam for all but 10% of the time and 419 K for all but 0.017% of the time.

The meeting of the Standard C earth station requirements may necessitate the use of a space diversity pair of receiving sites. For example, the use of a second receiving site at Lenox, WV, 35 km to the northeast of Etam, operated on a diversity basis with Etam, resulted in a reduction in L_2 to $0.4 + 2.5 = 2.9$ dB and a reduction of the noise temperature all but 0.017% of the time from 419 K to 307 K. Because an unacceptably high transmit power (22 kW) would have been required at Etam without diversity, the use of the diversity pair was implemented.

A nonstandard earth station requires approval on a case-by-case basis. A "station accessing leased facilities (transponders)" is one that produces less than 400 pWp on an adjacent satellite carrier (⩾3°away), in accordance with CCIR recommendation 465–1. Its emission limit is 20 dBW/4 kHz carrier. The abbreviation pWp means "picowatts psophometric" and refers to the use of a voice band weighting curve used by the CCIR. This curve approximates the response of a standard telephone set.

The side lobes produced by the antennas of the above earth stations must all meet the criterion $G_s = 32 - 25 \log \theta$, where θ is the antenna's half-power beamwidth in degrees. Many newer earth stations must meet more stringent sidelobe criteria, such as $29–25 \log \theta$ or $26–25 \log \theta$.

8.5. MAJOR OPERATIONAL COMMUNICATION SATELLITE SYSTEMS

With one exception most currently operational satellite systems are strongly oriented toward analog transmission at the present time, although digital transmission is starting to appear. The reason is that most systems had their origins in the mid-1970s, and built a sizable number of analog earth station facilities. The exception is Satellite Business Systems (SBS), owned by IBM and Aetna Life Insurance Company. SBS started later than the others, and has built an all digital system. Each of its satellites carries ten transponders of 43 MHz bandwidth with uplinks in the 14.0–14.5 GHz band and downlinks in the 11.7–12.2 GHz band. For the uplink, SBS has a G/T = 1.8 dB/K at the beam edge. For the downlink, it has an EIRP of 37.0 to 43.8 dBW at the beam edge.

Many other satellite systems are implementing digital transmission in one or more transponders, as well as at many of their earth stations. They include Telesat's *Anik*, RCA's *Satcom*, *Advanced Westar*, *Comstar*, American Satellite and *Intelsat*. U.S. domestic satellites which also may be used in the future for digital telephony include Hughes' *Galaxy*, GTE's *GSTAR*, SP Communications' *Spacenet*, AT&T's *Telestar*, and U.S. Satellite System's *USAT*.

8.6. FUTURE TRENDS IN COMMUNICATION SATELLITE SYSTEMS

In spite of the rapid development of communication satellite systems since the mid-1970s, and the increasing implementation of digital transmission via satellite since the early 1980s, many more developments are on the horizon. The use of on-board processing was described in Section 8.4.1.1. It is expected to allow implementation of the "switchboard in the sky" concept of controlled switching from the ground, as well as on-board controlled TDMA switching, possibly combined with scanning spot beams.[23] The addresses carried by individual packets might be used for beam and time slot control. Packet voice has been found to be feasible for telephone conversations, and this concept relates directly to the achievement of on-board controlled TDMA switching.

Satellites also may play a role in the achievement of nationwide mobile radio telephony, especially in connection with the use of error-correcting codes combined with digitized voice. Error correcting codes may allow the severe bit error rates (e.g., 10^{-1}) resulting from mobile radio multipath propagation to be reduced to values on the order of 10^{-3} to 10^{-4}, at which most voice coding techniques give satisfactory voice quality.

Intersatellite links involving an up, over and down mode of transmission have already been shown to be feasible through work done by the U.S. Air Force with its Lincoln Experimental Satellites 8 and 9 (*LES-8* and *LES-9*), which used a 36–38-GHz cross link. Such intersatellite links, when implemented commercially, may help to further global telephony as well as to provide more widespread use of land, maritime and aeronautical radiotelephony.

Early satellite system designs attempted to maximize satellite lifetime by keeping the satellite as simple, and therefore reliable, as possible, while allowing the earth stations to be relatively complex. The trend now, however, is toward more complex satellites but simpler earth stations. This trend is being aided by two major factors. One is the widespread use of large-scale integrated circuits, with their inherently high reliability. The other is the accumulation of many years of experience on the part of spacecraft designers. The complex satellite and the relatively inexpensive earth station make good economic sense, just as do the present complex central office and the relatively inexpensive end instrument. Many earth stations may use a single satellite. By putting more of the complexity into the satellite, the savings in earth station hardware are multiplied many times.

More complex satellites may provide not only higher EIRP (in the bands above 10 GHz where such EIRP is allowed) but also more narrow satellite beams for improved orbit-spectrum utilization. More on-board processing is another feature of the more complex satellite. Such on-board processing, in addition to the applications discussed in Section 8.4.1.1, can allow transmission to and from cities experiencing good weather in the 20–30-GHz bands, while using the 4–6-GHz bands to and from those cities experiencing precipitation conditions. The smaller earth station sizes will be advantageous primarily from an economic viewpoint.

Another future satellite system development may be the orbital antenna farm (OAF), in which a few giant space stations in geostationary orbit replace many present communication satellites.[24] Such an OAF could alleviate interference caused by the crowding of the geostationary orbit, thus permitting smaller earth stations. It could permit the mounting of very large and complex antennas and feed systems with very narrow spot beams. Other features would be the interconnection or "cross-strapping" of various satellite circuits and the achievement of economies of scale by providing common support functions (e.g., power, as well as propulsion for orbit corrections) for the multiple satellite systems involved. The OAF would be implemented with the help of the space transportation system (STS) or "shuttle" in its construction and maintenance.

The OAF would be serviced and maintained in operation through the use of high-energy upper stages operating from the STS.

Since the OAF concept was suggested in 1977, thinking has moved toward the idea of a platform consisting of a single STS-borne spacecraft with its mated transfer vehicle.[25] Such a scaled-down platform is now visualized as serving Intelsat requirements in the 1990s. Use of the STS can allow spacecraft deployment and checkout in low earth orbit where minor malfunctions can be corrected manually, something not yet possible at geostationary orbit.

PROBLEMS

8.1 Dual polarization is used commonly in the 4–6-GHz satellite bands, but not in the 11–14-GHz bands. Why?

8.2 Why is a satellite system's downlink budget more critical than its uplink budget?

8.3 A satellite transponder centered at 3950 MHz is used totally for TDMA digital transmission and has a 5 W power output. For earth stations that are at the beam edge (−3 dB) and that have a 45° latitude (satellite and earth stations at same longitude), what is the received level in a 50-dB gain dish if the satellite antenna has a 1.5 m diameter? (Neglect circuit and waveguide losses.) Assume a 55% antenna efficiency.

8.4 In the system of Problem 8.3 the effective earth terminal figure of merit is 20.0 dB/K. What is the available C/kT?

8.5 In the system of Problems 8.3 and 8.4, what symbol rate can be transmitted if the required $E_s/N_0 = 18$ dB, assuming losses and link margin total 4.0 dB?

8.6 You have a choice between placing a satellite earth terminal dish on a building roof top or near a parking lot on the ground. The path to the satellite is unobstructed in both cases. Which location would you choose and why?

8.7 Discuss the advantages and disadvantages of uniform carrier frequency spacing in a satellite transponder being used on an FDMA basis.

8.8 Explain how the operating level of a satellite transponder is actually determined on the ground.

REFERENCES

1. Keiser, B. E., "Wideband Communications Systems," Course Notes, © 1981.
2. Carr, F. S., "Aerosat-Current Status and the Test and Evaluation Program," *1975 Eascon Record,* p. 13 (Sept. 1975).
3. Anderson, R. E., "The Mobilesat System," *Satellite Communications,* Vol. 8, No. 3, pp. 16–18 (March, 1984).
4. Brown, R. J., et al., "Companded Single Sideband (CSSB) Implementation on Comstar Satellites and Potential Application to Intelsat V Satellites," *ICC '83 Conference Record,* IEEE Document 83 CH1874–7 pp. E2.1.1–E2.1.6 (June, 1983).
5. Edwards, Robert D., "An Improved Syllabic Compandor for Satellite Applications," *ICC '83 Conference Record,* IEEE Document 83 CH1874–7, pp. E2.3.1–E2.3.5 (June, 1983).
6. Helder, G. K., "Customer Evaluation of Telephone Circuits with Delay," *Bell System Technical Journal,* Vol 45, pp. 1749–1773 (Dec., 1966).
7. Benoit, A., "Signal Attenuation Due to Neutral Oxygen and Water Vapour, Rain and Clouds," *Microwave Journal,* pp. 73–80 (Nov. 1968).
8. Fang, D., "Attenuation and Phase Shift of Microwaves Due to Canted Raindrops," *Comsat Technical Review,* Vol. 5, No. 1, pp. 135–156 (Spring, 1975).
9. Smith, F. L., "A Nomogram for Look Angles to Geostationary Satellites," *IEEE Trans. Aerospace and Electronic Systems,* p. 394 (May, 1972).
10. Spilker, J. J., *Digital Communications by Satellite,* Prentice-Hall, Inc., Englewood Cliffs, NJ, 1977.
11. Puente, J. G., Schmidt, W. G., and Werth, A. M., "Multiple Access Techniques for Commercial Satellites," *Proc. IEEE,* pp. 218–229 (Feb. 1972).
12. McClure, R. B., "Analysis of Intermodulation Distortion in a FDMA Satellite Communication System with a Bandwidth Constraint," *Trans. IEEE International Conference on Communications,* 1970.
13. See note 10.
14. Babcock, W. C., "Intermodulation Interference in Radio Systems," *Bell Syst. Tech. J.,* Vol. 32, pp. 63–73 (Jan. 1953).
15. See note 10.
16. Assal, F., Gupta, R., Apple, J., and Lopatin, A., "Satellite Switching Center for SS-TDMA Communications," *Comsat Technical Review,* Vol. 12, No. 1, pp. 29–68 (Spring, 1982).

17. Pritchard, W. L., "Satellite Communication—An Overview of the Problems and Programs," *Proc. IEEE,* Vol. 65, No. 3, pp. 294–307 (March, 1977).
18. Kaneko, H., Katagiri, Y., and Okada, T., "The Design of a PCM Master-Group Codec for the Telsat TDMA System," *Conference Proceedings, ICC '75,* Vol. 3, pp. 44–6 to 44–10 (June, 1975).
19. Gruner, R. W., and Williams, E. A., "A Low-Loss Multiplexer for Satellite Earth Terminals," *Comsat Technical Review,* Vol. 5, No. 1, pp. 157–177 (Spring, 1975).
20. Huang, R. Y., and Hooten, P., "Communication Satellite Processing Repeaters," *Proceedings of the IEEE,* Vol. 57, pp. 238–251 (Feb., 1971).
21. See note 17.
22. Gray, L. F., and Brown, M. P., Jr., "Transmission Planning for the First U.S. Standard C (14/11 GHz) Intelsat Earth Station," *Comsat Technical Review,* Vol. 9, No. 1, pp. 61–89 (Spring, 1979).
23. Reudink, D. O., and Yeh, Y. S., "A Scanning Spot-Beam Satellite System," *Bell System Technical Journal,* Vol. 56, No. 8, pp. 1549–1560 (Oct. 1977).
24. Edelson, B. I., and Morgan, W. L., "Orbital Antenna Farms," *Aeronautics and Astronautics,* Vol. 15, No. 9, pp. 20–27 (Sept., 1977).
25. Cohen, N. L., and Stone, G. R., "DBS Platforms: A Viable Solution," *Satellite Communications,* Vol. 6, No. 13, pp. 22–27 (Dec., 1982).

9
Fiber Optic Transmission*

9.1. INTRODUCTION

Fiber optic transmission can be implemented wherever coaxial cable or wire-pair transmission is used. Its areas of present and future application extend all the way from the subscriber loop[1] to transoceanic cables.[2]

The basic concept in fiber optic transmission is illustrated in Fig. 9–1. As shown there, a modulated light source such as a light-emitting diode (LED) or a laser has its output coupled to a fiber which then transmits the light to a remote photodetector, which demodulates the light, thereby providing the desired signal output.

Fiber optic transmission has numerous advantages over the use of metallic cables.[3] First, the weight and bulk of the cable plant are reduced significantly, thereby allowing more efficient use to be made of duct space. A significant improvement in bandwidth also may be obtained, as is discussed in Section 9.2. This may allow future expansion without plant refurbishment. Accordingly, fiber optics allows significant cost savings in some applications, while having a cost comparable to that of wire cable in others. The cost of a 50 km transmission system using fiber optics has been shown to be less than the cost of microwave radio or coaxial cable transmission for a capacity of 240 duplex channels (15.44 Mb/s) or more. System costs, based upon a 30 km repeater spacing, have been estimated[4] by Future Systems, Inc., Gaithersburg, MD. The costs range from $10 500 for 45 Mb/s transmission to $13 600 for 270 Mb/s transmission. These estimates assume rural installation costs and one-for-one backup protection. Such factors as legal, property, maintenance and tax expenses have not been included because of their wide variability from one place to another.

Fiber optic systems are not troubled by short circuits or current leakage to ground, although open circuits can occur. Crosstalk is extremely low, and its effects can be eliminated completely by proper design. Problems of electromagnetic interference and noise do not occur along the optical portion of a fiber system, but such problems can exist on the wire side of the modems that may be used in getting a bit stream on and off a fiber. Correspondingly,

* Much of the material of this chapter is taken from B. E. Keiser, "Wideband Communications Systems," Course Notes, © 1981.

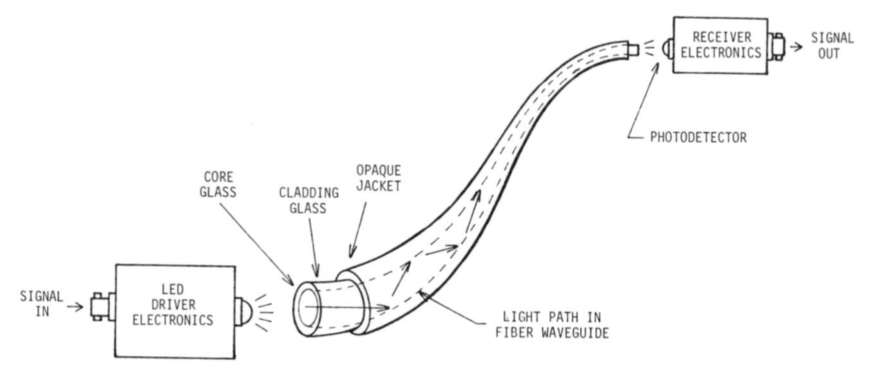

Fig. 9–1. Basic fiber optics communication system. (Courtesy Dr. Leonard Bergstein.)

there is no electromagnetic pulse problem along the fiber itself, although the fiber does exhibit a response to the ionizing radiation that directly accompanies a nuclear blast, as discussed in Section 9.2.

Optical fibers allow a very high degree of transmission security, which can be extended directly to the user. This results from the fact that optical fibers are very difficult to tap without the user becoming aware of the tap in the form of reflection level changes. In fact, for secure applications a triaxial fiber system can be built with a noninformation region on the outside to monitor the reflection level. This noninformation region can contain a pulse sequence used only for monitoring purposes, but designed to look like a real bit stream.

9.2. FIBER TRANSMISSION CHARACTERISTICS

The two major factors that limit the transmission distance along a single unrepeatered section of fiber are attenuation and dispersion. Dispersion is important in terms of its limiting effect on bandwidth, and is affected significantly by the process used in manufacturing the fiber. The vapor-phase axial deposition (VAD) process holds the record for the highest bandwidth multimode fiber,[5] having a bandwidth-distance capability of 9.7 GHz km at $\lambda = 1.31\ \mu m$. The modified chemical vapor deposition (MCVD) process is popular because it allows the mass production of commercial fibers.

9.2.1. Attenuation

The attenuation characteristics of both VAD and MCVD fibers[6] are shown in Fig. 9–2. Because the lowest attenuations are obtained at near infrared wavelengths, infrared emitting diodes (IREDs) are of major interest. The

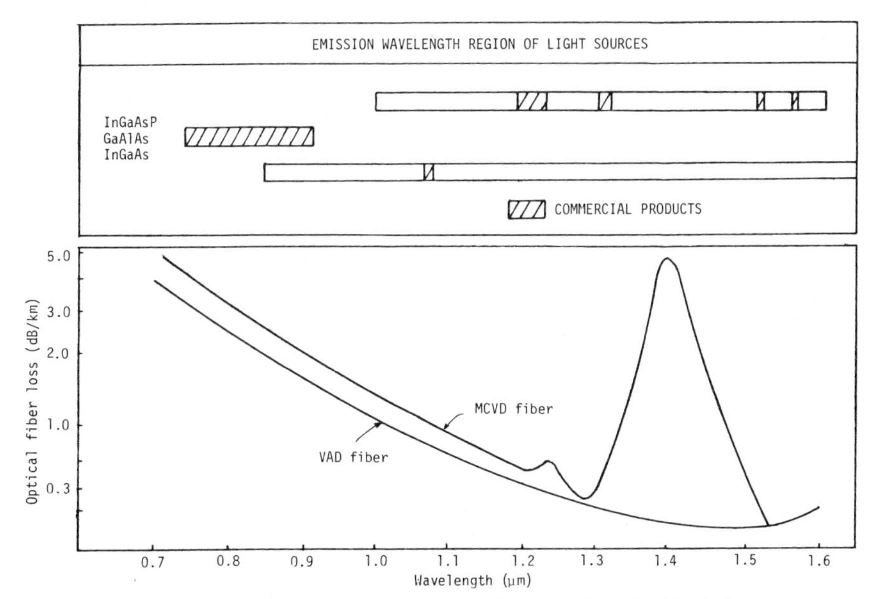

Fig. 9–2. Fiber attenuation. From I. Kobayashi [6], © IEEE, 1983.

attenuation peak at 1.4 μm is a hydroxyl (OH) absorption band. Just 1 ppm of OH produces 30 to 40 dB/km attenuation at 1.39 μm.

Figure 9–2 also shows the wavelengths of emission of some of the available light sources, which are discussed in Section 9.4.

Of concern in connection with defense programs is the behavior of an optical fiber in the presence of ionizing (nuclear) radiation. When a burst of such radiation occurs, there is a temporary generation of luminescent energy within the waveguide itself. This is followed by a temporary increase in attenuation and then by a smaller permanent amount of attenuation beyond the amount previously exhibited. The use of lead sheath around the fiber cable and its underground installation provide the best protection against this problem.

How does optical fiber attenuation compare with the attenuation of conventional coaxial cable? Figure 9–3 attempts to answer this question. To establish a comparison, however, note that the entire coaxial cable bandwidth up to a certain limit is used, whereas the optical fiber is operated at a specific wavelength at which it is modulated. Accordingly, the fiber optic attenuation appears flat as a function of bandwidth, whereas the coaxial cable attenuation increases steadily with bandwidth. Another difference is noteworthy. A relatively large outer diameter is required for a coaxial cable to exhibit low losses. On the other hand, the optical fiber has only a 0.4 mm nominal

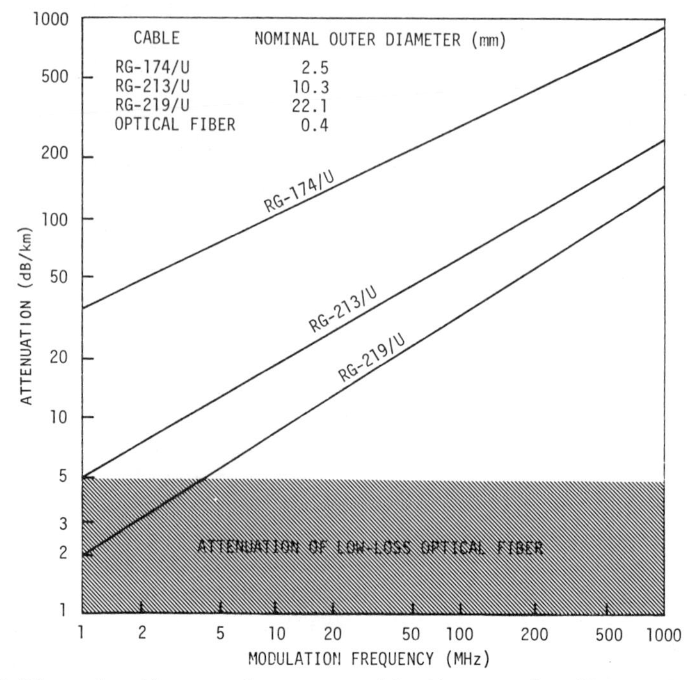

Fig. 9–3. Fiber optic cable attenuation versus coaxial cable attenuation. (Courtesy Dr. Leonard Bergstein.)

outer diameter, compared with a 2.5-mm outer diameter for the smallest (but highest loss) coaxial cable of Fig. 9–3.

9.2.2. Dispersion

If the optical energy passing through a fiber were to take only a single path along the length of the fiber, each light pulse would be received exactly as it was transmitted. However, pulse broadening occurs because of the multipath nature of wave propagation in a fiber. This problem is discussed in Section 9.3. Dispersion is measured in nanoseconds/kilometer of fiber length, and may be as low as 0.3 ns/km or as high as 10 ns/km, depending on the fiber type. A dispersion of 1 ns/km is equivalent to a 1 GHz bandwidth over a 1-km length of cable. A 10-km length of such a cable will have a 100-MHz bandwidth. The intersymbol interference resulting from dispersion thus may be a greater limitation on cable length than is attenuation.

9.3. FIBER TYPES

Optical fibers may be made, not only from glass, but also from plastic. The plastic fibers are useful only for short distances. They have diameters on the order of 1 mm or more and are well suited for use with LEDs. They are useful in rugged environments (e.g., automotive) but are not found, in general, in telecommunications applications. Plastic-clad silica fibers generally have a diameter less than 600 μm and are used with inexpensive emitters such as LEDs. Because of their large diameter, they allow a large number (>1000) of propagation paths, or modes, and thus are used only for relatively short distances.

Glass fibers have a diameter less than 200 μm. They are classified as multimode, graded index, and single mode, with the single mode fibers having a diameter on the order of only 5 μm.

An important factor in fiber optics is the index of refraction n of the fiber. It is defined as the ratio of the velocity of light in vacuum to the velocity of light in the medium. Thus if a fiber has an index of refraction of 1.5, it will carry light at a speed of $3 \times 10^8/1.5$ or 2×10^8 m/s. Since the velocity of propagation also equals $1/\sqrt{\mu\epsilon}$ where μ is permeability and ϵ is permittivity, the value of n also equals $\sqrt{\epsilon_r}$, where ϵ_r is the relative permittivity, or dielectric constant. In the foregoing example, the dielectric constant is 2.25.

Light wave propagation in a step-index fiber occurs as illustrated in Fig. 9–4. The core glass has an index of refraction of n_1 whereas the cladding has an index of refraction n_2. For total internal reflection at the core-cladding interface, the light flux must have an entrance angle θ_0 relative to the axis of the fiber that is less than the critical angle θ_c, where

$$\sin \theta_c = \sqrt{n_1^2 - n_2^2} = \text{numerical aperture (NA)} \qquad (9\text{–}1)$$

A fiber with a small NA thus requires a source with a narrow output beamwidth, such as a laser, for low coupling loss, whereas a fiber with a high NA can use an LED (broad output beamwidth) as a source. Single mode fibers tend to have low NAs while graded index and multimode fibers tend to have high NAs.

The path length of a ray that passes through the fiber's axis, called a

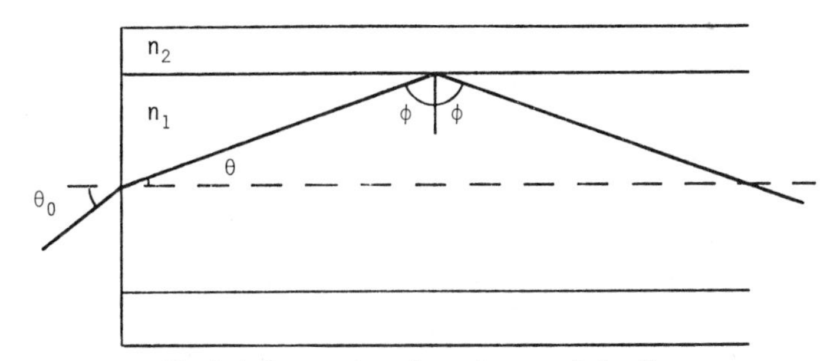

Fig. 9–4. Propagation of rays in a step-index fiber.

meridional ray, then is $L \sec \theta$, where L is the axial length of the fiber and θ is the entrance angle of the ray relative to the fiber axis. The problem of differential delay, and thus pulse dispersion, results partly from the dependence of path length on θ.

Figure 9–5 shows the cross sections of different fiber types as well as profiles of their refractive indexes. The single-mode step-index fiber has the lowest dispersion and is very useful for long-haul telecommunication applications. It works well with laser sources. The multimode step-index fiber is most useful for distances up to several km in systems using light-emitting diode

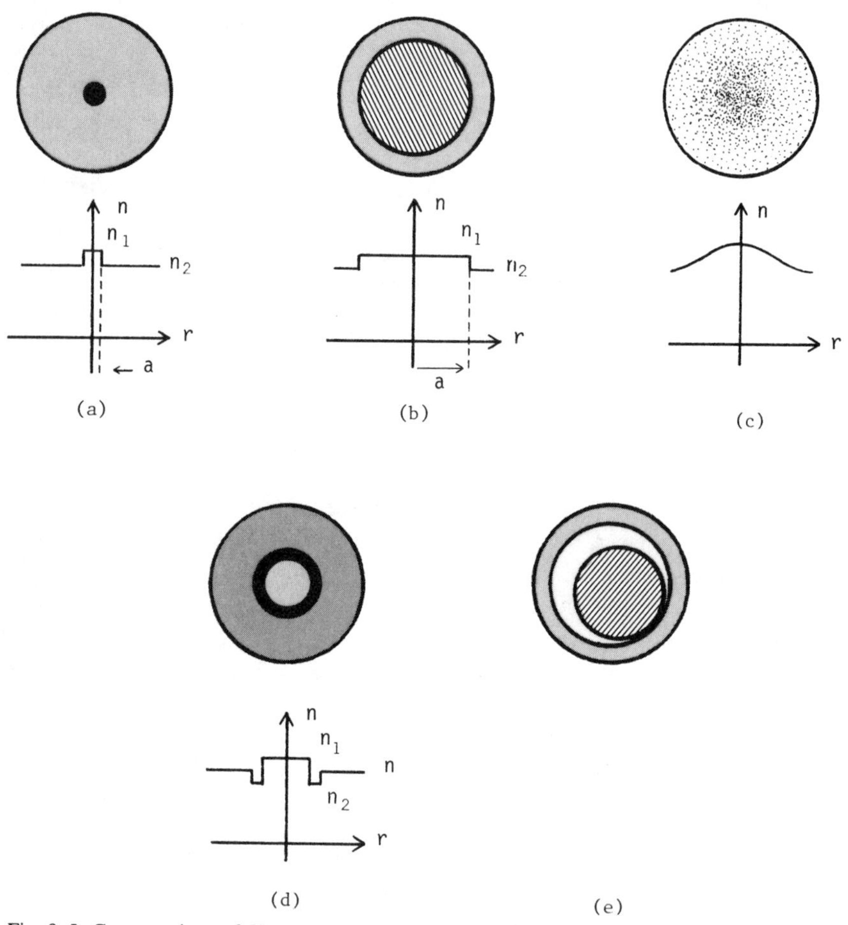

Fig. 9–5. Cross sections of fiber types and profiles of their refractive indexes. (a) Single mode step index fiber (b) Multimode step index fiber (c) Graded index (parabolic fiber (d) Double clad fiber (e) Plastic clad fiber. (Courtesy Dr. Leonard Bergstein.)

sources rather than lasers. The graded index fiber finds common use in systems from several km length to repeatered systems, such as cable television, up to 30 km or longer. The doubly clad fiber is useful in secure applications. The plastic clad fiber is used where rugged physical characteristics are important over short distances, as in automotive applications.

9.3.1. Multimode Fibers

The multimode fiber has a diameter in the 600-μm range. Figure 9–6 illustrates the light paths in a multimode step-index fiber. The group delay dispersion in a multimode fiber results from three factors: intermodal dispersion, material dispersion, and waveguide dispersion. Intermodal dispersion is the group delay spread that results from a variation in group delay between the different propagating modes. This is the dispersion that occurs because the path length of a ray that propagates at an angle θ relative to the fiber axis is $L \sec \theta$ rather than L.

Material dispersion is the group delay resulting from the nonlinear dependence of the fiber's refractive index on wavelength. This material dispersion is proportional to the spectral width of the optical source. Here again the laser is preferable to the LED because the laser has a more narrow spectrum and thus less material dispersion. Typical spectral widths are:

White light	4000 Å
LED	350 Å
Injection laser	15 Å

where 1 Å = 10^{-10} m.

Waveguide dispersion is the group delay resulting from the differing dispersions of each propagating mode. It is a relatively small effect.

The foregoing dispersion types also are found in single mode and graded index fiber, but intermodal dispersion and waveguide dispersion are less in magnitude than in the multimode fiber. Material dispersion is less for a laser source than for an LED.

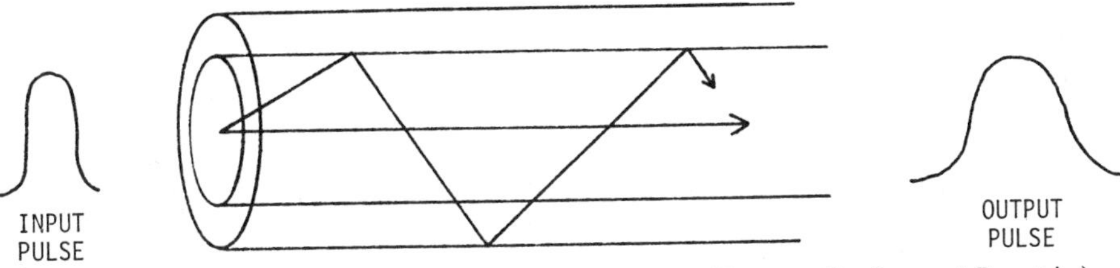

Fig. 9–6. Paths of light rays in multimode step-index fiber. (Courtesy Dr. Leonard Bergstein.)

9.3.2. Single-Mode Fibers

The single-mode fiber, also known as monomode, has a diameter of only 5 to 10 μm. This makes it more difficult to splice than the larger types. However, its dispersion is very low, allowing rates of 1 Gb/s and higher to be sent over distances well beyond 10 km. It has a very low NA, however, which means that a highly coherent high radiance source such as a laser or a laser diode must be used to drive it. Fig. 9–7 illustrates an idealized light path in a single-mode fiber.

Because the single-mode fiber has a greater data rate-distance product than the other fiber types, it is the preferred fiber for long-haul telecommunication applications. Loss as low as 0.2 dB/km has been achieved at 1.55 μm using a silica-based single-mode fiber.[7] The actual choice of wavelength, however, is governed not only by loss but also by dispersion characteristics and the availability of sources and detectors for a given wavelength. For example, a material dispersion null can be achieved at 1.30 μm, making this wavelength attractive even though the losses are somewhat higher than at 1.55 μm. A 100 Gb/s-km capability was achieved at Bell Telephone Laboratories[8] for laser sources with a 4-nm emission bandwidth over the 1.45–1.73 μm wavelength range. The diameter was chosen so that waveguide dispersion cancels material dispersion at predetermined wavelengths. The loss is as low as 0.4 dB/km at $\lambda = 1.6$ μm.

As an application example, the Nippon Telegraph and Telephone Public Corporation (NTT) installed a high capacity system, the F-400 M, over an 80 km route in 1980 with the characteristics[9] shown in Table 9–1. The repeaters operate with a −35 dBm received level and a −5.4 dBm transmitted level. Cable loss is 0.49 dB/km, but averages 0.58 dB/km because of splices every 1.3 km. The splicing technique used was fusion splicing with fiber position adjustment.

Single-mode optical fibers also are being planned into the submarine cable systems intended for installation across the Atlantic and Pacific Oceans in 1988. The Atlantic Ocean system is known as the SL Undersea Lightguide

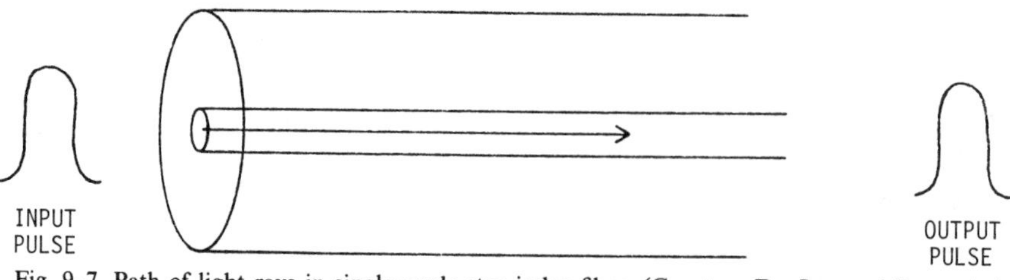

Fig. 9–7. Path of light rays in single mode step-index fiber. (Courtesy Dr. Leonard Bergstein.)

Table 9–1. NTT F-400M System Summary.

Information rate	397.200 Mb/s
Transmission rate	445.837 Mb/s
Transmission medium	Single-mode optical fiber cable
Wavelength	1300 nm
Repeater spacing	20 km maximum
Bit error rate	$<10^{-8}$ for 2400 km length
	$<10^{-11}$ per repeater
Timing jitter	<15° rms for 2500 km length
Supervisory span	280 km maximum
Power	Fed locally by commercial power

System,[10] or TAT-8, and will use single-mode fibers to carry 280 Mb/s at 1.30 μm with repeater spacings of 35 km. The use of DSI with ADPCM will allow over 35 000 two-way voice channels to be obtained. The cable will contain a total of 12 fibers. The Pacific Ocean system[11] is being planned by the Kokusai Denshin Denwa Company (KDD) and will extend from Japan to the west coast of the U.S.A.

9.3.3. Graded Index Fibers

The path of the rays in a graded index fiber is shown in Fig. 9–8. The key feature of the graded index fiber is essentially the same travel time for the various modes. Thus the effects of dispersion are minimized. This is achieved by causing the rays to travel faster near the edge of the fiber by making the refractive index lower there. The result is a light path that is almost sinusoidal. With a graded index fiber, the transmission of 1 Gb/s to a distance of one km is possible. Graded index fibers are made of pure silica and silica doped with boron, germanium, or phosphorus. The core diameter typically is 50–60 μm, with a cladding diameter of 125 μm. As many as 700 modes may propagate in such a fiber.

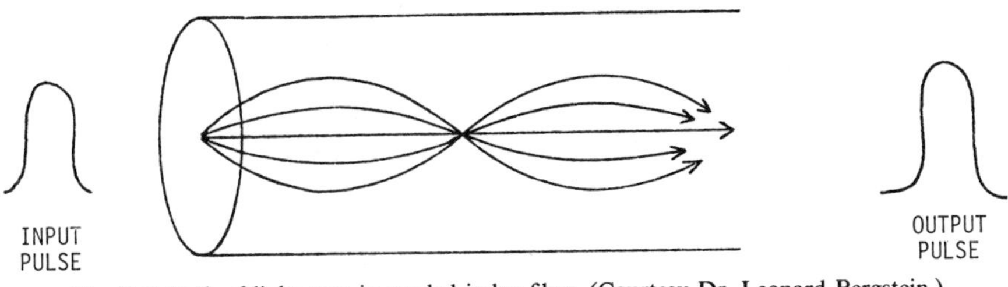

Fig. 9–8. Path of light rays in graded index fiber. (Courtesy Dr. Leonard Bergstein.)

9.4. OPTICAL SOURCES

9.4.1. Types

The most commonly used sources in optical fiber transmission are the laser and the light-emitting diode. Fig. 9–9 is a sketch of an injection laser. These devices are capable of output power levels of 5–50 mW on a continuous basis, and can be modulated at rates up to 2 GHz. Their line width is less than 20 Å. Their projected lifetime is on the order of 100 000 to 700 000 hours.[12] Lifetime depends on operating current density, with the degradation rate increasing with current density. Ambient temperature also is a factor. Degradation is a bulk phenomenon. Over a period of time, strain fields result in the formation of nonradiative recombination centers.

Injection laser diodes are available using the materials indicated[13] in Table 9–2.

The narrow beamwidth and narrow bandwidth output of the laser suit it well to long distance (> 5 km) and high data rate (> 100 Mb/s) applications.

Light emitting diodes (LEDs) are useful for short distance low data rate applications. Their beamwidth is appreciably larger than that of the laser. Table 9–3 compares lasers and LEDs with respect to several basic parameters for a wavelength of 800–850 nm. The ranges of values indicate variations among units of differing designs and output levels.

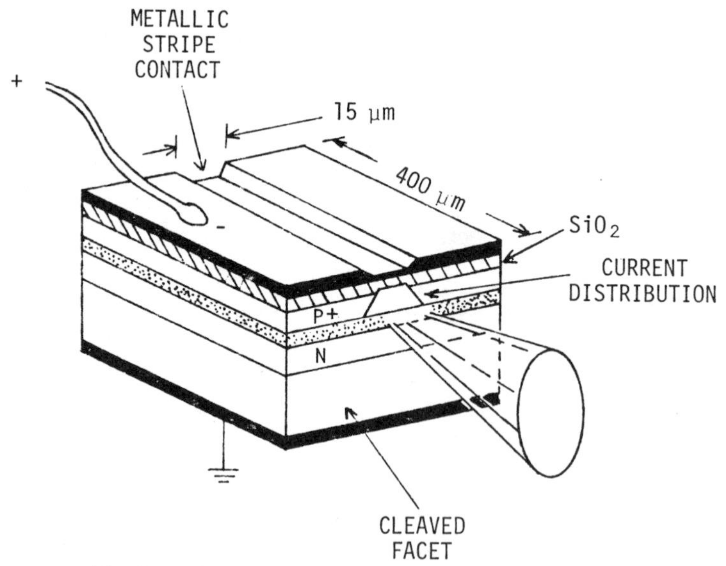

Fig. 9–9. Injection laser. (Courtesy Dr. Leonard Bergstein.)

Table 9–2. Injection Laser Diode Materials.

WAVELENGTH REGION	MATERIAL
800–900 nm	Alloys of gallium arsenide and aluminum arsenide
1300 nm	Allows of indium, gallium, arsenic, and phosphorus
1550 nm	Allows of indium, gallium, arsenic, and phosphorus

Table 9–3. Comparison of Laser and Light Emitting Diodes.

PARAMETER	LASER	LED
Power	5–50 mW	1–10 mW
Modulation	1–2 GHz	$\leq$ 200 MHz
Projected Lifetime	10^6 hours	10^6 hours
Coupling Loss to 0.14 NA fiber	−2 to −3 dB	−10 to −17 dB
Line width, $\Delta\lambda$	1–20 Å	200–400 Å

9.4.2. Modulation

Intensity modulation is used to transmit information digitally from optical sources. For the laser a threshold dc bias is used to obtain modulation response in the sub-nanosecond range. Thus two amplitude levels are used, one to represent 0 and the other to represent 1.

9.5. PHOTODETECTORS

A photodetector must have a high response to incident optical energy, an adequate instantaneous bandwidth to respond to the information bandwidth on the optical carrier, and a minimum of internal noise added to the detected signal. In addition, a photodetector should not be susecptible to changes in environmental conditions, especially temperature.

An optical detector may be a photoemitter, such as the photomultiplier; a photoconductor; or a photodiode, such as an avalanche device. The photodiodes are basically reversely biased *pn* diodes. The *p-i-n* type detectors convert optical power directly into electrical current, with responses on the order of 0.5–0.7 mA per mW. Avalanche photodetectors (APDs) provide an additional internal gain on the order of 10–100. Most telephony applications require high-sensitivity receivers and thus use APDs.

Detectors operating in the 800–1000 nm region usually are made of silicon, whereas detectors for wavelengths longer than 1000 nm are made from germanium, indium gallium arsenide (InGaAs), and mercuric cadmium telluride (HgCdTe) crystals. Fig. 9–10 shows the minimum detectable power for a bit error rate of 10^{-9}. The internal gain of the APD gives it a 7–15 dB advantage over the *p-i-n* diode. The quantum limit is for a noiseless receiver with 100% quantum efficiency. A field effect transistor (FET) or bipolar transistor is used to provide gain, as indicated in Fig. 9–10.

9.6. COUPLING OF SOURCES TO FIBERS

Figure 9–11 shows coupling loss from the light source to the fiber as a function of the numerical aperture (NA) for a step-index fiber. The injection laser, with its relatively narrow beam directed along the fiber axis, exhibits the smallest coupling loss. Edge and surface emitters have broader beams and thus higher coupling losses.

9.7. REPEATERS AND COUPLERS

Repeaters for fiber optic systems operate by detecting the optical pulses, regenerating them, and then modulating a new light source, as indicated in

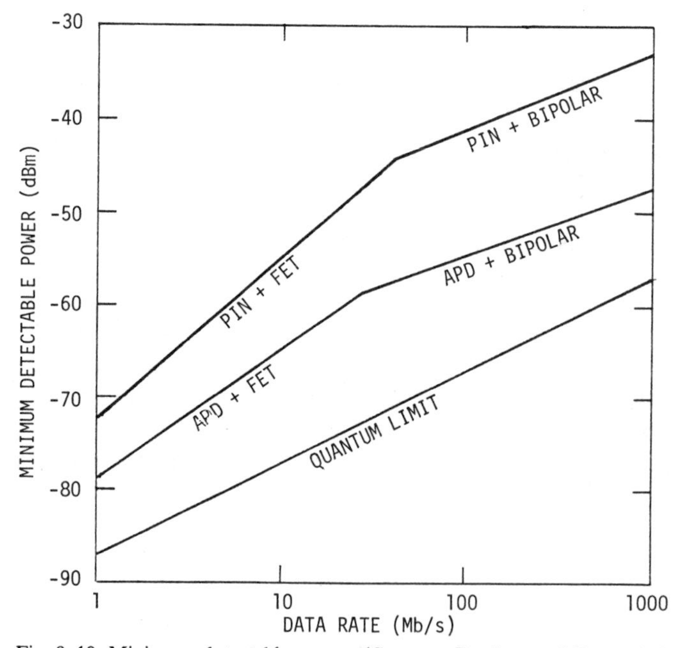

Fig. 9–10. Minimum detectable power. (Courtesy Dr. Leonard Bergstein.)

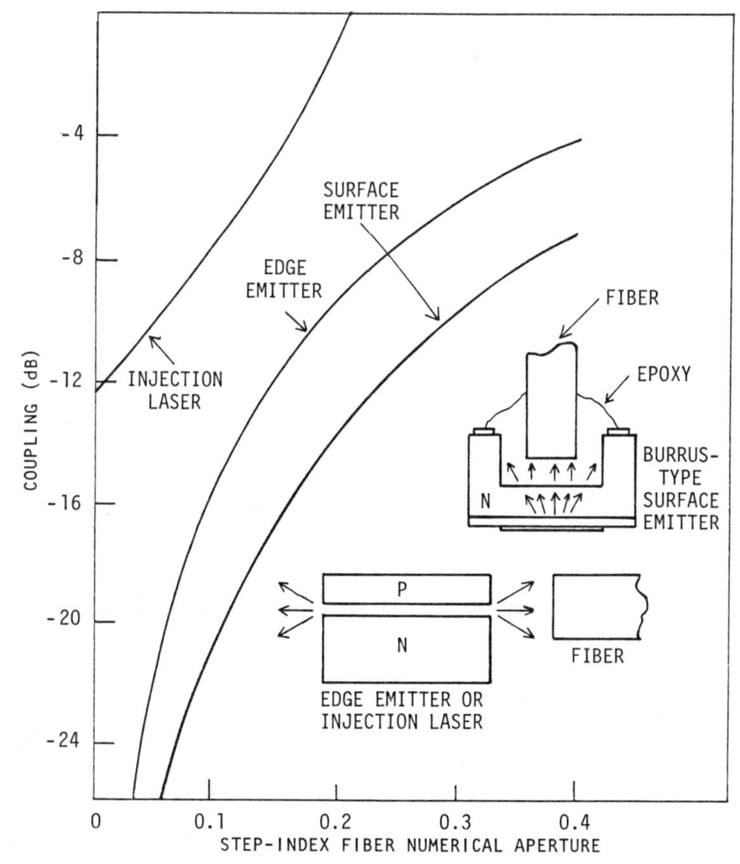

Fig. 9–11. Coupling losses. (Courtesy Dr. Leonard Bergstein.)

Fig. 9–12. Power for such a repeater must be obtained from the central office by means of wire cable paralleling the optical fiber, or may be delivered to the repeater site by the local power company, in which case an on-site battery back-up is required for use in the event of a power outage. Solar powering of the repeater is another alternative, but on-site solar arrays of sufficient size may be subject to vandalism. The distance between repeaters is governed by power budget considerations. Specifically, Figs. 9–2, 9–10, and 9–11, respectively, indicate the fiber attenuation, receiver sensitivity and coupling loss that may be encountered. Fig. 9–13 summarizes this information, illustrating the overall operating margin available for a 10^{-8} bit error rate. From this information, the maximum tolerable attenuation, and thus the maximum distance between repeaters, can be determined. For a high data

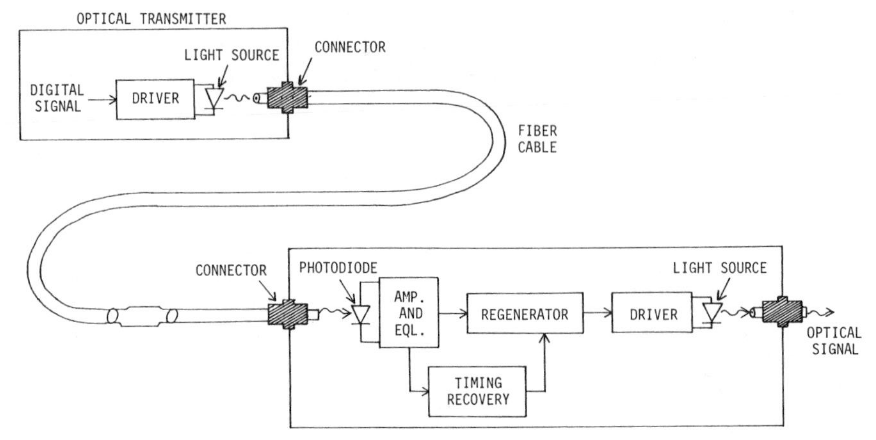

Fig. 9–12. Use of repeater in fiber optic link. (Courtesy Dr. Leonard Bergstein.)

rate system, a fiber's dispersion rating in GHz km also is a factor in establishing the tolerable distance between repeaters.

Optical link components can be built into coaxial connectors, as shown in Fig. 9–14. In addition, "tee" and "star" couplers have been devised. Their loss characteristics as functions of the number of terminals served are shown in Fig. 9–15.

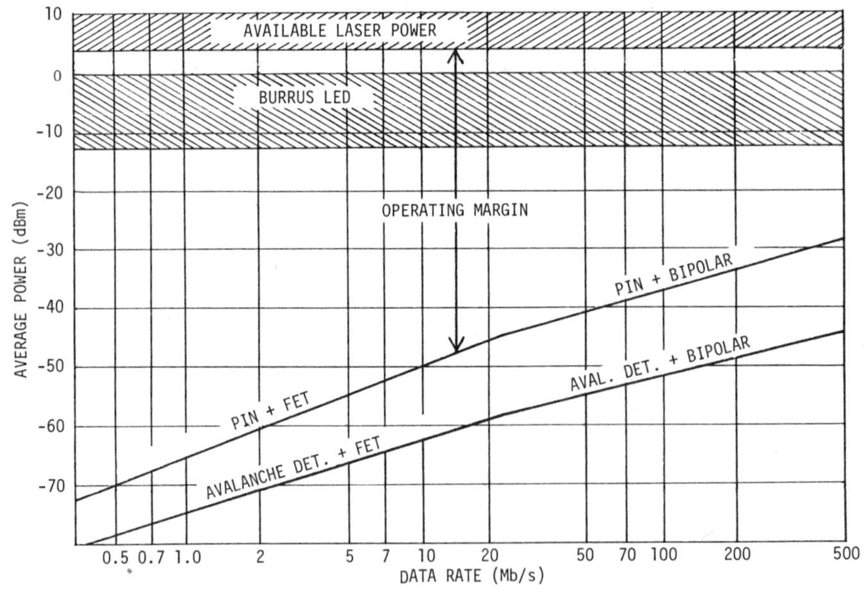

Fig. 9–13. Fiber optic link. Operational margin for b.er. = 10^{-8}.

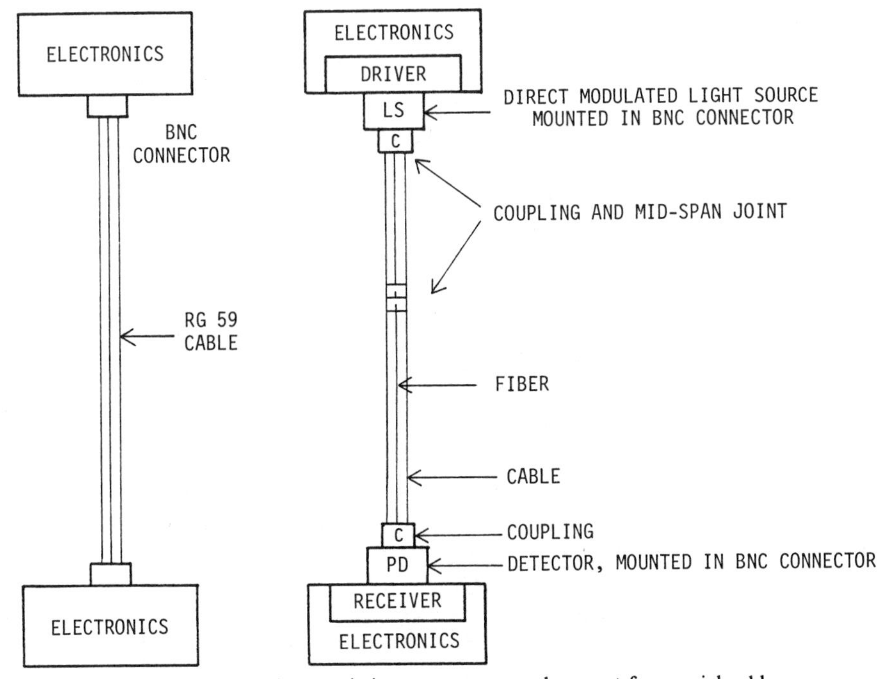

Fig. 9–14. Optical transmission system as a replacement for coaxial cable.

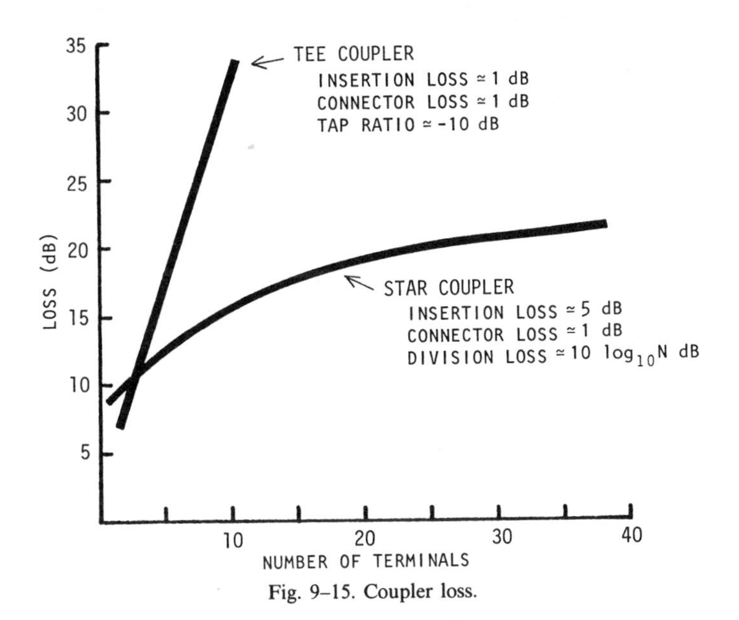

Fig. 9–15. Coupler loss.

9.8. NOISE SOURCES

Figure 9–16 illustrates the noise sources which are applicable to a direct detection optical communications receiver. The external noise sources are those which accompany the signal, and may have come from the optical source itself or may have entered along the transmission path. Internal noise includes "dark current," which is generated in the detector itself and constitutes a current that flows in the absence of any external light input. Such noise types as flicker ($1/f$) and thermal noise are common to all electronic amplifiers in one form or another.

9.9. OPERATIONAL AND PLANNED FIBER OPTIC SYSTEMS

Tables 9–4 and 9–5 list, respectively, systems being built[14] in the United States and internationally. Not all fiber optic links are compatible with the digital hierarchy, even though they are used for digital transmission. Many, instead, are designed to provide a connection between a host computer and its remote peripherals. As a result, they will be compatible with transistor-transistor logic (TTL) or emitter-coupled logic (ECL) rather than the bipolar-return-to-zero (BRZ) format used in telephony. Thus a fiber optic link's compatibility needs to be determined before it is ordered.

The Bell Northeast Corridor Network extends from Boston through New York City to Richmond, VA. It connects 19 digital switching systems and uses lasers as the sources. The transmission rate is 90 Mb/s. A 1300-nm wavelength is used.

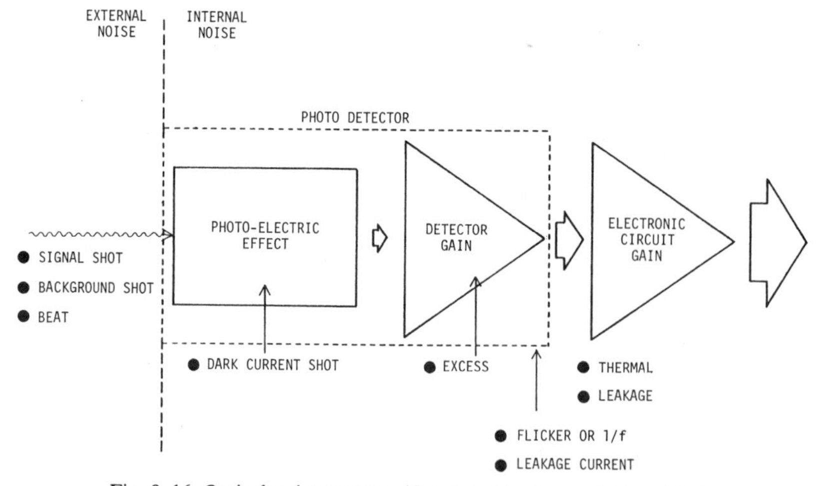

Fig. 9–16. Optical noise sources. (Courtesy Dr. Leonard Bergstein.)

Table 9–4. Long-Distance Fiber-Optic Communications Systems Under Construction in the United States

NAME OF SYSTEM	OWNER/ OPERATOR	ROUTE	TOTAL LENGTH, km	TYPICAL REPEATER SPACING, km	WAVE LENGTH, m	BIT RATE, Mb/s	ESTIMATED COMPLETION DATE	COMMENTS
Northeast Corridor	AT&T Communications	New York to Washington, D.C.	595	7	0.82	90	Complete	Built with multimode fiber; construction began 1982
		New York to Boston and Washington, D.C., to Richmond, Va.	646	7	0.82	90	Early 1984	Extensions to above system
California Lightguide Project	AT&T Communications	Sacramento to Los Angeles, with leg to Oakland and San Francisco	270	7	0.82	90	Complete	Phase I and II built with multimode fiber; phase III with single-mode fiber upgradable to 432 Mb/s; construction began June 1981
			384	7	0.82	90	Early 1984	
	Pacific Telesis		188	7	1.3	90	Early 1985	
	MCI Communications Corp.	New York to Washington, D.C.	366	24–32	1.3	405	Early 1984	Built with single-mode fiber; construction began spring 1983; follows Amtrak right-of-way. MCI has 5400 km of CSX right-of-way in eastern U.S.
FT3C-90	AT&T Communications	Camden, N.J., to Pleasantville, N.J.	83	26	1.3	90	Early 1985	Will be built with single-mode fiber; construction will begin 1984, pending approval of FCC
		Chicago to Plano, Ill.	96	26	1.3	90	Early 1985	
		Garden City to Rego Park, N.Y.	53	26	1.3	90	Early 1985	

Table 9–4. Long-Distance Fiber-Optic Communications Systems Under Construction in the United States (continued)

NAME OF SYSTEM	OWNER/ OPERATOR	ROUTE	TOTAL LENGTH, km	TYPICAL REPEATER SPACING, km	WAVE LENGTH, m	BIT RATE, Mb/s	ESTIMATED COMPLETION DATE	COMMENTS
	Electra Communications Corp. (Cable and Wireless [N.Y.] Inc. and Missouri-Kansas-Texas RR Co.)	Dallas to Houston, Texas, via San Antonio and Austin	880	35	1.3	432	Mid-1985	Will be built with single-mode fiber; construction will start 1984; route follows MKT rights-of-way; proposed extension to Kansas City
Lightnet	Southern New England Telephone CSX Corp.	Nearly every major city east of Mississippi River	8000	27	1.3	90	Early 1986	Will be built with single-mode fiber; first leg in Florida, scheduled for operation in 1984; route follows CSX rights-of-way; upgradable to 432 Mb/s
FT4E-432	AT&T Communications	Atlanta, Ga., to Greensboro, N.C.	646	26	1.3	432	Early 1986	Will be built with single-mode fiber; construction will begin 1984; FCC approval received
		Dallas to Houston, Texas	408	26	1.3	432	Early 1986	
		San Antonio to Seguin, Texas	67	26	1.3	432	Early 1986	
		Philadelphia to Pittsburgh, Pa. to Cleveland, Ohio	804	26	1.3	432	Early 1986	

SOURCE: *IEEE Spectrum*, Jan., 1984. © IEEE, 1984.

Table 9–5. International Long-Distance Fiber-Optic Communications Systems Under Construction

COUNTRY	NAME OF SYSTEM	OWNER/ OPERATOR	ROUTE	TOTAL LENGTH, km	TYPICAL REPEATER SPACING, km	WAVE-LENGTH, μm	BIT RATE, Mb/s	ESTIMATED COMPLETION DATE	COMMENTS
Belgium	—	RTT[1]	Lessive to Jemelle to Bulssonville	28	13 to 15	1.3	140	1984	Experimental operation from Lessive to Jemelle began January 1983
	—	RRT[1]	Veurne to Ieper to Roeselare to Gent to Ronse	157	13 to 18	1.3	140	1985	
	—	RRT[1]	Llege to Verviers to Stavelot	55	15 to 16	1.3	140	1985	
	—	RRT[1]	Kortrijk to Roeselare to Oostende	63	15 to 20	1.3	140	1985	
Canada	SaskTel Broadband Network	Saskatchewan Telecommunications	52 communities, major route is Yorkton to Regina to Saskatoon to North Battleford to Meadow Lake	3400	10 to 30	0.84 and 1.3	45	1984	Built with multimode fiber; 3000 km installed by end of 1983; distributes television programming as well as telephone communications

Table 9–5. International Long-Distance Fiber-Optic Communications Systems Under Construction (continued)

The Canadians pioneered the first commercial fiber-optic system, joining Regina and Yorkton, Sask., in 1980. In addition to the SaskTel Broadband Network in Saskatchewan, some 1800 km of optical fiber has been installed by Bell Canada in Ontario and Quebec, much of it as short links rather than as an integrated long-distance network. Another major fiber-optic system is operated outside the cities in Alberta by Alberta Government Telephone. Moreover, there are fiber-optic communications systems in all the other provinces of Canada except for Prince Edward Island, each one operated by the provincial telephone company. Space would not allow a comprehensive listing of all these systems.

COUNTRY	NAME OF SYSTEM	OWNER/ OPERATOR	ROUTE	TOTAL LENGTH, km	TYPICAL REPEATER SPACING, km	WAVE-LENGTH, μm	BIT RATE, Mb/s	ESTIMATED COMPLETION DATE	COMMENTS
Italy	—	Poste e Telecomunicazioni	Rome to Pomezia	31	25 to 30	1.3	140	1984	Single-mode fiber
	—	Poste e Telecomunicazioni	Milan to Turin	180	15	1.3	140	1984–86	Multimode fiber, upgradable to 560 Mb/s on some paths
	—	Poste e Telecomunicazioni	Genoa to Pisa to Livorno	220	15	1.3	140	1984–86	Multimode fiber
	—	Poste e Telecomunicazioni	Rome to Civitavecchia	65	25 to 30	1.3	140	1984–86	Single-mode fiber
	—	Poste e Telecomunicazioni	Sicily to Palermo to Catania	230	25 to 30	1.3	140	1984–86	Single-mode fiber
Japan	F-400M	Nippon Telegraph and Telephone Public Corp.	Length of Japan	2500	20	1.3	400	End of 1984	Is being built with single-mode fiber; field trial carried out 1980–82

Saudi Arabia	—	SCECO[2]	Huraymila-al to Muzahimyah	119	10.8	1.3	34	Complete	Multimode fiber
	—	SCECO[2]	Ghazlan to Damman	73	10	1.3	34	December 1983	Multimode fiber
	—	Aramco and SCECO[2]	Al-Kharj to Rhiada	95	15	1.3	8	April 1984	Multimode fiber
United Kingdom	—	British Telecom	London to Birmingham	202	8	1.3	34	Complete	Multimode fiber
	—	British Telecom	London to Birmingham	202	8	1.3	140	December 1983	Multimode fiber
	—	British Telecom	Manchester to Birmingham	170	24	1.3	140	December 1983	Single-mode fiber; upgradable to 565 Mb/s
	Project Mercury	Mercury Communications	London to Birmingham	213	21	1.3	140	June 1984	Single-mode fiber; upgradable to 565 Mb/s
	Project Mercury	Mercury Communications	Birmingham to Leeds to Manchester	297	25	1.3	140	July 1984	Single-mode fiber; upgradable to 565 Mb/s

It should be noted that in addition to the systems listed above, the British are also installing more than 4000 km of multimode fiber and 400 km of single-mode fiber along routes between 1 and 95 km long, to form a network that links virtually every major city in Great Britain.

Table 9–5. International Long-Distance Fiber-Optic Communications Systems Under Construction (continued)

COUNTRY	NAME OF SYSTEM	OWNER/ OPERATOR	ROUTE	TOTAL LENGTH, km	TYPICAL REPEATER SPACING, km	WAVE-LENGTH, μm	BIT RATE, Mb/s	ESTIMATED COMPLETION DATE	COMMENTS
International	—	RTT[1] and NLPTT[3]	Herentals, Belgium, to Breda, the Netherlands	55	21	1.3	140	1985	Multimode fiber
	—	RTT[1] and NLPTT[3]	Herentals, Belgium, to Breda, the Netherlands	55	25	1.3	140	1985	Single-mode fiber
	—	RTT[1] and NLPTT[3] DBP[4] and BTI[5]	Oostende, Belgium, to Broadstairs, UK	120	30	1.3	280	1984–87	Single-mode fiber, submarine cable; experimental operation scheduled to begin 1985, commercial operation 1987

[1] RTT is Régie des Télégraphes et des Téléphones.
[2] SCECO is Saudi Consolidated Electric Co.
[3] NLPTT is Netherlands Post, Telephone and Telegraph.
[4] DBP is Deutsche Bundespost.
[5] BTI is British Telecom International.
Fiber-optic systems are also being installed in West Germany; upon inquiry, however, the German source declined to give further information because it is proprietary to the Deutsche Bundespost.
SOURCE: *IEEE Spectrum,* Jan., 1984. © IEEE, 1984.

9.10. WAVELENGTH DIVISION MULTIPLEXING

All fiber optic transmission involves the modulation of an optical source whose unmodulated output is at a specific wavelength. In the case of a laser, this wavelength is very narrowly defined. Its counterpart in the frequency domain is a single frequency. Just as microwave and satellite systems use radio channels centered at different frequencies, so optical fibers can carry light of different wavelengths simultaneously. Each wavelength can be modulated separately from the others. This is called *wavelength division multiplexing* (WDM). Fig. 9–17 illustrates this concept. Wavelength division multiplexing is not yet in widespread use, but carries considerable promise for the future. Present limitations result from the fact that sources and detectors are only available for specific wavelengths. The couplers work on the prism basis by causing the different wavelengths to be spread out in angle and thus separated from one another or combined.

Experimental WDM systems have been built using LEDs operating at 1100, 1200, and 1300 nm. In some cases, 800 nm and 1300 nm are used. Applications include both trunks and local loops. At Elie, Manitoba, Canada, an experimental two-way system using LEDs at 830 nm for one direction of transmission and 920 nm for the opposite direction has been placed into operation. The two sources transmit bidirectionally on the same fiber using WDM couplers.[15]

9.11. FUTURE OPTICAL TELEPHONE NETWORK

Figure 9–18 illustrates an all-optical telephone network, in which not only the repeaters but also the switches operate entirely on an optical basis. Research now is underway to develop the components of such a system. Advantages will include not only extremely high switching speed, but low power operation and freedom from electrical interference of various types.

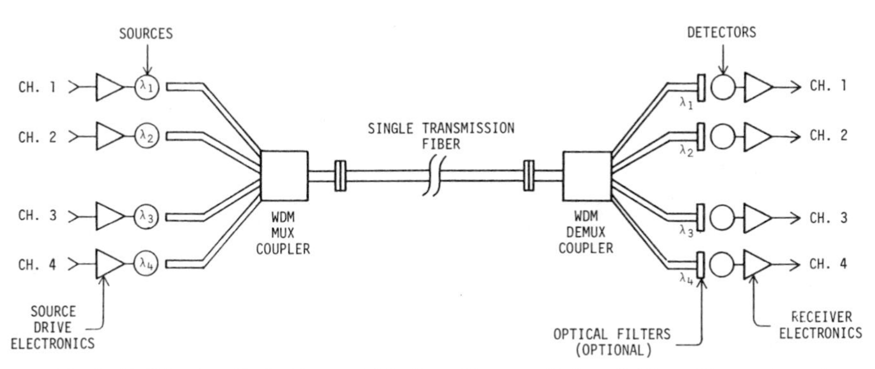

Fig. 9–17. Wavelength division multiplexing. (Courtesy Robert J. Hoss, Warner-Amex.)

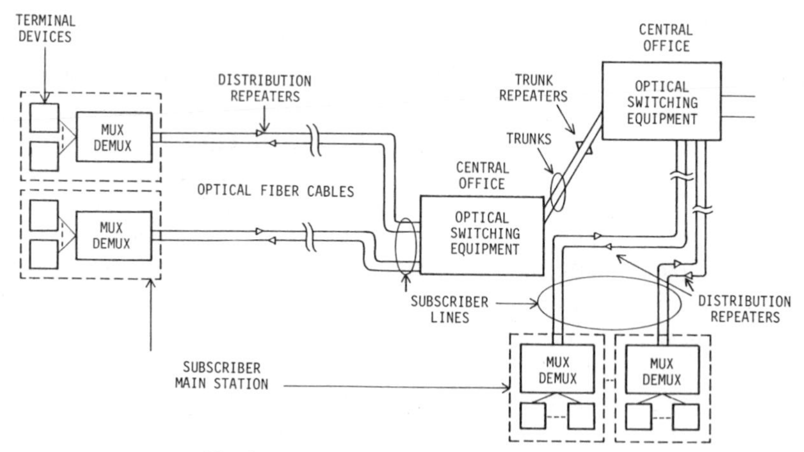

Fig. 9–18. Future optical telephone network.

PROBLEMS

9.1 An optical fiber with a 2.0 dB/km attenuation is to be used with a 10-km repeater spacing. The fiber is capable of 900 MHz-km. If digital video at 90 Mb/s per channel is to be transmitted, how many video channels can this system accommodate? If the repeaters were placed 5 km apart instead, what number of channels could be handled?

9.2 Optical fibers often are said to be immune to electromagnetic interference. Does this mean that the use of optical fibers in a telephone network will end all electromagnetic interference problems?

9.3 Lasers often are used with single mode fibers whereas LEDs often are used with graded index or multimode fibers. Will a laser work satisfactorily with a graded index or multimode fiber? Discuss.

9.4 With the common use of simple digital intensity modulation in fiber optic systems, what is the relationship of information rate to symbol rate in such systems?

9.5 Explain why a fiber optic system designed to interconnect a computer with its peripherals cannot be used directly in a telecommunications application.

REFERENCES

1. Kanzow, J., "BIGFON: Preparation for the Use of Optical Fiber Technology in the Local Network of the Deutsche Bundespost," *IEEE Journal on Selected Areas in Communications,* Vol. SAC-1, No. 3, pp. 436–439 (April, 1983).
2. Fitchew, K. D., "Technology Requirements for Optical Fiber Submarine Systems," *IEEE Journal on Selected Areas in Communications,* Vol. SAC-1, No. 3, pp. 445–453 (April, 1983).
3. Keiser, B. E., "Wideband Communications Systems," Course Notes, © 1981.
4. "Cost Comparisons for Long-Distance Communications," *Laser Focus,* Vol. 20, No. 3, p. 40 (March, 1984).

5. Li, T., "Advances in Optical Fiber Communications: An Historical Prospective," *IEEE Journal on Selected Areas in Communications,* Vol. SAC-1, No. pp. 356–372 (April, 1983).
6. Kobayashi, I., "Test Instruments for Fiber Transmission Systems," *IEEE Journal on Selected Areas in Communications,* Vol. SAC-1, No. 3, pp. 547–554 (April, 1983).
7. Nakagawa, K., "Second-Generation Trunk Transmission Technology,"*IEEE Journal on Selected Areas in Communications,* Vol. SAC-1, No. 3, pp. 387–393 (April, 1983).
8. Jang, S. J., et al., "Experimental Verification of Ultra-Wide Bandwidth Spectra in Double-Clad Single-Mode Fiber," *Bell System Technical Journal,* Vol. 61, No. 3, pp. 385–390 (March, 1982).
9. See note 7.
10. Runge, P. K., and Trischitta, P. R., "The SL Undersea Lightguide System," *IEEE Journal on Selected Areas in Communications,* Vol. SAC-1, No. 3, pp. 459–466 (April, 1983).
11. Niiro, Y., "Optical Fiber Submarine Cable System Development at KDD," *IEEE Journal on Selected Areas in Communications,* Vol. SAC-1, No. 3, pp. 467–478 (April, 1983).
12. Yonezu, H., "Reliability of Light Emitters and Detectors for Optical Fiber Communication Systems, *IEEE Journal on Selected Areas in Communications,* Vol. SAC-1, No. 3, pp. 508–514 (April, 1983).
13. Personick, S. D., "Review of Fundamentals of Optical Fiber Systems," *IEEE Journal on Selected Areas in Communications,* Vol. SAC-1, No. 3, pp. 373–380 (April, 1983).
14. Bell, T. E., "Communications," *IEEE Spectrum,* Vol. 21, No. 1, pp. 53–57 (Jan., 1984).
15. Mokhoff, N., "Fiber Optics," *IEEE Spectrum,* Vol. 19, No. 1, pp. 39–40 (Jan., 1982).

10
The Circuit Switching Environment

10.1. INTRODUCTION

There is a wide variety of circuit switching systems in the North American public telephone network. Existing systems vary from systems which were designed 50 or 60 years ago to the very latest digital switching systems. Signaling systems are the languages used by switching systems to "talk" to each other. Some signaling systems were designed to be used only on metallic conductors. Others employ signaling concepts suitable for use over derived circuits of carrier systems.

Existing systems also use various control concepts to control the establishment and disestablishment of telephone connections. These vary from the relatively simple direct, or progressive, control used in step-by-step systems to highly sophisticated stored program common control systems used in current state-of-the-art designs.

Talking paths are established through switching systems in different ways. Older designs use metallic paths which are established by electromechanical relay action. Later systems use solid state logic gates to form crosspoints in a space division matrix. Current state-of-the-art digital switching systems use time division techniques to establish talking paths repetitively many times each second.

The time required to perform each function in the establishment of switched connections varies with the system design. When systems with different timing characteristics need to "talk" to each other, selection of a signaling system, or "language," must consider the timing differences and make provision for them.

Because the preponderance of existing transmission systems is analog, new digital switching system designs must be quite versatile if they are to survive in the marketplace. They must be able to interconnect with most, if not all, of the existing switching systems via both metallic and derived circuits using both analog and digital transmission. Therefore, an understanding of the functional characteristics of existing switching systems is a prerequisite to the application of digital switching systems in the present circuit switching environment.

All Bell operating companies (BOCs) in the United States, except minority-owned Cincinnati Bell and Southern New England Telephone Company, were divested on January 1, 1984 by the American Telephone and Telegraph Company (AT&T) as a result of the settlement of a Government antitrust suit. The BOCs were formed into seven regional holding companies and continue to provide local telephone service as regulated monopolies. AT&T and other carriers provide competitive long distance services (see Chapter 13). Since AT&T still provides the bulk of such services, the circuit switching environment described in this chapter is primarily that of the former Bell System as the major portion of the North American network.

10.2. BASIC SWITCHING FUNCTIONS

The basic switching functions required to switch telephone calls are supervision, control, signaling, and provision of network paths.

10.2.1. Supervision

Supervision involves recognition of the busy or idle condition of circuits (lines and trunks) connected to the switching system. A transition from idle state to busy state is recognized as a demand for service requiring response by the switching system. A transition from busy to idle state is recognized as a termination of connection, or a "disconnect," requiring action by the switching system to restore all associated connections to idle state.

10.2.2. Control

The first control function is to recognize and respond to a caller's demand for service. The system control then prepares the system to receive the digits of the called number, or address, and returns dial tone. Upon receipt of the address digits, system control interprets the digits to determine the desired destination in terms of equipment terminations. It then examines the availability of a path through the switching system network to the equipment termination representing the destination. If a path is not found, the caller is so informed by a tone signal. If a path is found, system control causes a path to be established to the called line or to a trunk termination to another switching system. Ringing is applied to the called line. When the called line is answered by a transition from idle to busy state, ringing is discontinued. When conversation is completed, transitions from busy state to idle state are detected, and system control causes the network path to be released.

10.2.3. Signaling

Three types of signals are used in basic telephony. Supervisory signaling transmits the busy or idle state of lines or trunks as described in section 10.2.1. Address signaling is the transmission of the digits of the called telephone number by a caller to a switching system or by one switching system to another. Call progress signals are those signals transmitted to a caller to provide information relative to the establishment of a connection through the telephone network.

10.2.4. Switching Network

Each switching system has a network of talking paths used to connect lines to lines, lines to trunks, trunks to lines, and trunks to trunks. It also is used to provide access to peripheral equipment such as tone generators and digit receivers.

10.3. BASIC SWITCHING SYSTEM

Every switching system has three main functional groups of equipment and various auxiliary groups depending upon its network application. A block diagram of a typical local switching system is shown in Fig. 10–1. The Terminal Interface Group connects all lines and trunks to the switching system. The Switching Network provides talking paths, and the Control Complex controls all other functions. Service Circuits are peripheral equipment such as tone generators and digit receivers.

Switching systems have other equipment groups, such as power supplies, billing equipment, input/output devices, maintenance and administrative equipment, which support the primary switching functions but which are not shown on the block diagram.

10.4. CONTROL CONCEPTS

The three basic control concepts employed in telephone circuit switching are manual control, progressive control, and common control.

10.4.1. Manual Control

Manual control is the oldest type of telephone switching control. It is no longer used in the public switched telephone network in the United States except in a few toll switchboard applications. However, it is quite widely used for attendant positions in older types of private automatic branch exchange (PABX) switching systems.

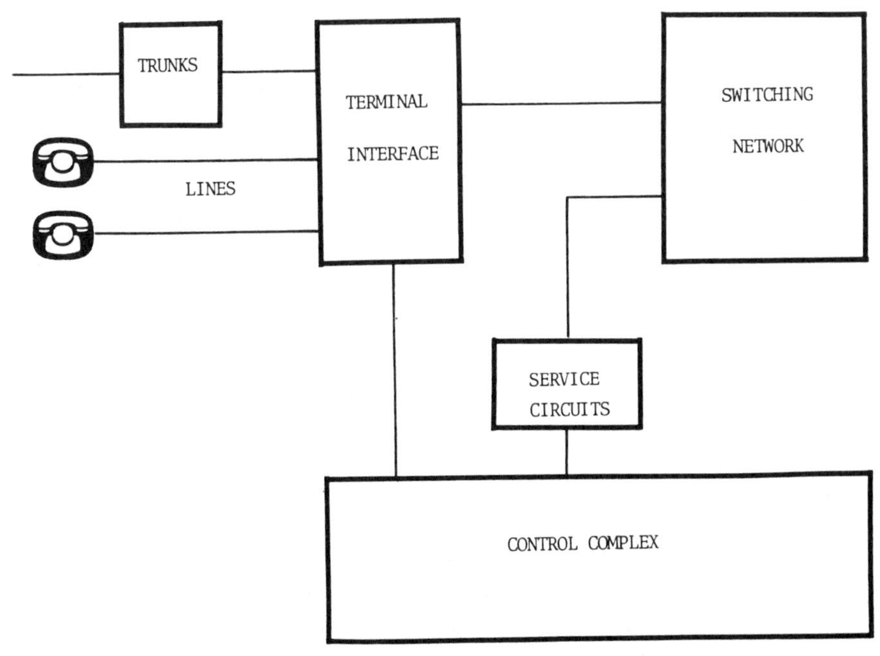

Fig. 10–1. Basic switching system block diagram.

The switchboard may use cords to connect lines and trunks, or it may be of cordless design. In either case, the switchboard and attendant perform all switching functions described in Section 10.2. Lamps (or metallic drops in older systems) indicate the supervisory state of lines and trunks. Transmission of address signals and call progress signals is by voice. Alerting, or ringing, is performed by operation of keys on the switchboard. The switching network is formed by cord circuits or contacts operated by pushbuttons.

Operation is relatively slow, depending upon the traffic load and the skill of the operator. Line and trunk capacity is very limited.

10.4.2. Progressive Control

In 1889, just 13 years after Alexander Graham Bell invented the telephone, Almon B. Strowger patented the first workable two-motion stepping switch which was controlled by pushbuttons on a telephone. After establishing a connection, the caller rang the called telephone by turning the hand crank of a magneto. The first operating system was installed at Fort Sheridan, Illinois, in 1891 and served 11 lines. The first commercial version was installed at La Porte, Indiana, the following year.

Improvements came rapidly. In 1895, the rotary dial was invented, and

Strowger's newly formed company developed an improved two-motion stepping switch quite similar to those in widespread use today. (Other designs of two-motion stepping switches are also in use.) Contacts are arranged in horizontal rows, called levels, in a semicylindrical bank. A wiper arm, attached to a vertical shaft, is actuated by dc pulses transmitted by the rotary dial on the connected telephone. Each pulse causes the shaft to elevate the wiper arm one level. After reaching the designated level, the wiper arm rotates horizontally across the selected level. Depending upon the application, the wiper arm rotates until it finds an idle circuit represented by one of ten pairs of contacts on the level, or it rotates to a specific pair of contacts directed by the number of pulses dialed. Thus, a connection is established progressively, one switch at a time, through a series, or train, of switches under direct control of the caller. Later improvements have incorporated a common control system to receive and translate digits. This is called *senderized operation.*

The stepping switch provides 100 terminations in a bank of contacts. A separate bank of control contacts has either 100 or 200 contacts depending upon whether one or two line banks are used. Control bank contacts are used to indicate busy or idle state of their associated line (or trunk) contacts.

Operation of two-motion stepping switches is controlled by multiple electromechanical relays which require meticulous adjustment. Maladjustment, caused either by wear of metal parts or by deficient maintenance techniques, can result in stepping errors. Some relays are quite large, and the resulting arcing can produce impulse noise in circuits, which causes errors in data transmission. Thus, step-by-step switching systems are not suitable for data transmission. There also is a traffic penalty associated with step-by-step switching. Each train of pulses causes a wiper arm to rotate over a row of ten pairs of contacts representing ten circuits. Large trunk groups must be divided into groups of ten for step-by-step switching. Thus, a call can encounter an all-trunks-busy condition in the group searched while trunks to the same destination are idle in other ten-trunk groups. Therefore, more trunks must be provided than would be required to carry the same volume of traffic in a single large trunk group to which each call has total access. Stepping of switches normally is at a rate of ten pulses per second, much slower than more modern switching systems.

10.4.3. Common Control

Although common control systems represent current state-of-the-art concepts, existing systems use different design technologies with different characteristics and capabilities. All common control systems, however, perform substantially the same functions.

10.4.3.1. Common Control Functions. Call processing in a local central office switching system typifies the common control functions performed in any switching system. The common control receives information from the terminal interface equipment that a caller's line has transitioned from idle (on hook) to busy (off hook), indicating a demand for service. The calling line is identified for billing purposes, and its equipment designation is transmitted to the common control. The common control establishes a path through the switching network from the calling line to an idle digit receiver and bridges a tone generator to the line to send dial tone to the caller. Upon receiving the first digit of the called party address, the dial tone is disconnected, and the digit is registered. The remaining digits of the address are received, the address is read by the common control, and the digit receiver is disconnected.

The common control translates the address into the designation of the equipment termination of the called line (or trunk if it is an interoffice call) and tests the called line to determine whether it is idle or busy. If busy, the common control connects the calling line to busy tone through the switching network and disconnects that connection when the calling party goes "on hook."

If the called line is found to be idle, the common control searches for an idle path through the switching network to its equipment termination. If one is not found, the calling line is connected through the network to reorder tone, and the common control proceeds in the same manner as if the line were busy. If an idle path is found, that path is reserved, the called line is connected to a ringing generator, and the calling line is connected to audible ringing tone. When the called party answers, the terminal interface equipment detects the off-hook condition and transmits that information to the common control. The common control then disconnects the ringing generator and audible ringing tone and connects the calling line to the called line through the network path previously reserved.

The network path remains established for the duration of the call. When either party goes on hook, that transition is detected by the terminal interface equipment, and the common control is informed. The common control disconnects the network path and restores the lines to idle state. If one party remains off hook, thereby indicating a possible demand for further service, the common control correctly interprets that condition and proceeds to serve the additional call.

If the original call were to a destination which required a connection through another switching system, the common control would search for an idle trunk to the distant (or an intermediate) office. If found, it would cause a transmitter, or sender, to send the address digits to the distant office and connect the calling line to the trunk. If not found, it would connect the calling line to busy tone and proceed as if it were a busy line. On-hook detection and disconnect would occur in substantially the same manner as for a local connection.

10.4.3.2. Types of Common Control. There are three basic types of technology used in common control systems, each reflecting the state of the art at the time of original design.

Electromechanical common control uses electromechanical relays which are hard-wired together in a fixed configuration to perform the control functions. Decisions are made as the result of the operation of particular relays. Variations in the operating program are implemented by wiring options. Many of these systems have not been modernized because of the difficulty and cost of changing the call-handling program. Because of the characteristics of electromechanical relays, timing functions are directly related to the operating time of the relay solenoids. Very few, if any, new systems of this type are being procured, but many will remain in service for a long time. Thus, new digital switching systems must be able to interact with them.

Wired logic common control systems use solid state technology. The solid state components are hard-wired together in such a way as to perform specified set of functions predicated upon recurring routines or contingent upon a finite set of inputs. Such systems require less space, less power, and less maintenance than electromechanical control systems. Memory is used for local data base information and for temporary record of calls being processed. A significant disadvantage is the necessity to modify the printed circuit boards in order to change the generic operating programs. While electromechanical systems can operate satisfactorily in uncontrolled temperature environments, wired logic systems generate and concentrate heat into a smaller space and generally require temperature controls.

Current state-of-the-art switching systems are designed around stored program common control systems. In these systems, both the generic operating program and the local data base information are stored in memory. The generic program normally is stored in read-only memory (ROM), while the data base information is stored in programmable random access memory (RAM). Call set-up information is stored in a scratch-pad memory. Some systems use a combination of wired logic and stored program control systems. Stored program control systems have lower capital and maintenance costs for large switching systems. Maintenance diagnostics are simplified and more elaborate, and power and space requirements are less than for other types of control systems. As in the case of wired logic systems, however, heat is concentrated into a small area, and environmental controls are needed to maximize the service life of components.

10.5. SIGNALING

Signaling was identified in Section 10.2.3 as one of the basic switching functions. Signaling concepts may be categorized in several ways. They may be classified as supervisory, address, and call progress signaling, as they are discussed in this section. They also may be referred to as customer line signaling and interoffice signaling, and the latter can be subdivided into ac and dc methods, into inband and out-of-band signaling, or, more recently, into in-channel and common channel systems.

10.5.1. Subscriber Loop Characteristics

The great majority of telephone subscriber lines consists of twisted pairs of copper alloy conductors called *tip* and *ring*. The transmission characteristics are determined by four electrical properties. These are the series resistance of the conductors, the inductance of the conductors, the capacitance between the conductors, and the leakage resistance, or conductance, between the two conductors. Applying the principles of Thévenin's Theorem,[1] a complex network can be represented by a simple equivalent circuit. Thus, a transmission line can be represented by a series of T-sections as shown in Fig. 10–2.

a.

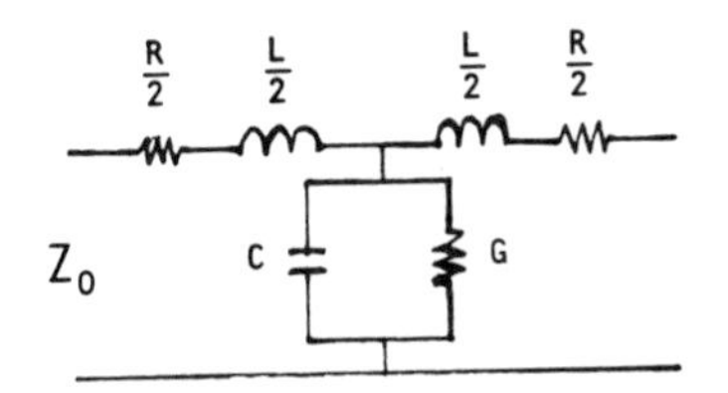

b.

Fig. 10–2. Subscriber loop characteristics.

Each T-section has the four properties of resistance, inductance, capacitance, and conductance, as shown in Fig. 10–2(a). Each combination of series and shunt properties comprise an impedance, which is repeated as additional T-sections are added to the line, as shown in Fig. 10–2(b).

Since the characteristic impedance of a transmission line of infinite length terminated in its characteristic impedance is expressed by the equation $Z_0 = \sqrt{Z_1 Z_2}$, the series impedance of a single section may be written as

$$Z_1 = R + j\omega L \tag{10–1}$$

The impedance of the parallel shunt branch is

$$Z_2 = \frac{1}{G + j\omega C} \tag{10–2}$$

Here, per unit length of line, R is the resistance in ohms, L is the inductance in henries, C is the capacitance in farads, G is the conductance in mhos, and $\omega = 2\pi f$, where f is the frequency in Hertz.

If all transmission lines were sufficiently long to approximate their characteristic impedance, and if all terminal equipment were to have the characteristic impedance of the lines, the transfer of power would be optimum. In the outside telephone loop plant serving subscribers, however, there are many variables. Loop resistance may vary from near zero to about 2000 ohms or up to about 3000 ohms with range extenders. On some long loops, loading coils insert lumped inductance at regular intervals to compensate for the additional capacitance. The use of different wire gauges causes impedance mismatches which affect the return loss. There may be one or more branch circuits, known as *bridged taps,* as shown in Fig. 10–2(b), which affect the impedance of the line.

Aerial cable is highly susceptible to power line interference which may reach magnitudes of up to 50 Vrms longitudinal and 5 Vrms metallic. Such interference contains primarily the odd harmonics of 60 Hz and contributes maximum noise between 420 and 660 Hz. Power noise, with other noise, can reach levels of about 20 dBrnc or more. There may be a difference in the ground potential at the telephone company central office and the subscriber's premises of as much as 3 V. Finally, lightning surges may induce voltages up to 5000 V peak and currents up to 1000 A peak with a maximum rise time on the order of 10 μs and a decay time of 10 μs or more.

All such characteristics impair the transmission, not only of voice and data signals, but also of signals used to control the telephone network. The effect of these impairments is increased when multiple sets of terminal equipment are bridged to the line at the premises of the subscriber. Telephone signaling systems must be designed to function in the presence of such impairments, and critical standards have been established to enable them to perform their functions satisfactorily.

10.5.2. Supervisory Signaling

Local supervisory signaling is used to determine the busy or idle condition of the subscriber's line. When the line is idle, it is said to be *on hook,* and a busy line is said to be *off hook.* When a subscriber desires to originate a call, the lifting of the handset causes a connection between the tip and ring conductors through the telephone set. The resulting current flow is detected in the central office and is interpreted as a demand for service. When a call has been completed, the subscriber returns the handset to its resting position, opening the tip and ring conductors. The central office interprets the on hook condition as idle and breaks down the connection. Most supervisory signaling systems supply negative battery across the tip and ring conductors at the central office; i.e., the ring conductor is more negative than the tip conductor. A typical battery voltage is −48 Vdc. Subscriber carrier systems provide ac signaling between carrier terminals and dc on the subscriber end.

Interoffice signaling systems may employ either dc or ac to provide supervision on trunks between switching systems. When dc signaling is used, it functions substantially in the same way as on subscriber lines except that the on-hook and off-hook conditions are established under the control of the switching equipment at each end of the circuit. Most interoffice trunks in the toll network are 4-wire circuits with a separate transmission path in each direction. These trunks generally use ac supervisory signaling. In such systems, a single frequency 2600 Hz tone is placed on each side of the trunk when the trunk is idle, or on hook. An off-hook condition is identified by removing the tone.

Both local and interoffice supervisory signaling are described in more detail in the following subsections.

10.5.2.1. Subscriber Line Supervision. There are two categories of supervision on subscriber lines: loop start and ground start.

Supervision of loop-start lines is described below. The negative central office battery is provided to the ring conductor through a current sensor, and the tip conductor is grounded. When the hookswitch in the telephone set is closed, current flow is detected by the sensor, and the line is interpreted as being off hook. No current flows in the on-hook state. Loop current varies inversely with the external circuit resistance. A loop with zero resistance will draw a minimum of about 40 mA, while a loop at maximum resistance range will draw a minimum of about 20 mA. Some types of central office terminating circuits can operate with higher loop resistance and lower line current. The maximum loop current should not exceed approximately twice the minimum values. External loop resistance higher than 10 000 ohms must be recognized as a disconnect.[2]

Lines which serve switching equipment at the subscriber's premises generally operate on a ground-start basis. The central office provides battery, negative with respect to ground, through a sensor to the ring conductor of the subscriber line. The tip conductor at the central office is open in the idle state.[3] To initiate an outgoing call, the subscriber terminal equipment seizes the circuit by grounding the ring conductor through a resistance. The central office detects current flow in the ring conductor and, when ready to receive address digits, responds by connecting the tip conductor to ground and returning dial tone. The subscriber terminal equipment detects the central office tip ground, changes to loop supervision, and sends address digits. Incoming calls are detected by the terminal equipment by detection of ringing signals or by detection of central office tip ground. Ground-start supervision has three major advantages over loop-start supervision for terminal equipment which dials automatically. First, the central office tip ground can be used as a start-dial signal, eliminating the need for dial tone detection circuitry. Secondly, removal of central office tip ground provides a positive disconnect signal to the terminal equipment. Thirdly, it tends to reduce the effect of simultaneous seizures.

Both loop-start and ground-start lines are susceptible to power line induction. If the battery supply for loop-start lines is grounded, the sensor should be balanced to provide immunity to 60 Hz induction. Ground-start arrangements are inherently unbalanced, and some method of assuring induction immunity should be applied.[4]

10.5.2.2. Loop Supervision on Trunks. There are several types of loop supervision used on trunks between switching systems. In all cases, dc signaling states are superimposed upon the same conductors used for voice transmission. To assure compatibility, the electrical characteristics, signaling protocols, and sensitivities must be maintained within certain ranges. The maximum loop resistance, or working limit, is dependent upon the sensitivity of a particular trunk circuit to dc pulsing and to the steady state dc signal. The working limit for any trunk group is that for the least sensitive trunk circuit at either end. Loop signaling trunk circuits should function with conductor leakage resistance as low as 30 000 ohms.

A reverse battery trunk, in its simplest form, is a one-way trunk which sends opens and closures from the originating end and battery and ground reversals from the terminating end. The originating end may substitute aiding battery and ground during dial pulsing to increase the working limit of a pulsing-limited trunk. At the originating end, the outgoing trunk circuit indicates an idle (on-hook) condition by maintaining an open condition with at least 30 000 ohms resistance across the tip and ring conductors. An off-hook condition is signaled by bridging the tip and ring conductors with not more than 500 ohms resistance. At the terminating end, the incoming trunk circuit signals an on-hook condition by connecting the tip conductor to ground and the ring conductor to −48 Vdc. To change to an off-hook condition, the incoming trunk circuit reverses the polarity of the tip and ring conductors. During the reversal, the electrical state is not defined. Therefore, the transition period should be kept as short as possible but not longer than 5 ms. To assure versatility, both incoming and outgoing trunk circuits should have options for reversing the tip and ring polarities.

There are several variations of reverse battery trunks. The high-low, reverse battery trunk substitutes a high-impedance polarity detector for the open state of the outgoing trunk circuit. In the on-hook state, the high resistance should be at least 30 000 ohms ±10%. The combination of this resistance in parallel with the leakage resistance of 30 000 ohms is sensed as 15 000 ohms by the incoming trunk circuit at the terminating end. Reverse battery, high-low trunks operate the same way, except that the originating end signals by polarity reversals, and dial pulsing is not permitted. High-low, reverse battery trunks can be arranged for dial pulsing in both directions.[5] Other types of dc signaling, such as composite (CX) and duplex (DX) systems, are in current use on metallic trunks.

10.5.2.3. Tone Supervision on Trunks. Inband signaling systems on carrier-derived channels employ single-frequency (SF) signaling units which generate a tone in the voice frequency band. The SF tone is injected into each transmit side of the 4-wire equivalent path. The presence or absence of tones is transformed into dc signals to and from the switching equipment trunk circuits. In modern systems, the single frequency is 2600 Hz. Formerly, other frequencies in the voice band were used. Some SF units could be arranged to send one frequency in one direction while another frequency was used in the other direction, thus enabling the units to be used on 2-wire metallic facilities as well as 4-wire facilities. Conventionally, the 2600-Hz tone is on when the trunk is on-hook and is off when the trunk is off-hook.

A 2600-Hz band elimination filter in the receive path blocks the tone so that the calling party does not hear the tone when his receive path is on hook. This permits call progress signals, such as busy tone and audible ring, to be heard by the calling party. The filter is inserted when tone is detected.

A significant problem with SF supervision is its susceptibility to mutual interference between voice transmission and signaling. SF units are subject to false operation, known as "talk-off," from voice sounds which are near the signaling frequency. The selection of 2600 Hz for the signaling frequency tends to minimize the probability of talk-off, but it does occur. Protective measures include specification of a minimum sustained duration of signaling tone to operate the unit and detection of voice-frequency energy other than the signaling frequency to block operation.[6]

10.5.2.4. E & M Lead Control. Many trunks use E & M lead signaling interfaces between switching system trunk circuits and their associated signaling equipment. The E & M designations are derived from their identification on circuit drawings. E & M leads normally are used only within a building and do not appear in the outside plant. They are separate from the transmission paths. E & M leads control the on-hook and off-hook supervision indications which do appear in the transmission paths. The M lead controls the signals from the switching equipment trunk circuit to the signaling unit, and the E lead controls the signals from the signaling unit to the switching equipment trunk circuit. Some have supposed that the lead designations were derived from the key letters in trans*M*it and rec*E*ive, and this is a good memory crutch regardless of its validity. See Fig. 10–3.

Originally, E & M lead signaling circuits used a single lead for each direction of transmission with a common ground return. Battery was supplied to both leads at the trunk circuit. In the signaling unit, the M lead was connected to ground through the windings of a relay while the E lead was connected to ground through the normally open contacts of another relay. Signaling protocol requires ground on the M lead when the trunk circuit is on hook and battery when the trunk circuit is off hook. The transition from ground to battery causes the relay in the signaling unit M lead to operate, passing the off-hook indication over the trunk to the distant switching equipment. The off-hook indication causes ground to be applied to the signaling circuit E lead, which completes a path through a relay in the trunk circuit E lead to battery. This causes the trunk circuit to recognize the off-hook state of the trunk. This is called a Type I interface. A disadvantage of this interface is the high return current through the office grounding system as a result of the battery for both leads being supplied at the trunk circuit. Additionally, the unbalanced signaling could cause interference to some solid state equipment, but the effect can be minimized by keeping the current below 50 mA through use of current limiters.

Five types of E & M lead signaling interfaces have been standardized. The original Type I interface is used with electromechanical switching equipment. The Type II interface, used with the No. 4 ESS toll switching system, uses one pair to signal in each direction between the trunk circuit and the signaling equipment. The signaling unit supplies ground on the M lead and battery on the SB (signal battery) lead while those leads are connected to

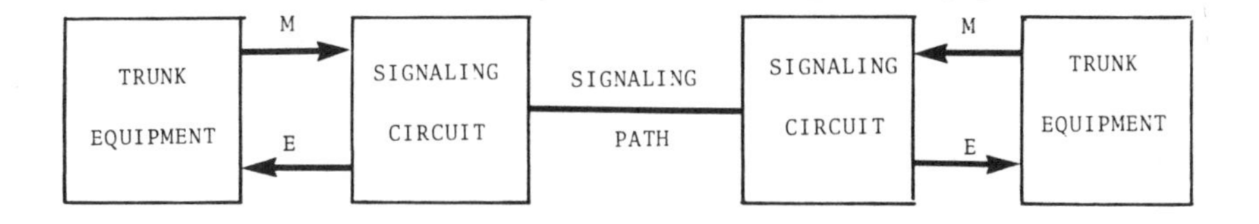

	SENT SIGNALING STATES					
	TRUNK TO SIGNALING CIRCUIT			SIGNALING TO TRUNK CIRCUIT		
TYPE	LEAD	ON-HOOK	OFF-HOOK	LEAD	ON-HOOK	OFF-HOOK
I	M	GROUND	BATTERY	E	OPEN	GROUND
II	M	OPEN	BATTERY	E	OPEN	GROUND
III	M	GROUND	BATTERY	E	OPEN	GROUND
IV	M	OPEN	GROUND	E	OPEN	GROUND
V	M	OPEN	GROUND	E	OPEN	GROUND

Fig. 10–3. E & M lead control signals.

normally open relay contacts in the trunk circuit. For the E lead, battery and ground are supplied at the trunk circuit, and the E and SG (signal ground) leads are connected to normally open relay contacts in the signaling equipment. E-lead protocol is the same as in the Type I interface, but the M-lead protocol requires the M lead to be open instead of grounded in the on-hook condition. The Type II interface greatly reduces the probability of interference to sensitive electronic devices, but it requires interface conversion when a Type II trunk circuit is used with a Type I or Type III signaling unit.

The Type III interface, used with the No. 1, 2, and 3 ESS family of switching equipment, operates with the same protocol as Type I. The E-lead arrangement is identical to that of the Type I interface, but the M lead uses two additional leads, SB and SG, for battery and ground. This provides complete separation of switching and signaling power supplies, reducing the likelihood of interference.

Type IV interface is an improved Type II arrangement with a fully symmetrical configuration but using a slightly different M-lead protocol. The signaling equipment supplies battery and ground on the M and SB leads, respectively. At the trunk circuit, the M and SB leads are connected to normally open relay contacts to signify an on-hook state. Those contacts are then closed, connecting ground to the M lead, in the off-hook state. The E-lead configuration is identical to that of the Type II interface.

The CCITT E & M lead interface, designated Type V by AT&T, is a 2-wire, symmetrical E & M lead arrangement. The signaling unit supplies battery to the M lead which is connected to ground through normally open relay contacts in the trunk circuit. Thus, the M-lead protocol requires the trunk circuit to send open while on-hook and ground while off-hook as in the Type IV interface. The E-lead protocol is the same as all other types.

Trunks between switching systems collocated in the same building can operate without signaling circuits if their E & M trunk circuits are compatible. Trunk circuits with Type II, IV, or V interfaces can be connected back-to-back metallically by cross-connecting E leads to M leads and, for Types II and IV, SB leads to SG leads. Trunk circuits with Type I or III interfaces can be interconnected through auxiliary links. E & M appliques can be used to interconnect trunk circuits with different types of signaling interfaces. E & M lead protocols are illustrated in Fig. 10–3.[7]

10.5.2.5. Control of Disconnect. Switched connections normally are held and disconnected under control of the calling party. This concept is commonly known as *calling party control.* Disconnect signals, however, may be initiated by either the calling party or the called party by returning to an on-hook condition. Switching equipment must be able to differentiate on-hook signals of different duration. During established calls, temporary on-hook indications can be caused by "hits" on a transmission facility, a legitimate flash to exercise a feature or recall an operator, or a disconnect followed by a reseizure. Hits should be ignored, and a flash and a disconnect should be treated differently. The timing requirements to differentiate hits, flashes, and disconnects are shown in Table 10–1. Any on-hook condition which endures longer than hits or flash timing is interpreted as a disconnect. Timers should begin timing as soon as possible after the on-hook state is detected by the line or trunk circuit.

Treatment of a disconnect depends upon the type of switching equipment, whether the calling or called party disconnects first, whether the disconnect affects a line circuit or a trunk circuit, the type of trunk circuit, and whether the other party remains off-hook. Time must be allowed to restore the line or trunk to idle state before making it available for reuse. This time varies from immediately to about 37 seconds depending upon the variables listed above. If both parties have disconnected, however, the guard timing to assure

Table 10–1. Disconnect Timing Requirements.*

DURATION OF ON-HOOK SIGNAL	INTERPRETATION
Timing without flash	
0–200 ms	hit
200–400 ms	hit or disconnect followed by reseizure
>400 ms	disconnect
Timing with flash	
0–200 ms	hit
200–300 ms	hit or flash
300–1100 ms	flash
1100–1550 ms	flash or disconnect followed by reseizure
>1550 ms	disconnect

trunk release before reseizure is measured in milliseconds according to the type of switching equipment and the round trip signaling time between switching offices.[8]

10.5.3. Address Signaling

When the digits of a called telephone number are transmitted to a switching center, the switching equipment must be prepared to receive and interpret them. It is not cost-effective for a switching system to have digit receivers permanently wired to all lines and trunks. Therefore, that equipment is grouped into a pool of common equipment which is connected to each line or trunk only so long as necessary to receive address digits. It then is disconnected and made available for use on another line or trunk. The switching center must signal the calling subscriber or distant switching center when it is ready to receive digits. The calling subscriber is informed by placing dial tone on the line. For trunk signaling, the distant office is notified by protocols described in Section 10.5.3.4. Subscribers transmit address digits to the central office switching equipment in either of two ways.

10.5.3.1. Dial Pulse Signaling. Dial pulses are transmitted by means of a rotary dial which opens and closes the tip and ring conductors at specified intervals. Some telephones use pushbuttons to automatically generate dial pulses, and some private automatic branch exchange (PABX) equipment automatically generates dial pulses toward the central office. As the line conductors are opened and closed, the line current is interrupted, deenergizing and then reenergizing a relay in the central office digit receiver.

The number of interruptions corresponds to the numerical value of the digit being dialed except that ten interruptions are used to designate the digit zero. To ensure correct interpretation of dial pulses (and all other address signals) in the presence of transmission line distortion, timing specifications are stringent. Pulsing speed is nominally ten pulses per second with a tolerance range of 8–11 pulses per second. The pulse period, therefore, is nominally 0.1 sec and comprises the open-circuit, or *break,* portion and the following closed-circuit, or *make,* portion. The structure of dial pulses in a digit train is illustrated in Fig. 10–4. The ratio of the break duration to the pulse period, called the *percent break,* is specified as a nominal 61% with a tolerance range of 58–64%. The interdigital on-hook interval must be long enough to distinguish between two series of pulses representing digits but short enough so that it is not interpreted as a disconnect. When pulsing into a step-by-step office, the interdigital interval must be long enough to allow the selector switch to step to the maximum level and also to allow time to condition a relay in the next switch. For manual pulsing by subscribers or operators, the interdigital interval is specified as 300 ms to 3 seconds except that the minimum time for step-by-step systems is 700 ms. (The normal human tendency in using a rotary dial is to space the digits at least one second apart.)[9]

Automatic outpulsing from PABX and other customer-premises equipment should conform to more stringent requirements. Pulsing should be at the rate of 9.8–10.2 pulses per second with a 58–62% break. The interdigital interval should be between 600 ms and 3 seconds with a desirable maximum of 1 second. When pulsing into step-by-step equipment, the minimum interdigital interval should be 700 ms. The first digit outpulsed should be delayed for 70 ms after receipt of dial tone to allow the dial pulse receiver to stabilize but must be outpulsed within 10 seconds.[10]

To overcome the distortion which is always present on cable pairs, central

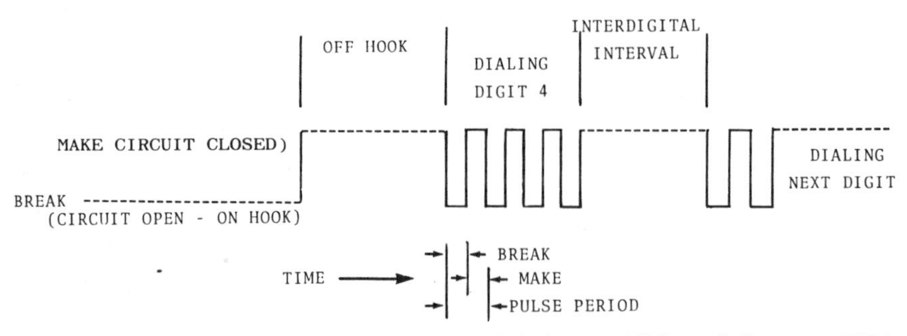

Fig. 10–4. Dial pulse address signals. (© American Telephone and Telegraph Company, 1980.)

described above. Receivers should be able to register dial pulses at speeds between 7.5 and 12 pulses per second under all acceptable conditions for outside cable plant.[11]

Dial pulse signaling also is used on some interoffice trunks, particularly those involving older switching equipment, such as step-by-step. Dial pulse transmitters should delay sending dial pulses for at least 70 ms after receiving a start-dial signal, and the distant office should not register any pulses for 30–70 ms after sending the start-dial signal to avoid registration of transients. Transmitters should send pulses at the rate of 9.8–10.2 pulses per second with a 58–62% break and an interdigital interval of 600–700 ms. Digit receivers should be able to register dial pulses at a rate of 7.5–12 pulses per second with a 40–80% break and a minimum interdigital interval of 180–300 ms.[12] Some of the more important dial pulse detection requirements for subscriber and interoffice signaling are summarized in Table 10–2.

Table 10–2. Dial Pulse Detection Requirements.

Central office impedance termination	900 ohms + 2.16 μF
Pulsing speed	7.5–12 p/s
Percent break	40–80%
Bridged ringers	0–5
Minimum interdigital interval	180–300 ms
External line circuit resistance	2000 ohms (3300 ohms with range extension)
Minimum line insulation resistance	10,000 ohms

10.5.3.2. Dual Tone Multifrequency Signaling. Dual tone multifrequency (DTMF) signaling is a form of address signaling using pairs of frequencies over the subscriber line. One of the frequencies is selected from a low group of four frequencies, and the other is selected from a high group of three frequencies. A fourth high-group frequency is used in certain private network applications and is reserved for future use in the public switched telephone network. Character representation of the tone pairs is shown in Table 10–3. The star (*) and number (#) symbols are used to activate special features.

Table 10–3. Dual Tone Multifrequency Signaling.

		NOMINAL HIGH GROUP FREQUENCIES (HZ)			
		1209	1336	1477	1633
Nominal Low Group Frequencies (Hz)	697	1	2	3	spare
	770	4	5	6	spare
	852	7	8	9	spare
	941	*	0	#	spare

DTMF transmitters used by subscribers may be powered by the subscriber loops or by a local power source. Table 10–4 contains the more important electrical requirements for signals from DTMF transmitters powered from a local power source.[13] The requirements for loop-powered transmitters are the same except that greater variation in levels is allowed because of the wide variation of loop current encountered on subscriber lines. If the transmitter is not equipped with a polarity guard, the oscillator in the station set requries the battery supply connected to the ring side of the line to be negative with respect to the tip conductor.[14]

Table 10–4. Subscriber DTMF Transmitter Requirements

Nominal level per frequency	−6 to −4 dBm
Minimum level per frequency:	
low group	−10 dBm
high group	−8 dBm
Maximum level per frequency pair	+2 dBm
Maximum difference between two frequencies in pair (high freq. ≥ low freq.)	4 dB
Frequency deviation	±1.5%
Automatic dialer pulsing rate:	
minimum duration of digit	50 ms
minimum interdigital time	45 ms
maximum interdigital time	3 s
minimum cycle time	100 ms

Central offices also can be equipped to send DTMF signals to subscriber PABXs arranged for direct inward dialing (DID). The DTMF requirements are only slightly different from those of subscriber equipment as shown in Table 10–5.[15]

DTMF receivers in central offices must operate with greater tolerances than transmitters. The line impedance should satisfy voice transmission requirements. The receiver should execute a code validity check to ensure that two frequencies are present, one and only one low-group frequency and one and only one high-group frequency. The receiver should register frequency

Table 10–5. Local Central Office DTMF Transmitter Requirements.

Same as subscriber key pad, *except:*	
Frequency deviation	±1.5%
Signal level per frequency	−7 dBm0 ±0.5 dB
Source impedance	600 or 900 ohms in series with 2.16 μF
Pulsing delay	≥ 70 ms after start signal
Extraneous frequency components > 500 Hz	≥ 20 dB below signal power

pairs in which each frequency is within 1.5% of nominal. The receiver should reject tone pairs having either frequency more than 3.5% from nominal. It may accept or reject signals between 1.5 and 3.5% of nominal. Speech and other nonsignal energy should be rejected.

Pulsing speed and duration of pulses must be confined within certain limits. The receiver should register signals with digits and interdigital intervals as short as 40 ms but should ignore those less than 23 ms duration. Between 23-ms and 40-ms duration, digit registration is optional.

Transmission impairments affect requirements for DTMF receivers. Signals between 0 and −25 dBm per frequency must be registered and may be registered at levels as low as but not lower than −55 dBm. If both frequencies are within the 0 to −25 dBm range, they should be registered if the amplitude of the high frequency is not more than 4 dB more or 8 dB less than that of the low frequency. Signal echoes which are delayed not more than 20 ms and which are at least 10 dB below the level of the primary signal should not preclude digit registration. Various impedances of subscriber loops (see 10.5.1) can cause the dial tone level at the input of the DTMF receiver to increase. Therefore, the receiver should register digits in the presence of dial tone up to 3 dB higher than that specified in Section 10.5.4. Noise tolerance also is a DTMF receiver requirement. Nominal value digits should be registered in the presence of 54 dBrn flat-weighted Gaussian noise in the 0–3 kHz band with an average error rate up to 1 in 10 000 and 83 dBrnc impulse noise with an average error rate not more than 14 in 10 000. Power-line-induced noise components of 60 Hz at 0.45 Vrms, 180 Hz at 0.13 Vrms, or 300 Hz at 0.013 Vrms should cause a change less than 1 dB in receiver sensitivity.[16]

DTMF signaling can experience problems when used for end-to-end signaling on either public or private networks. During call processing, some switching systems reverse polarity on the tip and ring conductors. If a polarity guard is not provided on the DTMF generator, the output is disabled. Many single-slot coin telephones use a positive rather than negative battery on the ring conductor to prevent fraud and to permit proper coin totalizer operation. Again, a polarity guard is needed or special operational features are required. When a call is made over either a public or private network trunk using echo suppressors and dial tone is sent from the distant switching system, the first DTMF tone pair can be attenuated. Depending upon the relative levels of the DTMF and dial tone signals and the type of echo suppressor used, the effect on the first digit may be blockage of the entire digit or attenuation of only the first portion.[17]

10.5.3.3. Multifrequency Signaling. While DTMF signaling generally is confined to subscriber lines, multifrequency (MF) signaling is the primary inband address signaling system used on interoffice, carrier-derived trunks. MF, like DTMF, employs two simultaneous frequencies, but MF tones are selected from a group of only six frequencies spaced 200 Hz apart. The 15 possible combinations represent the ten digits 0 through 9, signals indicating the beginning and end of pulsing, and three signals for special network use. The general-purpose representations used in the North American network for call set-up shown in Table 10–6.

The major advantages of MF signaling over dial pulsing are accuracy, speed, and signaling distance. Critical transmitter and receiver specifications ensure a very high degree of accuracy of address digits. Transmit power is held to within 1 dB of −7 dBm0. The power of extraneous frequency components should be at least 30 db below that of either signal frequency. Digits and interdigital intervals should be 58–75 ms duration, while the start-of-signaling (KP) pulse should be 90–120 ms. Frequency stability should be kept within 1.5% of nominal, and each frequency should start and end within 1 ms of each other. Longitudinal balance and return loss should be at least equal to the requirements for voice transmission, and the transmitter impedance should be the same as the switching equipment trunk circuit. The nominal digit rate of seven digits per second in the American network is increased to ten digits per second for international inband signaling using CCITT Signaling System No. 5. These signaling speeds permit significant economics in trunk usage compared to dial pulse signaling.

MF signaling is not only faster and more accurate than dial pulsing, it can be used over much greater distances. Distortion of the ac signals, however,

Table 10–6. Multifrequency Signaling.

DIGIT	FREQUENCIES
1	700 + 900
2	700 + 1100
3	900 + 1100
4	700 + 1300
5	900 + 1300
6	1100 + 1300
7	700 + 1500
8	900 + 1500
9	1100 + 1500
0	1300 + 1500
CONTROL SIGNALS	FREQUENCIES
KP—Preparatory for digits	1100 + 1700
ST—End of pulsing sequence	1500 + 1700

is largely a function of distance, and receivers must be designed to register distorted signals without sacrificing accuracy. A code validity check is required to assure that each pulse contains two, and only two, valid frequency components. The receiver must respond to a KP signal of at least 55 ms, and it may respond to a KP signal as short as 30 ms. Transmission distortion may cause the two frequencies to be shifted in time by as much as 4 ms. The receiver should accept pulses at up to 10 digits per second provided that each frequency component is at least 30 ms in duration and that the two components are coincident for more than 10 ms. Interpulse intervals should be at least 25 ms. This does not apply, however, to a series of pulses of 55-ms pulse period. Digits should be registered when received signal levels are between 0 and −25 dBm per frequency, but should not be registered if the signal level drops below −35 dBm per frequency. Received frequency stability is acceptable if each frequency is within 1.5% ± 5 Hz of nominal. The receiver should register address signals in the presence of circuit noise at levels of 63 dBrnc0 with compandors and 50 dBrnc0 without compandors and impulse noise at levels as high as 98 dBrnc0 with compandors and 81 dBrnc0 without compandors. It should also accurately register signals in the presence of power-line-induced noise at levels of 81 dBrnc0 at 60 Hz and 68 dBrnc0 at 180 Hz.[18]

The evolution of technology has resulted in a network comprised of some very slow-reacting switching systems and some very fast-reacting switching systems. When these are interconnected, timing considerations can be critical to effective exchange of signals. Certain types of switching equipment may cause the MF signaling transmitter to send a spurious KP signal followed by a normal KP signal. If the distant switching equipment reacts very quickly to the trunk seizure and returns a start-dial signal immediately, the spurious KP can be registered if it is of minimum specified duration. To protect against these and other transients, signaling protocols specify a delay in sending the start-dial signal, and receivers should be capable of registering MF signals received at least 35 ms after sending the start-dial signal.[19]

10.5.3.4. Control of Interoffice Address Signaling. As indicated in Section 10.5.3, time must be allowed for attachment of a digit receiver before address digits are transmitted over a trunk. The three basic protocols used to control address signals on trunks are immediate dial, delay dial, and wink start. A fourth protocol, stop-go, is used in certain network connections through a step-by-step or register-only tandem office to an office without immediate dial pulse receiving capability. On-hook and off-hook supervisory states are described in Sections 10.5.2.2 and 10.5.2.3. Seizure of a trunk by the calling office is signaled by changing its on-hook state to off-hook. The called office should recognize the seizure and prepare to receive address signals as rapidly as practicable with reasonable protection against registration of transients (see Section 10.5.3.6).

When the called office is a nonsenderized step-by-step switching system, the calling office should send the off-hook seizure signal for at least 150 ms, immediately followed by the entire dial-pulsed address. When a call is originated by a subscriber connected to a nonsenderized step-by-step office, the time of arrival of the first dial pulse in each digit at the terminating office will be under control of the calling party. The delay will be the calling party's interdigital time minus the step-by-step switching time for the previous digit. This time is variable, but the multilation of the first digit will be minimized if the called office is ready to receive digits within 65 ms after receipt of the trunk seizure. This type of control is called *immediate dial operation.*[20]

An important consideration in interoffice signaling involves a call-by-call integrity check to ensure trunk continuity and, on trunks with loop supervision, the presence of battery and ground with on-hook polarity. On trunks with tone supervision, the receipt of a start dial signal constitutes an integrity check.

The original method of controlled address signaling was *delay dial operation.* The calling office seizure is an on-hook-to-off-hook transition. Upon receipt, the called office immediately returns an off-hook, delay dial signal. The called office then sends a start dial signal by going on hook when it is ready to receive digits.

Originally, delay dial operation did not use an integrity check. The calling office, after sending a seizure signal, times for 75 ms or 300 ms, depending upon the trunk, and then examines the supervisory state from the called office. If it is on hook, it starts outpulsing; if it is off hook, it waits until the received supervisory state is on hook before outpulsing. Thus, the failure to receive a delay dial signal will permit the calling office to outpulse before a digit receiver in the called office is attached. This type of operation is unsuitable for synchronous satellite trunks and results in excessive call failures on terrestrial trunks. With the addition of an integrity check, the calling office will not outpulse address digits until it receives a delay dial signal followed by a start dial signal from the called office. For delay dial with integrity check, the delay dial signal must be at least 140 ms duration, and the start dial signal must not occur earlier than 210 ms after receipt of the seizure signal. If the calling office does not receive a start dial within a specified time interval, it should either route the call to reorder tone or retry on another trunk. For calls to be retried on another trunk, the time interval between seizure and receipt of delay dial signal should not exceed 4 sec, and the time interval between receipt of delay dial and start dial signals should not exceed 4 sec for the initial trial or 10 sec for a retrial. Second trial calls which exceed the 10-sec interval should be routed to reorder tone.[21]

Wink start operation is similar to delay dial operation except for the timing requirements. In response to a seizure, the called office sends a timed off-hook signal, called a *wink*, to the calling office. The wink should be from 140 ms to 290 ms, and the end of the wink should not occur earlier than 210 ms after receipt of the seizure signal. To compensate for transmission distortion, the calling office should recognize an off-hook signal of 100–350 ms as a wink. On two-way trunks, off-hook signals lasting longer than the intervals specified for delay dial operation can result from two conditions. Either a digit receiver is not available for attachment to the trunk or there is a near simultaneous seizure of the trunk from both ends. In wink start operation, a wink longer than 350 ms is an indication of either a near-simultaneous seizure on a two-way trunk or a malfunction of an outgoing trunk.[22]

Transients may occur on the trunk when the start dial signal is sent. To prevent registration of those possible transients when using dial pulse signaling, the called office should not register address signals for 30–70 ms after sending the start dial signal, and the calling office should delay sending address signals for at least 70 ms after receiving the start dial signal.[23]

10.5.3.5. Glare Detection and Resolution. Seizure of a two-way trunk by a calling office is received at the called office at a time determined by the propagation delay of the circuit. Time also is required for the called office to detect the seizure, mark the trunk busy in its memory, and attach a digit receiver. Then, and only then, is the called office ready to send a start dial signal to the calling office. If the called office seizes the same trunk for an outgoing call before it has marked the trunk busy as a result of detecting the seizure from the other end, each switching system will be sending an off-hook signal toward each other. Were it not for timing specifications, both switching systems would *glare* at each other indefinitely waiting for a start dial signal. Fortunately, timing requirements and network conventions enable common control switching systems to detect and resolve glare situations involving most address signaling control protocols.

Glare can be detected and resolved on E & M trunks. In wink start operation, a returned off-hook (wink) signal exceeding 350 ms duration is interpreted as glare. In delay dial operation, a returned off-hook (delay dial) signal longer than 4 sec on initial trial or 10 sec on retrial is interpreted as glare. To resolve a glare condition on E & M trunks, each common control switching office is assigned a *glare bit*, or equivalent electromechanical function, to identify one of the offices as the *control office*. When glare is detected, the control office should wait for the off-hook-to-on-hook transition (start dial signal) and then start outpulsing address digits. The other office should maintain the steady off-hook signal while retrying the originating call on another trunk and preparing to receive address digits on the glare trunk.

Then it sends a start dial (off-hook-to-on-hook transition) to the control office.

With reverse battery supervision, glare cannot be detected because the loop closure detector is replaced by a dry bridge (no battery) when the trunk is seized, but glare can be resolved. When an office seizes a reverse battery trunk, a timer times the waiting period for a wink or delay dial signal from the called office. It typically times out after 16–20 secs and retries the call on another trunk. If the office is not the control office, that timing is reduced to 4–8 secs. This enables the office to release the trunk before the control office times out, resolving the glare condition.[24]

10.5.3.6. Signaling Transients. The rather slow operation of electromechanical switching equipment permits it to be relatively immune to short changes of on-hook or off-hook state during signaling. However, the much faster switching operations performed by electronic switching systems increase their susceptibility to such transients, especially when the switching system determines the dc supervisory state by scanning or sampling. The transients described here are generated when dial pulse signaling is used, and all but one can be encountered only when receiving digits from a step-by-step trunk circuit. Care must be taken not to register transients.

Step-by-step trunk circuits and outgoing repeaters generate a spurious pulse just before the first pulse of each digit. The false pulse may start 6–10 ms before the first valid pulse and may last 2–4 ms. A pulse-repeating relay may cause a momentary "make" contact of about 10 ms during the normal break period, occurring generally about 5–10 ms after the valid make closure is released. Mechanical relays can generate contact chatter for 10–15 ms. Trunk circuits that use battery and ground dial pulsing can cause a false on-hook pulse as long as about 10 ms when it replaces battery and ground with an inductor bridge during the interdigital period. Step-by-step outgoing trunk circuits typically use an inductive off-hook bridge with loop reverse battery supervision. When another office returns answer supervision by reversing battery and ground, the drop in current caused by the inductance may be interpreted by the incoming trunk circuit as a short on-hook pulse.

One significant transient can be encountered when connected to any type of switching system if loop reverse battery trunk supervision is used. Some loop reverse battery outgoing trunks use an idle circuit termination consisting of resistor and capacitor in series across the tip and ring conductors. Although the capacitance should not exceed 0.5 μF, some older outgoing trunk circuits use as much as 2–3 μF capacitance. If the calling party disconnects first and the idle termination has been bridged across the tip and ring conductors when the called party disconnect is sent, the reversal of battery and ground by the incoming trunk circuit causes the idle circuit termination capacitive charge to reverse, giving the appearance of a short off-hook. This should not be interpreted as a new seizure.[25]

10.5.4. Call Progress Signals

A wide variety of audible tone signals is used in the network to provide call progress information to callers and operators. A precise tone plan, comprising four pure frequencies of 350 Hz, 440 Hz, 480 Hz, and 620 Hz, is used to derive over 20 separate indications by varying the tones used and their patterns. Each frequency should be controlled within 0.5% of nominal value, and power levels should not vary more than 1.5 dB from nominal. Interruption rates and cadence timing should be maintained within 10% of nominal values. Harmonics and extraneous frequency power should be at least 30 dB below the specified signal level. The call progress signals of primary interest are described in Table 10–7.[26]

10.5.5. Common Channel Signaling

The latest signaling concept to be employed in the public telephone network, and the trend for the future, is common channel signaling. The AT&T version is called Common Channel Interoffice Signaling (CCIS).[27] Other common channel signaling systems perform similar functions. CCIS is similar to, but differs in message format from, the international common channel signaling standard, CCITT System No. 6.

10.5.5.1. Principles of CCIS. In the CCIS system, supervisory signals, address signals, and other signals are exchanged between switching system processors over a network of signaling links instead of over the voice transmission paths. Signaling links utilize 4-wire voice frequency channels with modems and terminals. Initially, the VF channels operated at 2400 b/s, but

Table 10–7. Call Progress Signals

NAME	FREQUENCY (HZ)				LEVEL PER FREQUENCY	INTERRUPTION PATTERN
	350	440	480	620		
Dial tone	X	X			−13 dBm	steady
Line busy tone			X	X	−24 dBm	60 IPM (tone on 0.5 s and tone off 0.5 s)
Reorder/no circuit tone			X	X	−24 dBm	120 IPM (tone on 0.25 s and tone off 0.25 s)
Audible ring tone		X	X		−19 dBm	10 IPM (tone on 2 s and tone off 4 s)

all or most have been upgraded to 4800 b/s. Signaling messages are transmitted over one or more signaling links to a destination switching center (which may not be the final destination of the call) by means of packet switching technology. Voice paths are checked for continuity and connected through to complete the transmission path. An example of an associated signaling link is shown in Fig. 10–5. Since it would be uneconomical to establish such signaling links for all interoffice trunk groups, a network of common channel signaling links has been formed and is described in Chapter 13.

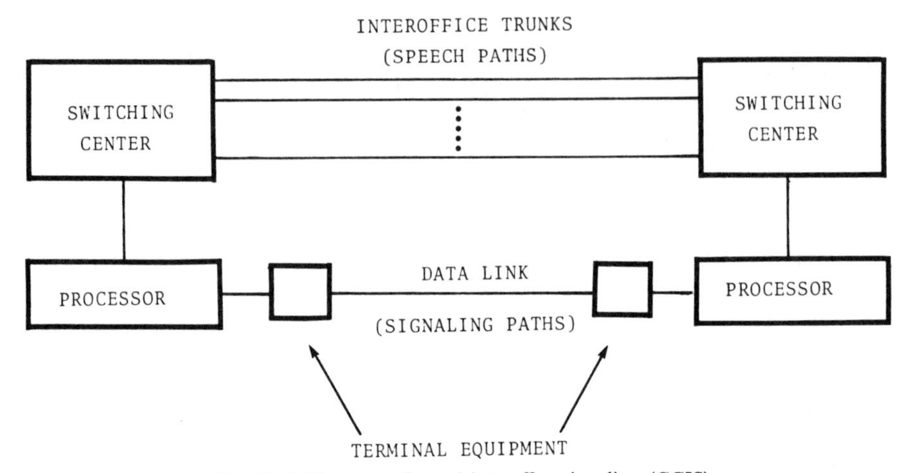

Fig. 10–5. Common channel interoffice signaling (CCIS).

10.5.5.2. Signaling Link Operation. Signaling messages are generated by switching system processors as signal units (SU). SUs contain 20 bits of signaling data each and are transmitted over signaling links in blocks of 12. SUs are sent by the switching system processor to a terminal access controller (TAC) which is custom-designed to be compatible with its associated processor. The TAC arranges the SUs according to priority and routes them to the proper signaling terminal. The terminal stores them in a transmit buffer until 11 have been accumulated. The terminal adds eight check bits to each SU and adds the twelfth SU as an acknowledgement signal unit (ACU) to complete the block of 12. After transmission, the block of SUs is stored until an acknowledgement is received indicating that all SUs in the block have been received correctly. Errors are corrected by retransmission. When no SUs containing signaling data are being transmitted, a synchronization signal unit is transmitted for link synchronization. Signaling reliability is enhanced by link redundancy with automatic transfer and diverse routing.

10.5.5.3. Call Setup with CCIS. To illustrate procedures for call setup by CCIS, an associated signaling link between the originating and terminating switching centers is assumed. The originating CCIS office sends an initial address message (IAM) consisting of an initial signal unit (ISU) and up to four subsequent signal units (SSU). Header information relates the ISU and SSUs to each other in the correct sequence. The ISU contains a trunk label consisting of a band number (9 bits) associated with a subgroup of 16 trunks and a trunk number (4 bits) identifying a specific trunk in that subgroup. Routing information and the digits of the called telephone address are contained in the SSUs.

Upon receipt by the destination switching office, the SUs are translated, and a continuity check circuit is connected to the designated trunk. That circuit simply loops the send and receive paths together through a lossless loop. The originating office connects a transceiver to the trunk, transmits a 2010-Hz tone to the terminating office, and measures the loss of the signal received over the return path. If the loss is out of limits, the failed trunk is locked out of service and subjected to a special test, and the call is reinitiated. If the loss is within acceptable limits, the originating office removes the transceiver and sends a continuity (COT) message over the CCIS link. The terminating office acknowledges the COT message by sending an address complete (ADC) message to the originating office. The call processing program then translates the called number and tests the called line for busy. If busy, it returns a subscriber busy (SSB) message to the originating office over the CCIS link. If the called line is idle, the terminating office applies ringing and returns audible ring tone over the incoming trunk which has been connected to the calling party by the originating office after receipt of the ADC message. When the called party answers, the terminating office connects a cross-office path between the called line and incoming trunk and sends an answer message over the CCIS link to the originating office which begins charge timing for billing. When the parties hang up, disconnect messages are sent over the CCIS link, charge timing is stopped, the cross-office paths are removed, and the trunks and lines are restored to idle condition after a specified guard interval.

10.5.5.4. Advantages of CCIS. The primary technical advantages of CCIS are increased signaling speed, reliability, and flexibility. CCIS passes signals at a much higher speed than inband signaling systems. This reduces post-dialing delay and the holding time of trunks. Increased accuracy is achieved by error correction of signaling messages. The increased capacity of signaling messages permits more flexibility in call routing, more efficient network operation, and enhanced subscriber services.

10.5.5.5. Signaling Message Formats. A comparison of signaling message formats used in CCITT System No. 6 and CCIS is shown in Fig. 10–6 which depicts three types of signal units. The principal differences are in the trunk labels and routing information.

Figure 10–6(a) shows the formats for a lone signal unit or an initial signal unit of a signaling message containing multiple signal units. CCITT System

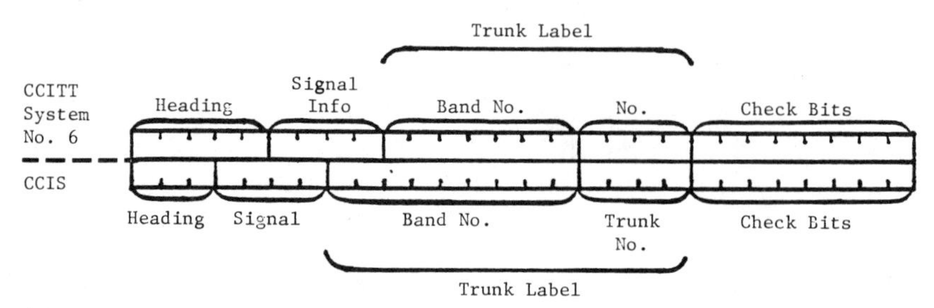

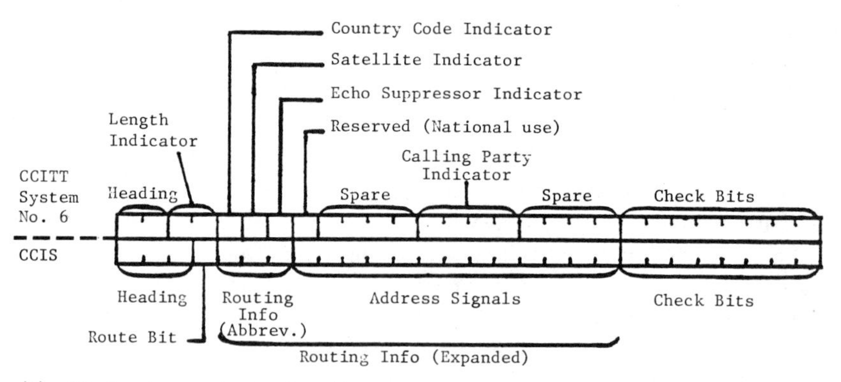

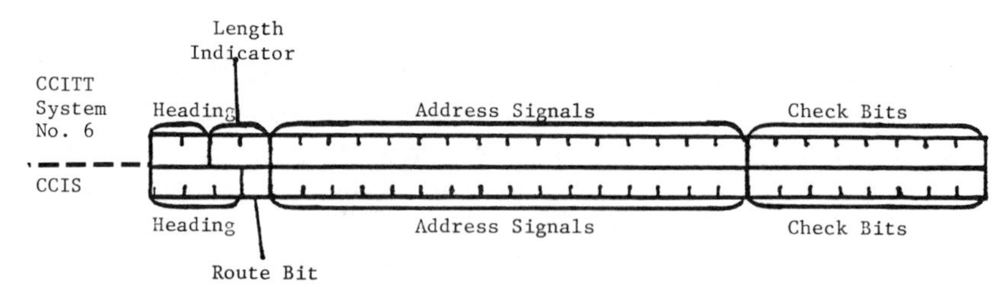

Fig. 10–6. Signaling message formats.

No. 6 uses a 5-bit heading, while CCIS employs only a 3-bit heading. In the international signaling system, blocks of heading codes are allocated for international, regional, and national use. The 11-bit trunk label in the CCITT System No. 6 permits identification of a total of 2048 trunks. The 7-bit band number identifies a specific group or subgroup of 16 trunks, and the 4-bit trunk number identifies a specific trunk. The 13-bit trunk label used in CCIS enables the identification of 8192 trunks to accommodate the larger trunk quantities in the AT&T domestic network.

The formats for the first subsequent signal unit in a signaling message are shown in Fig. 10–6(b). The CCITT System No. 6 includes a length indicator to indicate the number of subsequent signal units in the message. Three bits of route information indicate whether a country code, a satellite, or an echo suppressor is included, and a fourth bit is reserved for future use. The calling party indicator identifies the language spoken by a calling party and other characteristics of the call, such as whether it is a data call or a test call. The CCIS format uses 4-bit abbreviated or 16-bit expanded routing, and the choice is indicated by a route bit. If abbreviated routing is used, the first three address digits are included in the first subsequent signal unit in binary coded decimal format. Additional subsequent signal units are used for address digits as shown in Fig. 10–6(c).

Check bits 21–28 are used for checking the accuracy of signaling messages. Each check bit indicates whether the binary sum of a specific combination of 8–11 bits is odd or even.

10.5.5.6. CCITT Signaling System No. 7. A more advanced signaling system, based upon a modified Open Systems Interconnection architecture, is scheduled to replace both CCITT System No. 6 and CCIS for most network applications. It is being further refined by CCITT and is briefly described in Section 13.2.3.2.

10.6. SWITCHING NETWORK TECHNOLOGY

The switching network is a systematic collection of interconnecting transmission paths which permits speech connections to be established through a switching system between lines, between lines and trunks, and between trunks. The network also can be used to connect signaling equipment to external circuits. Historically, economic considerations have precluded the provision of sufficient paths such that a path between two terminals would always exist irrespective of the traffic load. Some lines, such as those in residences, generate very small volumes of traffic, while others, such as those in many business establishments and office complexes, originate and receive very large numbers of calls during the business day. Therefore, switching systems almost universally have used concentration in their switching networks; that is, there are fewer paths than would be necessary to allow all subscribers to participate in talking connections concurrently. Such networks are known as *blocking* networks.

When concentration is used, a large number of subscriber lines is concentrated into a smaller number of switching paths which are distributed in such a way as to be connectable to any desired terminal through expansion switches. Thus, blocking networks can be described as having *concentration, distribution,* and *expansion* stages, as illustrated in Fig. 10–7. As can be seen, a connection can be established between the originating and terminating appearances of all subscribers, but only three paths can be established concurrently. In addition, when subscriber No. 1 is talking, subscribers No. 2, 3, and 4 are blocked. Large switching networks can be assembled in this manner irrespective of the technology used.

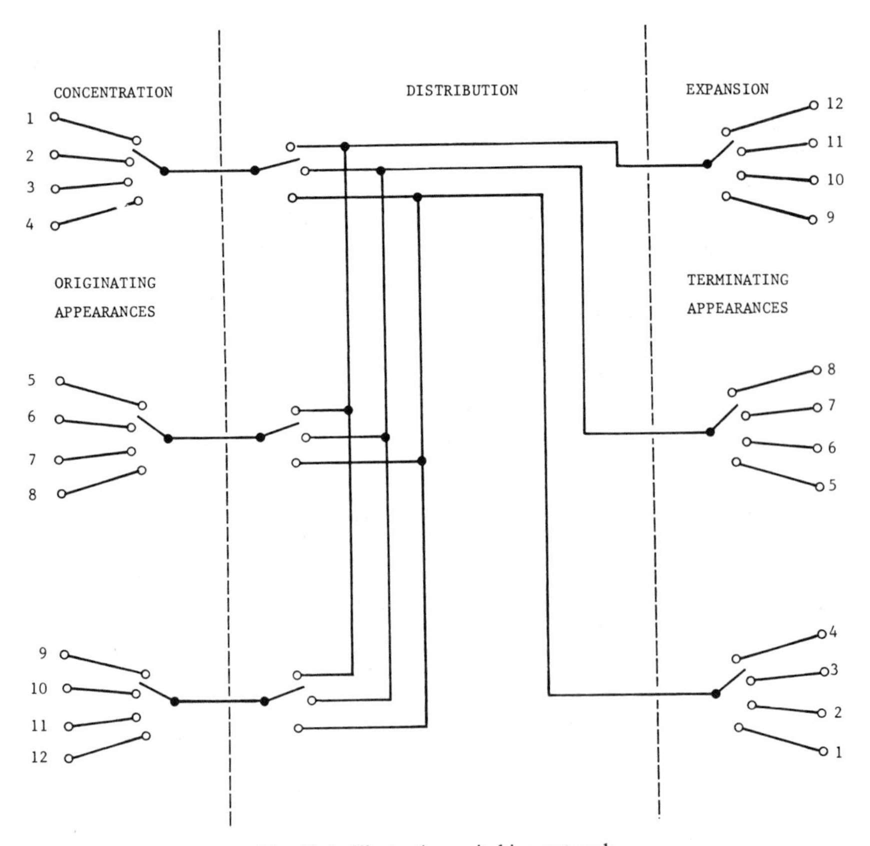

Fig. 10–7. Illustrative switching network.

10.6.1. Space Division Switching Technology

Conversation paths in a practical switching network can be separated from each other in space. The development of space division switching networks closely paralleled the development of control concepts.

10.6.1.1. Progressive Switching Networks. Progressive, or direct, control is described in Section 10.4.2 as it applies to the two-motion stepping switch employed in step-by-step switching systems. Establishment of a connection through the switching network is under the direct control of the calling party and occurs progressively as the address digits are dialed. There are many such switching systems in the North American network.

Using the Strowger two-motion stepping switch as an example, switches are connected in series to form a *switch train.* The first switch in the switch train is called a *linefinder,* and the last switch is called a *connector.* Subscriber lines have appearances on both linefinders and connectors. Intermediate switches are called *selectors.* Each subscriber line is connected to a group of connectors and to a group of linefinders through a line relay. When a subscriber initiates a call, the off-hook closure causes the line relay to extend battery and ground to specific portions of a preselected linefinder in a linefinder group. The linefinder steps vertically and then rotates the shaft horizontally until the wiper arms find the calling line. Each linefinder is mated to a first selector which returns dial tone. The caller then dials the address digits. First selectors are arranged to drop one, two, or three digits of the 7-digit telephone number depending upon the number of digits required to route the call. Fig. 10–8a illustrates the connection established by the last four dialed digits, 2738, of a local call.

When the digit "2" is dialed, the first selector steps vertically to the second level and rotates its wiper arms horizontally. When its control arm finds an idle path to a second selector, usually indicated by battery on the control contact, its other wiper arm clamps onto contacts, extending the voice path to the second selector. The digit "7" causes the second selector to step in a similar manner to the seventh level and connect to a path to a connector having a terminating appearance of the called line. The last two digits are dialed into the connector. The digit "3" causes the connector to step vertically to the third level, and the digit "8" causes the shaft to rotate its wiper arms to the eighth set of contacts. Finding the line idle, the wiper arm clamps onto the contacts, and ringing voltage is applied to the line. Upon answer, the call is established.

Each linefinder in a group connects to either 100 or 200 lines. Each selector has 100 tip-and-ring paths, and each connector has either 100 or 200 lines connected to it. If the connector serves 200 lines, three digits are dialed

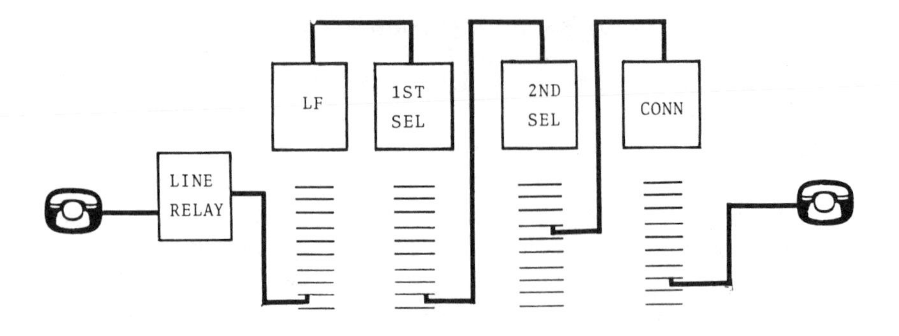

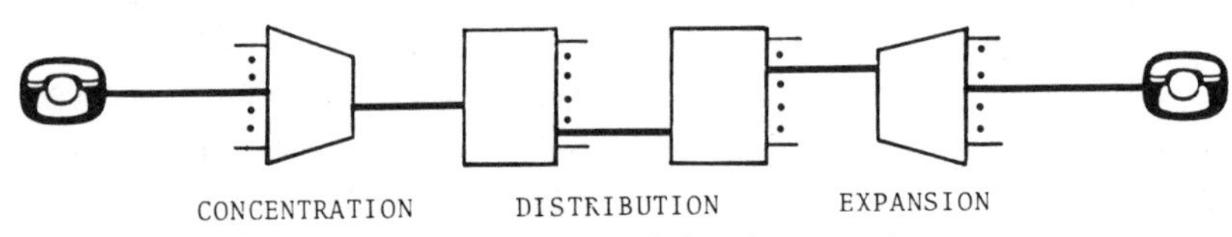

Fig. 10–8. Step-by-step switch train connection.

into it—the first digit being a "1" or "2" to select the upper or lower bank of 100 line terminations.

The linefinders comprise the concentration stage of the switch train, concentrating 100 or 200 lines into a smaller number of paths. The connectors form the expansion stage, expanding the selected path to connect to 100 or 200 lines. The distribution stage is formed by the selectors. The symbolic representation of these stages is shown in Fig. 10–8(b).

Step-by-step switching systems are characterized by low initial cost, minimum postdialing delay, durability, simplicity, and lack of an environmental control requirement. However, they are feature-limited, require large floor space, and consume large amounts of power. Changes and expansion require major rewiring operations, and directory number assignments are inflexible. The operation of the large stepping magnets and relays cause electrical impulses which interfere with data transmission, and relay maladjustments are a major source of troubles. One of the most undesirable characteristics of step-by-step networks is the traffic penalty which results from the limited availability of paths out of each selector switch. For example, in a group of second selectors, there may be several idle switches but the ten switches connected to the dialed level of the first selector could be busy, thereby blocking the call.

10.6.1.2. Crosspoint Switching Networks. Whereas in progressive networks, the selection of a path proceeds stage by stage with no knowledge of conditions ahead, development of common control systems made it possible to identify the input and output terminals and then examine the entire switching network for possible paths between them. Crosspoint networks, sometimes known as grid networks or coordinate networks, were developed to take advantage of the characteristics of common control systems.

A crosspoint switching network is an assembly of individual coordinate switches arranged into a switching array. Switches may be square with the same number of inlets and outlets, or they may be rectangular with different quantities of inlets and outlets. Individual switches are arranged in the form of a matrix with inlets and outlets on the vertical and horizontal axes. Transmission paths run vertically and horizontally, and connections are made by closing crosspoints at the intersection of the selected inlet and outlet as shown in Fig. 10–9.

In a nonblocking crosspoint switch, there must be a sufficient number of paths and associated crosspoints to ensure that every inlet can be connected concurrently to a separate outlet. This is illustrated in Fig. 10–9. In Fig. 10–9(a), three inlets are connected to three outlets in a square matrix. As can be seen, the six crosspoints could satisfy the criteria for a nonblocking, three-line switching system. If the number of lines is doubled, as in Fig. 10–9(b), the required number of crosspoints is increased to 30. For a nonblocking square matrix, the number of crosspoints required can be calculated by $N_x = N(N - 1)$, where N is the number of inlet/outlet pairs.

It is readily apparent that nonblocking crosspoint switching networks are prohibitively expensive in crosspoints except in extremely small sizes. Most crosspoint networks consist of multiple stages. Even then, nonblocking arrays are not economical. For example, a 100-line, nonblocking, three-stage crosspoint network would require a minimum of 5257 crosspoints according to $N_x(\text{min.}) = 4N(\sqrt{2N} - 1)$, where N is the number of lines (ports).[28]

Therefore, most crosspoint networks are blocking networks. Two-stage blocking networks can be used for switching systems of small size. Fig. 10–10 illustrates a simplified two-stage network with concentration and expansion stages. The six-line network, shown by solid lines, can accommodate three simultaneous conversations for certain combinations of lines and requires 24 crosspoints. However, if inlet 1 is connected to outlet 4, inlets 2 and 3 are blocked from outlets 5 and 6. When it is increased to nine lines by the addition of concentration and expansion stages represented by dash lines, more conversation paths exist, and 54 crosspoints are required, but the network still is blocking. Although the two-stage array requires fewer crosspoints than nonblocking networks of the same size, they are not very efficient switching vehicles.

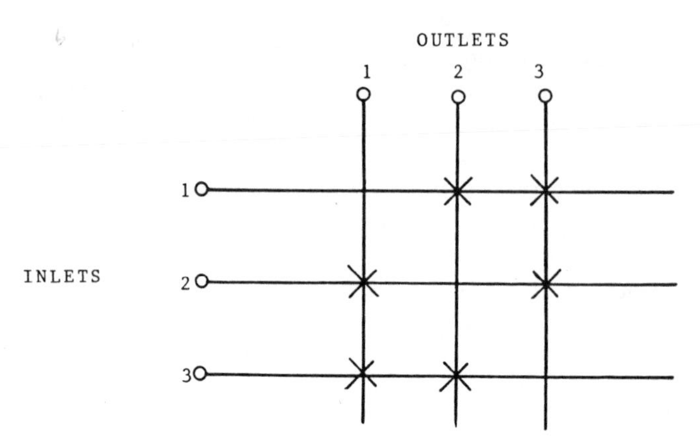

a. Three lines, six crosspoints

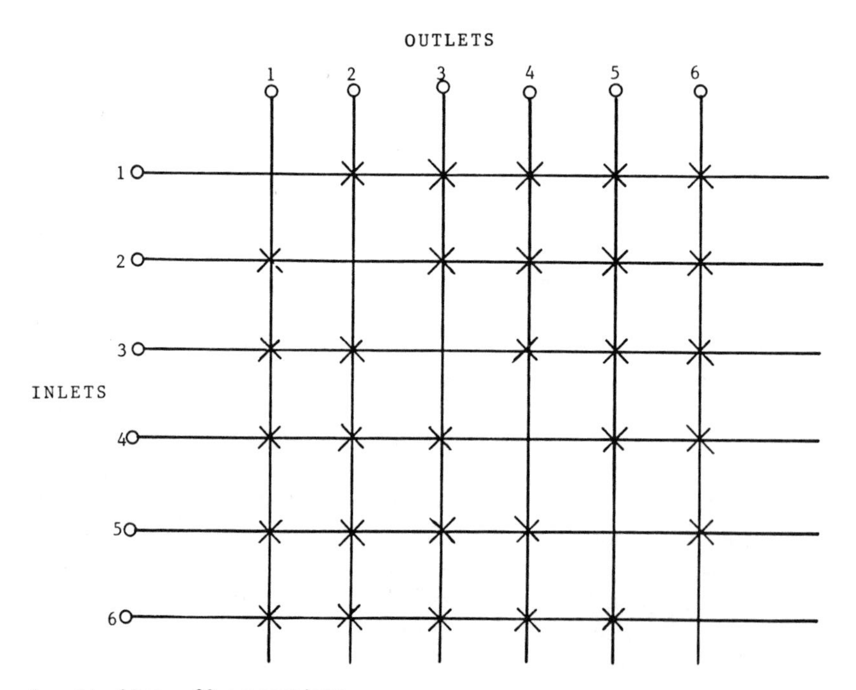

b. Six lines, 30 crosspoints

Fig. 10–9. Illustrative crosspoint switching matrices.

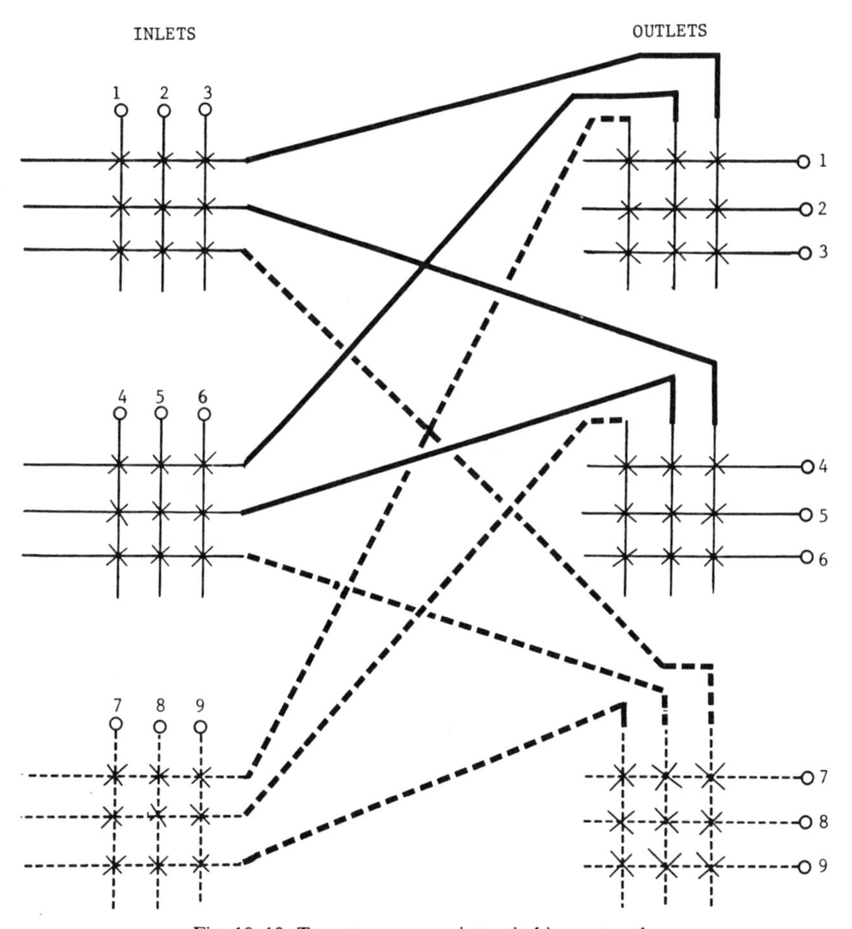

Fig. 10–10. Two-stage crosspoint switching network.

By adding a center distribution stage, the network shown in Fig. 10–11 can accommodate the same number of simultaneous connections as that shown in Fig. 10–10 under certain conditions for both six-line and nine-line configurations with fewer crosspoints, 16 and 27 compared to 24 and 54, respectively. A practical, three-stage crosspoint switching network, capable of serving 100 lines, is shown in Fig. 10–12. In this example, the A stage is composed of ten rectangular matrices to concentrate ten inputs each into N outputs, where N is the number of center stages. Center (B) stage arrays are 10×10 square matrices to distribute the concentrated paths to the C stage switches. C (expansion) stage switches are the mirror image of those in the A stage and expand the concentrated paths to reach any of the 100 terminations. The traffic capacity and the number of crosspoints required are dependent

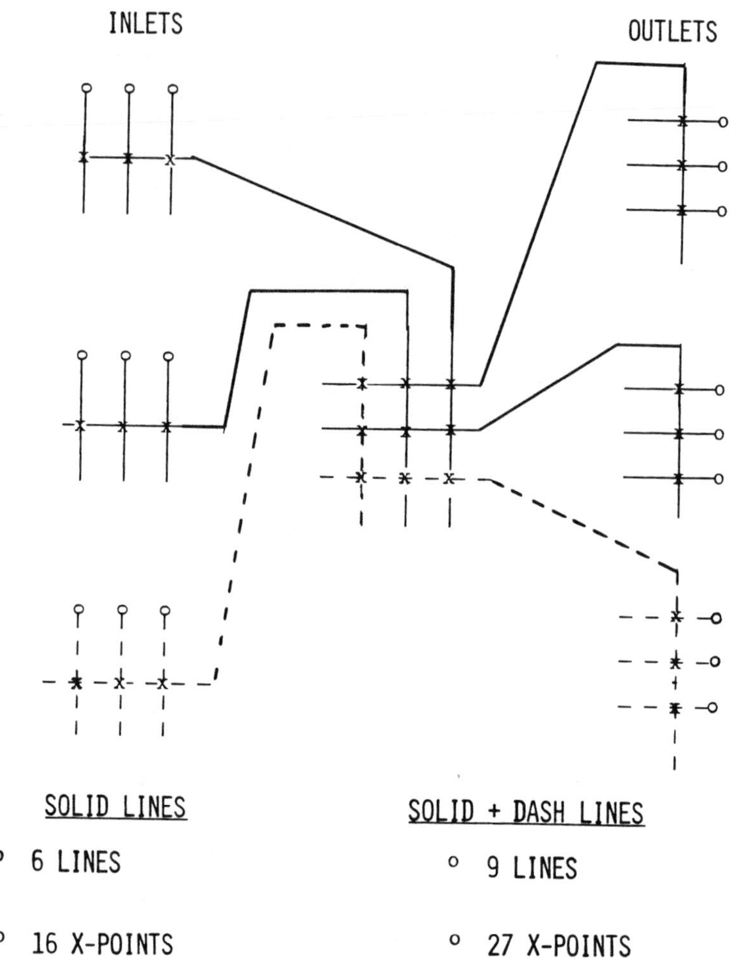

Fig. 10–11. Three-stage crosspoint switching network.

upon the number of crosspoint switches in the center stage. The number of crosspoints[29] required for a blocking, three-stage network is found by

$$N_x = 2Nk + k\left(\frac{N}{n}\right)^2 \tag{10–3}$$

where N = number of lines (ports)
n = number of lines in each concentration and expansion switch
k = number of center stage arrays.

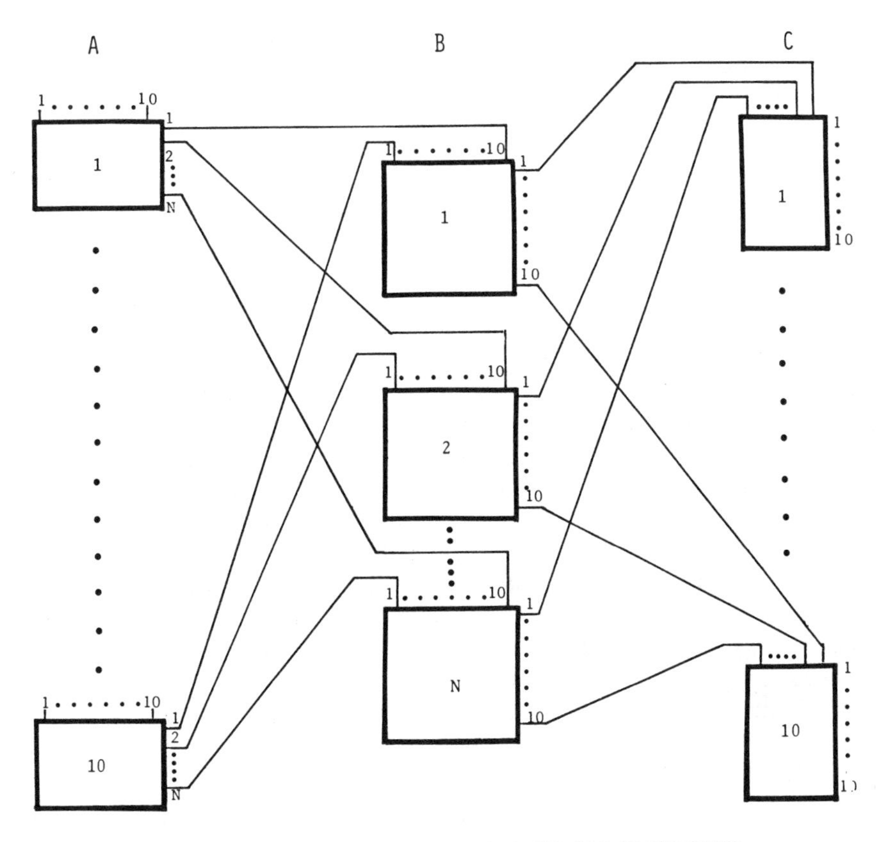

Fig. 10–12. 100-line, 3-stage switching network.

A multistage crosspoint network may have several center stages. In planning a system, however, the number of center (distribution) stages should be fixed in the initial installation. The number of individual switches in each center stage may be increased, along with those in the concentration and expansion stages, as the switching system grows, but the basic architecture fixes the maximum size of the system.

10.6.1.3. Crosspoint Technology. In the evolution of switching technology, three distinct generations of crosspoints have become widely used in North America along with some use of other types. Each type, coupled with its own common control system, affects timing considerations during call setup and disconnect operations over trunks to other switching systems.

Crossbar switches generally are configured in 10×10 square or 10×20

rectangular arrays. Each switch has 10 horizontal paths and 10 or 20 vertical paths. A combination of magnets closes and holds connections between vertical and horizontal paths, providing a total of ten possible connections simultaneously on any switch. Five selecting bars are mounted horizontally across the front of each switch, and each bar has one flexible selecting finger attached to it for each vertical path. Each vertical unit has ten groups of contacts, one group of three to six pairs of contact springs for each horizontal path. Each group of contact springs, when closed by a selecting finger and hold magnet, is considered to be a single crosspoint. Thus each crosspoint can switch from three to six conductors. To operate a crosspoint, a select magnet causes one of the horizontal selecting bars to be rotated slightly up or down, moving all selecting fingers up or down against a stop, a hold magnet slightly rotates a vertical holding bar, associated with a selected vertical unit, against five selecting fingers (one per horizontal selecting bar), moving them to one side. The selecting finger associated with the horizontal selecting bar which is operated is pressed against one group of contact springs on the vertical unit, closing the crosspoint. The other selecting fingers merely swing into a neutral position between the contact springs. After the hold magnet operates, the select magnet releases, returning all but one of its selecting fingers to normal position. When the hold magnet releases, the operated selecting finger returns to normal position, releasing the connection through the crosspoint. A crossbar switch is illustrated in Fig. 10–13.[30] Crossbar switches are grouped together to perform the concentration, distribution, and expansion functions of a blocking network.

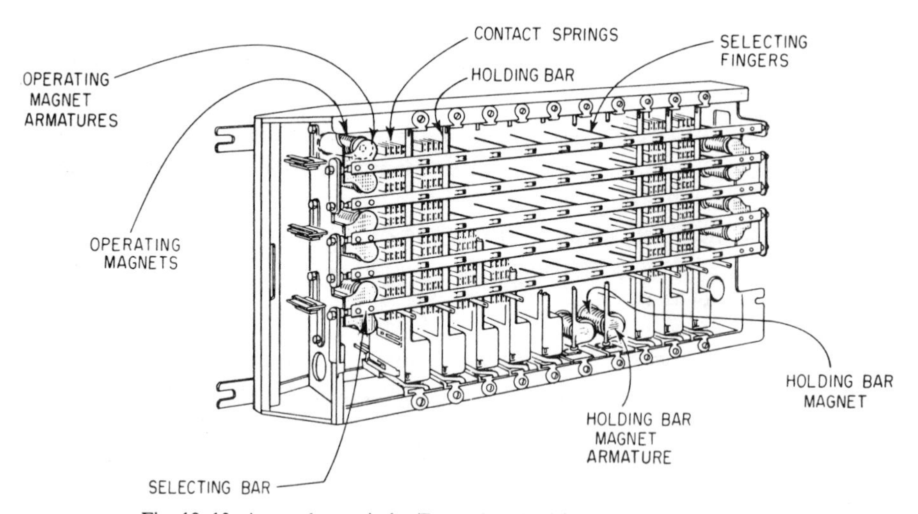

Fig. 10–13. A crossbar switch. (Reproduced with permission of AT&T.)

A second generation of crosspoint technology involved the development of the dry reed switch consisting of two magnetic reeds sealed in a glass capsule and mounted between plates of a two-state magnetic alloy. When the alloy is subjected to a short pulse of current, the reed contacts close and remain magnetically closed until another pulse of current removes the residual magnetism, opening the contacts. Other techniques use crosspoint coils with operate and holding windings to operate the contacts. Reed crosspoints are arranged into crosspoint arrays. Their use, along with electronic wired logic or stored program common control systems, permitted the design of switching systems with larger capacity, greater reliability, and faster switching times, but requiring less space and less power than crossbar systems.

Space division switching also employs solid state logic devices to switch telephone calls. Diodes, silicon controlled rectifiers, transistors, and integrated circuits have been used as crosspoints. Logic gates are arranged in switching arrays in the same manner as reed relays. Gates are closed to establish a connection and opened to disconnect. Solid state crosspoints are characterized by long life, low power consumption, high reliability, and very fast switching times.

10.6.2. Time Division Switching

Analog signals, continuously varying in amplitude, must be altered in order to be switched in time. The speech signals must be changed into pulses which can be multiplexed into a digital stream, each pulse or group of pulses representing one amplitude sample of speech. Four types of pulse modulation currently have application in telephone switching. In all cases, the speech signals are sampled at a controlled rate, each sample pulse representing the amplitude of the signal at the moment of sampling. In Pulse Amplitude Modulation (PAM), the amplitude of the pulse is the same as the amplitude of the signal. In Pulse Width Modulation (PWM), the amplitude of the signal sample is represented by the width of the pulse. In Pulse Code Modulation (PCM) and its variations, the amplitude of the PAM pulse is quantized and encoded as explained in Chapters 3 and 4. In Delta Modulation, only the difference in amplitude between two adjacent speech samples is encoded as explained in Section 4.4. PAM, PWM, and Delta Modulation schemes are used in several time division PABX systems, but only PCM systems serve the public network. Time division multiplexed (TDM) signals can be switched either in a buffer memory or by means of space division logic gates.

10.6.2.1. Time Switching in Memory. In this concept, a time division data stream, such as that used in T-carrier systems (see Section 6.3), is fed into an *information memory* with each traffic channel occupying a separate time slot position in the buffer. A *control memory* contains the same number of words, or channels, as there are time slots in the information memory, but the content of each word in the control memory is the number of a time slot, or channel, in the information memory. Speech channel information is read into and out of the information memory under control of clock pulses. A switching processor, under control of the same clock, controls call setup by reading into the control memory a correlation between input and output time slots in the information memory.

Conventionally, the speech time slots are read into the information memory sequentially and read out asequentially in the sequence specified in the control memory. In the illustration in Fig. 10–14, the input switches are operated sequentially, but the output switch sequence determines the sequence of the readout. For example, if the first time slot read out is channel 3 during an entire switched connection, then the speech in channel 3 has been switched to channel 1 of the output data stream.

Figure 10–15 depicts a time slot interchange (TSI) such as is used in a PCM switching system. In this example, the bus, or PCM highway, has 24 time slots per frame. The 24 voice channels in each frame are read into the information memory in the sequence shown, but the readout sequence is in accordance with the sequence of channels written into the control memory by the control processor. Assume that the illustration represents a 24-port

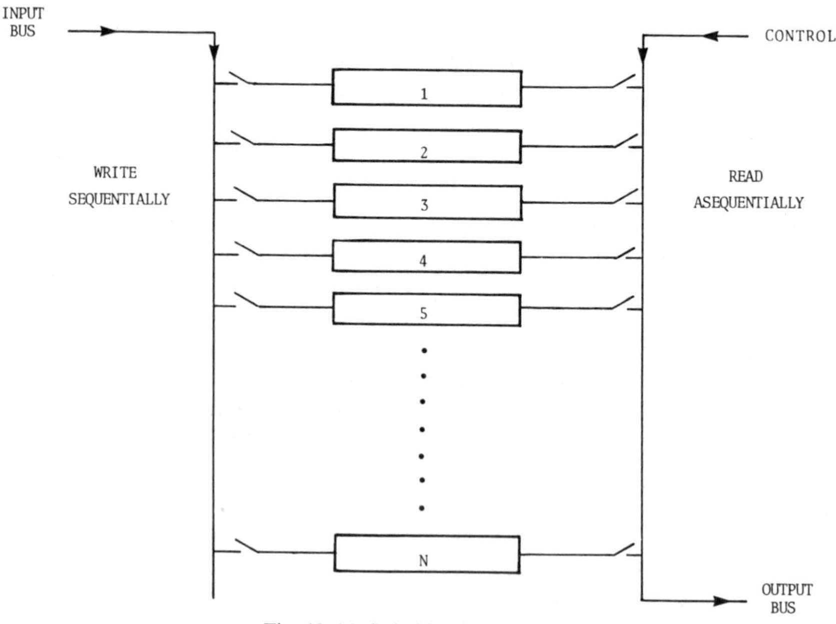

Fig. 10–14. Switching in memory.

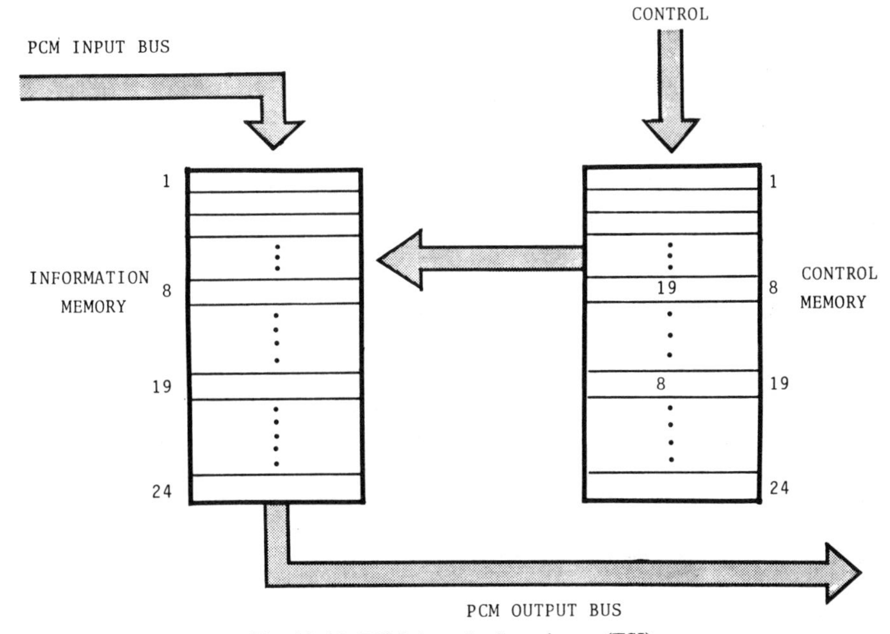

Fig. 10–15. PCM time slot interchange (TSI).

switching system and that ports 8 and 19 are to be connected to each other. When the 8th time slot in the output frame is read out to the output bus, the information memory spills out the word in its time slot 19. Similarly, 19th time slot in the output frame will contain the PCM word from time slot 8 in the information memory. However, since time slot words are being read in and out at clock rate, output time slot 19 will contain the PCM word read into time slot 8 in the following frame. Since this word represents only one amplitude pulse, the difference is not detectable.

10.6.2.2. Time Switching in Space. A time-multiplexed signal can be switched in space. One method sometimes used with PAM switching is *resonant transfer.* The charge on a capacitor at the sending end is transferred to a capacitor at the receiving end by closing a solid state switch under clock control.

A preferred method, one commonly used with PCM systems, employs solid state logic gates as crosspoints arranged in a square or rectangular matrix. A time-multiplexed PCM highway is connected to each inlet, and an output bus is connected to each outlet. Both input and output highways are controlled by a clock. Under direction of a call processor, the control memory causes the logic gates to be opened and closed at clock rate and in the desired sequence to rearrange the time slots in the input and output highways.

10.6.2.3. Sampling and Coding Rates. Sampling rates in PCM switching systems range from 8000 to 16 000 times per second. Higher sampling rates provide higher fidelity and require less complex filtering but cannot interface standard PCM transmission systems directly. Therefore, 8 kHz sampling has been adopted as the worldwide standard for PCM switching and transmission.

As discussed in Chapter 3, 8-bit PCM with $\mu = 255$ companding is the current North American standard. Other techniques are used. CCITT has standardized both the north American system and the A-law companding method used in Europe and most other countries. Some PABX systems employ more quantization steps and linear coding to facilitate digital signal processing. A disadvantage, of course, is that conversion is required to directly interface a digital transmission system using standard 64-kb/s PCM.

10.6.3. Digital Switching Applications

Since the first digital central office was installed in France in 1970, digital switching systems have found application as local switching systems, toll (transit) switching systems, and PABX systems. In North America, stored program, space division, electronic switching systems have been replacing older electromechanical systems in the larger metropolitan areas since about 1966. It was natural, therefore, that local digital systems would first replace older systems in smaller communities. This did, in fact, occur as hundreds of digital systems were placed in service in the 1976–1980 period, primarily outside the former Bell System. AT&T placed initial emphasis on the toll network. Digital systems have replaced older electromechanical systems in a large portion of the toll network. In the 1976–1980 period, over 150 digital switching systems were placed in service in toll switching applications, 56 of which were No. 4 ESS.[31] The opening of the terminal equipment market in the United States to competition stimulated the development of digital PABX systems, and a variety of products have found wide acceptance. Although digital transmission systems are heavily used for relatively short-haul circuits, development of long-haul digital transmission systems has not kept pace because of economic considerations. The paradoxical result in the United States is that most digital switching systems are in smaller communities where transmission systems are predominantly analog, and most digital transmission systems are in large metropolitan area complexes where switching systems are predominantly analog. The probable result is that reed relay switching systems in metropolitan areas will switch telephone traffic metallically for several years.

10.7. WHY DIGITAL SWITCHING?

Considering the refinements made in stored program control space division switching in the last 15 years, why is the interest in digital switching at such a high level? The concept of digital switching is not new. An experimental digital switching system was constructed by Bell Telephone Laboratories in 1959. The system employed a PAM remote concentrator switcher connected to a central PCM switching network by means of a 24-channel PCM transmission system over two cable pairs. The entire system, less common control, used solid state components. The PCM switching network was a space division matrix with AND gates. The model, known as Experimental Solid State Exchange (ESSEX), was the first known practical demonstration of the integration of digital transmission and switching systems.[32] With the level of solid state technology at that time, however, the system was not economically competitive with analog switching.

The revolutionary advances in solid state technology in the last decade have radically changed those cost relationships as they relate to switching systems. The development of large and very large scale integration has enabled powerful microprocessors and coder/decoders (codecs) to be designed on single integrated circuit chips. The cost of semiconductor memories has continued to drop as larger and larger memories are placed on single chips. The overall advance in solid state technology, with the resulting cost reductions, is the primary factor in the development of cost-effective digital switching systems.

10.7.1. Advantages of Digital Switching

Digital switching has technical, economic, and operational advantages over analog switching. Switching speed is faster than relay operation, and there is an improvement in transmission quality resulting from the switching of discrete signals impervious to Gaussian noise. Signal processing is less complex and more accurate when signaling is in digital format. A significant advantage in both transmission qualtiy and cost is the integration of switching and transmission systems, including digital pair-gain systems, made possible by PCM technology. Quality is improved by reduction of A/D and D/A conversions, and cost reductions are achieved by the elimination of hardware components.

Other economies result from a reduction in floor space requirements, efficient use of distributed control, increased line and trunk capacities, and reduced manpower requirements. Digital switching, combined with digital transmission and common channel signaling, opens the door for a wide variety of new subscriber services which increase the profitability of common carriers.

The switching of discrete signals, along with digital stored program control, makes possible improvements in performance monitoring, maintenance diagnostics and repairs, and system operational management. Administration is simplified with the interfacing of computerized service order procedures to digital switching systems.

10.7.2. Disadvantages of Digital Switching

Although the advantages far outweigh the disadvantages, the digital switching balance sheet is not totally one-sided. During the transition to an all-digital network, extra equipment is required for A/D and D/A conversion. The transition, if fully implemented in a short span of years, must inevitably involve retirement from service of both switching and transmission systems long before their economical service lives are exhausted. Much of the long-haul network in North America is composed of microwave radio links. To fully realize some of the advantages of digital switching, the interoffice transmission systems should be digital, but digital transmission requires several times the bandwidth used in analog transmission systems. As this is written, there is a regulatory restriction on approval of frequencies for digital transmission systems with low spectrum efficiency. While analog switching systems can function satisfactorily with independent timing systems, digital switching systems interconnected by digital transmission links require their clocks to be synchronized. In spite of these disadvantages, the trend toward digitalization of the telephone network is irreversible.

PROBLEMS

10.1 Why must digital switching system design be influenced by characteristics of other switching systems to which it will be connected?

10.2 What are the three main functional groups of any switching system?

10.3 Why is stored program control superior to other types of control systems?

10.4 Name four electrical properties of local telephone cable plant, and explain how they affect voice transmission quality and signaling.

10.5 What are bridged taps and how do they affect the impedance of subscriber loops?

10.6 What are the advantages and disadvantages of ground-start supervision on PBX trunks relative to loop-start supervision?

10.7 Discuss the relative merits of dc and ac supervision on interoffice trunks.

10.8 Explain the functional relationship between switching system trunk circuits and single frequency signaling equipment when used with E & M lead control.

10.9 What factors affect the use of DTMF signaling on end-to-end connections, and how may they be overcome?

10.10 When a digital toll switching system initiates a call to a local crossbar office using inband signaling with E & M lead supervisory control and wink-start control of address signaling, how should the digital system respond to receipt of an off-hook signal of 490 ms?

10.11 How does common channel interoffice signaling ensure improved transmission quality?

10.12 Why is a traffic penalty inherent in step-by-step switching systems?

10.13 How many crosspoints would be required for a three-stage space division switching network to serve 500 ports in a nonblocking configuration? How many would be required to serve the same 500 ports in a blocking configuration which has 20 ports in each concentration and expansion array with 15 center stage arrays?

10.14 How is time division switching performed in a space division network?

REFERENCES

1. C. F. Myers, ed., *Principles of Electricity Applied to Telephone and Telegraph Work,* American Telephone and Telegraph Company, New York, 1961, pp. 151–154.
2. *Local Switching System General Requirements,* PUB48501, American Telephone and Telegraph Company, Basking Ridge, N.J., 1980, Section 6.2 (hereafter cited as *LSSGR*).
3. Ibid., Section 6.2.1.1.
4. Ibid., Section 6.2.
5. Ibid., Section 6.3.
6. *Notes on the Network,* American Telephone and Telegraph Company, 1980, Section 5, pp. 67–79 (hereafter cited as *Notes*).
7. Ibid., Section 5, pp. 44–58.
8. Ibid., Section 5, pp. 11–15; *LSSGR,* Section 6.2.
9. *Telephone Instruments with Loop Signaling for Voiceband Applications, EIA Standard RS-470,* Electronic Industries Association, Washington, D.C., 1981, pp. 33–41 (hereafter cited as *RS-470*).
10. *Private Branch Exchange (PBX) Switching Equipment for Voiceband Applications, EIA Standard RS-464,* Electronic Industries Association, Washington, D.C., 1979, pp. 59–66 (hereafter cited as *RS-464*).
11. *LSSGR,* Section 6.2.
12. Ibid., Section 6.3.
13. *RS-464,* pp. 75–79.
14. *RS-470,* p. 45.
15. *LSSGR,* Section 6.4.2.
16. Ibid., Section 6.2.6.2.
17. *Notes,* Section 5, pp. 153–154.
18. *LSSGR,* Section 6.4.1.
19. Ibid., *Notes,* Section 5, p. 87.
20. *LSSGR,* Section 6.3.4.2.
21. Ibid., Section 6.3.4.4.
22. Ibid., Section 6.3.4.3.
23. Ibid., Section 6.3.4.6.
24. Ibid., Section 6.3.4.9.
25. Ibid., Section 6.3.6.
26. Ibid., Section 6.4.3.

27. *Notes,* Section 6.
28. Bellamy, John C., *Digital Telephony,* John Wiley & Sons, New York, 1982, pp. 225–227.
29. Ibid., pp. 223–225.
30. *Local Dial Switching Equipment Reference Guide,* American Telephone and Telegraph Company, New York, 1957, Vol. 4, pp. 2–3.
31. Joel, Amos E., Jr., ed., *Electronic Switching: Digital Central Office Systems of the World,* IEEE Press, New York, 1982, Appendix C.
32. H. E. Vaughan, "Research Model for Time Separation Integrated Communications," *Bell Sys. Tech. J.,* Vol. 38, pp. 909–932 (July 1959).

11
Digital Switching Architecture

11.1. INTRODUCTION

In digital switching system design, architectural considerations involve much more than simply replacing an analog switching network with a digital switching network. Several fundamental questions must be answered, and each must then be followed by a large number of decisions, most of which involve trade-offs. Some of the most prominent questions which require major decisions up front are:

1. What is the intended application and maximum size of the switching system?
2. What type of voice encoding should be adopted?
3. How should lines and trunks be interfaced and conditioned for the switching network?
4. What network architecture should be selected?
5. How should signaling and service circuits be handled?
6. What control concept should be used?
7. How should maintenance diagnostics, traffic management, and switching system administration be accommodated?
8. What recent technological advances should be considered for implementation in the system?
9. What future potential innovations should the design be planned to accommodate?

Not all of these can be answered with finality in the beginning, but all should be addressed, and tentative decisions should be made and then reexamined as the design process continues. The application and approximate maximum size of the system are determined by a perceived view of the marketplace. Recent advances in solid state technology should be evaluated, and those most promising for the intended application should be identified for close scrutiny.

If the intended application is for use as a PABX, several encoding schemes are available for consideration. However, a public network switching system should be able to accommodate the world standard of 64 kb/s PCM. The geographic area of the anticipated market will determine whether μ-law or A-law PCM is selected. For a world market, both should be available as options.

Basic architectural decisions must consider all operational features and capabilities to be provided by the switching system, not only in its initial design capability but also for retrofit of possible future innovations such as provision for data switching.

11.2. TERMINAL INTERFACE TECHNIQUES

The functions required to be performed in interfacing lines and trunks with the switching network apply to all digital circuit switching systems, but implementation techniques vary widely. The trend, however, is clearly visible.

11.2.1. Terminal Interface Functions

Referring to the block diagram in Fig. 10–1, the terminal interface area of a digital switching system includes all equipment components required to interface lines and trunks and to condition the speech paths for presentation to the switching network. An analog subscriber line is terminated in the tip and ring terminations of a line circuit.

Line circuit functions are typically defined by the acronym BORSCHT, derived from the first letters of the seven functions described below.

B = Battery voltage supplied to the line. Most central office battery supplies are nominally negative 48 Vdc with a range of negative 42.5–52.5 Vdc, although the voltage applied by some central offices at certain stages of switching can exceed 78 Vdc and, in rare cases, can exceed 100 Vdc.

O = Overvoltage protection to protect the line circuit equipment from lightning and power line induction. The protector is placed between the tip and the ring to ground and must be designed to protect the sensitive solid state components in the line circuit.

R = Ringing circuits to apply controlled ringing voltages to the line to activate the ringer in the called telephone.

S = Supervisory circuitry to detect on-hook and off-hook conditions and dial pulsing.

C = Coder/decoder (codec) equipment to convert analog speech to PCM words. Codecs either contain or are associated with low-pass filters in the transmit and receive paths.

H = Hybrid transformer to convert the two-wire line to a four-wire line consisting of separate transmit and receive paths. A balance network reduces undesired feedback from the receive path to the transmit path.

T = Test circuitry which connects test equipment to the two-wire line to test the line for continuity, shorts, and impedance characteristics.

The terminal interface area also contains equipment to terminate analog and digital trunks. The functions performed for trunks are similar in some respects to those performed for lines but are modified for the differences in signaling and supervision.

Since most subscriber lines generate low volumes of traffic, lines generally are concentrated for economic reasons. An initial concentration stage of switching is usually included in the terminal interface area. There may be one or two stages of concentration, and it may be performed on the two-wire or four-wire side of the hybrid network or on the PCM side of the codec. Finally, one or more stages of digital multiplexing are included to increase the number of time slots in the PCM stream entering the switching network. Trunk circuits usually bypass the terminal equipment concentrator and directly interface the multiplexor.

11.2.2. Implementation Considerations

As digital switching systems began to be designed, implementation architecture was influenced by two major factors. Concepts which had been proved effective in electromechanical and electronic space division systems tended to be carried over to digital design. Also, the relative costs of different solid state implementations influenced trade-off decisions which were subsequently changed as larger scale integration became feasible.

11.2.2.1. Analog Line Interface. Basic line interface functions are shown in Fig. 11–1. The line circuit (LC) provides access to the line for battery feed, overvoltage protection, ringing voltage, supervisory signaling, and test circuits. The hybrid network is shown here as a separate component, and the coder/decoder functions are shown separately.

One major architectural question has been the location of the line concentrator (not shown in Fig. 11–1). In the early designs, the cost of the discrete solid state components in codecs mitigated in favor of shared codecs following the concentrator. Therefore, some terminal interface designs in the 1970s placed two-wire, analog, space division concentrators between the line circuits and the hybrid networks and used shared codecs between the hybrids and the PCM multiplexors. Other implementations placed a four-wire concentrator between the hybrid network and the shared codecs. These decisions have been influenced by the state of the art of solid state crosspoint technology,

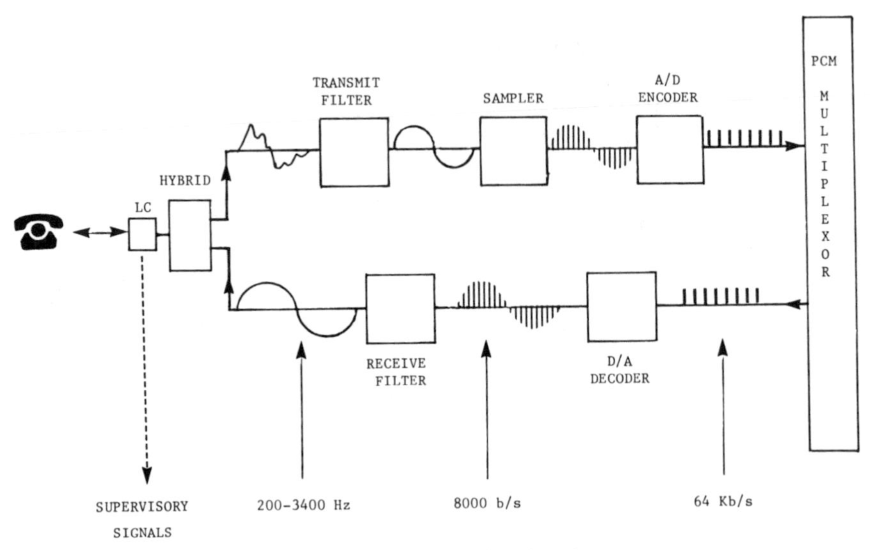

Fig. 11–1. Terminal interface functions.

especially the limitation on voltages that can be switched through them. As larger scale integration progressed, and as more functions came to be implemented with low voltage techniques, it became feasible to combine many of the line interface functions into a smaller number of components, including single-chip codecs. High voltage silicon technology now permits the passage of higher voltages used for ringing. Current technology permits line circuits to include all line interface functions except concentration and multiplexing using a minimum of components as shown in Fig. 11–2.[1] This is followed by PCM bus formatting and concentration.

The concentration function, if performed prior to A/D conversion, is a space division network. If performed after A/D conversion, switching may be performed in memory (time division), through TDM crosspoints (space division), or both. A functional diagram of a line interface module is shown in Fig. 11–3. The speech paths of 32 lines are converted by the line circuits into PCM streams which are time multiplexed by the line access time switch. The outputs of 20 line access cards, appearing as 20 terminal links, are concentrated through a time-multiplexed space switch into 4 network links at a serial bit rate of 2.56 Mb/s. Each of the 32 time slots contains 10 bits, 8 of which are used for standard PCM words and 2 of which are used for signaling and channel control. The 4 network links are interfaced to the main switching network.[2]

11.2.2.2. Analog Trunk Interface. Analog trunks may be terminated at a switching system in any quantity via metallic paths or in 12-channel

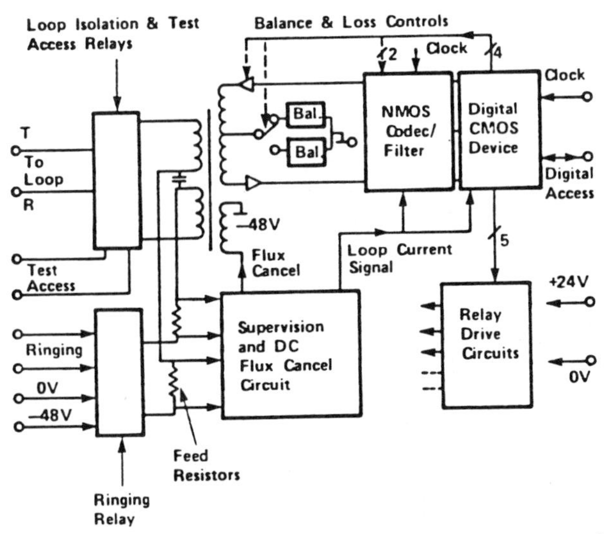

Fig. 11–2. DMS-100 line circuit block schematic. (From J. Terry, D. Younge, and R. Matsunaga, "A Subscriber Line Interface for the DMS-100 Digital Switch," *NTC '79*, © IEEE.)

groups via analog carrier systems. The basic treatment is similar to that accorded subscriber lines; however, the implementation is somewhat different. Inband and common channel signals must be processed separately, and integrity checks are performed as described in 10.5. Speech paths are converted to PCM signals and multiplexed to conform to the PCM bit rate required by the switching network. Codecs may be used on a per-channel or multichannel basis. Since trunks carry high traffic volumes compared to lines, they normally are not concentrated in the terminal interface modules.

11.2.2.3. Digital Trunk Interface. Digital trunks arrive at the switching system in basic groups of 24 or 30 channels. The North American T-carrier group of 24 channels terminates in a digital terminal unit. Under clock control, signaling is extracted from the least significant bit in every sixth frame and processed separately by a signal processor. The 24 speech channels can be multiplexed with other T-1 channels and placed on a PCM highway to the switching network.

11.2.3. Implementation Trends

The recent advances in large scale integration (LSI) and very large scale integration (VLSI) have established a rather clear trend toward consolidation of line interface functions in the line circuit itself. Filters and codecs already had been manufactured on a single LSI chip. Subscriber line interface circuits (SLIC), containing all BORSCHT functions, now are being produced on a

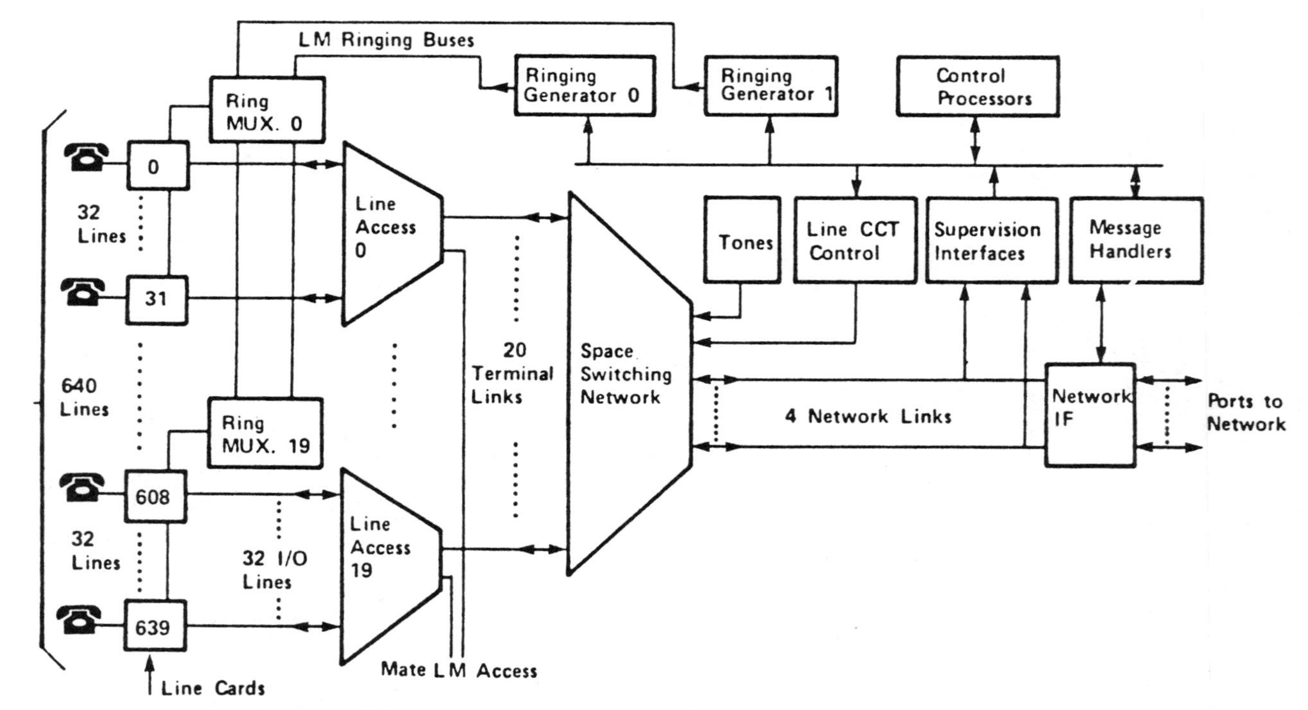

Fig. 11–3. DMS-100 line module functional diagram. (From J. Terry, D. Younge, and R. Matsunaga, "A Subscriber Line Interface for the DMS-100 Digital Switch," *NTC '79,* © IEEE.)

single VLSI chip.[3] The addition of a remote-controlled time slot assignment circuit could enable digital telephones to be connected to digital transmission facilities without the traditional first stage of digital multiplexing.

With consolidation of functions in the subscriber line interface circuit, concentration will be performed increasingly on a digital rather than analog basis. The entire terminal interface function can be located either in the local central office or in a remote location. Remote switching units are connected to the host digital switching system by digital transmission facilities over existing cable pairs. The number of connecting digital facilities is determined by the capacity of the concentrator and the traffic load.

11.3. SWITCHING NETWORK CONSIDERATIONS

Selection of switching network architecture involves several considerations. A small or medium-size PABX can be designed with only a single switching stage, whereas an economical design for a larger system requires multiple stages. Time and space switching have somewhat different traffic-handling characteristics. The effect of switching network architecture on control complexity, packaging, and expansion should be considered. The intended application, whether local or tandem switching, and its ultimate size, influence architectural selection. The fact that there are so many different switching network architectures in the marketplace is a clear indication of the many trade-offs which must be considered.

11.3.1. Multistage Digital Switching

Multistage digital switching networks generally involve both time and space division switching. While some switching systems do use multiple time networks only, the time delay involved in time slot interchanges does place a practical limit on the number of time stages used in a large switching system. All PCM switching networks are symmetrical about a central point.

11.3.1.1. Time Switching Considerations. The principle of time switching in memory was described in Section 10.6.2.1. A major factor in time switching in memory is delay. In a 24-time slot system, speech can be delayed up to 23 time slots. In a 32-channel system, speech can be delayed up to 31 time slots. Some time switching networks use 256 time slots or more in a 125-μs frame. Therefore, the maximum delay of speech in a single time switch is in the range of 120–124.5 μs. When added to other delays inherent in transmission and switching systems, it can add to the effect of echo on the talkers. Consequently, most large switching systems include both time division and space division switching stages.

The number of time slots that can be switched in memory during a 125-μs frame time is a function of the cycle time of the information memory. Each time slot requires a write cycle and a read cycle. With 8 kHz sampling, the maximum number of time slots, or channels (C), that can be switched in memory is

$$C = \frac{125}{2t_c} \tag{11-1}$$

where 125 is the frame time in microseconds and t_c is the memory cycle in time in microseconds.[4] From this, it can be seen that a 256-channel TSI requires a memory with a minimum cycle time of 244 ns, and a 1024-channel TSI requires a memory with a cycle time of 61 ns. Thus, memory cycle time places a practical and economic limit on the size of TSIs.

The impact of TSI size on the control memory should also be considered. The control memory, like the information memory, requires one word for each channel switched in time, but the word length is a function of the number of channels. The number of bits (B) required in each word in the control memory is calculated by

$$B = \log_2 C \tag{11-2}$$

where C is the number of channels.[5] Thus, the maximum number of time slots that can be switched in time with a control memory using 8-bit words is 256, and a 1024-channel TSI requires 10-bit words in the control memory. The number of outgoing time slots in the TSI may be equal to, greater than, or less than the number of incoming time slots, but the most economical arrangement of control memory requires that it control the side of the information memory having the greater number of time slots.[6]

11.3.1.2. Space Switching Considerations. Time division space switching involves a matrix which switches m incoming time-multiplexed lines (TML) to n outgoing TMLs having the same number of time slots. The crosspoints in the switching matrix are formed by logic gates controlled by a control memory. The number of outgoing TMLs may be equal to, greater than, or less than the number of incoming TMLs. The control memory controls the logic gates to allow specific time slots in incoming TMLs to be connected to specific time slots in outgoing TMLs.[7] Time slots are not delayed in space division switching as they are in TSI switching. In space division switching, incoming and outgoing TMLs are synchronized by the same clock such that channel 1 of an incoming TML is always switched to Channel 1 of one of the outgoing TMLs.

11.3.1.3. Time-Space-Time (TST) Structure. In a time-space-time switching network architecture, multiple time switches are used as the input and output switching stages, and space switches are used as the center stage. The number of internal time slots (i.e., the time slots between the input and output stages) is the major factor in the probability of blocking through the switching network. If the number of internal time slots is twice the number of input/output time slots minus one, the network is nonblocking. To achieve a good grade of service (i.e., low probability of blocking), the number of input time slots must be significantly less than the number of output time slots in the first time stage. The third stage is a mirror image of the first stage.

In time division switching with multiple stages, the arrangement of concentration, distribution, and expansion functions shown in Fig. 10–6 is reversed. Expansion is always performed in the first stage, and concentration is always performed in the last stage. The distribution (center) stage or stages always switch on a 1:1 ratio.[8] For example, the No. 4 ESS input time stage expands seven TMLs of 120 time slots each into eight TMLs of 105 active time slots each for switching. The output time stage then concentrates them back into seven TMLs for transmission.[9] This arrangement does not preclude concentration in the terminal interface module.

Operation of a TST network, shown in Fig. 11–4, involves finding an idle time slot through the space stage which can connect an input time stage to the desired output time stage. The first TSI delays the information in the input time slot 5 until the selected idle space switch time slot 38 appears at the output. Space switching connects the output multiplexed line of the first TSI through the space switch to the input of the third-stage TSI. That time slot information then is switched in memory to the desired output of the third-stage TSI in time slot 20. A similar connection is established for the other conversation path.[10]

11.3.1.4. Space-Time-Space (STS) Structure. The STS network has a center time stage to effect distribution between the first and third space stages. Time-multiplexed lines function as inputs and outputs to both space stages. The time stage is symmetrical and serves to interconnect time slots in the output TMLs of the first stage to time slots in the input TMLs of the third stage.

Operation of an STS network is shown in Fig. 11–5. The first space stage functions as an expansion stage, having more output TMLs than input TMLs. The last space stage is reversed and is used to concentrate distribution TMLs into a smaller number of output TMLs. To establish a connection, the path controller must find a time switch in the center stage which has an idle input time slot and an idle output time slot corresponding to the time slots

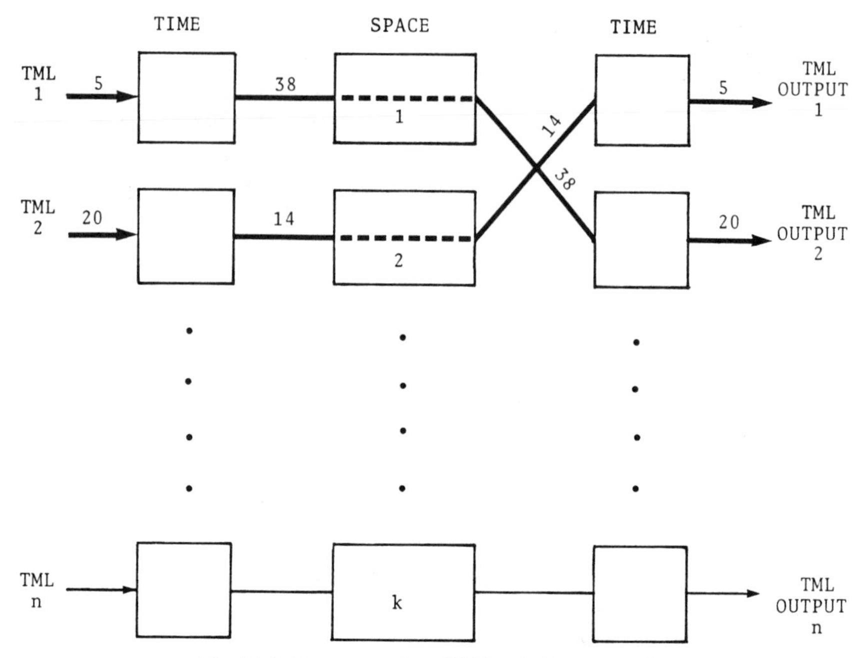

Fig. 11–4. Time-space-time (TST) switching network.

used in the space stages. The first stage space switch closes a crosspoint path to an internal TML connected to a TSI with an idle time slot 3. The TSI then delays the information until time slot 16 in the frame. The third stage then connects that internal TML to TML OUTPUT 2. In the same manner, time slot 16 in TML 1 is connected to time slot 3 in TML OUTPUT 1 to complete the other half of the conversation.

11.3.1.5. Combined versus Separated Switching. The examples in Figs. 11–4 and 11–5 depict unidirectional paths through the switching networks. Actually the speech highways can be designed to carry two-way traffic. In Fig. 11–4, the two speech paths are shown to use separate time switches, and the two internal time-multiplexed lines are switched through separate arrays in the space stage. This concept is called *separated switching,* and control memories must be provided for each switching array. However, if both directions of speech are selected identically, only one control memory is needed for the bidirectional paths.

In *combined switching,* connections can be established between any two highways connected to a single switching array. Paths may be selected symmetrically or quasi-symmetrically. Symmetrical path selection, in a TST network, uses identically numbered time slots through the center space stage

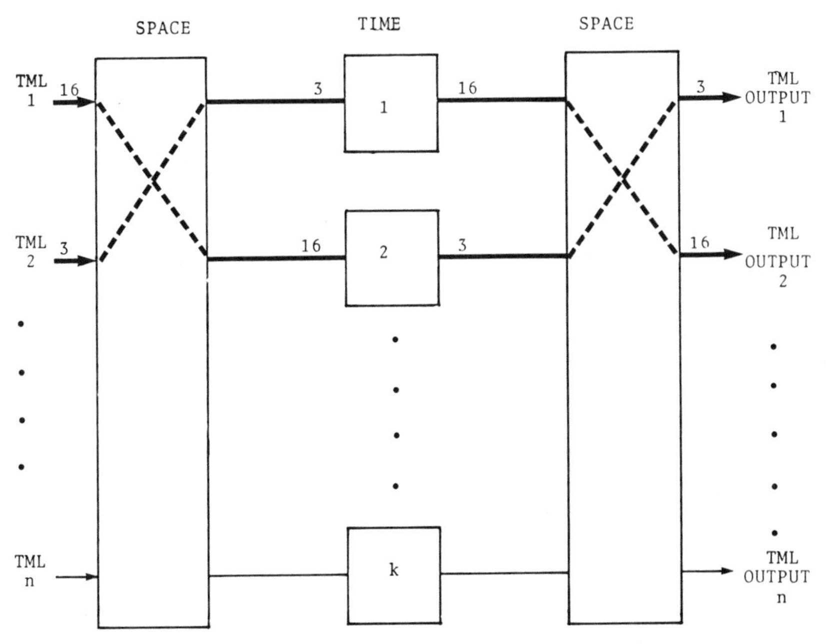

Fig. 11–5. Space-time-space (STS) switching network.

for both speech directions but cannot connect speech paths in the same highway. In an STS network, the two speech directions are switched symmetrically through the same TSI in the center stage. With quasi-symmetrical selection, the two speech directions use two related TSIs controlled by one control memory in an STS network, while they conventionally use even/odd time slots through the space stage of a TST network, permitting connections within the same highway.[11]

11.3.2. Economic and Traffic Considerations

The economics of PCM switching network design must always be weighed against the traffic-carrying capacity of the network. One trade-off is the cost of crosspoints for space switching versus the cost of memory for time switching. For several years, the cost of memory decreased at a faster rate than the cost of logic gates. Currently, however, large scale and very large scale integration is significantly reducing the cost of gates, although the effect is limited by the cost of pin arrangements to access the speech paths. Because of the rapid advances being made in solid state technology, the cost ratio between logic gates and memory depends upon the technological state of the art at the moment and upon the relative costs of control. Time switches capable of switching 4096 time slots have been designed.[12] Generally, it appears that time switching is still more economical than space switching. It must be remembered, however, that time switching produces transmission delay.

Studies have shown that separated switching generally is more economical than combined switching. However, there are some exceptions. Certain multistage configurations with space switching in the first and last stages are more economical if combined switching is used.[13]

The number of time slots (TS) per time-multiplexed line has an effect upon both cost and traffic capacity. Generally, in multistage networks, if the number of terminations and traffic capacity per termination are held constant, switching network costs are reduced if the number of time slots per multiplexed line are increased. This enables the number of multiplexed lines to be decreased, saving logic gates by making the space switches smaller. This also reduces the number of bits required in the space switch control memory but increases the number of bits required in the TSI control memory. This is true for both TST and STS networks. The probability of blocking is reduced in TST networks but is increased in STS networks as the number of time slots per multiplexed line is increased. To retain the same probability of blocking, an STS network must be expanded when the number of time slots per multiplexed line is increased. Such action still results in a slight decrease in comparative cost at traffic capacities of about 0.8 erlang per termination or higher.[14] At lower traffic loads, however, a lower loss probability is achieved with fewer time slots per multiplexed line and more multiplexed lines.[15]

Another architectural judgment involves the number of center stages in the switching network. With a relatively small number (e.g., 30) of time slots per multiplexed line, a three-stage PCM switching network is comparable in cost to one having four or more stages up to about 3000 terminations. With 120 time slots per multiplexed line, comparability extends up to about 20 000 terminations. With more terminations, however, multiple center stages are more economical with fewer time slots per multiplexed line.[16] With more time slots per multiplexed line, costs of three-stage and multiple-center-stage networks are roughly equivalent, but the additional center stages may be needed to reduce the probability of blocking.

The most easily recognizable architectural issue is that of the relative merits of TST (or T—T) and STS (or S—S). This decision is influenced by the factors identified above and others, such as whether data is serial or parallel. There is no clear judgment independent of other factors. Generally, for a given grade of service, expansion in the center stages is somewhat easier and less expensive in T—T networks. However, since the cost of the switching network rarely exceeds ten percent of the cost of the switching system, other factors generally control the decision on the type of network.[17]

11.3.3. Digital Symmetrical Matrices

The foregoing discussion of switching network architecture was in terms of separate modules of time switching and space switching. Successful switching system designers have designed complete switching networks consisting of identical modules of combined time and space switching. Each switch element contains time switching, space switching, and microprocessor control for switching any of 30 channels in any of 16 incoming PCM links (time-multiplexed lines) to any of 30 channels in any of 16 outgoing PCM links. The PCM links are bidirectional, thus serving 16 two-way PCM ports with 30 multiplexed speech channels each. Switch elements can be arranged in multiple stages.[18]

Combined time and space switches have been designed on a single chip with large scale integration (LSI) and are called digital symmetrical matrices (DSM). Port capacity is limited by the number of pins on IC cans, chip complexity, and power consumption. Using a 28-pin DIL can, a DSM can support eight bidirectional PCM links and dissipate about 200 mW of power.[19] With very large scale integration (VLSI), the potential application of DSMs in digital switching networks appears to be very promising. Analyses have shown that networks composed of DSMs can be designed to be nonblocking and are most cost effective if the least possible number of stages is used.[20]

11.4. SERVICE CIRCUIT TECHNIQUES

Service circuits are common or shared equipment units, implemented in hardware or software, associated with communication paths during the progress of calls. They often are referred to as *pooled common equipment.* Examples of service circuits include tone generators, tone receivers, signaling transmitters, signaling receivers, ringing generators, recorded announcements, conference circuits, and echo suppressors (cancelers).

In analog switching systems, service circuits generally are hardware units which are physically switched into and out of lines and trunks to provide specialized functions. In digital switching systems, service circuits may be implemented in hardware or in memory, or mixed. Incoming tone address signals may be picked off in the terminal interface module and sent to a signal processor or they may be switched through the digital network to hardware units or to memory locations for interpretation and registration. Outgoing tone signals may be generated by hardware oscillators or in memory.

Since most local digital central office switching systems have or will have direct digital interface to remote digital switching units via digital transmission facilities, the trend is to implement service circuits digitally whenever practicable.

11.4.1. Tone Generation

Digital generation of audible and signaling tones in PCM switching systems is constrained by 8-kHz sampling and 8-bit quantization according to the μ-255 companding law. Fundamentally, digital tone generation is accomplished by storing digitally encoded tones in read only memory (ROM) and then reading out the contents to be decoded for the listener. The PCM constraints result in a unique relationship between the frequency or frequencies of the desired tone signal and the number of samples which must be read out to reproduce the exact tone. One cycle of a presynthesized waveform, representing a single tone or a mixed tone, is stored in ROM. When read out under control of a counter incremented at an 8-kHz rate and decoded either locally or at the distant end of a circuit, the analog tone signal is reproduced. For example, eight words in memory are required to reproduce the digital equivalent of a 0-dBm, 1000-Hz test tone applied at the zero transmission level point. This tone is known as a *digital milliwatt.* Digital tone generators have several advantages over analog tone generators. They do not require impedance matching or amplification, and they do not drift in frequency.[21] If tones require interruption, such as busy tone, they are interrupted by turning the ROM on and off at the precise intervals required to produce the desired cadence.

11.4.2. Tone Reception

In a digital switching system, it is desirable, from a control viewpoint, to treat all incoming signals in the voice path uniformly, bringing all incoming tone address signals through A/D conversion and switching them through the PCM system. Given the conversion of the analog tone signals to PCM, it is equally desirable to interpret and register them digitally to save the hardware and D/A conversion necessary to use conventional tone receivers. Several approaches to digital tone detection have been used with varying degrees of success, but most designers have settled on either digital filtering or the discrete fourier transform (DFT).

Digital signal detection involves single frequency (SF) tone signals used for supervision (Section 10.5.2.3), DTMF address signals on subscriber lines (Section 10.5.3.2), and MF address signals on trunks (Section 10.5.3.3). Analog signal receivers have rather wide tolerances to compensate for distortion caused by aging transmitters, variations in subscriber keying characteristics, and transmission line impairments. Such distortion compounds the problem of digital recognition.

In the digital filtering method, the PCM samples are linearized and passed through a series of digital bandpass filters centered at each signaling frequency. The filtered signal power at each frequency is measured repeatedly to detect the presence or absence of power for each tone. A signal processor then interprets the signals and translates them for call setup. This method requires a substantial amount of memory and rather elaborate arithmetic calculations.[22]

The DFT method also extracts a measure of signal power at each signaling frequency but must exercise care to detect signals of minimum specified duration. When used with tone supervision, protection against speech power operation must be provided. One system which can handle tone supervision for 128 analog channels and satisfies CCITT standards contains 333 ICs and consumes 44 watts of power. The same principle can be used for recognition of DTMF and MF address signals, but must be far more elaborate to cope with the multiplicity of signals.[23]

While both of the methods described above are used for digital recognition of tone signals, some designers prefer to pass the signals through the digital network and a decoder to conventional analog tone receivers. Others, using analog concentration in the terminal interface area, connect the voice channels to tone receivers at that point. With increasing advances in LSI and VLSI, however, the trend appears to be toward fully digital recognition of tone signals.

11.4.3. Digital Conferencing

Analog conference calls typically involve a conference bridge which adds all signals together so that each conferee is heard by all others. With PCM switching systems, the same conference bridge can be used, but it requires converting all signals back to analog and then encoding the composite conference signals for transmission to all conferees. With two-wire transmission on subscriber lines, all signals are passed through a hybrid network for conversion to separate transmit and receive paths required for the digital switching system. When several conferees are thus connected through a digital switching network to an analog conference bridge, the multiplicity of hybrid network impedance mismatches can cause severe instability. This can result in severe echo or even singing, irrespective of the introduction of some attenuation. Therefore, in a digital switching system, digital conferencing is preferred.

There are two basic methods of performing digital conferencing in a companded PCM system. The first method involves an instant speaker algorithm in a switching-type bridge. During conversation, the probability of speech power in any random 8-kHz sample is quite low. If the speech level in each sample of several conferees is compared to identify the sample having the largest binary number, the sample is generally that of the active speaker. The conference processor then transmits that sample to all conferees during the next frame while blocking all other samples. This method involves rather simple circuitry for three-way conference bridges but becomes much more elaborate for larger conferences. The effect of the switching-type conference bridge is similar to that of a long trunk connection with echo suppressors.[24]

The second method involves a summing-type conference bridge using a summing algorithm. Since standard PCM uses a companding algorithm, simple binary addition is not possible. The compressed PCM samples first must be expanded to linear samples. Then all samples in a single conference frame can be summed arithmetically and compressed for transmission to conferees. During the summing process, the sample representing each speaker's own voice signal is removed from the summed data before it is compressed and sent back to that conferee's receiver. Depending upon the number of conferees, some attenuation may be added to the composite signal before compression. The effect of echo is reduced if the sign bits of one-half the incoming samples are inverted, producing a phase inversion to give the conference better stability.[25]

One designer combined elements of the two conferencing methods described above as shown in Fig. 11–6. A loudness analysis circuit estimates the loudness of each conferee's signal. The loudest signals are expanded and summed in pairs before recompression and transmission to conferees. Attenuation over a range of 0–20 dB is selected for each speaker based upon that signal's actual loudness as well as its relative loudness compared to others. This is designed to provide a gradual blending of voices when one speaker interrupts another and to provide zero loss for the path from the conference speaker to other conferees.[26]

11.4.4. Digital Recorded Announcements

In the public telephone network, recorded announcements must be available to many subscribers simultaneously. Analog announcement machines generally use a magnetic disk or a continuous magnetic tape. Digital recorder-announcers are an outgrowth of the same techniques used in digital tone generators and digital conference circuits. Speech segments are stored in memory and read out under processor control. Since standard announcements contain substantial redundancy of expression, memory can be conserved by recording the announcements in segments which can be arranged in the sequence needed. Memory requirements can be reduced further if the speech

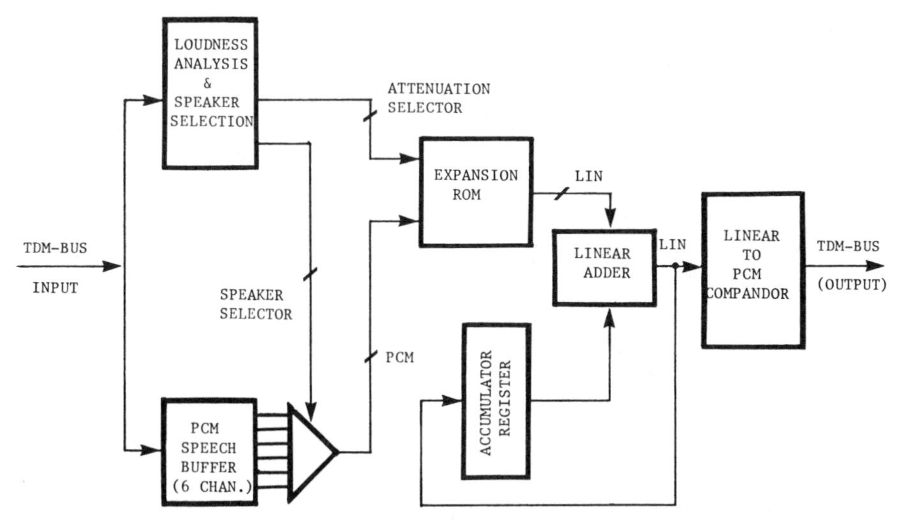

Fig. 11-6. Digital conference circuit. (From E. A. Munter, "Digital Switch Digitalks," *IEEE Communications Magazine,* Nov. 1982, © IEEE.)

samples are stored in 4-bit ADPCM instead of 8-bit PCM. In such case, the speech is converted from PCM to ADPCM when it is recorded. An ADPCM-to-PCM decoder is used to provide the PCM-to-ADPCM encoder with prediction values and scaling factors. Speech is read out in segments, including phrases, subphrases, and pauses to multiple output channels.[27]

11.4.5. Digital Echo Suppressors

Echo suppressors or echo cancelers are used to control echo on network trunks having a round trip delay of over about 45 ms and on certain other trunks. For many years, echo suppressors have been provided on a per-trunk basis at each end of the trunk. Now, digital technology permits echo suppressor modules to be provided in digital switching systems and inserted into trunk connections as needed.

In normal operation, voice-actuated echo suppressors attenuate signals in one direction (echo) while permitting signals in the other direction (talker) to pass through unattenuated. They can be overridden by both talkers speaking simultaneously. Suppressors can be disabled for data communciations by transmission of a 2100 Hz tone in each direction followed by continued presence of a strong data signal.

Digital echo suppressor modules can be provided on a pooled basis and switched into and out of trunk connections as required under processor control. The processor prevents connection of multiple echo suppressors in tandem. The echo suppressor terminal in the No. 4 ESS toll switching system, for example, can serve up to 1680 trunks simultaneously. It requires only a small fraction of the space and uses only a tenth of the power of conventional analog suppressors.[28] Other digital echo suppressors are packaged in various configurations.

When satellite circuits are involved in telephone connections, conventional echo suppressors do not provide sufficient suppression of echo, and the resulting speech clipping is not acceptable to the general public. Digital echo cancelers (see Section 5.7.2) enable satellite circuits to be used in general-purpose telephone connections. As integrated circuit technology permits continuing cost reductions, echo cancelers are expected to be used increasingly on terrestrial circuits.

11.4.6. Digital Pads

In telephone networks, the level of signals is closely controlled between the end switching systems by a system of amplifiers to increase signal level and pads to attenuate it. In analog implementation, amplifiers and pads are separate devices. Pads are simple resistors arranged to attenuate signals a specified amount. In digital implementation, digital pads can either amplify or attenuate signals by specified amounts.

A digital pad is a lookup table in read only memory in which all possible levels of PCM samples are stored except the sign bit. Input PCM samples address locations in the same ROM, and a control input designates the amount of attenuation or gain required. Correlation of the two inputs produce the attenuated or amplified output. Digital pads are shared devices switched into connections which require loss or gain according to a transmission plan.[29]

Since digital pads actually obliterate the original bit stream, digital data communication is not possible where they are used. Therefore, digital pads are used only in PBX applications and in local central offices where their use can be restricted to nondata transmissions. Digital pads should not be used in analog-to-analog or analog-to-digital connections but may be used in digital-to-digital and digital-to-analog connections. Digital pads do result in reduced signal-to-distortion performance comparable to that caused by the use of back-to-back digital channel banks.[30]

11.4.7. Provisioning of Service Circuits

Three different performance criteria are used in the North American network for the provisioning of service circuits. The Ten High Day Busy Hour (THDBH) is the clock hour which has the highest average traffic load for the ten highest traffic days of the year, excluding Mother's Day, Christmas,

Table 11–1. Provisioning of Service Circuits.

SERVICE CIRCUIT ELEMENT	TIME FRAME	PROBABILITY OF ALL CIRCUITS BUSY
DTMF receivers	HDBH[1]	<5%
Interoffice receivers	ABSBH[2]	<1%
Interoffice transmitters	THDBH[3]	<1%
Tone circuits	HDBH	<0.1%
Ringing circuits	HDBH	<0.1%
Dial tone delay	ABSBH	Max. 1.5% > 3 seconds
	THDBH	Max. 8% > 3 seconds
	HDBH	Max. 20% > 3 seconds
Conference circuits	ABSBH	<1% of 3 port attempts
Announcement circuits	ABSBH	<1% with overflow to tone trunks

[1] High Day Busy Hour
[2] Average Busy Season Busy Hour
[3] Ten High Day Busy Hour

and high traffic days attributable to unusually severe weather or catastrophic events. The High Day Busy Hour (HDBH) is the busy hour of the one day of the ten which has the highest traffic during the busy hour. The Average Busy Season Busy Hour (ABSBH) is the clock hour which has the highest average traffic in the three highest traffic months of the year. The three months of the "busy season" do not have to be consecutive, and Christmas, Mother's Day, and high traffic days attributable to unusually severe weather and catastrophic events are excluded.

The standards of performance for most service circuits in the North American network are shown in Table 11–1.[31] Tone circuits and ringing circuits have the most stringent requirement and are provisioned more liberally than any other service circuit. The general engineering guidelines for pooled common equipment require that the most liberal provisioning apply to equipment necessary to terminate a call during call setup. Tone circuits, providing call progress tones, and ringing circuits fit that requirement. The least liberally provisioned equipment is that required to originate a call, following the principle that calls which already have entered the network should be given precedence over calls just being attempted. The dial tone delay requirements and DTMF receiver standards illustrate this principle.

11.5. CONTROL ARCHITECTURES

Three basic control system architectures, each with variations, are used in digital switching systems. All systems use stored program control, but some use wired logic processors in specialized applications. Central control, shared control, and distributed control systems have evolved with several variations, including hybrid systems, which make them almost indistinguishable as basic architectures.

11.5.1. Control Workload Distribution

A stored program control system is an aggregation of several separate programs which work together to process calls. Generally, there is an executive program which coordinates and schedules the work of specific application programs. The tasks performed by application programs are grouped in many ways. One program may be concerned only with the on-hook or off-hook status of lines or trunks, or both. Another may control path setup through all or only a portion of the switching network. A separate program may control service circuits, while another controls call routing. When all control functions are combined into one processor, the executive program controls and schedules the work of the application programs.

Not all of the work performed by control systems is related to call processing. The control system may be working at 30–50% or more of its capacity on housekeeping functions even when there is no traffic load at all. Processor time required to run the executive program and to perform the essential functions related to maintenance and administration is known as *fixed overhead.* That time is the same for all switching systems of a specific type irrespective of the level of traffic and the amount of equipment in the system. *Variable overhead time* is the time spent on tasks performed at a constant rate but at a lower priority than fixed overhead tasks. This time includes such tasks as detecting and responding to line and trunk originations.

The time spent in processing calls is highly variable and is dependent upon the level of traffic. This time is zero if there are no demands for service. Once initiated, however, call processing is performed on a priority basis ahead of variable overhead time. The remainder of the work required by a control system is *deferrable work time.* This includes useful tasks which do not have critical timing requirements, such as administration, audits, and routine maintenance. Audit programs check the status of resources such as hardware subsystems, data blocks, and overall system operation. If the switching system is operating at full traffic capacity, the control system performs no deferrable work. As a peak traffic load subsides, however, the control system uses the increasingly available time to perform deferrable work. Deferrable work is sometimes known as *fill time.*[32]

11.5.2. Central Control Systems

In a central control system, all work described in Section 11.5.1 is performed by a single processor. In a small or medium-size PABX, this generally is a microprocessor or minicomputer. In a large central office, however, a very powerful computer is required.

Tasks may be scheduled in different ways. In a timed schedule, tasks are grouped into classes and assigned to fixed time increments, or slots. Scheduled classes in each slot are arranged in priority sequence. The last class of tasks in each slot is comprised of fill work. During each time increment, scheduled classes of tasks are performed sequentially. When all scheduled tasks have been completed, the remainder of the slot time is occupied with fill work. If traffic becomes so heavy that there is insufficient allotted time in the slot to complete all scheduled tasks, the executive program must resolve that situation and control whether to complete scheduled tasks in the slot or to proceed to the next slot of tasks.

Central control systems can follow a cyclic schedule of tasks. All classes of tasks are arranged sequentially by priority and by required frequency of performance. Thus, one class of tasks may be performed several times as often as another class. There is no fixed cycle time. At low traffic loads, the time required to cycle through all tasks is short. As traffic builds up, cycle time increases and fewer complete cycles are executed in a given time period. If a particular task takes an exceptionally long time to complete, the time between executions of higher frequency tasks could increase to a point at which it would exceed a threshold established to assure a particular quality of service. In such cases, the executive program can limit the amount of work done in a class so as not to jeopardize system operation. For example, during traffic overloads, a maximum limit can be placed on the number of new call originations that will be processed during the task of recognizing new calls to be processed. This maximum limit can be set so that new calls that are recognized are processed efficiently, enabling the system to reduce the backlog of work in other critical tasks. As the backlog of work decreases, the task of recognizing new call originations will occur more frequently, and more new calls will be allowed into the system per unit of time. Cyclic schedules function well in heavy traffic and require low overhead.

Hybrid schedules can be designed so that tasks having critical timing requirements are executed according to a timed schedule while other tasks not having such timing requirement are executed according to a cyclic schedule. Various combinations are possible. One arrangement is to perform cyclic tasks when there is no other work to be done and interrupt them at each time interval when timed scheduled tasks are to be executed. Another arrangement involves grouping tasks into classes according to their required frequency of performance as in the cyclic schedule. In this case, however, a minimum time between executions of each class is specified. Thus, the time schedule controls task execution during light traffic periods. During heavy traffic, the cyclic schedule controls task execution with interrupts to conform to the timed schedule. This arrangement assures timely execution of tasks having critical timing requirements while taking advantage of the attractive characteristics of cyclic schedules in heavy traffic.[33]

The complexity of central control systems increases with size, and failures can be catastrophic. Redundancy is essential for high reliability in large switching systems and is highly desirable in any system. The central control generally is fully duplicated, and continuously monitoring maintenance programs effect automatic switchover in case failures exceed a threshold. Even then, central control systems have been known to fail completely and to require many hours to restore service.

11.5.3. Shared Control Systems

In a shared control system, multiple processors (generally two) share the load. Each processor works at only 50% capacity while the switching system is operating at full rated capacity. This concept reduces the probability of processor failures during periods of extremely heavy random traffic, and the reserve processor capacity is able to more easily absorb peak traffic loads without jeopardizing critical functions. Each processor can monitor the performance of the other and can handle the full rated switching system load in case the other processor fails. Processors can be triplicated, in which case two processors share the load while the spare monitors their operation and is automatically switched into service if one fails. The processor complexity, however, is the same as in central control systems because each processor handles all control functions.

11.5.4. Distributed Control Systems

The definite trend in switching system architecture is toward distributed control systems. Processors may perform specialized tasks only, or they may perform all or virtually all control tasks associated with a group of lines and trunks.

11.5.4.1. Distribution of Control by Function. A popular method of distributed control is to allocate a group of specialized tasks to a controller which does nothing but execute those tasks repeatedly. The application programs for such tasks are relatively simple and are ideally suited to microprocessor technology.

A typical example of a specialized processor is that of a line processor which continually detects on-hook and off-hook conditions of a group of lines, sends that information to another processor, and starts and stops ringing. Some central control systems have allocated signal processing tasks to a separate, stored program control processor to perform such tasks as line and trunk scanning, reception of incoming address signals, and control of outgoing address signals.

11.5.4.2. Distribution of Control by Block Size. There is an increasing trend toward allocation of all or most processing functions to a controller for a group, or block, of lines and trunks. In a typical distributed control system, a central controller maintains orderly operation of a number of block controllers and may also engage in path selection, maintenance diagnostics, and administration. In this case, a terminal module processor performs all tasks associated with a group of lines and trunks, including path selection within the module, while a main processor manages centralized data base information, performs path selection between terminal modules, and executes other centralized programs. A major advantage of terminal module control is that a switching system can grow modularly without incurring an extremely high initial cost for a powerful central computer designed for ultimate system size. A second advantage is that virtually autonomous terminal modules can be located in outlying areas and operated as remote switching modules with host control of centralized functions.[34]

11.6. MAINTENANCE DIAGNOSTICS AND ADMINISTRATION

In addition to call processing, a digital switching system must have programs to perform diagnostics, collect billing and traffic data, provide for program and data base changes, and make traffic load analyses for network management purposes. In electromechanical and even electronic analog switching sytems, adjunctive devices performed many of these tasks. With integrated circuit technology, however, these programs should be designed into the system from the conceptual design stage onward.

11.6.1. Maintenance Diagnostics

An ideal design objective is a self-healing switching system which can detect and correct all faults without human intervention. In the absence of such perfection, diagnostic programs must be designed to detect and correct faults whenever possible and to alert maintenance personnel to the existence of uncorrected faults. System effectiveness and maintainability are directly attributable to the consideration given to them during architectural design. Functional units which are critical to maintaining continuity of service should be duplicated so that a unit can be removed from service without impairing call processing capability. Traffic-sensitive common equipment, such as service circuits, generally is provided on the basis of $n + 1$ redundancy.

Equipment should be grouped in such a manner as to facilitate maintenance procedures. This enables a group of hardware units to be removed from service without affecting other portions of the system. This includes partitioning of power supplies such that power supply faults affect only a limited group of equipment.

11.6.1.1. Maintenance Phases. System maintenance involves seven basic phases: *fault detection, fault analysis, fault isolation, fault reporting, fault localization, fault clearance,* and *service restoration.*[35] While functionally separate, these phases may overlap.

There are two common methods of fault detection, and both are used concurrently. Hardware and software may be designed to activate alarms and initiate diagnostic tests when call failures or other fault conditions occur. Other faults are detected by on-line tests performed routinely as part of the operating program, or they may be performed as audit routines to check blocks of hardware and software.

Ideally, fault detection processes should also isolate the fault to a particular group of equipment units. In case a fault involves more than one equipment group or, for some other reason, is not readily identifiable with an equipment group, further analysis is needed. Failure of path setup, for example, may result from a fault in any of several equipment groups. Analysis may be performed by logical deduction, by historical comparison, or by specific triggering of diagnostic tests.

Once the fault is identified with a specific equipment group, action must be taken to isolate the fault from the rest of the system to avoid further degradation. Traffic may be diverted to a redundant equipment group if one exists, or a faulty group may be locked out of service.

Statistical reporting of faults is essential for effective maintenance. Printed records of fault characteristics and the results of automatic diagnostic tests serve two major purposes. They assist technicians in further actions to clear the fault and restore full service, and they provide a historical record which can be used to develop engineering changes to make the system more reliable. Visual and audible alarms may accompany printed reports when service impairment exceeds a preset threshold.

The next maintenance phase is fault localization. Its purpose is to identify specific faulty equipment within the equipment group. A fault may be localized automatically to a specific printed circuit board by the diagnostic process, which is the ideal objective, or it may require action by a technician. This is a trade-off between the ideal objective and the cost to achieve it. On-demand diagnostic programs can be initiated by technicians when needed to localize the fault.

Fault clearance involves physical replacement of the defective board or boards. When a board is replaced with a spare, a test program should be run to test the spare board before returning the faulty equipment unit to service. Faulty boards are sent to a repair shop where off-line tests are performed to ascertain the exact trouble and to effect repairs.

The final phase is restoration of service. After the equipment unit with its replacement board is tested, the entire equipment group is functionally tested. When this test is satisfactorily passed, the group can be returned to service in the switching system.

11.6.1.2. Diagnostic Methods. A major advantage of using a 32-channel PCM format in a switching system is that 30 channels can be used as traffic channels and two channels are available as maintenance and control channels. Some of the 16 bits thus made available in each frame can be used as control messages. Some can be used for parity checks while others can be used for integrity checks.

On-line tests can be used to check control paths between a controller and the equipment units controlled. Detailed checks can be made to verify order execution, data validity, and transmission accuracy. Orders can be looped back and compared. Data can be written into memory, then read out and compared. Network paths can be set up and checked for both continuity and data integrity. Codecs can be checked separately and in connection with network path checks. Signaling transmitters and receivers can be checked by connecting them through the switching network and comparing sent and received signals.

Numerous system-specific parameters are counted as events occur and compared against an alterable threshold. Such events as path check failures, parity failures, signaling timeouts, clock alarms, power fluctuations, loss of framing alignment, and many others are counted and recorded. When a threshold is exceeded, the data are printed out and may be accompanied by audible and visual alarms depending upon the predetermined urgency of that parameter. Some thresholds, when exceeded, may trigger automatic actions to initiate diagnostic tests, switch to spare equipment units, or curtail certain types of traffic. Periodic polling of equipment units can be performed on either a positive-response or negative-response basis. The positive-response concept looks for a response only from a defective unit. However, the unit may be so disabled that it cannot respond or it may be so busy with traffic that a response is seriously delayed. A negative-response concept looks for a positive response from all equipment units which are operating within established parameters. This more positively identifies faulty equipment units but does place an additional workload on the affected control systems. Both concepts are productive when judiciously used.

Since the mean time between failures of any equipment unit in a well-engineered switching system is much greater than the mean time to repair that failure, most maintenance strategies are designed around an assumption that a trouble is the result of a single fault. In most cases, the single-fault assumption is correct, but not always. The prioritization of application programs (Section 11.5.1) often results in many maintenance programs being defined as deferred workload. Therefore, most of the maintenance logic of those programs is tested only during low traffic periods. Units under test are removed from service, tested, and returned to service. When a fault occurs during a high traffic period, the logic is called upon to isolate the fault while fully duplicated equipment is on line. This usually involves removing a unit from service before testing, which removes it from the effects of heavy traffic when actually a unique combination of heavy random traffic may have uncovered a program error which would occur only during such rare conditions. Multiple faults and transient errors do occur and do complicate the maintenance programs.[36]

The most critical maintenance programs are those concerned with severe degradation of control systems which require restart of executive programs to recover system sanity. A common cause of system recovery is the mutilation of memory data. Other causes include loss of a vital function, loss of a major facility, problems in control software, multiple failure of redundant units, clock failure, and excessive failures of various types. Most systems have system-dependent recovery phases. When a system is first activated, initialization of all control programs is required. Afterward, required restarts apply only to portions of the system which are in trouble unless the system is shut down entirely. The first recovery phase generally saves all established calls and reinitializes temporary memory. The second phase may lose some calls but saves most of them; calls in process of setup are generally lost. The third phase reinitializes major portions of the system; all or most calls are lost. The fourth phase involves reinitialization of the entire system, and all calls are lost. The sequence and effect of the various recovery phases varies with the system.[37]

11.6.2. Administration

Switching system administration includes data base management, generic program changes, and collection of data for billing, traffic engineering, service evaluation, and network management.

11.6.2.1. Data Base Management. The purpose of data base management is to provide the capability of administering the switching system data efficiently on a continuing basis. All insertion, deletion, or modification of data is controlled by the data base management system. The data used in the switching system is organized into formats suitable for processing and for use in system memories, but not for administration. Data base information must be available in clearly readable form. Therefore, the data base management system provides translation between the administrative view of the data and the machine language formats.

The switching system data base contains all configuration data pertaining to switching system equipment and connectivity including data on subscribers, lines, trunks, service circuits, switching network equipment, call routing, features, traffic data collection schedules, and billing information. Access to the data base is provided by the data base management system for both administrative and maintenance purposes. The data base management system is provided in one or more of the processors associated with the switching system. An important prerequisite is that it must be designed into the switching system and not added as an afterthought.[38]

11.6.2.2. Generic Program Changes. A generic program is the set of executive program instructions that controls call processing, maintenance, and administrative functions. A means must be provided to modify the program to correct program errors, to provide new features or feature enhancements, to provide system performance enhancements, or to consolidate accumulated program changes.

Program updates are changes to small portions of a generic program to correct program errors or to make minor changes in call processing. Program retrofits are major revisions in a generic program, which should be implemented in a working system with minimum impact on subscriber service.[39]

11.6.2.3. Data Collection. Switching systems generate large quantities of statistical data in connection with call processing. Some of the data is used for billing purposes. Billing options for local calls vary according to tariffs which are effective in each locality. Generally, message accounting systems used for toll billing can also be used for billing local timed message-rate calls. For toll billing, call data includes calling party identification, called telephone number, date, time call is answered, duration of call, and time call is disconnected. Switching system designs must include not only the capability to record all data required for long distance calls but also options to record data needed for billing of services under many tariff variations and for special studies. Timing of calls must be adjusted for inaccuracies in timer operation, path setup, time delays after called party answer, and clock variations. If conference bridges are provided for three-way calling or larger conferences, timing must include the usage of conference bridges in billing data. Since billing is tariff-dependent, the data collection system must be easily alterable.

Data collection also includes service measurements which are used to evaluate the quality of service being provided. Service measurements include total switching system traffic counts segregated by type and disposition of calls, customer access statistics, switching system irregularities, network service statistics, and maintenance measurements. While many of the measurements are standardized, some are dependent upon switching system design. Schedules for data collection vary from every 30 seconds to every 24 hours.

11.6.3. Traffic Administration

Much of the collected data is used for multiple purposes, and traffic administration is one of the dominant users. Traffic administration comprises those actions necessary to assure that equipment and circuit quantities are provided in sufficient number to provide a specified grade of service for assumed traffic loads and includes both traffic measurements (data collection) and traffic engineering.

Traffic is simply the flow of messages through a communications system. Teletraffic theory is based upon statistical probabilities and certain assumptions concerning telephone traffic. Four universal assumptions which are dominant in teletraffic theory are:

1. Traffic originated by a large number of sources is random and independent of all other subscribers.
2. The holding times (or duration) of individual calls and trunks are exponentially distributed.
3. The holding times of certain types of common equipment (i.e., signaling units) of a given type are constant.
4. There are random variations of traffic loads hourly, daily, and seasonally.[40]

There are two types of traffic measurements used in traffic administration: peg counts and usage. *Peg counts* measure the cumulative number of occurrences of events during a specific time period. *Usage* measures the duration of a specific condition such as the busy state of a line, trunk, or unit of equipment, called a *server*. Peg counts should be 100% accurate. Usage may be measured as actually occurred in seconds, the most accurate but generally the most costly method, or it may be calculated from a peg count obtained by scanning or sampling at specified intervals. The scan rate is related to the average holding time of the servers in the group being measured.[41] If the scan rate is at least one-half the average holding time, the result should be highly accurate. A scan rate of one per second is commonly used for service circuits, while a scan rate of one per 100 seconds for interoffice trunks produces acceptable results.

Generally, both peg counts and usage are required for the same system components or events when both measurements are meaningful. Total switching system measurements are recorded for significant parameters pertaining to originated calls, incoming calls, outgoing calls, terminating calls, calls initiated but not completed, feature utilization, and processor real time usage allocation. Standard measurements for trunks and traffic-sensitive components, such as service circuits, include peg count of total attempts, peg count of attempts which find all servers busy, usage of servers available for traffic, and usage of servers out of service for maintenance. The traffic data collection system also should compute dial tone delay and delay for service circuits. For network management purposes, selected parameters are accumulated for predetermined periods on a continuous basis and are available for readout on demand or when thresholds are exceeded.[42]

Selection of sensing points for traffic measurements requires a knowledge of how the data will be used. In teletraffic theory, offered traffic load consists of random demands for service from a large number of sources to be connected to a finite number of servers via a switching network. Servers may consist of a group of trunks or service circuits. If all servers are simultaneously busy, the next demand for service, or call attempt, is either blocked or delayed, depending upon the function of the servers. If the servers are a group of outgoing trunks, the attempt is blocked, or lost, unless there is an alternate route, in which case the attempt overflows to the alternate group of trunks. If the servers are subscriber digit receivers, the attempt will queue and wait until a server becomes available. All offered traffic which is not carried immediately by the server group is lost traffic, overflow traffic, or delayed traffic. Traffic relationships are illustrated in Fig. 11–7.

In determining the quantity of servers to provide in a group, the following factors must be considered:

1. Whether the servers will handle traffic on a blocking basis or on a delay basis;
2. The type of access to servers—full or limited availability;
3. The statistical characteristics of the offered traffic in terms of randomness and holding times;
4. The required grade of service; i.e., the amount of loss or delay that can be tolerated.[43]

Grade of service for blocking systems is expressed as probability of loss. For delay systems, it generally is expressed in terms of probability of delay exceeding *t* seconds and average delay of offered or delayed calls. The grade of service is determined by the offered load, the number of servers, and the underlying assumptions of traffic characteristics. Three basic formulas in

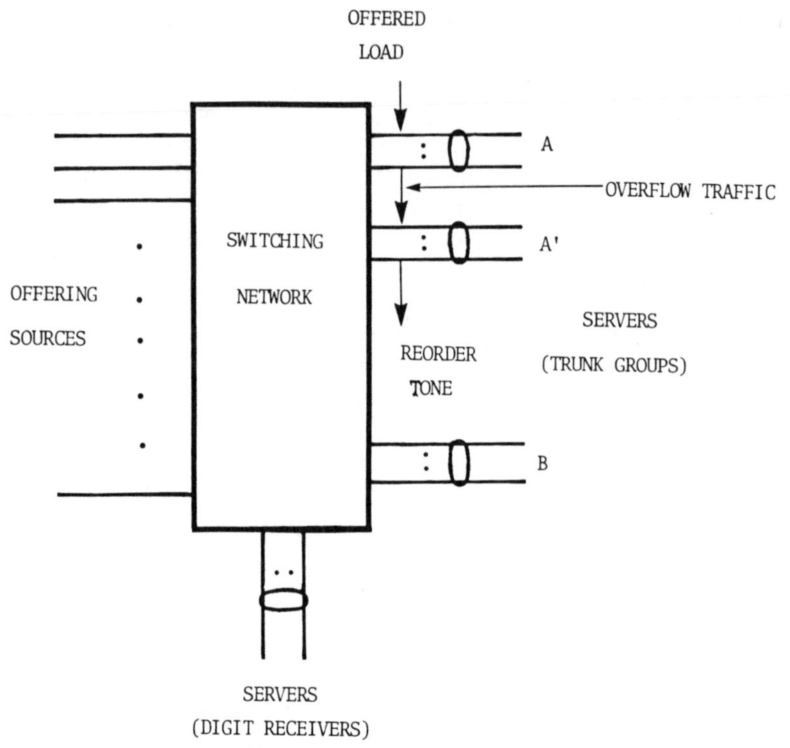

Fig. 11–7. Traffic relationships.

prominent use in North America are based upon three different assumptions as to the behavior of call attempts finding all servers busy.

The Erlang formula of the first kind, known as Erlang B formula or Erlang loss formula, is based upon the assumption that calls finding all servers busy will be cleared from the system and will have no further effect on it. This is known as the *blocked calls cleared* assumption. The probability of blocking for a call in traffic load a offered to s servers is expressed by

$$B_{s/a} = \frac{a^s/s!}{\sum_{k=0}^{s} \left(\frac{a^k}{k!}\right)} \tag{11–3}$$

where a is the traffic load in erlangs (one erlang = 3600 sec). In Fig. 11–7, traffic offered to trunk group A behaves according to the assumptions underlying the formula because calls finding all trunks in group A busy will overflow and be offered to trunk group A′. However, traffic offered to trunk group B does not overflow and likely will be retried within a short time or even immediately. For a two-way trunk group, traffic loads offered from each end are added to obtain the total offered load a.[44]

The underlying behavioral assumption for the Poisson formula is that a call finding all servers busy will wait for its intended holding time, then depart the system, and will not reappear for the remainder of the study period, usually an hour. If a server becomes idle while the call is waiting, the call will seize the server and hold it for the balance of its intended holding time. This is known as the *blocked calls held* assumption. The probability of blocking for a call in traffic load a offered to s servers is defined by

$$P_{s,a} = e^{-a} \sum_{j=s}^{\infty} \frac{a^j}{j!} \tag{11–4}$$

where a is the traffic load in erlangs and e is the natural or Naperian logarithmic base, a constant 2.71828.[45] Although the queue discipline of the Poisson formula is unlike any actual physical behavior of waiting traffic, it does offer some allowance for retrials and just happens to give a slightly better solution for blocking on a trunk group which has no alternate route for blocked calls. Though retrial habits of different groups of people vary widely according to circumstances, this formula does happen to interpolate between the Erlang B formula and the delay formula, Erlang C.[46]

The Erlang formula of the second kind, known as the Erlang C formula or Erlang delay formula, is based upon the assumption that offered calls which find all servers busy will wait indefinitely for a server to become available and then will occupy that server for its full intended holding time. This is known as the *blocked calls delayed* assumption and implies that the offered load is equal to the carried load. Perhaps the Erlang C formula can best be understood by its relationship to the Erlang B formula. Using the terms of Eq. (11–3), the probability of delay $P(>0)$ for a call finding all s servers busy can be expressed by

$$P(>0) = \frac{B_{s,a}}{1 - \frac{a}{s}\left(1 - B_{s,a}\right)} \tag{11–5}$$

where a is the offered load in erlangs. This formula is valid only when the offered load is less than the number of servers, when calls are handled in the order of arrival, and for exponentially distributed holding times. For constant holding times, formulas were developed by A. K. Erlang for only single-server applications, but graphs have been prepared for multiple-server groups for both exponential and constant holding times. Values of $P(>0)$

for constant holding times are only slightly lower than the corresponding values for exponential holding times. For holding times which are neither constant nor exponential, the data for exponential holding times should be used because they yield a slightly better grade of service.[47]

A comparison of the results of the three traffic formulas is shown by the curves derived from them in Fig. 11–8. The abscissa denotes the load in erlangs offered to a group of ten trunks, and the ordinate reflects the probability that a call attempt finds all servers busy. For the Erlang C curve, the ordinate reflects the probability of delay $P(>0)$. For low blocking probabilities up to about 1%, the Erlang C and Poisson curves yield the same results, and there is only a slight difference in results obtained from Erlang B. For example, at an offered load of 4 erlangs to a 10-trunk group, the Erlang B curve reflects a blocking probability of 0.5% while the Poisson and Erlang C curves show 0.9% probability of blocking. At higher traffic loads, however, there is some divergence of the Erlang C and Poisson curves because of the Erlang C assumption that calls wait indefinitely while the Poisson assumption is that they wait only a limited time. The greater divergence of the Poisson and Erlang B curves results from the Erlang B assumption that blocked calls disappear from the system immediately. An offered load of 6 erlangs to the 10-trunk group results in approximately 4% blocking under the Erlang B assumption and over 8% blocking under the Poisson assumption.[48] The true blocking probability is a function of retrial habits.

The assumptions that offered traffic is purely random and that it is from a large number of sources are not always true. For example, in Fig. 11–7, the traffic overflowing from trunk group A to trunk group A′ is from a single source and has a peakedness characteristic. This can be seen by examining the pen recording of the number of simultaneously busy trunks in a trunk group during a 45-minute period, as shown in Fig. 11–9. If the traffic shown represented the characteristics of the traffic offered to trunk group A in Fig. 11–7, and if trunk group A had 15 trunks, the overflow traffic offered to trunk group A′ would consist of spurts, or peaks, of traffic overflowing trunk group A at about 9:46, 9:58–10:01, and 10:23. If one were to assume that first-routed traffic, in addition to overflow traffic from other trunk groups, is offered to trunk group A′, it can be seen that *average* traffic intensity would not be a true representation of the total offered traffic load. The peakedness factor can be quantified by relating the peakedness of overflow traffic to that of random traffic. If random traffic has a peakedness of one, overflow traffic can have peakedness up to about four.[49]

Two theories, producing about the same results, were developed at approximately the same time to solve the problem of nonrandom traffic. R. I. Wilkinson, of Bell Telephone Laboratories, published an *equivalent random theory* in 1956, and G. Bretschneider published a *traffic variance method* the same

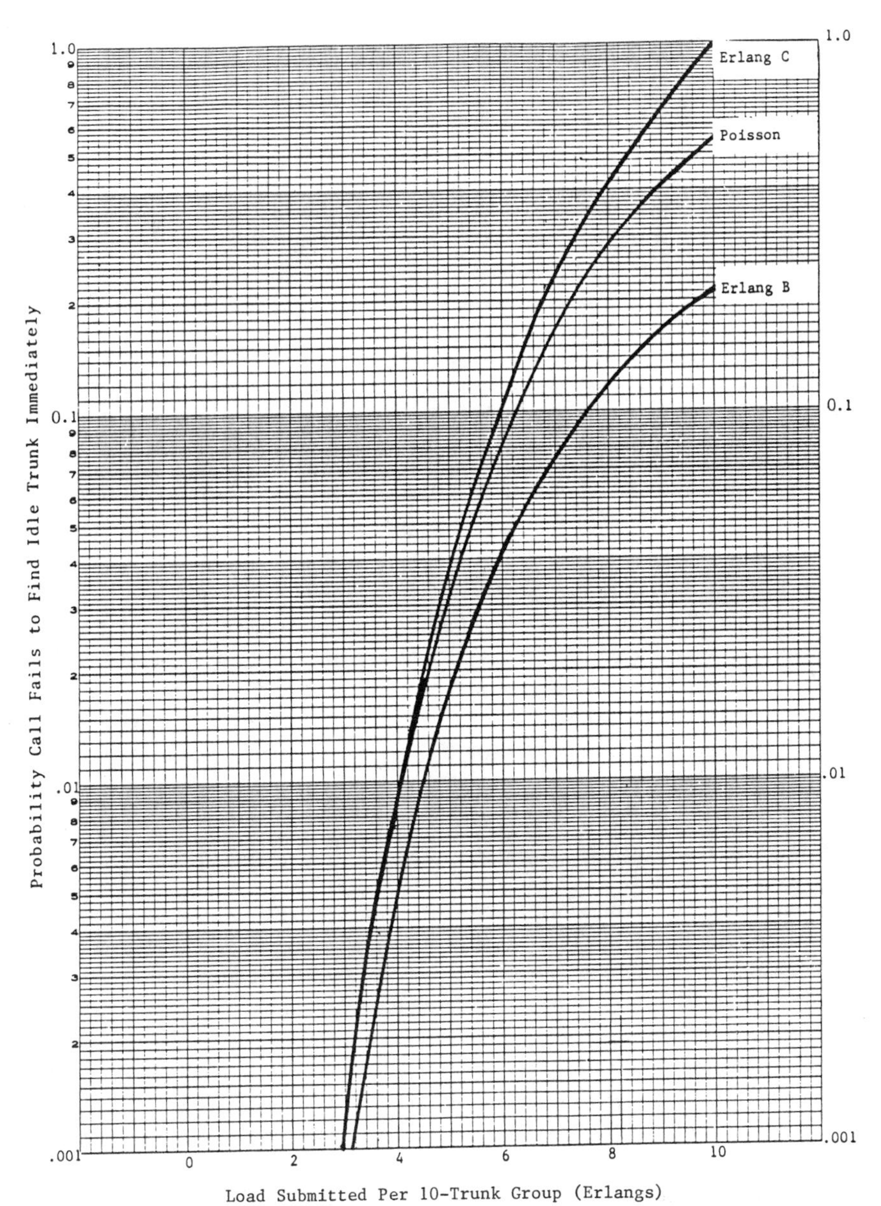

Fig. 11–8. Comparison of major traffic formulas. (From *Defense Communications System Traffic Engineering Practices,* Vol. XII.)

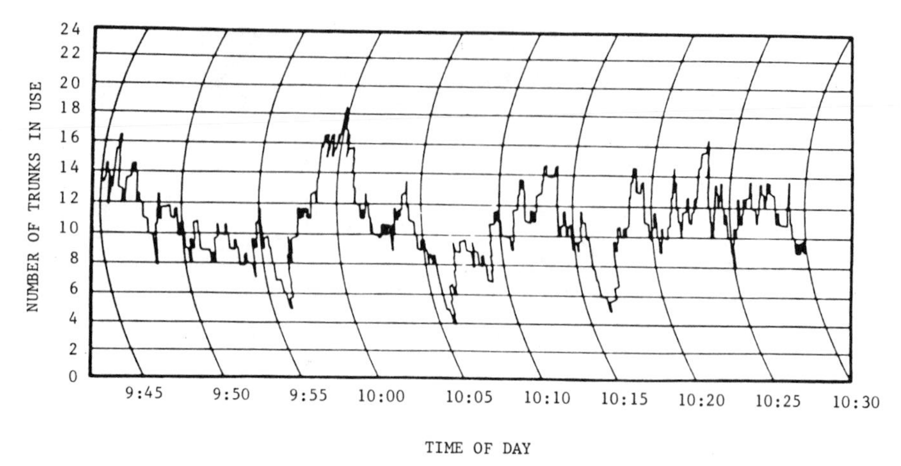

Fig. 11–9. Continuous load distribution during the busy hour. (From *Defense Communications System Traffic Engineering Practices,* Vol. I.)

year. The principles are the same; only the methods vary slightly in handling the variance. The principle involves finding a pseudotrunk group with random traffic input which is equivalent to the actual trunk group with peaked traffic. Wilkinson's method uses a series of curves based upon the mean and variance of the traffic; while Bretschneider's method uses tables and curves based upon the mean and random factor characteristics of the traffic to produce the same results.[50]

11.6.4. Network Management

A modern telephone network using extensive alternate routing and common control switching makes efficient use of transmission facilities but is subject to degradation during periods of heavy traffic overloads or major equipment failures. As traffic increases beyond engineered capacity, the use of alternate routing involves more trunks and switching systems per call. Under those conditions, more and more calls receive all-trunks-busy signals, and retrials add to the traffic demands on service circuits and control systems. This effect is much more severe with inband signaling than when common channel signaling is used.

The most degrading effect is switching congestion. If an incoming call finds all digit receivers busy, it will queue until a receiver becomes available or until the transmitter at the other switching system times out and routes the call to reorder tone or an announcement. Transmitter holding time has increased, and the control system has had to use real time to route the call twice—once to the trunk group and once to reorder. Under exceedingly heavy traffic overloads, the mean holding time of service circuits and the average call processing time increase. As the overload continues to feed on itself, ineffective attempts build up, switching congestion spreads, and the network carried load decreases. Finally, so much processor time is spent on ineffective attempts that trunks begin to be idle and trunk holding times become shorter.

This effect can be seen in Fig. 11–10, which shows the results of an event simulation of a network of about 50 switching systems. As the offered load built up toward the engineered load of 1600 erlangs, the carried traffic in-

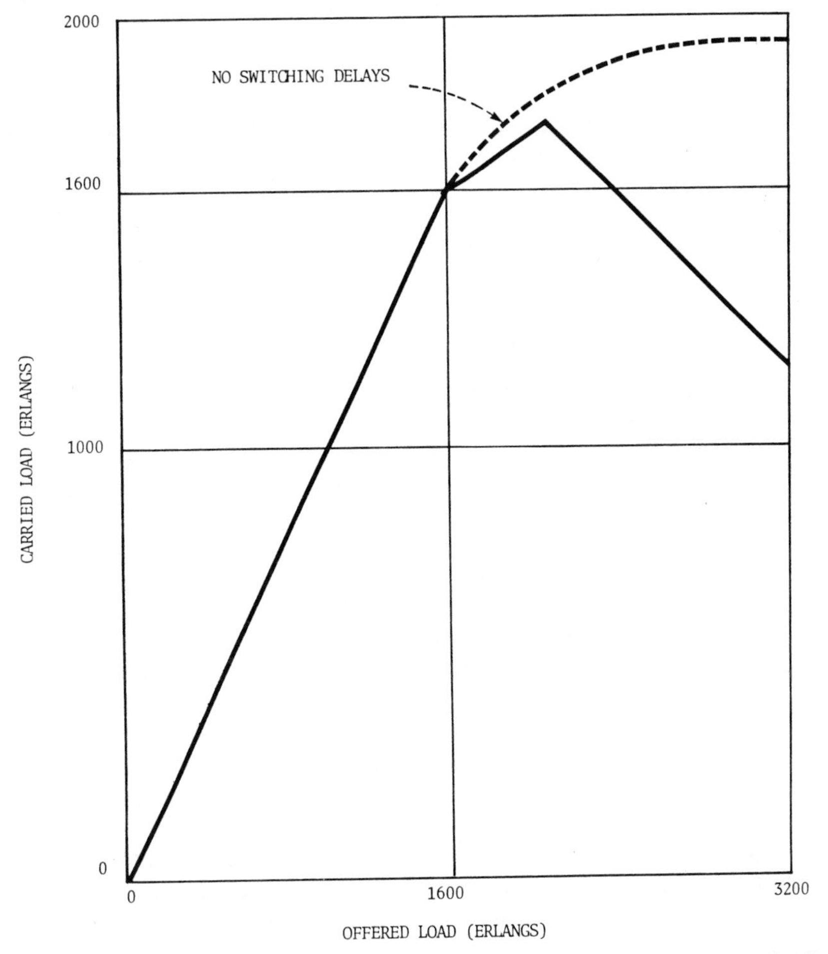

Fig. 11–10. Effect of switching congestion. (Copyright 1977 Bell Telephone Laboratories. Reprinted by permission.)

creased linearly. The dash line shows how the carried load would be affected with no switching congestion as the offered load increased to twice the engineered load. The solid line, however, shows that the carried load did continue to increase for awhile but at a slower rate. A point was reached—it generally occurs at about 20–30% above engineered load—where the carried load began to decrease sharply and continued to decrease even though the offered load continued to rise.[51]

Network management is real-time surveillance and control to enable a degraded network to operate as efficiently as possible. Network management is an art—not an exact science. Traffic analysis is used by network managers to take actions to implement controls, but there are an infinite number of possible situations that can be encountered in a large network, and each one must be treated individually.

11.6.4.1. Principles of Control. The overall objective of network management is to enable as many calls as possible to be completed under any condition of degradation. Certain principles have proved valid for large networks. The overriding principle is to *use the minimum controls necessary to enable the network to operate as normally as possible.* Other important principles are:

1. Inhibit switching congestion.
2. Use all available network capacity.
3. Keep available circuits filled with traffic which has a high probability of completing calls.
4. When all trunks are exhausted, give priority to calls needing only a minimum number of trunks to complete connections.[52]

11.6.4.2. Principal Controls Available. There are two major categories of controls available for network management. *Expansive* controls are employed to divert some traffic from a portion of the network which is overloaded to call destinations via other portions which are underloaded. *Restrictive* controls remove some traffic from overloaded portions of the network when that traffic has a low probability of completion. Most controls are of the restrictive type. Controls may be applied manually or automatically, and switching systems should have the capability to implement controls using both methods as appropriate. Dynamic controls generally are more effective than manual controls because they can be applied and removed instantaneously based upon traffic thresholds. The controls described in the following paragraphs apply to the AT&T network and have the greatest impact on switching system design.[53]

Alternate route cancellation is an effective control to protect overloaded switching systems. It may be implemented manually or automatically, but is always applied selectively. All traffic overflowing a specific route may be prevented from advancing to its next alternate route, or all overflow traffic may be prevented from advancing to a specific route. Both variations may be implemented manually, and the latter may be applied automatically upon receipt of a signal from a congested switching system.

Trunk directionalization changes two-way trunks into one-way operation. It has two major applications. In case of severe switching congestion or a natural disaster which attracts focused heavy calling, trunks can be partially directionalized in favor of the affected switching system. This will allow calls to depart the system, but the number of calls entering the system will be reduced. In addition, during periods of exceptionally heavy traffic throughout the network, such as exists on Mother's Day and Christmas, trunk groups can be partially directionalized in favor of traffic proceeding down a network hierarchy. That traffic may already have used several switching systems and trunks and may need only one or two more links to complete. In electromechanical switching systems, trunk directionalization is applied dynamically using manually set thresholds. In stored program systems, it can be applied through software.

Code blocking is used when one or more locations are virtually isolated from the network, generally because of equipment failure, and traffic to the affected destinations has little or no chance to complete. Calls to affected codes should be diverted to a recorded announcement at a switching location as near to the point of origin as practicable.

Recorded announcements are an effective way to increase the spacing between retrials. In one subjective study of calling habits, it was found that the average spacing between retrials was approximately doubled when attempts were diverted to announcements rather than reorder tone.

Call gapping limits the number of calls routed to a specific code or to a specific number. It is very effective in controlling mass calling by blocking some of the calls at or near the point of origin rather than at or near the destination switching system. An adjustable timer in stored program control switching systems, set at one of several time intervals from 0 to 600 seconds, allows one call to have access to the network and then blocks all others for the duration of the time interval, after which it allows another call to go through and recycles itself. A call gapping time interval set to infinity functions as a code blocking control.

Reroutes allow calls to be routed in other than the normal manner. This permits heavy traffic in one portion of the network to use idle capacity in another portion. Reroutes are particularly effective in cross-routing through different time zones and in routing calls around a major facility failure.

Common channel signaling systems reduce the need for controls which have their greatest application in a network with predominantly inband signaling. However, the CCIS network, irrespective of its redundancy, can be affected by traffic overloads, failures, and switching congestion. Dynamic overload controls are effective in controlling CCIS link overloads and in reducing the traffic load on congested CCIS processors.

11.6.4.3. Switching System Requirements. All switching systems connecting to the AT&T network should have automatically applied controls to maintain a high level of call processing efficiency in the presence of brief periods of peak traffic loads and manually applied controls to protect the system when it is subjected to extended periods of overload. Traffic measurements and overload indicators should automatically alert maintenance technicians to service degradation. Indicators and measurements should be designed to identify as accurately and as soon as possible the causes of overload.[54]

Requirements for implementing network controls and for furnishing network management surveillance data vary with the size of the switching system and its position in the network. Small local systems serving fewer than about 5000 lines and switching fewer than about 5000 trunk-to-trunk calls in the busy hour require only internal overload protection. Larger systems need varying network management capabilities depending upon size and function. Such systems must be capable of providing certain specified surveillance data at 30-second and 5-minute clock intervals to network management centers.[55]

Large and medium capacity local switching systems should be capable of applying code controls and trunk group controls manually, and large systems should be able to administer manually a hard-to-reach code list generated by No. 4 ESS toll switching systems and disseminated through CCIS.[56]

Automatic controls are divided into four categories. Toll and tandem switching systems should provide full capability for all categories. Local system requirements vary according to size capacity and network configuration. Dynamic overload control (DOC) thresholds comprise two levels of switching congestion and total switching failure. Those conditions need to be sensed and transmitted to connecting offices via CCIS or via dedicated circuits for up to 16 trunk groups with inband signaling. Large and medium capacity systems should be able to receive and respond to up to three levels of DOC signals within one second. Responses should cover a range of predefined controls which affect variable percentages of traffic destined for or via the affected switching system according to the severity of the condition. They also should be capable of implementing trunk reservation on both one-way and two-way trunk groups in favor of the switching system transmitting the DOC message. Responses must be flexible to reserve from 0 to 15 trunks for inward-only traffic. Large switching systems should be capable of receiving and automatically administering at least 128 hard-to-reach codes.[57]

PROBLEMS

11.1 What are BORSCHT functions and why are they potentially hazardous to digital switching systems using solid state technology?

11.2 What memory cycle time is required for the information memory in a time slot interchange to switch 512 time slots? How many bits are required in the control memory?

11.3 When time-multiplexed lines are switched through a space division stage of a digital switching system, why do input PCM code words retain the same time slot number in the output?

11.4 How does the number of time slots per time-multiplexed line affect the probability of blocking in time-space-time and space-time-space digital switching network architectures?

11.5 What factors complicate digital recognition of analog tone signals?

11.6 How are digital pads used to impart gain to a digital signal?

11.7 What is the effect when traffic increases to a level which precludes a processor from having sufficient real time available to process all demands for service and what can be done to alleviate the condition?

11.8 What are the advantages of distributed control over central control architecture?

11.9 What advantage does a 32-channel PCM format have over a 24-channel format in a switching system?

11.10 Compare the validity of the underlying assumptions of the Poisson, Erlang B, and Erlang C traffic formulas and explain the effect of retrials on traffic measurements.

11.11 Explain how a traffic overload can cause switching congestion, and how it can degrade a network when all equipment and facilities are intact. Name two controls that are the most effective in inhibiting switching congestion.

REFERENCES

1. Terry, J. B., Younge, D. R., and Matsunaga, R. T., "A Subscriber Line Interface for the DMS-100 Digital Switch," *National Telecommunications Conference Record, 1979,* IEEE Press, 1979, pp. 28.3.1–28.3.6.
2. Ibid.
3. Caves, Terry and McWalter, Ian, "Filter Codec and Line Card Chips: the New Generation," *Telesis,* No. 4, pp. 2–7 (Ottawa, Bell-Northern Research, 1983).
4. Bellamy, John C., *Digital Telephony,* John Wiley & Sons, New York, 1982, p. 246 (hereafter referred to as *Digital Telephony*).
5. Ibid.
6. Rothmaier, Klaus, and Scheller, Reinhard, "Design of Economic PCM Arrays with a Prescribed Grade of Service," *IEEE Transactions on Communications,* p. 925 (July 1981). (Hereafter cited as "Economic PCM Arrays.")
7. Ibid.
8. Ibid.
9. Huttenhoff, J. H., et al., "Peripheral System," *Bell Syst. Tech. J.,* pp. 1037–1041 (Sep. 1977).
10. Pitroda, Sam G. "Telephones Go Digital," *IEEE Spectrum,* p. 51 (Oct. 1979).

11. "Economic PCM Arrays," p. 926.
12. Gotoh, Kazuhiko, and Itoh, Masahiko, "Design Concepts of a Digital Switching System for Higher Performance," *National Telecommunications Conference Record, 1980,* IEEE Press, 1980, pp. 19.2.1–19.2.5.
13. "Economic PCM Arrays," p. 932.
14. Ibid., pp. 930–931.
15. Lotze, Alfred, Rothmaier, Klaus, and Scheller, Reinhard, "TDM Versus SDM Switching Arrays—Comparison," *IEEE Transactions on Communications,* p. 1455 (Oct. 1981).
16. Ibid.; "Economic PCM Arrays," pp. 931–932.
17. McDonald, John C., "Techniques for Digital Switching," *IEEE Communications Society Magazine,* p. 11 (July 1978).
18. Richards, Philip C., "Technological Evolution—The Making of a Survivable Switching System," in Joel, Amos E., Jr., ed., *Electronic Switching: Digital Central Office Systems of The World,* IEEE Press, New York, 1982, p. 196.
19. Charransol, Pierre, et al., "Development of a Time Division Switching Network Usable in a Very Large Range of Capacities," *IEEE Transactions on Communications,* p. 982 (July 1979).
20. Jajszczyk, Andrzej, "On Nonblocking Switching Networks Composed of Digital Symmetrical Matrices," *IEEE Transactions on Communications,* p. 2 (Jan. 1983).
21. Pitroda, p. 59; Munter, Ernst A., "Digital Switch Digitalks," *IEEE Communications Magazine,* p. 15 (Nov. 1982).
22. Munter, p. 18.
23. Ikeda, Yoshikaza, and Norigoe, Masamitsu, "New Realization of Discrete Fourier Transform Applied to Telephone Signaling System CCITT No. 5." *IEEE Global Telecommunications Conference Record, 1982,* IEEE Press, 1982, pp. D8.1.1–D8.1.6 (hereafter cited as *GLOBECOM '82*).
24. Munter, p. 19.
25. D'Ortenzio, Remo J., "Conferencing Fundamentals for Digital PABX Equipments," *IEEE International Conference on Communications Record, 1977* IEEE Press, 1977, pp. 2.5–29 to 2.5–36 (hereafter cited as *ICC '77*).
26. Munter, pp. 19–20.
27. Ibid., pp. 20–23.
28. Talley, David, *Basic Electronic Switching for Telephone Systems,* Hayden Book Company, Inc., Rochelle Park, N.J., 1982, pp. 276–277.
29. Munter, pp. 17–18.
30. *Local Switching System General Requirements,* PUB48501, American Telephone and Telegraph Company, Basking Ridge, N.J., 1980, Section 7.4.11.1 (hereafter referred to as *LSSGR*).
31. Ibid., Sections 11.2 and 11.4.
32. Brand, Joe E., and Warner, John C., "Processor Call Carrying Capacity Estimation for Stored Program Control Switching Systems," *Proc. IEEE,* p. 1342 (Sep. 1977).
33. Ibid., pp. 1344–1345.
34. Penney, Brian K., and Williams, J. W. J., "The Software Architecture for a Large Telephone Switch," *IEEE Transactions on Communications,* pp. 1369 (June 1982).
35. Treves, Sergio R., "Maintenance Strategies for PCM Circuit Switching," *Proc. IEEE* p. 1363 (Sep. 1977).
36. Willet, R. J., "Design of Recovery Strategies for a Fault-Tolerant No. 4 Electronic Switching System," *Bell Sys. Tech. J.* pp. 3019–3040 (Dec. 1982).
37. Penney and Williams, p. 1372; Meyers, M. N., Routt, W. A., and Yoder, K. W., "Maintenance Software," *Bell Syst. Tech. J.* pp. 1139–1167 (Sep. 1977).

38. *LSSGR,* Section 8.5.
39. Ibid., Section 8.6.
40. *Switching Systems,* American Telephone and Telegraph Company, New York, 1961, pp. 91–98, 105.
41. *LSSGR,* Section 8.2.2.1.
42. Ibid., sec. 8.2.
43. *Telephone Traffic Theory—Tables and Charts, Part 1,* Siemens Aktiengesellschaft, Munich, 1970, pp. 15–17 (hereafter referred to as *Telephone Traffic Theory*).
44. Cooper, Robert B., *Introduction to Queueing Theory,* The Macmillan Company, New York, 1972, pp. 65–71.
45. Ibid., pp. 77–80.
46. *Engineering and Operations in the Bell System,* Bell Telephone Laboratories, Inc., 1977, p. 484 (hereafter cited as *Engineering and Operations*).
47. *Telephone Traffic Theory,* pp. 351–359.
48. *Defense Communications System Traffic Engineering Practices,* Defense Communications Agency, 1969, Vol. XII, pp. 2–3 (hereafter cited as *Traffic Engineering Practices*).
49. *Engineering and Operations,* p. 485.
50. *Traffic Engineering Practices,* vol. XII, p. 5.
51. *Engineering and Operations,* pp. 493–495.
52. *Notes on the Network,* American Telephone and Telegraph Company, 1980, Section 11, p. 2.
53. Ibid., Section 11, pp. 6–9.
54. *LSSGR,* Section 5.3.8.
55. Ibid., Section 16.0.
56. Ibid., Section 16.2.
57. Ibid., Section 16.1.

12 Operational Switching Systems

12.1 INTRODUCTION

An appreciation of the differences in digital switching system architectures can be gained by examining the architectures employed in operational systems. Although some size capacities and other operating parameters are presented for the purpose of perspective, the emphasis in this chapter is on overall system architecture (review Section 10.3 and Fig. 10–1).

All systems discussed are in operation in various countries. The omission of many other systems is not intended to imply inferiority or any operational or maintenance deficiency. The systems selected represent the three main categories of switching systems (PABX, local, and toll) but are discussed noncategorically because of overlapping functions and because each has some distinctive architectural characteristics of interest. The scope of information presented on each system is dependent upon its availability and its applicability to the architectural focus. Equipment terms frequently are manufacturer-specific terms representing generic functions.

12.2. ROCKWELL 580 DSS

A block diagram of the Rockwell 580 DSS PABX in its largest size is shown in Fig. 12–1. This formerly was known as the Wescom 580 DSS. Although manufacture-discontinued by Rockwell, its unique architecture is of interest as an example of distributed control. The 580 employs a four-wire nonblocking PCM time division network. Analog line circuits and trunk circuits terminate on dedicated channels of 24-channel codecs. The outputs of four codecs are multiplexed into a 96-channel format, and eight of the 96-channel buses are further multiplexed to form a 772-channel bus, four of which channels are used for framing. A direct digital interface connects to 24-channel PCM lines.

The network consists of four time switching blocks of 772 channels each. One spare network block is provided and is automatically switchable in case of failure of an active block. All incoming channels are written and stored in each network block to provide full availability.

Conventional DTMF receivers are provided. Incoming tone address signals

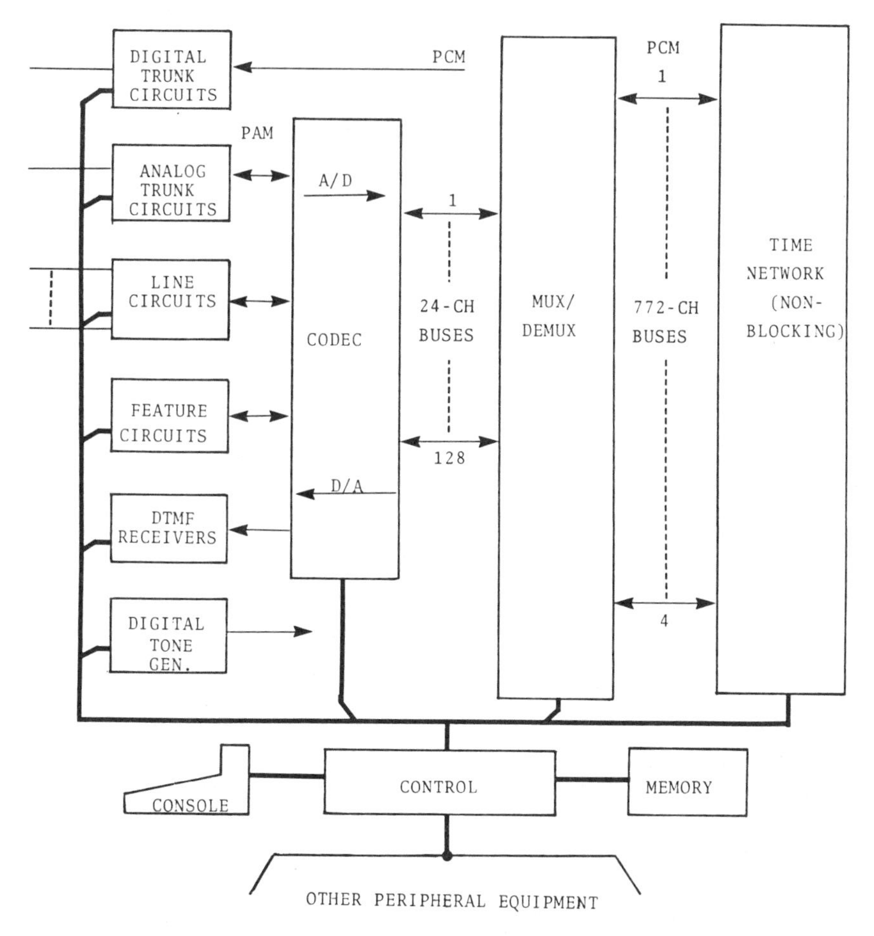

Fig. 12–1. Rockwell 580 DSS block diagram.

are switched through the time network to the DTMF receivers. Duplicated digital tone generators produce the 16 tones used in the system.

System control is distributed among six specialized controllers, each employing identical hardware and each having its own memory. Each controller is duplicated, and each processor unit uses two 8080 microprocessor chips. Each controller operates asynchronously and has its own autonomous maintenance subsystem. About 90% of processor real time is allocated to call processing, and about 10% to maintenance. Communications between processors is via interprocessor buffers with 16 bytes of memory. Communication among processors is shown in Fig. 12–2. The state processor communicates with

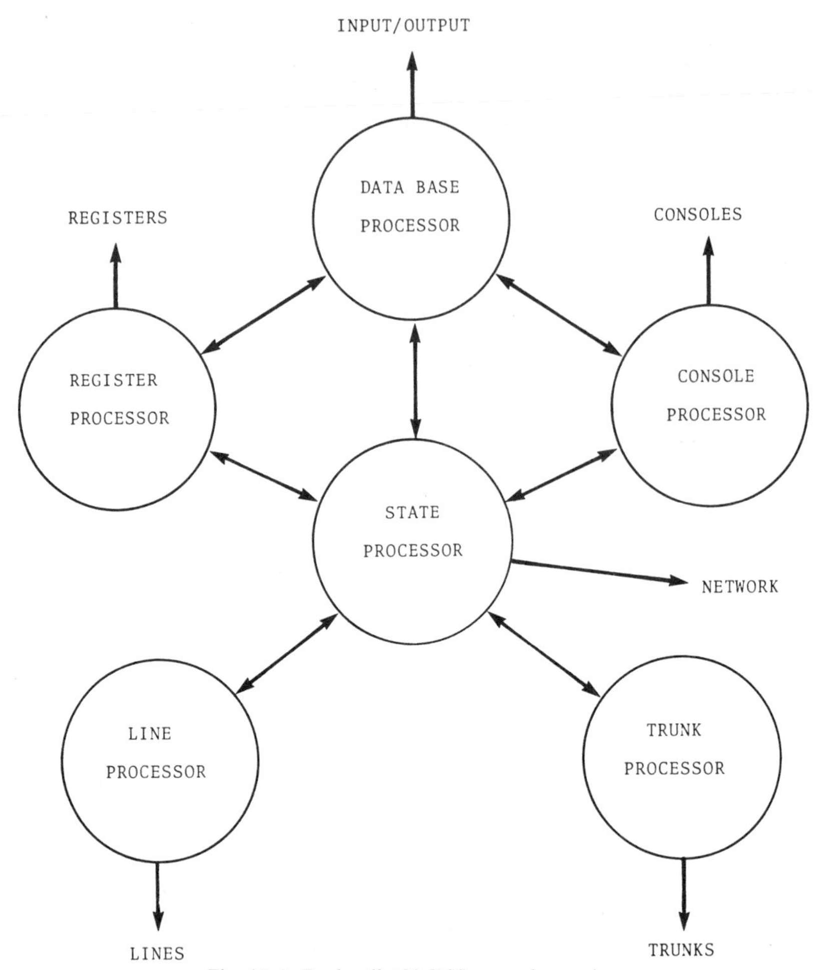

Fig. 12–2. Rockwell 580 DSS control complex.

all other elements of the control system. The register processor and console processor exchange information with the data base processor. The line processor and trunk processor communicate only with the state processor.

The line processor connects to a sense point and a control point on each line circuit and performs five repetitive functions. It detects on hook, off hook, and hookswitch flashing through the sense points by continuous scanning, and it starts and stops ringing by writing a "1" or "0" to the line control points. When a change in condition is detected, the line processor communicates this fact to the state processor.

The trunk processor interfaces both analog and digital trunks with four sense points and four control points per trunk. It continuously scans up to 600 trunks and reports supervisory transitions to the state processor. Since there is a wide variety of trunk types, the program is more complex than that for the line processor. Trunk signaling is controlled by the trunk processor in response to messages from the state processor.

The data base processor analyzes either dial pulse or DTMF address digits received by the combination dial pulse and DTMF receivers. The first digit is forwarded to the data base processor for translation and determination as to the number of digits to be expected. When all digits have been received, the register processor sends them and the line identifier to the data base processor.

The console processor can interface up to 16 attendant consoles, analyze signals from them, and return signals to them. It sends status information and commands to the state processor and executes commands from the state processor. It does not control any voice path.

The data base processor contains in its memory user class-of-service information and translation tables. It also contains all system configuration information as to lines and trunks in service, and uses about 60% of the call processing memory. It sends class-of-service information upon request and sends dialed number translations to the register processor. Maintenance messages and configuration changes are processed through input/output devices connected to the data base processor.

The state processor is the most complex processor in the system. It uses about 25% of the overall memory storage and maintains a record of the current state of each line, trunk, and register in the system. It uses messages from other processors to make decisions as to what should be done next and sends commands to the other processors to change the state of devices under their control. It is the key processor in the system and contains the generic executive program for call processing.

The Rockwell 580 system has been enhanced by adding a capability for simultaneous voice/data communications through its network ports. Voice is switched at 64 kb/s, and data up to 9.6 kb/s can be switched in the same connection. This is accomplished by switching 10-bit words rather than the normal 8-bit words. For data speeds above 9.6 kb/s up to 64 kb/s, the data must use a separate PCM port. Data ports require new interface equipment and software to provide an RS-232-C interface to compatible devices.[1]

12.3. STROMBERG-CARLSON SYSTEM CENTURY DCO

The first Stromberg-Carlson digital central office was placed in service in 1977. The initial architecture was TST with analog concentrators used to concentrate low-traffic lines into the 1920-port system. Four such systems could be combined into a 7680-port size. Subsequent enhancements have enlarged the System Century DCO to a maximum size of 32 400 lines and 4000 trunks with a capability of processing over 114 000 call attempts in the ABSBH.[2]

The basic architecture of the 1920-port system is shown in Fig. 12–3. Line and trunk circuits and the concentrator are not shown. The input to each port circuit is from the tip and ring leads of a subscriber line or the tip and ring plus signaling leads of a trunk. The port circuits filter the analog input and covert it to parallel PCM code words.

A multiplexer-demultiplexer and the port group control (PGC) logic arrange the PCM words, to which control bits have been added, into a 30-channel serial data stream of 256-bit frames. The data rate of the port group highway input to the time slot interchange is 2.048 Mb/s. A direct digital interface with T1 carrier can replace a port group control. Up to eight port

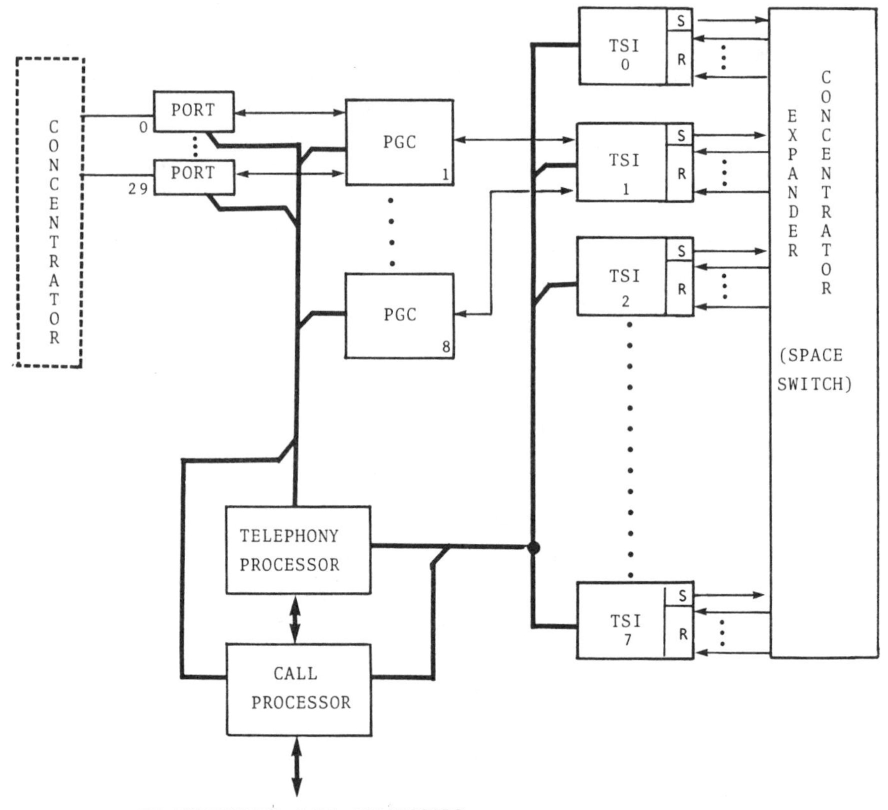

Fig. 12–3. Stromberg-Carlson System Century DCO, 1920-port system.

group controls can access each TSI, and one call processing system can control up to eight TSIs.

The 240 PCM words of the eight port group highways are converted back to parallel by a dual data conditioner and are read into the TSI memory sequentially. Under control of the call processor, each word is assigned to one of the 128 time slots on a cross-office highway at the output of the TSI. The cross-office highway connects the send logic of a TSI to the receive logic of that TSI and all other TSIs. Under call processor control, the receive logic in each TSI accepts the PCM words directed to it by operating logic gates, thus establishing a connection through the TST network. The cross-office highway system comprises the central stage space switch.

In the enhanced version, the previous ports and port group control units for lines have been replaced by a local line switch as shown in Fig. 12–4. Up to 12 line groups, of 90 analog lines each, are processed into 30-channel PCM streams and transmitted via 36 line group highways to a duplicated

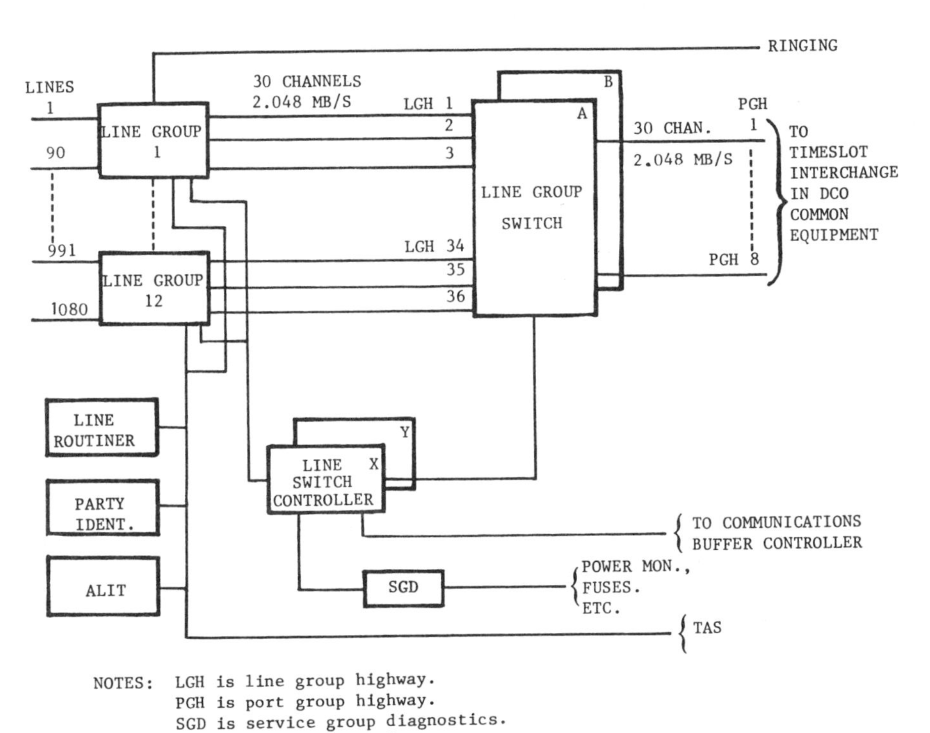

Fig. 12–4. System Century local line switch. ("DCO" and "SYSTEM CENTURY" are registered trademarks of Stromberg-Carlson Corporation. Copyright 1982, Stromberg-Carlson Corporation. Reproduced by permission of Stromberg-Carlson Corporation.)

line group switch. The line group switch concentrates the 1080 lines into 240 time slots by means of solid state crosspoints and transmits them on eight port group highways to the TSI. The output of the TSI has been doubled to 256 time slots to achieve a 6.7% expansion in the first time stage. After being switched through the center space stage, the 256 paths are concentrated back to 240 time slots in the last time stage.

By the addition of DS-1 modules and a message assembler, the local line switch can function as a remote line switch. It is connected to its host DCO by 2–8 T1 carrier lines of 24 voice channels, each, as shown in Fig. 12–5. The common control system in the host DCO controls path setup, and all calls to the outside world are connected through the host. Two options are available. An intranodal switch can be provided to permit lines served by the remote switch to be connected to each other without holding T-carrier paths during the connection. The host establishes the connection and, after the called party answers, drops the connection through the line group switch and performs a time slot interchange through the intranodal switch. A second option provides a controller for local switching in the event the transmission system to the host DCO is interrupted.

The control system is shown in Fig. 12–6. The telephony processor is a preprocessor that performs specialized control functions. It stores call state, port identification, and event code data for each port and communicates this information to the call processor (CPU). The call processor continually polls the telephony processor for line and call states. It contains the system executive program and controls path selection and setup. All control elements, line group switches, TSI, control memories, and communication buffer controllers are duplicated. Both systems operate in an on-line mode with the standby system duplicating in unison all functions performed by the primary system. The maintenance and administration processor monitors the entire operation. When a fault is detected, it reverses the appropriate elements of the duplicated systems, placing the standby unit in service with no lost calls. It also performs diagnostics on faulty components, program audits, circuit testing control, administration, traffic data collection, and provides input/output interfaces for program load and data base changes.

12.4. NORTHERN TELECOM DMS-100

The DMS-100 is a large, local central office digital switching system which also can perform toll switching. A block diagram of the system is shown in Fig. 12–7. The DMS-100 has substantial commonality with the DMS-200 toll switching system and the DMS-250 system.

Line modules are arranged in reliability pairs. Each line module frame concentrates 640 analog line circuits into 60–120 digital channels, depending

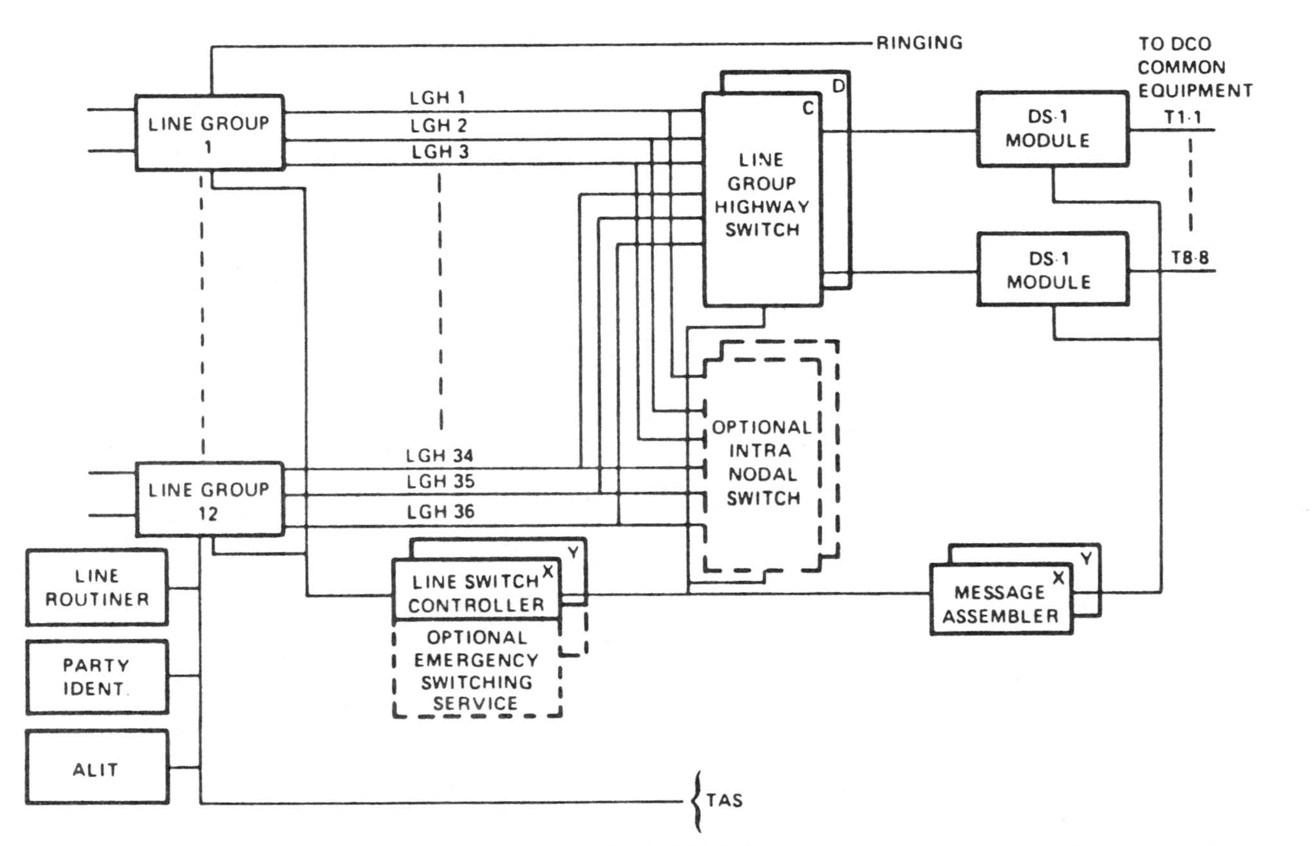

Fig. 12–5. System Century remote line switch. ("DCO" and "SYSTEM CENTURY" are registered trademarks of Stromberg-Carlson Corporation. Copyright 1982, Stromberg-Carlson Corporation. Reproduced by permission of Stromberg-Carlson Corporation.)

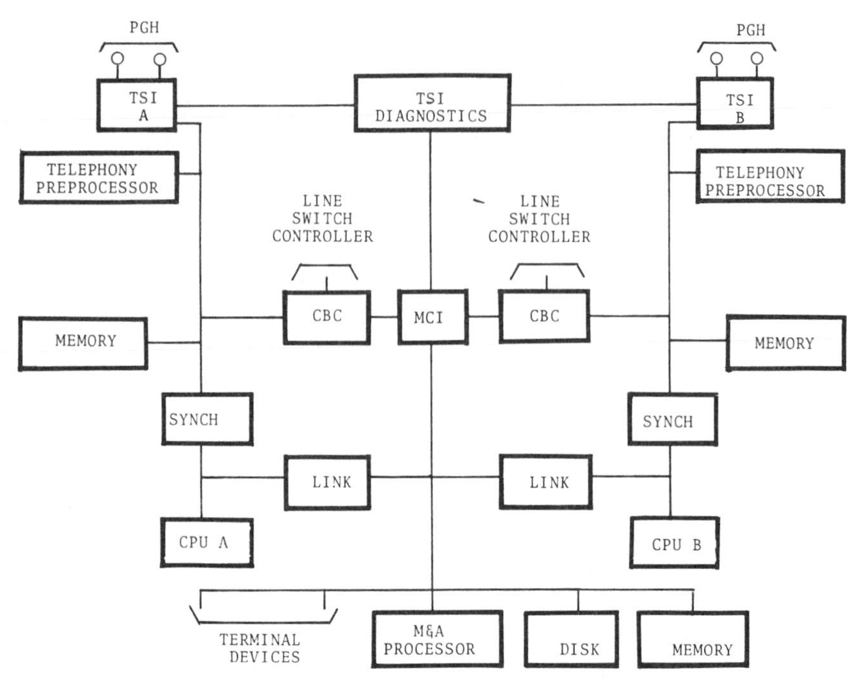

Fig. 12–6. System Century control system. ("DCO" and "SYSTEM CENTURY" are registered trademarks of Stromberg-Carlson Corporation. Copyright 1982, Stromberg-Carlson Corporation. Reproduced by permission of Stromberg-Carlson Corporation.)

upon traffic load, through a space switching network (Fig. 11–3). The reliability pair can terminate 1280 lines. Each line module uses an unduplicated common control in a load-sharing mode, each controller handling 640 lines. If the controller in one line module should fail, the controller in the mate module can control 1280 lines.[3]

In addition to the line modules, the peripheral subsystem contains analog trunk modules and digital carrier modules. Each analog trunk module provides supervisory control, PCM conversion, and signaling for 30 analog trunks. Each group of 30 trunks is multiplexed and fed to a network module port. Each digital carrier module can convert five 24-channel T-carrier circuits into four 30-channel PCM streams to be fed to four network module ports.

The switching network consists of identical network modules arranged into two planes. Each network module contains sixteen 8 × 8 time-space switch matrices arranged to provide two stages of time-space switching. Each module switches between 64 peripheral ports, each terminating one time-multiplexed line composed of 30 voice channels and two signaling channels, and 64 junctor ports. The serial data rate through the switching network is 2.56 Mb/s. Each module has 64 two-way junctors which connect to other

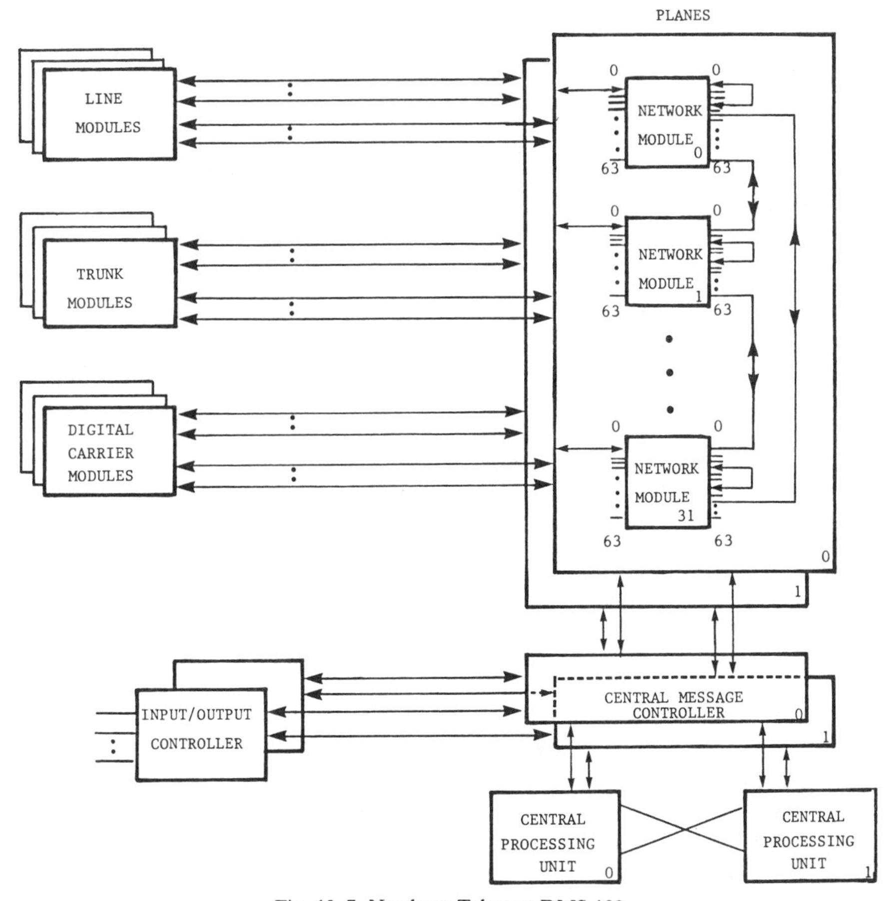

Fig. 12–7. Northern Telecom DMS-100.

modules in its plane and to itself. The network thus provides four stages of time switching. Fig. 12–8 illustrates the matrix arrangement of a network module.[4]

Each peripheral module transmits its time-multiplexed lines to one of 32 network modules in each network plane, and the peripheral subsystem control selects one network plane to connect to its receive section on a per-channel basis. In the event of a fault in the network plane being used, the peripheral module control autonomously switches to the other network plane. The switching load is shared by the two network planes, and each network plane is capable of carrying the full load in the event of a fault in one of them.[5]

The DMS-100 has control distributed over the four major subsystems. Central control has responsibility for the overall control and sanity of the

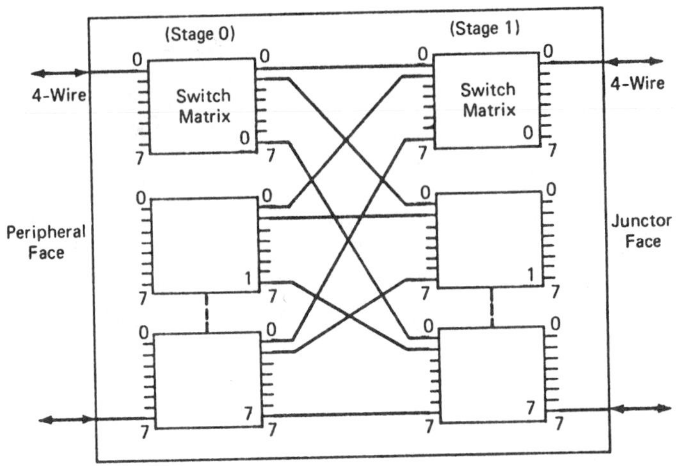

Fig. 12–8. DMS-100 network module switching matrix. (From J. Terry, H. Krausbar, and J. Hood, "DMS-200 Traffic Peripherals," *ICC '78*, © IEEE.)

switching system. Duplicated central processing units (CPUs) and their memories operate with one active and one standby. The standby CPU tracks the active CPU at each step in the control process. At the completion of each processor microcycle, each CPU matches its results with those of its mate. If either CPU detects a mismatch, it triggers an interrupt for self-diagnostics, after which information is exchanged between the CPUs, and fault recovery procedures are executed to isolate the faulty CPU. Control messages from and to central control are exchanged with the control elements of the network modules and input/output controllers through a pair of central message controllers (CMCs). The CMCs operate in a load-sharing mode. The active CPU transmits control messages to both CMCs, and both CPUs receive messages from both CMCs. This enables the standby CPU to maintain current status of the switching system at all times. One CMC can carry the entire load should the other fail. Each CMC is connected to 2.56 Mb/s serial transmission links to all input/output controllers and network modules.[6]

Duplicated input/output (I/0) controllers connect the CMCs to a maintenance access position and to tape and disk drives. They provide man/machine interfaces and a means of automatic or manual memory reload when required.

Network modules are controlled by two specialized microprocessors. One interfaces the data buses to the CMCs. It receives all messages on the bus, examines the address, and executes those intended for its network module. The other microprocessor controls the message channels of the 64 ports by loading the time switch control memories.

Peripheral subsystems are controlled by multiple processors. The line module has a master processor and three subordinate processors. The master processor controls ringing, coin telephone functions, and automatic number identification, performs hardware monitoring, and executes fault detection routines and audits. The central control message processor controls the exchange of messages between the line module and the network. A peripheral processing message processor inserts and extracts channel supervision messages in each speech channel for the master processor. It also checks parity on speech paths. The signaling processor performs all other telephony functions associated with the line circuits including supervisory scanning, digit collection, real-time ringing control, and control message decoding.[7]

12.5. GTE GTD-5 EAX

A simplified block diagram of the GTD-5 EAX is shown in Fig. 12–9. The GTD-5 EAX is a local switching system with toll switching capability. The small version serves up to about 20 000 lines, while the large version can

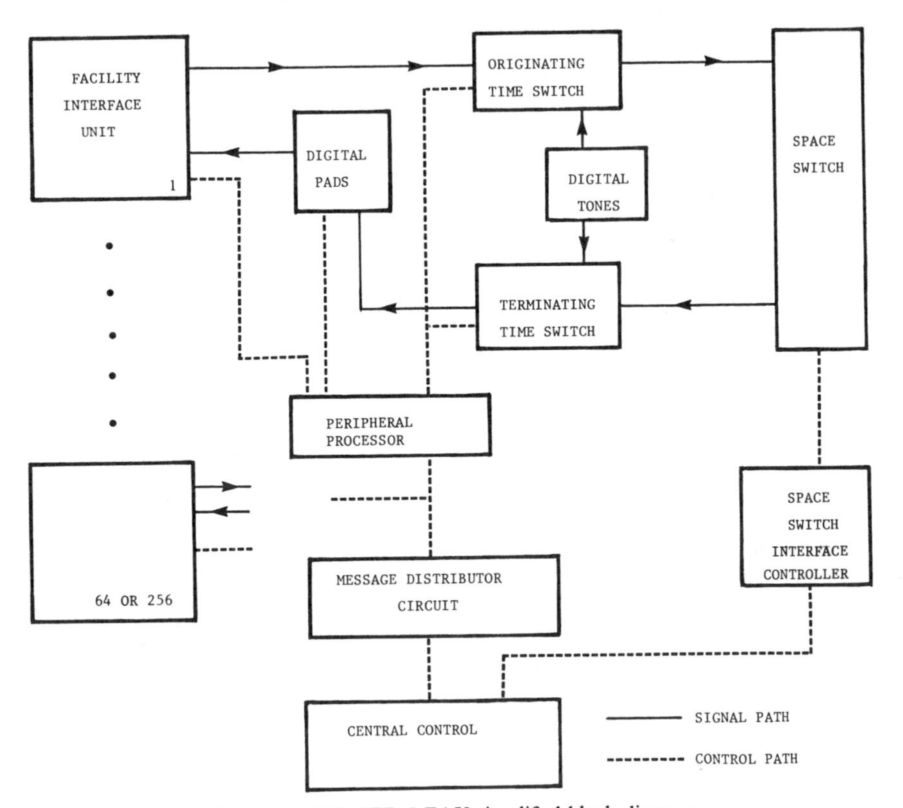

Fig. 12–9. GTE GTD-5 EAX simplified block diagram.

accommodate up to about 145 000 lines. There are three major equipment groups: peripheral, network, and common control.[8]

The peripheral equipment group interfaces lines and trunks to the switching network by means of various types of facility interface units (FIU). An analog line FIU interfaces 768 analog lines to 192 PCM channels. Analog line signals are converted to PCM by single-channel CODECs and multiplexed into 24-channel, 8-bit serial buses. The serial buses are converted to 12-bit parallel buses, and three supervisory bits and one parity bit are added. The parallel buses are then concentrated onto 192 of 193 channels, the 193rd channel being used for maintenance. Analog trunk FIUs interface 192 analog trunks to the 193-channel bus, and digital trunk FIUs interface eight 24-channel T-carrier lines to the digital bus. A digital three-port conference FIU interfaces up to 64 three-port conferences to 192 channels. Each FIU module has a duplicated control unit which performs all FIU functions on an active/hot standby basis.

The switching network is a three-stage TST network which operates unfolded. The first time stage is composed of originating time switches, and the last time stage is composed of terminating time switches. A time switch and peripheral control unit (TCU) consists of an originating time switch, a terminating time switch, a digital tone source, programmable digital pads, and a peripheral processor. Each TCU is fully duplicated and can interface up to four FIUs or a total of 772 PCM channels. Each duplicated copy interfaces with separate control units in the FIUs. The peripheral processor controls the active/hot standby status of the FIU control units and controls and monitors the time stages.[9]

The originating time switch performs time-slot interchanging between four 193-channel buses and two 386-channel buses called rail A and rail B. The rails remain separated as they are switched through duplicated space switches, but their channels can be interchanged in the terminating time switch which distributes the switched channels among the FIUs to which it is connected. The time slots are read into the time switch duplicated information memories sequentially and read out asequentially under control of the central processor and duplicated control memories.

The space switch is a duplicated 32 × 32 solid state switching matrix to form a space switch of 772 time slots divided between rail A and rail B. The matrix can be expanded to 64 × 64 to interface with 64 originating and 64 terminating switches. The two matrix copies per rail operate in an active/hot standby mode. The space switching matrices employ buffer memories. The PCM samples are read into a buffer memory and read out two time slots later. A space interface controller interfaces the space switch to the central processor and writes and reads the call path setup data into and out of a common memory.

An administrative processor complex and up to seven telephony processor complexes comprise the central control. The administrative processor provides access to input/output modules, and the telephony processors perform the call processing functions in a load-sharing mode. Hence, the control system uses a hybrid shared/distributed concept. Communication with the peripheral processor is via a duplicated message distributor circuit.[10]

Three-way conferences are established by switching each conferee's transmit path through the TST network to the conference FIU. The conference circuit uses a modified instant-speaker algorithm for conference control, and the conference path is switched back to each conferee via the TST network. Thus, each path is switched through the network twice.

Analog line units can be remoted with the addition of a link selector bypass circuit and T-carrier terminal equipment to provide access to the host for 768 lines at 4:1 concentration. A remote switching unit (RSU) also is provided to serve small, outlying communities requiring up to a maximum of 3072 lines. The switching elements are similar to the host unit, employing a TST network with separate originating and terminating time switches.[11]

The GTD-5 is equipped for common channel signaling using the CCIS system. A CCIS data link controller interfaces the CCIS data links to the message distributor circuit.[12]

12.6. AT&T NO. 5 ESS

The No. 5 ESS is a local digital switching system designed by Bell Telephone Laboratories. A toll switching capability is planned for future addition to the system. A simplified block diagram is shown in Fig. 12–10. The basic architecture involves a central processor, a solid state space switch, and multiple microprocessor-controlled interface modules which provide line and trunk interfaces and switching functions. The first No. 5 ESS was placed in service in March 1982, in a single-module configuration.

The interface module (IM) is the basic building block of the No. 5 ESS. Line units, trunk units, and digital line trunk units comprise the interface portion of the IM. Each line unit can interface up to 512 lines and PABX trunks. BORSCHT functions (see Section 11.2.1) are performed by channel circuits which are shared through a concentrator designed into the line unit. The concentrator is a two-stage space switch using high-voltage silicon integrated circuit technology. The bipolar gated diode crosspoints can handle ringing and testing voltages.[13] Concentration ratios of 8:1, 6:1, and 4.5:1 are available as options. The concentrator produces an output consisting of a time-multiplexed PCM signal with 64 time slots.

A trunk unit provides terminations, supervision, and address signaling

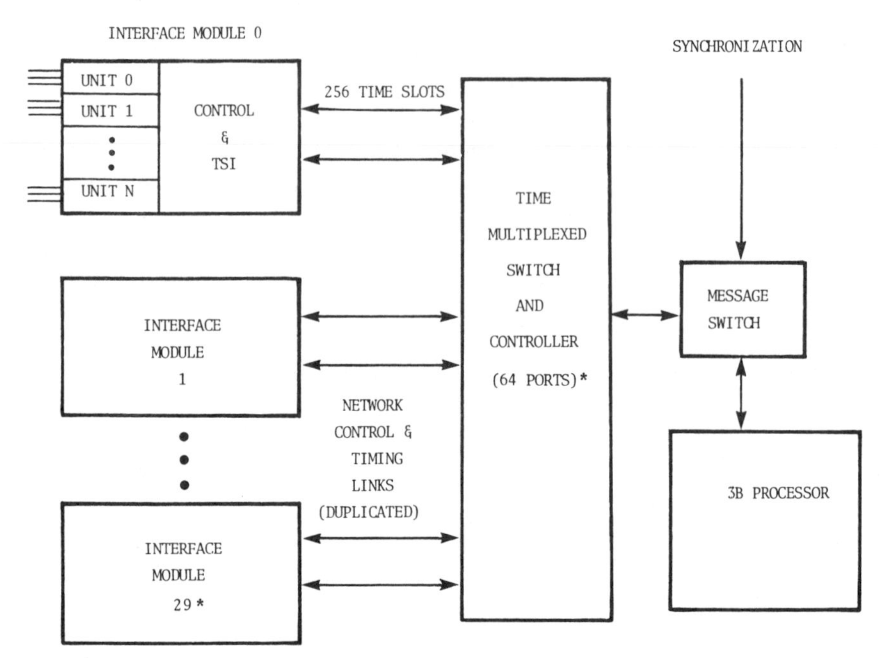

Fig. 12–10. AT&T No. 5 ESS switching system.

for up to 64 analog trunks. Either dial pulse or multifrequency (MF) address signaling may be used with immediate dial, delay dial, or wink start control. The analog signals are converted to PCM and are not concentrated.

A digital line trunk unit can terminate up to ten 24-channel T-carrier lines. Each digital T-carrier line terminates in a separate digital facility interface. Under processor control, the digital speech paths are processed for the first stage of time switching.

The interface module can terminate up to 4096 lines and trunks and switch them through its time-slot interchange (TSI). Digital service circuits are provided for all lines and trunks terminating on the IM. A module control unit provides microprocessor control for calls switched through the module. The time-slot interchange can connect any of its 512 peripheral time slots to other peripheral time slots or to any of 256 time slots on each of a pair of network control and timing links to a time-multiplexed space switch. Thus, a switching system with a single IM switches all calls through the single TSI, and calls between analog lines are switched through the concentrator, the time switch, and the concentrator again for a network configuration of SSTSS.

In a multimodule system, two network control and timing links connect each interface module to a time-multiplexed switch (TMS) which functions as a single-stage switch. The No. 5 ESS then functions as a TST network plus the two space stages of concentration used with analog lines. The TMS has 64 ports for time-multiplexed lines and will support 30 IMs. It is planned to increase the TMS to a three-stage, 256-port space switch to support 127 IMs in the future. The TMS and its control processor are fully duplicated.

The central control uses a 3B20D processor and the duplex multienvironment real-time (DMERT) operating system to perform overall control of the No. 5 ESS.[14] (The processor and the operating system are described in extensive detail in *The Bell System Technical Journal,* Vol. 62, No. 1, Part 2, January 1983.) The central processor is the hub of a star network configuration of controllers. The 3B20D directly interfaces an input/output processor, which connects to external devices, and moving head disk files containing infrequently used programs and data as well as recovery programs. All communication with the TMS and IM controllers is via a packet data message switch using "nailed up" connections, and optical fiber transmission links. Data bases and data base management are distributed among processors throughout the system. Call processing tasks are terminal oriented and are decentralized to the extent practicable.[15] Control messages transmitted from the central processor to subordinate processors are queued at the recipient processors' "mailboxes" until their work activity permits reading. Thus, the processors do not function in a master-slave relationship. Administrative functions, such as billing, plant measurements, service evaluation, and network management, are performed by the central processor in conjunction with the input/output processor.[16]

Maintenance routines are resident in the distributed processors and comprise four major categories of maintenance software: switch maintenance, terminal maintenance, system integrity, and human-machine interface. Functional responsibilities relate to the standard maintenance functions described in Section 11.6.1.1.[17]

By incorporation of some additional components, an interface module can function as a remote switching module (RSM), serving a maximum of 4096 lines at 8:1 concentration. Intramodule calls are switched only by the RSM, but all other calls are switched through the host No. 5 ESS. Connection to the host is via 4–20 24-channel T-carrier links. If all host links should fail, intramodule calls can still be processed by the RSM. The RSM can be located up to about 100 miles from its host.[18]

12.7. ITT SYSTEM 1240

The ITT System 1240 is a fully distributed switching system. There is no central control and no network map in memory. The system is composed of terminals, terminal selection units, and the switching network as shown in Fig. 12–11. Control functions are distributed throughout the system.[19]

Terminal selection units (TSU) contain a pair of access switches and up to eight terminal interfaces (TI), each with its own microprocessor and memory. Separate TSU modules are used to terminate each type of terminal connected to the system. Fig. 12–12 illustrates the various types of terminals which can access the switching network. An analog line TSU can terminate 480 subscriber lines in eight groups of 60 lines each. Under TI microprocessor control, each line circuit performs all BORSCHT functions and, in addition, adapts battery feed to loop resistance, allocates power according to telephone set type (standard or electronic), selects line balance networks, switches attenuation pads according to the connection characteristics, and performs a variety of other functions. Each group of 60 line circuits is arranged into a pair of bidirectional 30-channel (plus two control channels) PCM lines connected to a terminal interface. The TI connects to digital tone sources and has a 32-channel PCM port to each of two access switches.[20] All PCM links in the switching network operate asynchronously in a serial bidirectional mode and contain 30 voice channels and two control channels, each with 16-bit words and optional A-law or μ-law companding. The data rate is 4096 kb/s. Each access switch has one PCM port to each of four planes of switching elements comprising the switching network.

An analog trunk TSU contains four trunk TIs which interface 30 trunks each. Trunk modules perform functions similar to those of line modules except for the characteristic differences between lines and trunks. Service circuits are accessed through the switching network as are clock pulses, tones, certain data bases, and maintenance/administration peripherals.[21]

The switching network and the access switches in the TSUs are composed of identical switch elements. Each switch element has 16 bidirectional PCM ports interconnected by a 39-line parallel TDM bus containing seven subfields for clocks, control messages, and data. Each port is a solid state integrated circuit of 11 500 transistors on a single LSI chip. The switch element can switch any time slot of the receive section of any port to any time slot of the transmit section of any port. Both time switching and space switching are performed in each switch element, and channel selection in the transmit portion of the PCM link can be performed to minimize time delay through the switch element.[22]

The switching network contains four planes of switch elements in up to three switching stages. Each stage can contain up to 16 groups of eight switch elements each. For a very small switching system, only one switch element per plane is necessary to establish single-stage connections. For somewhat higher-capacity systems, a second stage of switching can be added just by connecting second-stage switch elements to the unused ports on the net-

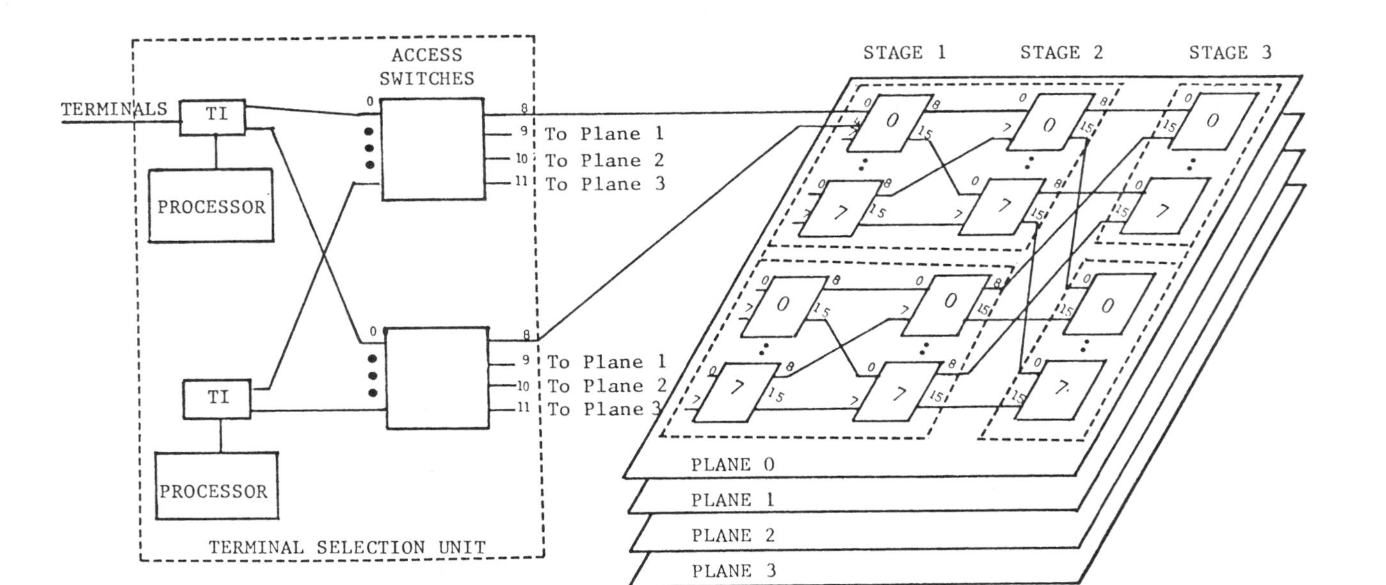

Fig. 12–11. ITT System 1240 switching system. (From J. Cotton et al., "An Expandable Distributed Control Digital Switching Network," *ICC '80,* © IEEE.)

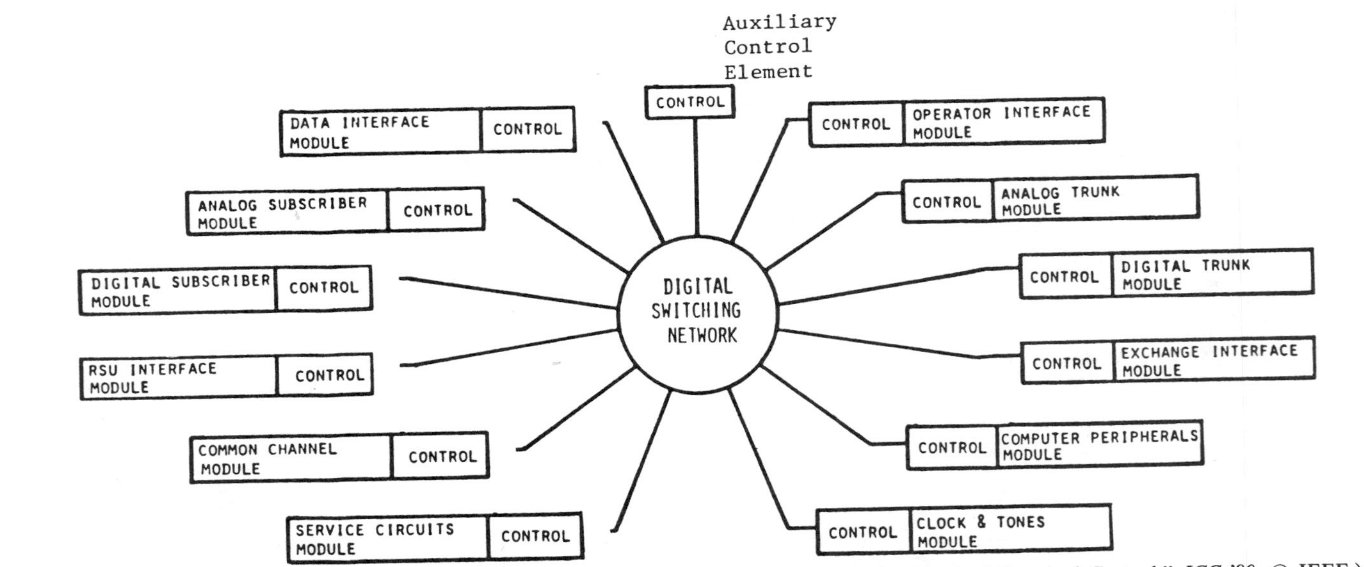

Fig. 12–12. ITT System 1240 digital exchange modules. (From R. Chea et al., "Circuit Terminations and Terminal Control," *ICC '80,* © IEEE.)

work side of the first-stage. As the system grows, additional first-stage switch elements are added to handle the additional terminals, and second-stage elements also are added to switch between first-stage switch elements. For still larger systems, a third stage is added to interconnect groups of second-stage switch elements. Two planes of switch elements are necessary for reliability. If four planes are provided, traffic is switched on a load-sharing basis, and the network is nearly nonblocking for all lines with an average of 0.25 erlang per terminal. In such case, line traffic is concentrated by 2:1 and trunk traffic is expanded by 1:2.[23]

Port and channel selection within a switch element can be performed in different ways. Selection options include a specific channel of a specific port, any channel of a specific port, any channel of any port (8–15), any channel of a low-numbered port (8–11), or any channel of any high-numbered port (12–15). When a specific channel is not specified, the first free channel in the selected port will be used to minimize the absolute delay across the switch element. Since a specific connection involves seven switch elements if the third stage is used, plus additional delays in the TSI in the terminal interface, switching delay is an important consideration as a contributing cause of echo. Design studies show that, at a capacity of 200 000 lines or 60 000 trunks and at maximum design load of 0.55 erlang per internal link, the average round trip delay through the three-stage switching system would be about 370 μs, and that 99% of connections would encounter less than 500 μs delay. Below 25 000 lines and at a normal maximum load of 0.5 erlang per link, the average delay of less than 250 μs is less than that encountered in a typical TST switching system.[24]

A control element consists of a microprocessor, its memory, and a terminal interface. There are two categories of control elements in the ITT 1240. Terminal control elements are directly connected to a group of terminals, and auxiliary control elements provide additional processing and data base storage capacity. Terminal control elements are directly connected to small groups of lines, trunks, service circuits, or other types of terminals. Their programs and memory data are related only to the control of their terminals. Some auxiliary control elements contain call processing logic associated with specific groups of lines or trunks, and others contain system-level data and operate in an inquiry/response mode. Liberal sparing is used for reliability.[25]

Path setup is under control of the terminal control elements associated with the originating and terminating lines. Each terminal has a unique network address. The originating terminal control element establishes a simplex path to the terminal processor responsible for the terminating line and sends a packet-switched message giving details of the connection to be established and requesting the setup of a return simplex path. Thus, both paths are established independently by separate terminal processors. The switching system is configured and functions in a manner similar to a public network. Connections are established between terminals having specific network addresses. Signaling is similar to common channel network signaling. Control is accomplished by a combination of nailed-up connections and packet-switched messages. The terminal address concept enables any second- or third-stage switch element to establish a path to the addressed terminal without additional software translation, the path selection sequence being defined by the network address.[26]

Reliability is achieved through redundancy, load sharing, and fault detection and recovery techniques. Terminal control elements are not duplicated because failure would affect only 60 lines or 30 trunks or service circuits. Critical elements are duplicated or are provided with a liberal quantity of spares. Fault detection and reaction programs are distributed over all control elements to reduce the service impact, to reduce detection and recovery delays, and to make the switching system less vulnerable to the loss of any single control element. Programs which control recovery from less critical failures are assigned to a pair of peripheral and maintenance control elements which isolate faulty equipment, activate standby units, and alert maintenance personnel as appropriate.[27]

The architecture of the ITT System 1240 has the flexibility for use over a wide range of system sizes and applications. A remote digital subscriber unit can serve 6–90 lines by connecting one terminal interface to a host system by means of one 32-channel PCM link using a simplified version of CCITT Signaling System No. 7. A satellite unit can serve 60–2000 lines with full availability trunking to the host. In case of isolation from the host, limited line-to-line connections can be established by the satellite unit. A supervised exchange contains both lines and trunks and can serve from less than 300 to 100 000 lines. It is dependent on its host for remote operations, maintenance control, and mass memory storage. As it grows, it can easily be converted to an independent switching system providing full features and the capability to act as host to supervised exchanges. The ultimate growth of the system is limited only by the traffic grade of service that can be provided by four full network planes.[28]

12.8. GTE GTD-3 EAX

The GTD-3 EAX (formerly known as the No. 3 EAX) is a PCM trunk switching system arranged in a space-space-time-space-space (SSTSS) configuration. A basic sector can serve up to 15 000 trunks, and four sectors can be combined to serve up to 60 000 trunks. Although primarily designed as a toll switching system, the GTD-3 also can be arranged to serve local subscribers in a combined local/toll system. The basic sector contains three major categories of equipment: a processor complex, a peripheral and telephony complex, and a maintenance and test complex. A block diagram is shown in Fig. 12–13.[29]

The basic input is 24-channel T-carrier PCM. Analog trunks must be converted to that format prior to termination on the system. Group equipment multiplexes eight 24-channel PCM streams onto a 192-channel parallel PCM bus for presentation to the first switching stage.[30]

The switching network in a sector is composed of two stages of space switching and a central stage of time switching as shown in Fig. 12–14.[31] The line group equipment frame (LGEF) contains 32 units of line equipment, four units of group equipment, and a space switch which functions as the first and last stages of space switching. The core of the switching network is an STS configuration called a *highway junctor frame* (HYJF). The HYJF has an A-side space switch, a B-side space switch, and a central time switch.

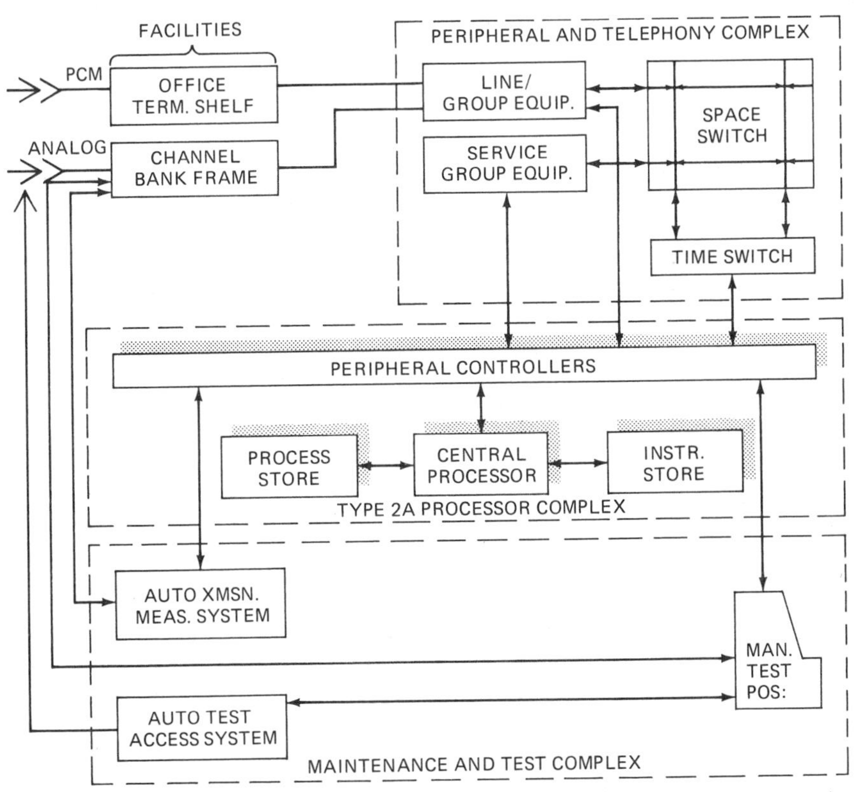

Fig. 12–13. GTE GTD-3 EAX block diagram. (From R. Holm, "No. 3 EAX System Description," *ICC '77*, © IEEE.)

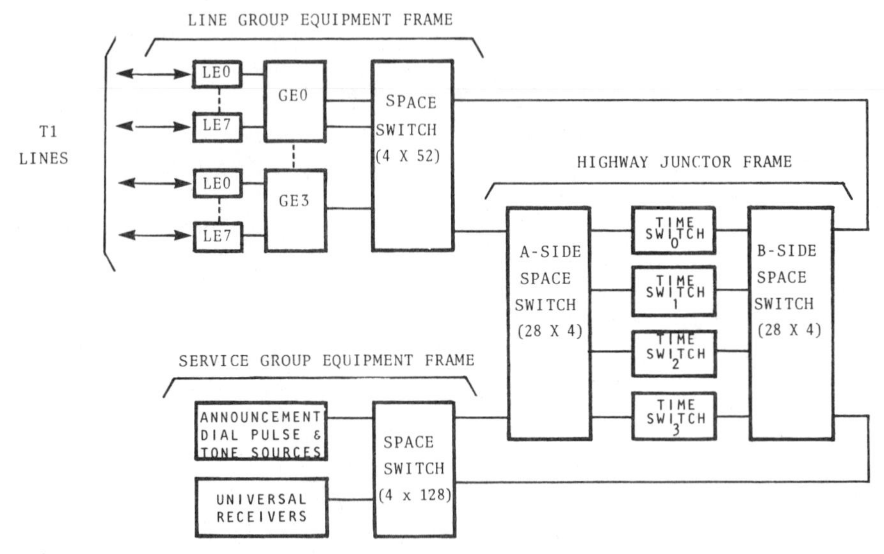

Fig. 12–14. GTE GTD-3 EAX switching network. (From J. Bagley, "The No. 3 EAX Digital Switching System," *Electronic Switching: Digital Central Office Systems of the World,* 1982, © IEEE.)

Incoming trunk calls are switched through the LGEF, the A-side space switch, the time switch, the B-side space switch and the LGEF serving the outgoing trunk. In a multisector configuration, each sector contains an additional HYJF for intersector switching.

The central time switch, called a *highway junctor,* contains five memories. The A-side and B-side line switching memories control the crosspoints in the A-side and B-side space switches, respectively. The channel switching memory (CSM) is a control memory for the two data transfer memories (DTM) which are the information memories for the time switch. There are four highway junctors in a sector. Time switching is timed by a time slot counter. During even-numbered time frames, voice samples from the A-side space switch are stored sequentially in DTM1, and samples from the B-side space switch are stored asequentially in DTM2 under control of the CSM. During odd-numbered time frames, the previously stored samples in DTM1 are read out asequentially to the B-stage space switch under control of the CSM, and the previously stored samples in DTM2 are read out sequentially to the A-stage switch. At the same time, the next samples are stored in reverse order; i.e., the second voice sample from the A-side space switch is stored in DTM2, and the second sample from the B-side space switch is stored in DTM1. These samples are read out during the succeeding even-numbered frame. This process continues for the duration of the connection.

The A-side information memory voice samples are always loaded/unloaded sequentially under time slot counter control, and the B-side information memory voice samples are always loaded asequentially under CSM control.[32]

Two service groups per sector each provide 384 receiver channels and 768 sender channels for trunk signaling. The service groups function in the same architectural position as the line group equipment and contain a space switching stage. Analog announcements, after A/D conversion, are fed through the service group space switch to the network.[33]

The processor complex contains a central processor, peripheral controllers, input/output equipment, a manual control panel, and associated memories and interprocessor channel equipment. The central processor is a general purpose computer which controls the operation of a sector. Peripheral controllers provide communication between the high-speed central processor and the slower-speed equipment in the peripheral and telephony complex and the maintenance and test complex. All except the manual control panel and the input/output access equipment are duplicated and interconnected by duplicated interprocessor buses. In a multisector configuration, sector processors are interconnected via an interprocessor communications link. The processor which controls the incoming trunk also controls the intersector highway junctor frame associated with an intersector call.[34]

The maintenance and test complex utilizes general diagnostic techniques as failure detectors. Each PCM frame in the switching system contains 193 time slots, 192 of which consist of speech time slots containing an 8-bit word and an odd parity bit. The remaining time slot is used for maintenance. Data problems are detected by parity checks. During the time slot that is being used by the T-carrier systems for framing, the maintenance software can set up a path through the switching network for verification of path continuity and data integrity. Pattern generators in the group equipment and highway junctors insert data test patterns which are checked for accuracy. Automatic test equipment can test transmission quality and signaling integrity on trunks.[35]

12.9 AT&T NO. 4 ESS

The No. 4 ESS toll switching system was placed in service in 1976. It is a large, centrally controlled PCM switching system employing a time-space-space-space-space-time (TSSSST) switching network with a capacity of 107 520 four-wire trunks and service circuits.[36] A simplified block diagram of the initial version of the system is shown in Fig. 12–15.

The switching network is composed of two types of switch frames: a time slot interchange (TSI) frame which performs both time and space switching, and a time-multiplexed switch (TMS) frame which performs two stages of

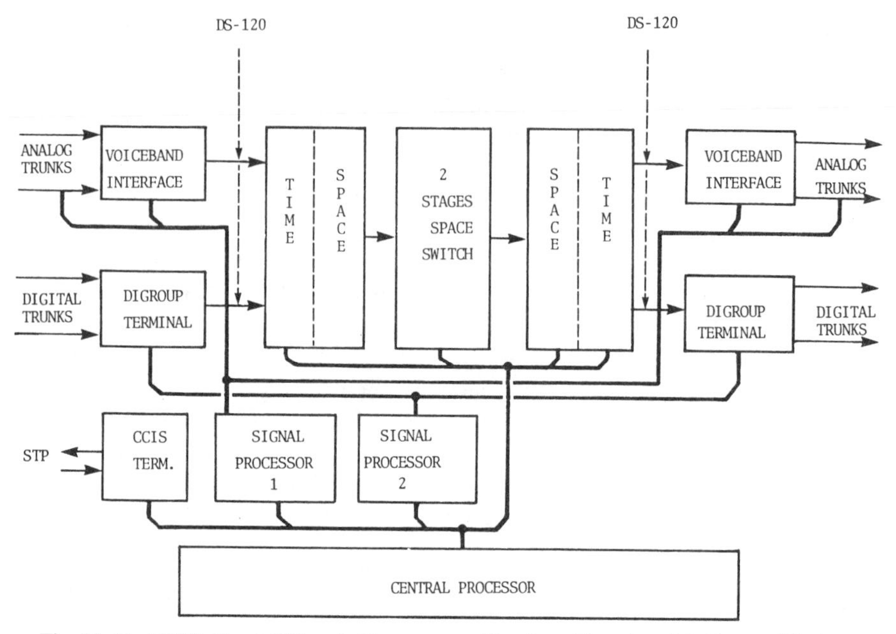

Fig. 12–15. AT&T No. 4 ESS switching system. (Based on Figs. 1 and 2, Bruce, Giloth, and Siegel, "No. 4 ESS—Evolution of a Digital Switching System," *IEEE Transactions on Communications,* July 1979, © 1979 IEEE.)

space switching. A maximum-capacity switching system contains 64 TSI frames and 8 TMS frames. The switching network operates in a folded configuration but is shown unfolded in Fig. 12–15 for clarity of illustration. All network inputs and outputs are T-carrier-compatible DS-120 carrying 120 trunks and 8 maintenance channels in each 128-time-slot PCM frame.

The TMS frames operate as mated pairs with each frame performing the same switching functions to facilitate automatic takeover without loss of calls in case one should fail. Each TMS frame is composed of 16 × 16 arrays of crosspoint logic gates arranged as a 256 × 256 switch array. Up to four pairs of TMS frames can be provided, and a full-sized system thus is configured as a 1024 × 1024 switch array providing 1024 unidirectional paths in each of 128 time slots, or a total of 131 072 paths. Since no retiming is done in the TMS, delay is controlled by precisely cut lengths of coaxial cable between the TSI and TMS frames and between those frames and the network clock.[37]

The time slot interchange frame contains fully duplicated equipment in two identical half-frames. The TSI provides initial time-space switching and final space-time switching for 14 DS-120 PCM links. A later version has doubled the capacity to 28 DS-120 links. Each DS-120 link arrives at the receive side of a switching and permuting circuit (Fig. 12–16) where the 8-bit PCM data are converted to 9-bit parallel format with an added parity bit and stored in buffer memory A. Retiming occurs when the data are read out under control of a time slot counter and passed through a decorrelator to buffer memory B. The decorrelator spreads the traffic in seven buffer memory A's equally over eight buffer memory B's such that the 120 channels of each DS-120 link are distributed equally among the eight buffer memory B's. This accomplishes two purposes. It deloads the network inputs' possible effects of correlation of trunk seizure and holding times which might result from a large trunk group arriving on a single DS-120 link. It also expands the traffic for switching through the network by distributing the 840 channels of seven DS-120 inputs over 960 channels (excluding maintenance channels) of eight outputs, thereby reducing the probability of blocking.

The contents of buffer memory B are read out asequentially under control of the time slot (control) memory, loaded by the central processor via the peripheral unit bus, and switched through an 8 × 8 space switch to the TMS frame. After proceeding through the two space switching stages of the TMS, the PCM samples are returned to the same or another TSI frame where they arrive at the transmit side of a switching and permuting circuit (Fig. 12–17). The samples are switched through the final stage of space switching (another 8 × 8 switch), passed through a timing detector and register to correct slight timing differences, and stored in buffer memory C under control of another time slot memory. The data then are read out sequentially by the time slot counter, passed through a recorrelator which concentrates the eight PCM streams back into seven streams, and converted from parallel to serial DS-120 links by PCM transmitters.[38]

Control is vested in a duplicated central processor with certain repetitive

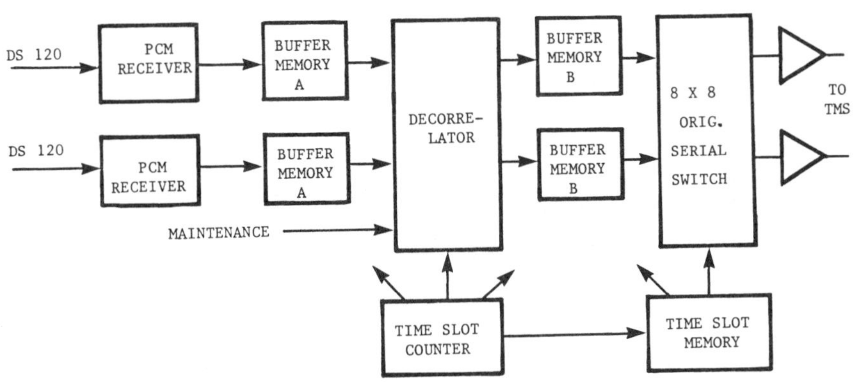

Fig. 12–16. No. 4 ESS switching and permuting circuit (receive side). (Reprinted with permission from *The Bell System Technical Journal.* Copyright 1976, AT&T.)

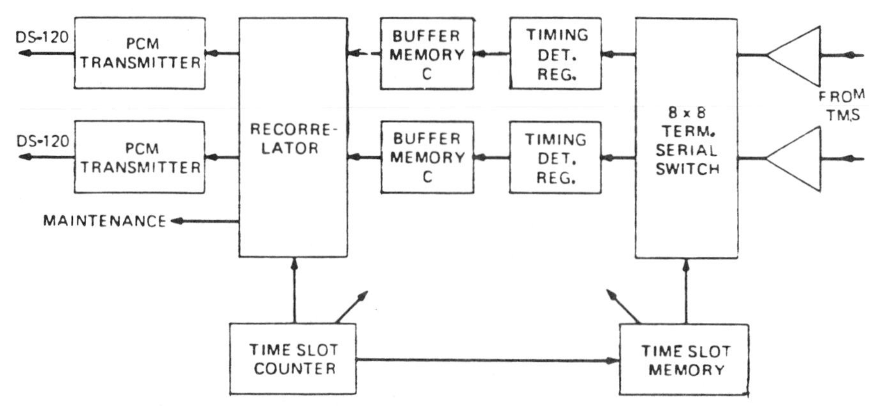

Fig. 12–17. No. 4 ESS switching and permuting circuit (transmit side). (Reprinted with permission from *The Bell System Technical Journal.* Copyright 1976, AT&T.)

functions distributed to peripheral equipment. The central processor performs executive control and audits, call processing tasks, maintenance routines and diagnostics, system initialization when required, and input/output interfaces for administration and network management.

In the initial version of the No. 4 ESS, analog trunks enter the system through a voiceband interface frame containing seven voiceband interface units (VIU), a switchable spare, and a duplicated controller. Each VIU terminates 120 4-wire analog trunks and combines them with eight maintenance channels to form four 32-channel buses. These are sampled and multiplexed into a pair of 64-channel PAM buses. Each 64-channel PAM bus passes through a PCM coder, the output of which is a parallel data stream with 64 PCM words. The outputs of two coders are combined in an access circuit and converted to a DS-120 serial stream for presentation to the switching network. In the reverse direction, the same functions are performed inversely to convert a DS-120 stream into 120 analog trunks and 8 maintenance channels.

Digital trunks are terminated in a digroup terminal containing eight digroup terminal units (DTU), a switchable spare, and two controllers. Each DTU interfaces five 24-channel T-carrier streams to a single TSI port. Each T-carrier bit stream is synchronized to the No. 4 ESS clock.[39]

Two signal processors interface the analog and digital trunks to the central processor. Signaling Processor No. 1, a wired logic processor, handles signaling for up to 4080 analog trunks, and Signal Processor No. 2 handles signaling for up to 3840 digital trunks. Both signal processors perform supervisory scanning, collection and transmission of address digits, and interprocessor communication with the central processor.[40] A CCIS terminal group interfaces the No. 4 ESS control processor to data links to CCIS signal transfer points.[41]

Several enhancements have been made to the No. 4 ESS switching system. Central processor enhancements include increasing the program store and call store memories from 64K ferrite core arrays with 1.4-μs read/write cycle time to 256K semiconductor memories using 16K integrated circuit memory chips with a 700-ns read/write cycle time. The capacity of each TSI has been increased from 14 to 28 DS-120 links without significantly increasing power consumption by using medium scale integrated circuit chips. The TMS frames have been reduced in size and consume one-third less power. A new synchronization unit using a phase-locked loop automatically synchronizes the No. 4 ESS clocks to a reference frequency.

The most significant architectural enhancements involve a redesign of the peripheral equipment (Fig. 12–18). A new digital interface frame (DIF) combines the functions of four digroup terminals and one Signal Processor No. 2 and contains a duplicated controller and a duplicated peripheral unit bus interface. The DIF also contains 32 digital interface units and two spares, each active unit interfacing five DS-1 signals containing a total of 120 digital voice channels. A fully equipped DIF can terminate 3840 digital voice channels and feed them in DS-120 format to the TSI. The duplicated controller

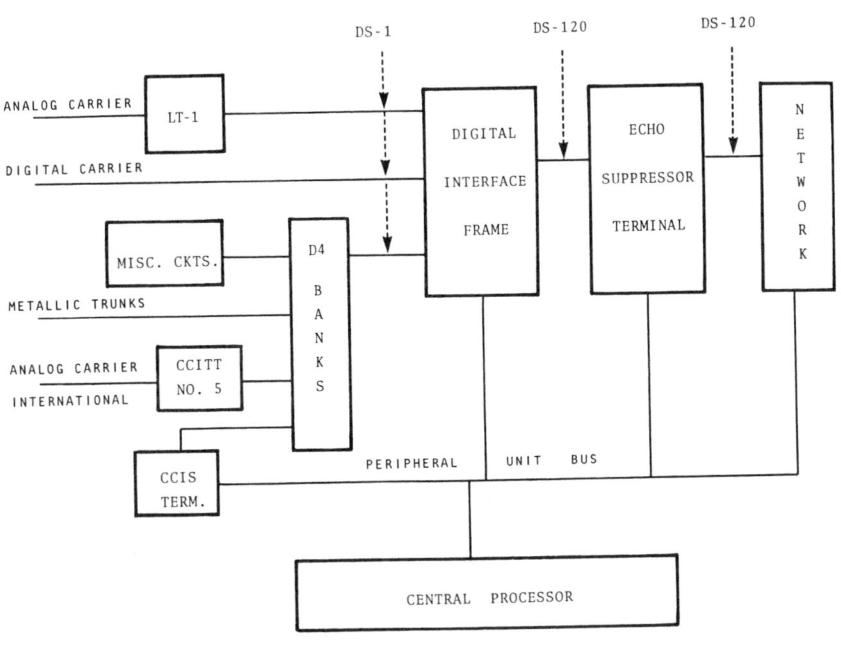

Fig. 12–18. No. 4 ESS enhanced architecture. (Based on Fig. 4, Bruce, Giloth, and Siegel, "No. 4 ESS—Evolution of a Digital Switching System," *IEEE Transactions on Communications*, July 1979, © 1979 IEEE.)

performs the functions formerly the responsibility of the Signal Processor No. 2 and performs maintenance functions. The voiceband interface and Signal Processor No. 1 have been eliminated. Analog trunks now encounter a transmultiplexor, called an LT-1 connector, which performs conversion between two 12-channel analog carrier groups and a 24-channel DS-1 signal. The DS-1 signals are terminated on the DIF along with other DS-1 signals received via digital transmission systems. Metallic trunks, analog international carrier trunks, and conventional service circuits are converted to DS-1 format by D4 channel banks.

A digital echo suppressor terminal has been designed to operate on DS-120 signals traversing paths between the digital interface frame and the network. Using improved suppression algorithms, the time shared echo suppressor can control echo on 1680 trunks simultaneously on a per-trunk basis under control of the central processor.

The enhancements described above represent No. 4 ESS systems placed in service after 1979. New frames are incorporated into older systems as they are expanded.[42]

PROBLEMS

12.1 Discuss the advantages of the control architecture used in the Rockwell 580 DSS.

12.2 Explain how the cross-office highway system of the Stromberg-Carlson DCO constitutes a space switch. Construct an equivalent crosspoint switch with eight TSIs.

12.3 Describe the switching network of the DMS-100 in terms of time (T) and space (S) switching.

12.4 Fig. 12–9 shows the GTD-5 in a folded TST configuration, but the system must operate unfolded. Explain why this is so.

12.5 How does the control system in the No. 5 ESS differ from a conventional master/slave relationship?

12.6 Explain how the ITT 1240 is able to switch calls through multiple stages of time and space switching without a central control to determine connection paths through the switching network.

12.7 Examine Fig. 12–14, and explain why the GTD-3 EAX can be said to function in both a folded and unfolded configuration simultaneously.

12.8 Describe the decorrelator functions in the No. 4 ESS and their effect on the probability of blocking through the switching network.

REFERENCES

1. S. G. Pitroda, "Integrated Voice/Data Capabilities of the Rockwell 580 DSS," *National Telecommunications Conference Record, 1981* IEEE Press, 1981, p. F3.2.1.
2. *System Century DCO System Description,* Stromberg-Carlson, Issue 1, Feb. 1982.

3. Younge, Dale, "The Line Module," *Telesis,* No. 4, pp. 13–15 (Ottawa, Bell-Northern Research, 1980).
4. Terry, J., Krausbar, H., and Hood, J., "DMS-200 Traffic Peripherals," *International Conference on Communications Record, 1978,* IEEE Press, 1978, pp. 32.3.1–32.3.6.
5. Bourne, John, Perry, John, and Workman, Rick, "The DMS-100 Distributed Control System," *Telesis,* No. 4, pp. 6–12 (Ottawa, Bell-Northern Research, 1980).
6. Ibid.
7. Seviora, Rudolph, "Control of Peripheral Modules," *Telesis,* No. 4, pp. 24–26 (Ottawa, Bell-Northern Research, 1980).
8. Esperseth, M. H., and Mnichowicz, D. A., "GTD-5 EAX: A Family of Digital Switches," *Electronic Switching: Digital Central Office Systems of The World* (Amos E. Joel, Jr., ed.) IEEE Press, New York, 1982. (Hereinafter referred to as *Digital CO Systems*), pp. 107–111.
9. Puccini, S. E., and Wolff, R. W., "Architecture of GTD-5 EAX Digital Family," *International Conference on Communications Record, 1980,* IEEE Press, 1980, pp. 18.2.1–18.2.8 (hereafter cited as *ICC '80*).
10. Jinbo, W. S., and Magnusson, S., "GTD-5 EAX Hardware Description," *ICC '80,* pp. 18.3.2–18.3.7.
11. Jackson, D. L., and Kelly, M. J., "GTD-5 EAX RSU Survivability," *National Telecommunications Conference Record, 1981,* IEEE Press, 1981, pp. D5.4.1–D5.4.5.
12. Krikor, K., Jackson, D. L., and Dempsey, J. B., "GTD-5 EAX Common Channel Interoffice Signaling," *National Telecommunications Conference Record, 1981,* IEEE Press, 1981, pp. A4.3.1–A4.3.6.
13. Smith, William Bridges, "No. 5 ESS: Bell's Local Digital Switch," *Telephone Engineer & Management,* pp. 41–45 (Jan. 1, 1982).
14. Mitze, R. W., et al, "The 3B20D Processor & DMERT as a Base for Telecommunications Applications," *Bell Syst. Tech. J.* Vol. 62, No. 1, Part 2, pp. 174–175 (Jan. 1983).
15. Barclay, D. K., Byrne, E. R., and Ng, F. K., "A Real-time Database Management System for No. 5 ESS," *Bell Syst. Tech. J.* Vol. 61, No. 9, Part 2, pp. 2423–2437 (Nov. 1982).
16. Duncan, Tom, and H. Huen, Wing, "Software Structure of No. 5 ESS—A Distributed Telephone Switching System," *IEEE Transactions on Communications,* pp. 1379–1385 (June 1982).
17. Beuscher, Hugh J., "No. 5 ESS Maintenance Software," *IEEE Transactions on Communications,* pp. 1386–1392 (June 1982).
18. Smith, pp. 44–45.
19. Cotton, J. M., et al., "An Expandable Distributed Control Digital Switching Network," *ICC '80,* p. 46.2.5.
20. Chea, R., et al, "Circuit Terminations and Terminal Control," *ICC '80,* pp. 46.3.1–46.3.7.
21. Ibid.
22. Cotton et al., pp. 46.2.2–46.2.4.
23. Richards, Philip C., "Technological Evolution–The Making of a Survivable Switching System," *Digital CO Systems,* pp. 200–201.
24. Cotton et al., pp. 46.2.3–46.2.6.
25. Becker, G., et al., "Call Processing in a Distributed Control System," *ICC '80,* pp. 46.4.3–46.4.4.
26. Cotton et al., pp. 46.2.4–46.2.6.
27. Guilarte, W., and Wang, F. C., "Maintenance Advantages for a Distributed System," *ICC '80,* pp. 46.5.1–46.5.3.
28. Cox, J. E., et al., "A Digital Switch for Wide Range of Application," *ICC '80,* pp. 46.1.4–46.1.7.

29. Holm, R. K., "No. 3 EAX System Description," *International Conference on Communications Record, 1977,* IEEE Press, 1977, pp. 27.1–194 to 27.1–195 (hereafter cited as *ICC '77*).
30. Litterer, J. E., and Tam, L. K., "No. 3 EAX Trunks and Service Circuits," *ICC '77,* pp. 27.3–204 to 27.3–206.
31. Bagley, John M., "The No. 3 EAX Digital Switching System," *Digital CO Systems,* p. 49.
32. Bieszczad, E. S., et al., "The No. 3 EAX Network and Master Clock," *ICC '77,* pp. 27.2–199 to 27.2–201.
33. Bagley, p. 49.
34. Holm, pp. 27.1–195 to 27.1–196.
35. Ray, D. P., Bernasek, B. H., and Hupcey, J. A., "No. 3 EAX Maintenance and Test Facilities," *ICC '77,* pp. 27.5–214 to 27.5–218.
36. Richie, A. E., and Tuomenoska, L. S., "System Objectives and Organization," *Bell Syst. Tech. J.,* pp. 1017–1027 (Sep. 1977).
37. Huttenhoff, J. H., et al., "Peripheral System," *Bell Syst. Tech. J.,* pp. 1029–1037 (Sep. 1977).
38. Ibid., pp. 1037–1041.
39. Boyle, J. F., et al., "Transmission/Switching Interfaces and Toll Terminal Equipment," *Bell Syst. Tech. J.,* pp. 1062–1081 (Sep. 1977).
40. Huttenhoff et al., pp. 1041–1049.
41. Croxall, L. M., and Stone, R. E., "Common Channel Interoffice Signaling: No. 4 ESS Application," *Bell Syst. Tech. J.,* pp. 361–362 (Feb. 1978).
42. Bruce, R. Allen, Giloth, Paul K., and Siegel, Eugene H., Jr., "No. 4 ESS—Evolution of a Digital Switching System," *IEEE Transactions on Communications,* pp. 1001–1011 (July 1979).

13
Evolution of the Switched Digital Network

13.1. INTRODUCTION

The North American public switched telephone network is evolving from a fully analog network to one which is expected to approach a fully digital network. As the evolution progresses over a period of many years, digital switching systems and digital transmission facilities are being added at the discretion of some 1500 separate telecommunications companies.

The Canadian portion of the network is operated by a group of telephone companies which independently provide local and intraprovincial services and which associate together as Telecom Canada to provide interprovencial and cross-border services. Competitive services are provided by CNCP Telecommunications.

In the United States, approximately 80% of the telephones are served by the operating companies which were part of the Bell System prior to January 1, 1984. On that date, 22 wholly owned Bell Operating Companies (BOCs) were divested as a result of a consent decree which settled a government antitrust suit against the American Telephone and Telegraph Company (AT&T). The divested BOCs were formed into seven regional holding companies. AT&T was reorganized, and the former AT&T Long Lines Department, which handled interstate and international services, became AT&T Communications. The remaining telephones are served by some 1400 or more telephone companies, many of which are part of separate holding companies.

In this chapter, Section 13.2 describes the North American network in terms of numbering, routing, signaling, and transmission parameters as a purely analog network. Section 13.3 then discusses the gradual evolution of that network from analog to digital with emphasis on transmission and synchronization. Section 13.4 describes the local telephone networks of the divested BOCs in terms of network configuration, signaling, and provision of access to the networks of competing long distance carriers.

13.2. THE NORTH AMERICAN ANALOG NETWORK

The public telephone networks of the United States and Canada, although subject to many different ownerships and different national interests, function essentially as one large network. The combined networks utilize unified numbering and traffic routing plans, compatible signaling protocols, and common transmission standards. Although parts of Mexico and certain Caribbean islands are part of the North American Numbering Plan, the contents of this chapter are intended to be applicable only to the United States and Canada.

13.2.1. NETWORK NUMBERING PLAN

The numbering plan consists of two parts: a numbering plan area (NPA) code and a 7-digit telephone number. The NPA code is a 3-digit number in the form *N* 0/1 X, where N is any digit, 2 through 9, and *X* is any digit, 0 through 9. The middle digit is either a 0 or a 1. All N11 codes were excluded so as not to conflict with local service numbers. The remaining 152 NPA codes, or area codes, will be used until exhausted for assignment to geographic areas. Seven-digit telephone numbers consist of a 3-digit central office (CO) code plus a 4-digit station number. Central office codes, originally assigned in the form *NNX,* allowing a total of 640 codes per NPA, have been exhausted in some areas. Future growth may exhaust NPA codes. Therefore, a system has been devised to use the form *NXX* interchangeably for both NPA and CO codes. All 63 *NN*0 codes have been designated to be the last CO codes assigned in any NPA and will be assigned as both CO and NPA codes in reverse sequence. Ultimately, all toll calls likely will need to be preceded by the digit 1 for direct-dialed station-to-station calls and by the digit 0 for calls which require operator assistance.[1]

13.2.2. Network Routing Plan

Central office switching systems which switch subscriber loops to each other and to the network are called *end offices* and are designated as class 5 offices. Switching systems which perform the first stage of concentration of traffic from end offices to the network and perform final stage distribution of traffic to end offices are called *toll centers* or *toll points* and are designated as class 4C or class 4P systems, respectively. The only difference between a center and a point is that a center provides an inward operator function while a point does not. Higher-class switching systems constitute *control switching points* (CSP) and are designated as *regional centers* (class 1), *sectional centers* (class 2), and *primary centers* (class 3).

There are ten regions in the United States and two in Canada, each served by a regional center. Each region is subdivided into sections, each served by a sectional center, and each section into primary areas, each served by a primary center. The network is a hierarchical arrangement of switching systems with class 1 offices at the top of the hierarchy and proceeding downward through class 2, 3, and 4 offices to class 5 end offices. Each switching system other than regional centers is connected to an office in the next higher level by a final trunk group, liberally sized to block no more than about 1% of the traffic offered in the ABSBH. Each regional center has a final trunk group to every other regional center. A switching system connected by a final trunk group to an office in the next higher level is said to *home* on that office. A switching system may perform a switching function other than that of its designation but is classified according to the highest function performed. For example, a class 2 office which has a class 4 office homed on it also performs a class 3 switching function. Hierarchical homing arrangements are illustrated by the solid trunk group lines in Fig. 13–1.[2]

In addition to final trunk groups connecting hierarchical levels, high usage (HU) trunk groups can be provided between any two switching systems, subject to certain constraints, when justified by traffic volume and economics. In some cases, HU trunk groups are sized according to the same criteria as used for final trunk groups, but generally they are sized to overflow a substantial portion of offered traffic. Justification for establishment of interregional HU groups is based solely upon first-routed traffic; i.e., traffic which would be offered first to the potential HU group. A "one-level limit" rule, however, limits the establishment of interregional HU groups to those which can be justified by first-routed traffic between switching *functions* which differ by no more than one level. The purpose of this rule is to keep traffic at the lowest possible level in the hierarchy. Once an HU group is justified by first-routed traffic, however, any traffic proceeding to that region may be offered to it. Interregional traffic may not be offered to intraregional HU groups progressing up the hierarchy of the home region because this tends to cause additional switching of calls.[3]

Network routing doctrine is illustrated in Fig. 13–1. Assume that a call originating at end office A is destined for a subscriber served by end office J. Since there are no interregional HU groups below the class 3 level in this example, the call must progress up the hierarchy via final trunk groups through toll center B to primary center C. Primary center C has several possible routes to the destination region. The call will be offered first to HU trunk group C–L. If all trunks in that group are busy, alternate routes C–M, C–N, and C–D will be searched in that sequence for an idle trunk. Observe that the call will not be offered to HU trunk group C–E because of the doctrine requiring calls to be kept as low as possible in the hierarchy. Observe also that the HU group C–N can be justified only because of first-routed traffic volume between C and N as a result of regional center N's performance of a class 3 switching function for the class 4 office homed on it. Once established, however, other traffic is permitted to use it. Sectional

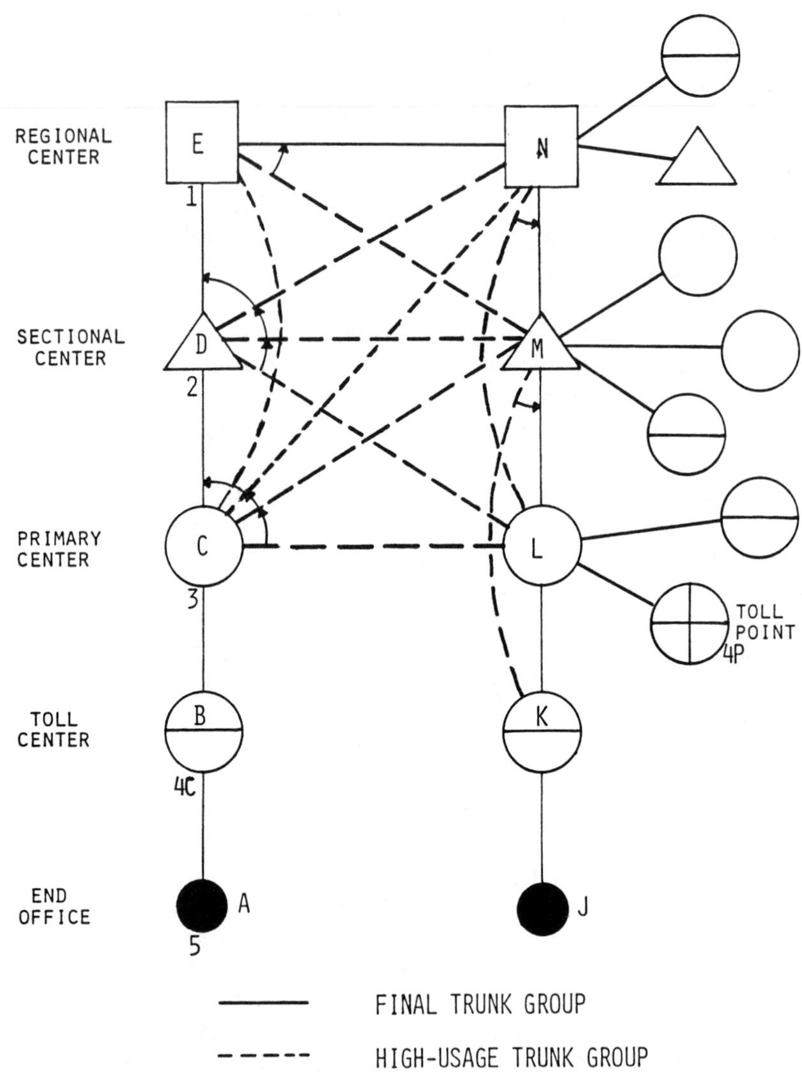

Fig. 13–1. North American public network hierarchical routing.

center D can use any of its routes to the destination region and the final route to regional center E if all Hu groups are busy. Proceeding down the destination region hierarchy, the call will use the most direct route available to the destination switching center. "Skip-level" HU trunk groups are used first with blocked traffic overflowing to final trunk groups.[4]

13.2.3. Network Signaling

The fundamentals of both inband and common channel signaling are set forth in Chapter 10. This subsection discusses the application of those systems to domestic and international networks.

13.2.3.1. Inband Signaling. Inband signaling on carrier-derived trunks employs single frequency (SF) supervisory signals with a tone-on-idle protocol. Address signaling is dial pulse (DP) or multifrequency (MF). MF signaling on domestic calls is at the nominal rate of seven pulses per second for most switching systems.

Signaling on international trunks employs CCITT signaling system No. 5 between foreign and North American gateway switching systems. The system uses compelled two-frequency supervisory signals and pulsed MF address signals at the nominal rate of ten pulses per second, using the same six frequencies as used in domestic signaling. A seizure is indicated by transmitting a 2400-Hz tone to the called gateway switching system. When the called system is ready to receive address digits, it responds by sending a 2600-Hz tone to the originating switching system which then proceeds to outpulse the address digits. The two frequencies also are used for other purposes during call setup and disconnect sequences. This use of compelled supervisory signaling accomplishes two purposes. One, it automatically enables detection of simultaneous seizures by distinguishing between the start dial signal and the seizure signal. Second, it enables inband signaling to be used on circuits employing time assignment speech interpolation (TASI), a system which increases utilization of expensive transoceanic cable trunks by using voice-operated high-speed switching to trunks only when speech is present, taking advantage of the silent periods during conversations. When used with TASI, the seizure signal looks like a speech spurt and is connected to an available TASI channel. The signal persists until the distant switching sytem returns a start dial signal. The originating system removes the seizure signal and outpulses address digits before the TASI channel can be released.[5]

Domestic switching systems, at which international calls are originated or switched through, must outpulse more digits than are required for domestic calls. Therefore, the address digits are outpulsed in two stages. The first stage outpulsing routes the call to an international gateway switching system. At each switching node, digits may be added or deleted for routing purposes. After a transmission path is established between the originating and gateway switching systems, the second stage outpulsing occurs on an end-to-end basis. After the gateway office receives the ST pulse at the end of the first stage outpulsing, it delays at least 700 ms to allow time for switching transients to dissipate and then sends a second-start-dial signal, an off-hook signal of at least 400 ms duration. The second-start-dial signal is immediately followed by a 480-Hz tone which persists until the ST signal is received or until the MF receiver times out. Since many international calls are dialed by operators, the purpose of the tone is to alert operators to start sending the second-stage address pulses.[6]

13.2.3.2. Common Channel Signaling. Two common channel signaling systems have been defined for worldwide use. CCITT System No. 6 uses analog voiceband transmission while CCITT System No. 7 uses the standard 64-kb/s digital transmission link. Common channel interoffice signaling (CCIS), used in the United States, is a variation of CCITT No. 6. The main difference is increased label size because of the large sizes of trunk groups used. There are other format differences as well.[7]

Basic CCIS is described in Section 10.5.5. All intertoll trunks (trunks connecting toll switching systems) are served by CCIS. CCIS had been extended to some local switching systems prior to the divestiture of the BOCs.[8] Divested BOCs plan to implement CCITT System No. 7 beginning in 1986.

Additional stored program controlled processors have been added to the CCIS network to support new or enhanced network services. Data bases have been established at locations called *network control points* (NCP). NCPs are collocated with selected *signal transfer points* (STP). They use the 3B20D processor and DMERT operating system used in the No. 5 ESS. Certain switching centers, called *action points* (ACP), are the focal points of subscriber-dialed calls requiring special handling. The subscriber dials a special code and number which causes the call to be routed to an ACP which must be served by CCIS. The ACP holds the call and sends an inquiry message via CCIS to a designated NCP for call routing information. The NCP returns routing instructions to the ACP which routes the call accordingly. The CCIS messages are routed like a datagram in packet-switched networks. The NCPs are grouped into mated pairs, widely separated geographically, each sharing the load but able to carry the entire load in case one fails or is isolated from the network. These direct signaling messages are routed through the CCIS network based upon the message address.

Figure 13–2 illustrates the arrangement of the CCIS network. Two STPs are located in each hierarchical region and operate as mated pairs. Each toll switching system served by CCIS is connected to the pair of regional STPs by data links, called A links, operating with modems on analog circuits at 2.4 or 4.8 kb/s. Each regional STP is connected to every other regional STP by B links, and the mated pairs are connected to each other by C links. In regions with very heavy calling volumes, area STPs have been established to augment the regional STPs. Because of capacity limitations at some

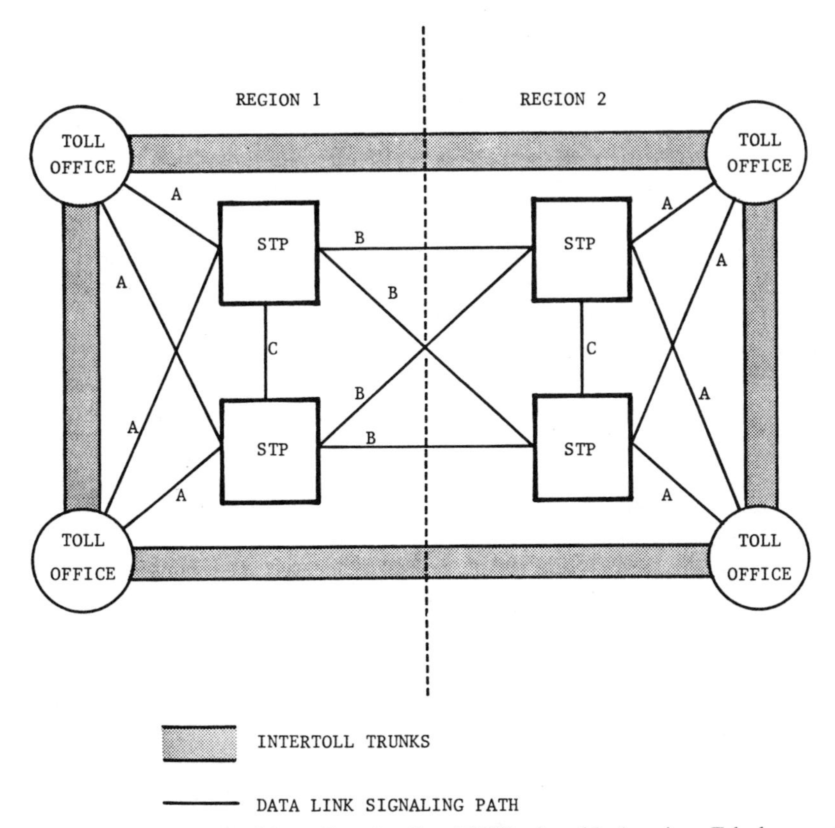

Fig. 13–2. Common channel interoffice signaling (CCIS) plan. (© American Telephone and Telegraph Company, 1980.)

regional STPs, some NCPs have required the activation of specialized STPs to handle signaling associated with use of the NCP data bases. The STP schedule provided for an end-1982 United States network of 20 regional STPs, 6 area STPs, and 4 NCP-STPs providing CCIS signaling for 222 switching systems, 135 toll operator systems, and 28 NCPs.[9]

During 1986, the CCIS network is planned to be replaced by new No. 2 STP equipment having a higher capacity and signaling link data speeds of 56 kb/s in addition to the present 2.4 and 4.8 kb/s. The transition plan calls for replacing mated pairs of STPs one-for-one. Concurrently, a modified CCITT System No. 7 is scheduled to be phased into the network signaling topology.[10]

CCITT System No. 7 is a signaling system designed for use in digital networks. It is composed of two basic sections: a message transfer part (MTP) and multiple user parts (UP). The MTP has three levels of signaling, which provide standards and protocols for a 64-kb/s data link or analog circuit, signaling functions over a single data link, and network functions for switching signaling messages through a packet-switched network. The UPs are functional procedures for the exchange of messages through the signaling links or network irrespective of the modes of transmission or routing. User Parts provide for basic telephone service and circuit switched data.

CCITT No. 7 signaling messages can be of variable quantities of 8-bit bytes. Errors are detected by analysis of a 10-bit cyclic code. Errors are corrected by retransmission of the message and all subsequent messages. This assures that the signaling messages are properly sequenced. Channel 16 of the 32-channel PCM format is used. When used with T1 carrier, signaling must occupy a full voice channel, reducing the T1 message channels to 23.[11]

CCITT No. 7 has several advantages over CCIS. Error correction assures proper message sequencing. Synchronization is simpler, and there is less delay in the system. By using 64-kb/s signaling links, CCITT No. 7 can handle much higher busy hour traffic loads. Studies have indicated that CCITT No. 7 is less costly than CCIS. CCIS does have some advantages. It was designed for low overhead, short addressing labels, and minimum retransmission to correct errors. However, the higher data rate of CCITT No. 7 more than offsets those advantages. Canada has opted for CCITT No. 7 in its digital network.[12]

CCIS is to be replaced by a modified version of CCITT No. 7 in the United States in the 1986 time frame. AT&T's System No. 7 is planned for 56-kb/s data links. The No. 2 STPs, which will replace the existing STPs used with CCIS, will have to have the capability of supporting both System No. 7 and CCIS formats. Initially, System No. 7 likely will use only the message transfer parts, with user parts to be added later.[13]

13.2.4. Transmission Requirements

All transmission systems are imperfect and impair the quality of information transmitted. A major objective of transmission system engineering is to control impairments in such a way as to achieve an acceptable compromise between transmission quality and the cost of providing the system.

13.2.4.1. Transmission Impairments. There are several characteristics of transmission systems which degrade the quality of signals. Among these are noise, loss, amplitude/frequency distortion, crosstalk, echo, and delay distortion. The effects of some impairments can be minimized by quality design of equipment components and circuitry. Others, such as loss, noise, and echo, are inherent in telephone transmission systems and can be dealt with only by a carefully controlled transmission plan.

Noise is any signal or interference on a circuit other than the signal being transmitted. There are several types of noise: thermal noise, random noise, impulse noise, induced noise, interchannel noise, quantizing noise, and internal noise resulting from unbalanced conditions. Noise always has amplitude and is amplified or attenuated along with the signal in analog systems. The level of inherent circuit noise increases with circuit length; i.e., when the length of a circuit is doubled and the noise per unit length is constant, the total noise level is increased by 3 dB. Design control keeps inherent noise levels well below signal levels. In the event of a weak signal, however, a lower signal-to-noise ratio reduces the quality of the received signal. Noise is measured with a meter which accounts for the interfering effect of noise at different frequencies, the overall interfering effect of combined noise elements, and the effect of noise bursts of different duration. Noise measurements are referred to a reference noise level, defined as 10^{-12} watt, or −90 dBm. A 1000-Hz tone at a level of −90 dBm would give a meter reading of 0 dBrn. Noise measurements using C-message frequency weighting are expressed in terms of dBrnc and, when referred to 0 TLP, are expressed in terms of dBrnc0.[14]

Loss is another impairment which is inherent in all transmission systems. Loss is the attenuation of power which occurs in traversing a circuit and generally is proportional to circuit length. Amplification can overcome loss, but amplification must be carefully controlled to provide satisfactory end-to-end communications. As signals proceed along a transmission path, they are attenuated, and amplifiers are inserted at intervals to restore the proper levels. If signals arrive at a transmission junction or a switching system at levels which are too high, they are attenuated to a predetermined level by the insertion of attenuation pads. Metallic subscriber loops have variable amounts of loss depending upon the length and gauge of the conductors used. On extremely long loops, lumped inductance, called a *loading coil,* is inserted in the loop at specified intervals to counteract loss due to the capacitance between the conductors.

The third impairment discussed in this subsection is *echo.* Echo is a reflected signal returned with sufficient magnitude and delay to be perceived as distinct from the signal directly transmitted. In telephony, echo occurs generally because of an imperfect impedance match between a 2-wire circuit and a 4-wire circuit. This is illustrated in Fig. 13–3. Figure 13–3(a) depicts a typical toll connection proceeding from telephone A through local central office B, toll offices C and D, and local central office E to telephone F. The subscriber loops and toll connecting trunks (TCT) are 2-wire circuits, although TCTs may be either 2-wire or 4-wire circuits. In this example, a single 4-wire

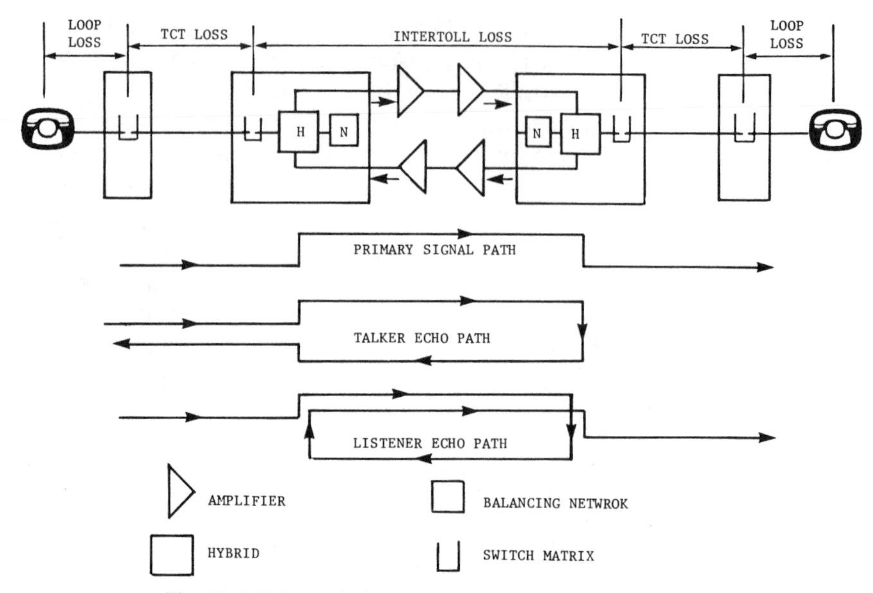

Fig. 13–3. Echo paths in long distance telephone connections.

intertoll trunk is shown between the toll switching systems. This requires 2-wire/4-wire conversion at each of the toll offices. The hybrid transformers habitually use a compromise balancing network which does not perfectly balance the impedance of the 2-wire path. Since the impedances of toll connecting trunks are closely controlled, the impedance mismatch at the hybrid transformer in a toll office is less than if it were located at the serving central office because of the wide variance of subscriber loop impedances. The primary signal traverses the path shown in Fig. 13–3(b). When the signal arrives at toll office D, theoretically half the signal power proceeds down the 2-wire path while the other half is dissipated in the balancing network N. In actual practice, however, the imperfect impedance match permits some of the signal power to flow across the hybrid transformer and return to the talker, as shown in Fig. 13–3(c). If the impedance mismatch at toll office C is significant, part of the echo signal power can cross the hybrid transformer and be heard by the listener, as shown in Fig. 13–3(d). Since the hybrid transformer divides the signal power equally between two impedances, the hybrid loss is 3 dB. Transformer core and copper losses are typically about 0.5 dB. Although there are other sources of echo, the principal source is hybrid unbalance.[15]

The effect of echo primarily is a function of transhybrid loss and delay, and delay is a function of the distance between the hybrid transformers and of the propagation velocity of the transmission facility. Tolerance of echo varies with individual people and is influenced by echo level and delay. If circuit amplification is too high or if transhybrid losses are too low, echo signals can circulate around the listener echo path and cause circuit instability, or singing. A near singing condition causes speech to have a hollow sound as if talking into a barrel.[16]

13.2.4.2. Control of Impairments. It should be apparent from the foregoing discussion that transmission impairments must be controlled. A conflicting relationship exists. Amplification overcomes loss but it also increases the effect of noise, and loss controls echo. Since echo is the most objectionable impairment encountered in telephone networks, control of echo must receive primary consideration. Echo can be kept to a tolerable level through control of loss and transmission delay. The amount of loss needed to control echo increases with increasing delay, but the amount of loss must not reduce the signal level below an acceptable level. Since an amplifier affects both signals and noise, noise must be kept sufficiently low that amplification does not cause interference with the signal.

13.2.4.3. Via Net Loss Plan. The plan developed in the early 1950s to allocate loss for echo control is called the via net loss plan. *Via net loss* (VNL) is the loss in dB assigned to a trunk to compensate for its added propagation delay, terminal delay, and loss variability, and is calculated for each trunk by

$$\mathrm{VNL(dB)} = 0.102D + 0.4 \qquad (13\text{–}1)$$

where 0.102 is the one-way incremental loss in dB per ms (reduced to 0.1 for practical calculations) and D is the echo path delay in ms. Since the echo path delay is a function of circuit length and velocity of propagation, a VNL factor (VNLF) for different transmission media has been calculated by

$$\mathrm{VNLF} = \frac{2 \times 0.102}{v} \qquad (13\text{–}2)$$

where v is the velocity of propagation in kilometers per ms. Since the factor 0.102 applies to one-way transmission, it is necessary to multiply by 2 to obtain a value applicable to round-trip delay. The VNLF for two-wire exchange cable is 0.04 and for all carrier systems is 0.00093. Hence, for toll plant, via net loss is calculated by

$$\mathrm{VNL(dB)} = 0.00093D + 0.4 \qquad (13\text{–}3)$$

where D is the one-way distance in kilometers. The factor 0.4 is added to compensate for loss variability.[17]

Loss can compensate for echo only to the point at which loss itself begins to degrade circuit quality below acceptable limits. Current engineering practices use loss alone on trunks up to 2979 kilometers in length. Longer high usage trunks and certain interregional final trunks are equipped with echo suppressors or echo cancelers and are operated at 0 dB loss. Certain trunks are excepted from the use of echo suppressors. Between class 5 offices, carrier circuits which are over 2979 kilometers long are operated at 8.9 dB net loss.[18]

13.2.4.4. Analog Network Transmission Plan. Losses in a subscriber-to-subscriber toll connection can be associated with three connection categories: the originating and terminating subscriber loops, all switching system networks, and the interconnecting trunks. Loops include the circuitry between the terminal equipment and the serving central office main distributing frame at which the loop is terminated. Only the originating switching network is considered separately. The networks in all other switching systems are considered to be part of the interconnecting trunks. Trunk losses are calculated from the output side of the switching network of one switching system to the output side of the network of the next switching system in the connection.[19]

Subscriber loops in Bell operating company territories have an average loop loss of about 3.7 dB. Some loops have near-zero loss while a small portion have losses in the range of 8.5–10 dB. In some rural areas, losses can exceed 10 dB. Current design standards for Bell operating companies limit loop losses to 8 dB.[20] Losses through the switching network of the originating switching system are very small, generally less than 1 dB, and vary with switching system technology.

Because loop and originating switching losses are highly variable, network transmission plans focus on controllable trunk losses. The analog transmission plan is illustrated in Fig. 13–4. The loss objective for connections from originating end office to terminating end office is VNL + 5 dB. The 5 dB portion is allocated to toll connecting trunks between the end offices and the connecting toll offices. The VNL portion is the sum of the VNLs for all intertoll trunks in the dialed connection. Trunks less than 322 kilometers long, which directly connect two end offices, are designed for a loss of 3 dB if they have gain and for a loss of 0–5 dB if metallic without gain. Such trunks over 322 kilometers long are designed for a loss of VNL + 6 dB up to 2979 kilometers and a loss of 8.9 dB if over 2979 kilometers.[21] It can be seen that connections between local offices can range from about 6 to 9 dB, and losses between subscribers can range from about 14 to 17 dB on

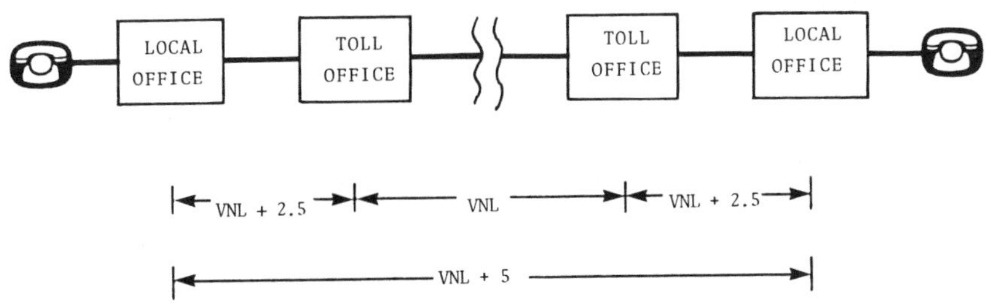

Fig. 13–4. Analog network transmission plan.

the average, but some connections can be well over 20 dB. Connections losses of about 14–16 dB are considered optimal.

Although trunk losses can be controlled by pads and amplifiers, connection losses are variable, and it is important that switched connections have as much uniformity as possible within economic limits. Therefore, test signals of all trunks are adjusted to the power level specified at individual switching offices. This is accomplished by measuring test tones and comparing them to a reference level. This process employs the *transmission level point* (TLP) concept. Any point in a circuit, at which signal levels can be measured, can be designated as a transmission level point. The transmission level at any point in a transmission system is the ratio of the signal power at that point to the power of the same signal at an arbitrary reference point called the *zero transmission level point* (0 TLP). TLPs are usually designated as having a numerical value such as −2 TLP or −3 TLP. Since transmission losses vary with frequency, the TLP is assigned for specific frequencies, usually 1004 Hz for voiceband circuits.

The TLP concept enables signal levels to be brought to a common level for switching, enables circuit gains and losses between two TLPs to be accurately determined, and provides a convenient way for signals and interference to be expressed in values referred to the same TLP.

The term *level* in transmission level point should not be confused with signal power level. There is no direct relation between power level and TLP. Only when a test signal is applied to a properly adjusted circuit at a power level which is numerically the same as the specified TLP at which it is applied will the measured power level of the test signal at any other TLP in the circuit correspond to the designated TLP value. It also is important to recognize that the TLP concept applies to individual trunks and not to switched connections. Commonly used transmission level points are 0 TLP at the outgoing side of a local switching system network and −2 TLP at the outgoing side of an analog switching system network to which an intertoll trunk is connected. Test signals normally are applied at power levels below the TLP to avoid overloading transmission systems. When referred to the 0 TLP, signal power can be expressed by the unit dBm0, meaning dBm at the 0 TLP.[22]

13.3. THE EVOLVING DIGITAL NETWORK

The transition of the public telephone network from analog to digital switching and transmission is an evolutionary process. Digital transmission systems have been finding their way into the network for over 20 years, while digital switching systems have been in the process of being added to the network for over 9 years. At the present or even a greatly increased rate of introduction, the network will not be entirely digital for many years.

13.3.1. Digital Network Transmission Considerations

Although realization of a fully digital network is many years away, many metropolitan areas have a high density of short-haul digital transmission systems, and some long-haul digital systems have been placed in service. A transmission plan for the fully digital network was necessary before a transmission plan could be developed for the mixed analog/digital network which exists now and will exist during transition.

13.3.1.1. Loss and Level Considerations. Digital transmission and switching involve loss and level considerations which were not factors in a purely analog network. The TLP concept, which functions well in an analog environment, requires modification for use in a digital environment. In an analog system, test tones used to line up (adjust levels in) voiceband channels generally are applied at a power level of one milliwatt (0 dBm). Digital systems, however, involve the encoding and decoding of analog signals, and the direct use of the one-milliwatt analog test tone is not feasible. Therefore, a digital milliwatt has been standardized. For 8-kHz sampling and the standard 64-kb/s channel rate with $\mu = 255$ companding, the digital milliwatt is defined by the CCITT through the bit stream of an eight-word sequence shown in Table 13–1.[23] A consequence of this definition is that the bit stream represents a 1000-Hz sine wave at a power level 3.17 dB below the overload level of the encoder.

In order to utilize the TLP concept in digital systems to the extent possible, encode and decode level points have been defined at the inputs and outputs of encoders and decoders, respectively. The encode level in dB at the encode level point (ELP) is the power level in dBm of a sine wave at that point which will result in coded values equivalent to the digital milliwatt. The decode level in dB at the decode level point (DLP) is the power in dBm of

Table 13–1. Digital Milliwatt.*

	BIT NUMBER							
	1	2	3	4	5	6	7	8
Word Number: 1	0	0	0	1	1	1	1	0
2	0	0	0	0	1	0	1	1
3	0	0	0	0	1	0	1	1
4	0	0	0	1	1	1	1	0
5	1	0	0	1	1	1	1	0
6	1	0	0	0	1	0	1	1
7	1	0	0	0	1	0	1	1
8	1	0	0	1	1	1	1	0

NOTE: Bit numbers 4, 6, and 8 individually duplicate the system framing sequence used in T-carrier channel banks. Therefore, to prevent a receiving channel bank from framing on bit numbers 4, 6, or 8 without giving an alarm, the digital milliwatt should not be transmitted over in-service digital facilities.

a sine wave at that point which will result from decoding the digital milliwatt. For example, at 0 EL, a 0-dBm signal produces the digital milliwatt, and encoder overload occurs at 3.17 dBm. (It is common practice to use 3 dBm rather than the more accurate 3.17 dBm.) A D/A converter operating at 0 DL decodes the digital milliwatt into a 1000-Hz sine wave at a level of 0 dBm.[24] A/D and D/A converters can be operated at levels other than zero. By changing the reference voltage of the encoder, a −3 dBm signal can be encoded into a ditital milliwatt, resulting in a 3-dB relative gain. This is equivalent to encoding at zero level and inserting 3 dB of analog gain before the encoder. Both result in a 3-dB gain. One could also use 3 dB of digital gain following a 0 EL encoder, but this would destroy bit integrity for transmission of digital data. A decoder also can operate at levels which will result in a specified amount of loss. For example, a −3 DL decoder will decode a digital milliwatt at −3 dBm, equivalent to a 3-dB loss. In summary, an encoder operating at 0 EL results in no gain or loss. An encoder operating at −3 EL, for example, results in a 3-dB equivalent gain. A decoder operating at 0 DL results in no gain or loss. A decoder operating at −5 DL, for example, results in a 5 dB equivalent loss.

13.3.1.2. Digital Network Transmission Plan. In an all-digital network, loss is not calculated on a per-trunk basis. Rather, it is allocated on a connection basis. Current plans specify a fixed loss of 6 dB between end offices. Intertoll digital trunks operate at 0-dB loss. Direct digital trunks between end offices are allocated loss of 3 dB if less than 322 kilometers in length and 6 dB if longer than 322 kilometers. In an all-digital network, the voiceband channel is encoded once and decoded once. All loss is inserted at the receiving end of the connection. The DL at the terminating office has no universal value but corresponds to the amount of loss needed for each type of connection. In the connection shown in Fig. 13–5, the terminating office would use a −6 DL. Class 5 offices will operate at 0 EL and have sufficient DL options so that the proper amount of loss can be inserted for any type of connection.[25] Echo cancellation will be used on long digital trunks; however, the additional delay in an all-digital network could require reduction of the present 2979 kilometers echo suppression rule.

13.3.1.3. Mixed Analog/Digital Transmission Plan. As the network evolves from analog to digital, digital conversion of switching is influenced at least as much by area economic judgment as by systematic planning. Both local and toll digital switching systems are interspersed throughout the network. Digital transmission systems have been used for over 20 years primarily in pair gain applications and for route augmentation, principally in metropolitan areas and along high-density routes. As a result, some trunks in the toll network are digital, some are analog, and some are combination analog and digital. Trunks are defined as analog, digital, or combination depending upon the type of switching and transmission facilities involved. A *digital trunk* is one that is derived on digital facilities and terminates digitally on a digital switching system at each end. A *combination trunk* is one that is derived on digital facilities and terminates in a digital interface at one end and in an analog interface at the other end. All other trunks are defined as *analog trunks* whether derived (1) on analog facilities, (2) on digital facilities which terminate in an analog interface at both ends, or (3) on a mixture of analog and digital facilities.[26] Figure 13–6 illustrates various configurations of digital, combination, and analog trunks.

The mixed analog/digital transmission plan is shown in Fig. 13–7. Digital toll offices operate at −3 TLP, but analog toll offices operate at −2 TLP.

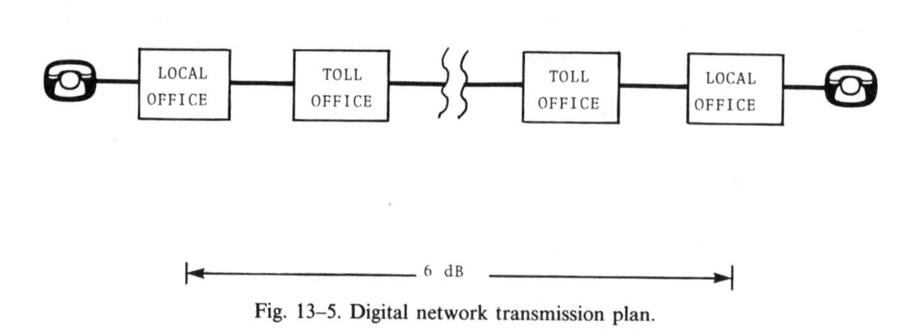

Fig. 13–5. Digital network transmission plan.

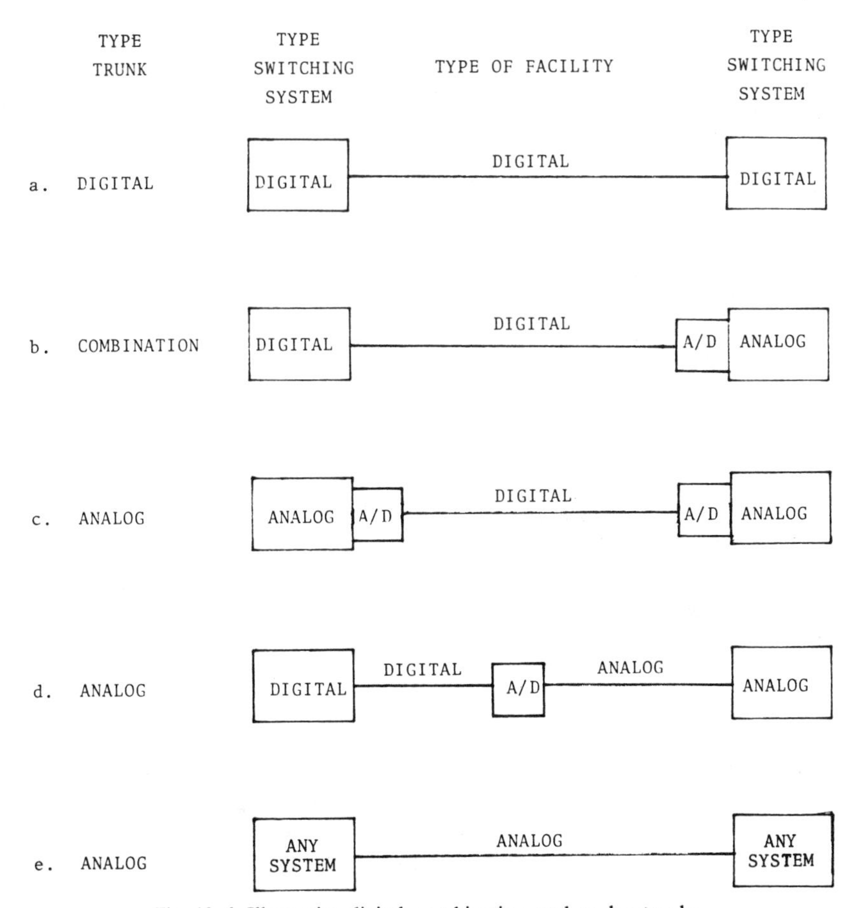

Fig. 13–6. Illustrative digital, combination, and analog trunks.

Both analog and digital end offices operate at 0 TLP. The only new element in this plan is the combination trunk. By definition, the only difference between a digital trunk and a combination trunk is that a digital trunk is terminated in a digital switching system at both ends while a combination trunk is terminated in a digital switching system at one end and in an analog switching system at the other end. Both are derived on digital facilities. The TLP levels shown in Fig. 13–7 apply to transmit levels of test tones used in circuit line-up procedures. To illustrate, a 0-dBm test tone applied to a combination intertoll trunk at an analog toll office produces the digital milliwatt in the digital facility. That signal is decoded by a −3 DL decoder at the digital toll office. An additional 3-dB pad reduces the tone further to −6 dBm for measurement at the test equipment in the digital toll office. In the reverse

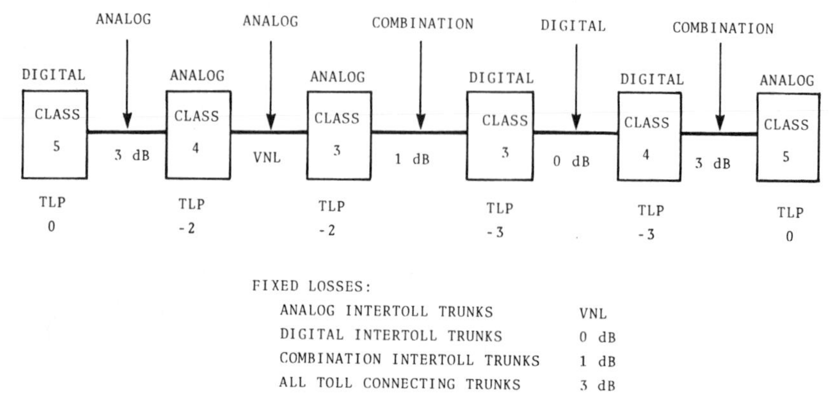

Fig. 13–7. Mixed analog/digital network transmission plan.

direction, a test tone inserted at the digital toll office is padded to −3 dBm and encoded at −3 EL to produce the digital milliwatt in the digital facility. The tone is reduced to a receive level of −4 dBm at the analog toll office and is further reduced by a 2-dB pad to−6 dBm for measurement at the test equipment.[27]

13.3.2. Digital Network Synchronization

A single digital switching system, operating in isolation from other digital switching systems and digital transmission facilities, functions under control of its own internal clock. As digital switching systems are interconnected by digital facilities, however, a need for synchronization and for a plan to satisfy that need become apparent.

13.3.2.1. The Need for Synchronization. Digital transmission can be either asynchronous or synchronous. In asynchronous transmission, such as teletypewriter code, each character has a start signal to indicate the beginning of the character bits and a stop signal to indicate the end of the character bits. In synchronous transmission, the receiver derives timing from the incoming bit stream. In PCM transmission, transmitters and receivers must operate at the same bit rate to avoid losing bits. In addition, a frame aligning bit enables a receiver to align itself to the timing of the fixed-format frame to depict the separate time slots.

The need for synchronization can be seen by examining the configurations illustrated in Fig. 13–8. In Fig. 13–8(a), a digital carrier channel bank transmits a bit stream at clock rate F_0. The other channel bank's receiver receives at the same bit rate but transmits a clock rate F_1. That rate is received

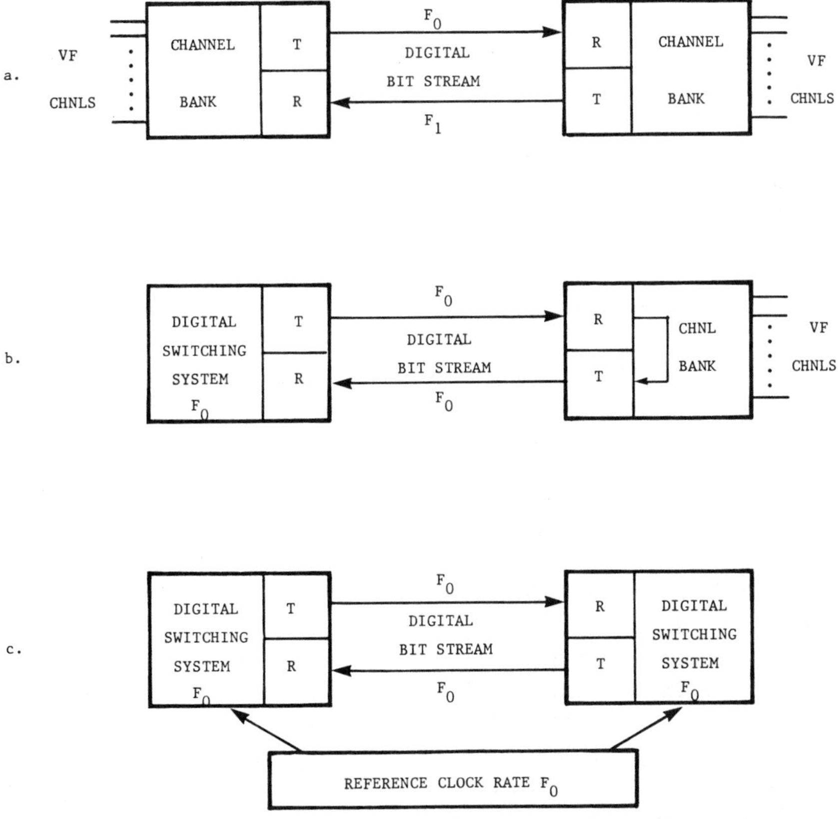

Fig. 13–8. Need for digital synchronization. (© American Telephone and Telegraph Company, 1980.)

satisfactorily by the receiver because it aligns itself with the timing of the transmitter at the other end of the system. Thus, an isolated digital transmission link can use different clock rates in each direction of transmission without impairing transmission quality.

If a digital switching system is substituted for one of the channel banks, as in Fig. 13–8(b), the digital switching system transmits at a bit rate determined by its internal clock, F_0, and needs to receive at the same bit rate. If the distant transmitter sends at a bit rate which is higher than the switching system clock, the receiver buffer will eventually overflow, causing one frame to be lost. If the received bit stream is at a lower bit rate, the buffer will underflow, causing a frame to be repeated. Either occurrence is called a *slip.* In Fig. 13–8(b), slips are prevented by forcing the distant channel bank transmitter to operate at the same clock rate as its receiver which derives its timing from the bit stream transmitted at the switching system clock rate. This is called *channel bank loop timing.*

Now assume that the remote channel bank also is replaced by a digital switching system as shown in Fig. 13–8(c). If each digital switching system transmits at its own internal clock rate, slips will occur if the two clocks are not running at precisely the same rate. When only two digital switching systems are interconnected, synchronization can be achieved by operating the two systems in a master/slave mode whereby one system derives its timing from the other. In a network, however, slips can better be prevented by forcing all digital switching systems to use a common reference clock.

Clock synchronization assures that transmissions from digital nodes have the same average line rate, and buffer storages are provided to absorb the differences between the average line rate and the actual line rate. Actual rates can vary because daily temperature variations alter the electrical length of the digital transmission facility conductors and because of phase jitter. Accumulation of jitter is prevented by buffering, but the buffer storages need to be greater than the 125-μs frame time to absorb jitter.[28]

13.3.2.2. The Network Synchronization Plan. The synchronization plans for North America involve hierarchical synchronization networks in Canada and the United States with special cross-border arrangements. In the United States, the master clock is the AT&T Communications reference frequency (ARF) standard, composed of three cesium-beam frequency standards, each having greater accuracy than 1×10^{-11}, with output frequencies of 20.48 Mhz and 2.048 MHz. Outputs of the three standards are compared in order to identify any malfunction. If one standard malfunctions, it is isolated. If the two remaining outputs ever differ, both are removed from service until the malfunctions are corrected. As a prime standard, there is no other backup provided.

Clocks have been classified into four strata for hierarchical ranking. The ARF is the stratum 1. Primary stratum 1 standards must have minimum accuracy of 1×10^{-11}. Stratum 2 clocks must have minimum stability of 1×10^{-10} per day and either have, or be adjustable to, a minimum 20-year accuracy of 1.6×10^{-8}. They must be able to pull themselves into synchronization with any other stratum 2 clock and must have a synchronization unit capable of tracking a DS-1 or 2.048-MHz signal. Stratum 3 clocks must have a minimum 20-year accuracy of 4.6×10^{-6} and a short-term accuracy of 3.7×10^{-7} per day or better. They must be able to pull themselves into synchronization with any other stratum 3 clock and be capable of switching automatically from primary to secondary DS-1 reference. Stratum 4 clocks must have a 20-year accuracy of 32×10^{-6} and be able to pull into synchronization with any other stratum 4 clock. Switching systems having clocks which do not meet these requirements pose special problems until their clocks can be upgraded.

The hierarchical synchronization network is shown in Fig. 13–9. The primary reference frequency of 2.048 MHz is transmitted via baseband over existing broadband analog facilities from the ARF standard at Hillsboro, Missouri to selected stratum 2 nodes. Any stratum 2 node which has analog facilities used for transmission of the ARF reference frequency must derive synchronization from the ARF. Otherwise, it must derive synchronization from some other stratum 2 node which is synchronized by the ARF. All nodes which are not connected to the ARF normally will derive their synchronization from DS-1 signals from digital nodes in the same or higher stratum. Below stratum 2 nodes, each digital node has access to a primary and a secondary frequency reference. In case of failure of the primary reference facility, the node pulls in synchronization over the secondary reference facility.

The Canadian network also uses a cesium-beam frequency standard equivalent to the accuracy of the ARF. Each network will operate plesiochronously. CCITT defines two signals as plesiochronous if their corresponding significant instants occur at nominally the same rate with any variation being constrained within specified limits. Digital nodes deriving synchronization from standards of equivalent accuracy can operate synchronously with each other. The CCITT plan for international digital synchronization assumes that national digital networks interconnected by digital links will each be controlled by national reference clocks which meet the accuracy requirements specified for stratum 1 clocks.

The slip objective for end-to-end connections is one slip or less in five hours. For international connections, the objective is not more than one slip in 70 days.

For the AT&T circuit switched digital capability, which will provide alternate voice and end-to-end 56-kb/s full duplex data, a digital office timing supply (DOTS) has been developed for the No. 1A ESS local office, which qualifies as a stratum 3 clock. Digital access and crossconnect systems (DACS) also qualify as stratum 3 clocks. Digital PABXs, D4 channel banks, and remote digital switching units are classified as having stratum 4 clocks.

Local telephone companies and the long distance networks of other common carriers (OCC) can derive synchronization from sources traceable to the ARF. If OCCs prefer, they can provide clocks with stratum 1 accuracy and operate plesiochronously.[29]

13.3.3. Digital Implementation Progress

Quality of switched telephone connections is a highly subjective consideration, depending not only upon circuit clarity but also upon talker characteristics,

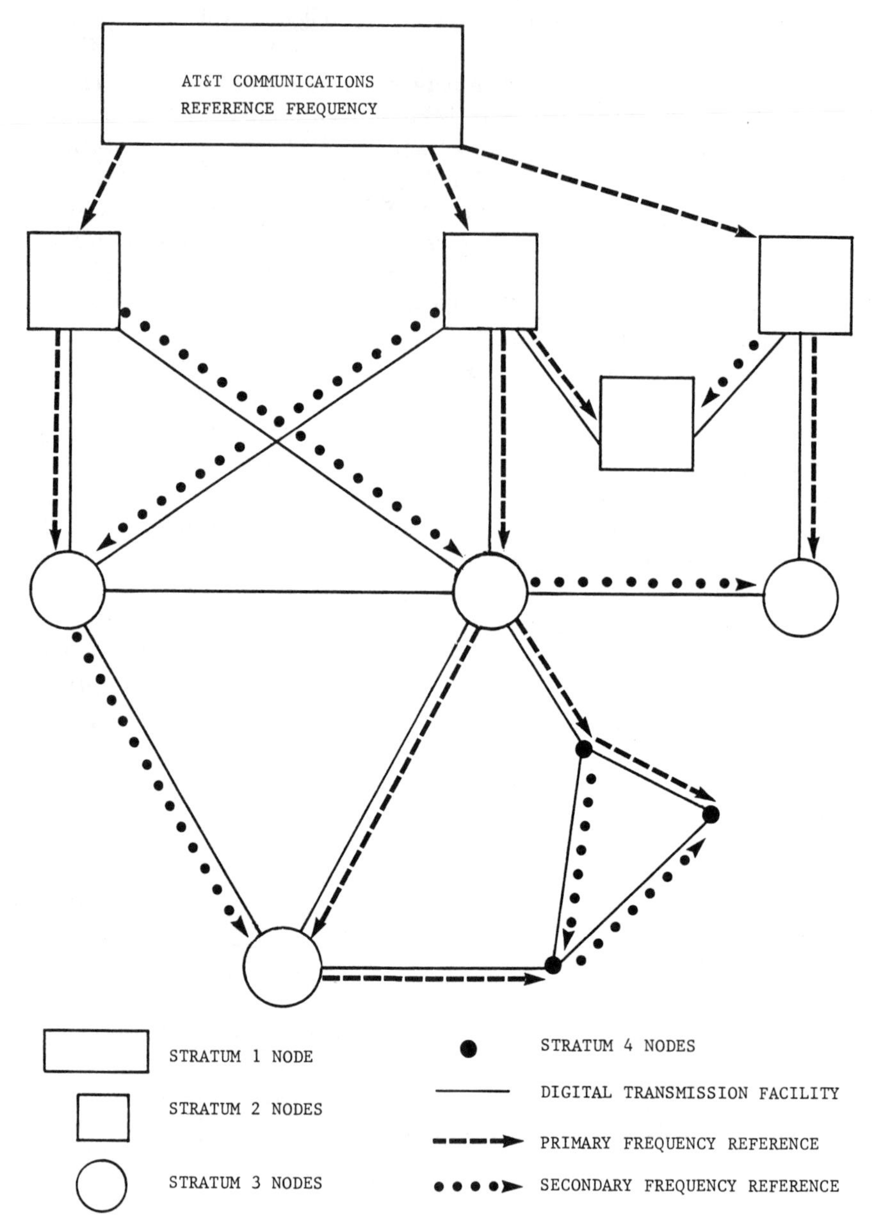

Fig. 13–9. Network synchronization plan. (Reproduced with permission of AT&T.)

hearing efficiency, and room noise. Therefore, any subjective evaluation of circuit or connection quality should be tempered by noncircuit-related factors. However, subjective studies are helpful and can be made more meaningful by using either the same evaluators or a large population of evaluators. By measuring actual circuit and connection parameters during a large-scale subjective study, data have been correlated such that projections using only technical parameters can produce reasonably reliable results.

Subjectively the quality of toll connections is evaluated by having listeners rate each connection as excellent, good, fair, poor, or unsatisfactory. Fig. 13–10 shows the results of projected performance as the network evolves from analog to digital. The curves show the projected percentages of toll connections that would be rated good or excellent as the percentage of intertoll trunks increases from all analog to all digital. The flat projection for the all-digital network over distance is based upon the uniform allocation of 6 dB of loss for all connections irrespective of distance.[30]

Most digital transmission facilities introduced into the network have involved relatively short systems, primarily in metropolitan areas. Some long-haul systems have been placed in service mainly on selected routes necessary to establish a thin-route digital network. On the other hand, the penetration of digital switching systems has focused on large toll switchers and relatively

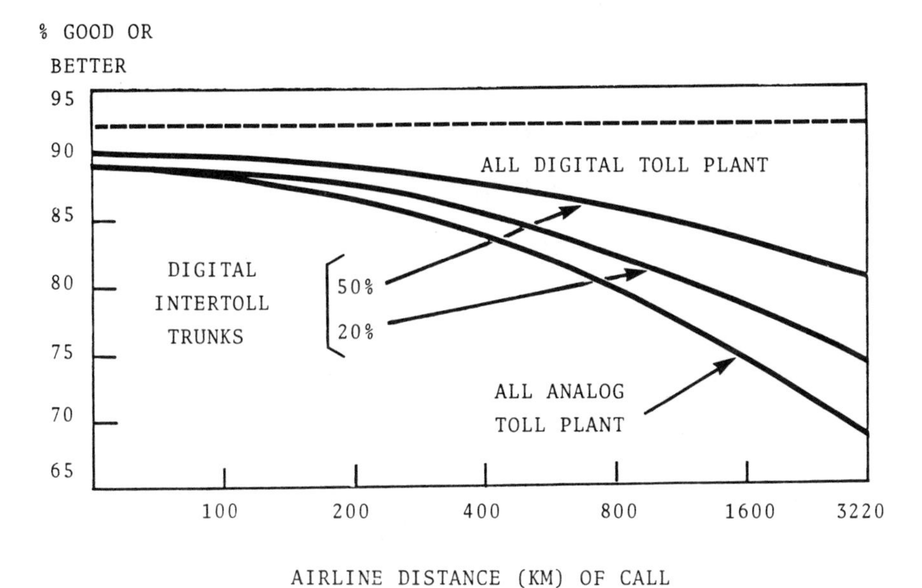

Fig. 13–10. Call connection quality during transition of network from analog to digital. (From W. L. Ross, "Transmission Planning for the Evolving Digital Network," *ICC '77*, © 1977 IEEE.)

small local switchers. There are some exceptions, of course, but the economic incentives have not been sufficient to justify wholesale replacement of serviceable plant sustantially before expiration of its life cycle. Consequently, the network likely will continue to use mixed analog/digital technology at least through the rest of this century.

13.4. INTRA-LATA NETWORKS

Concurrently with the diverstiture of the BOCs, a new network configuration was implemented for handling local area calls and connections between local and competing long distance carriers. The consent decree required separation of exchange and interexchange functions. Former exchange areas were reconfigured into local access and transport areas (LATAs) in BOC territories. Calls within a LATA typically are handled by the BOC serving that LATA. Calls between LATAs are carried by inter-LATA carriers (ICs). Non-BOC local telephone companies generally are not required to conform to these requirements but may participate on an optional basis. General Telephone is subject to similar requirements resulting from its own consent decree in connection with its acquisition of an interstate long distance carrier.[31] It is expected that most local companies will participate in the LATA concept, and they may be required by the Federal Communications Commission to do so.

13.4.1. Local Access and Transport Areas (LATAs)

The continental United States is divided into 181 LATAs including 163 LATAs comprising the divested BOC service areas and 18 comprising the service areas of non-BOC service areas and some small BOC areas. Additionally, there is one LATA each covering Alaska, Hawaii, and Puerto Rico.[32] Each LATA is roughly equal to a standard metropolitan statistical area (SMSA) or a standard consolidated statistical area (SCSA). Among the guidelines established by the consent decree, each LATA must include at least one AT&T class 4 switching system. Deviation from SMSA and SCSA boundaries is permitted primarily to preserve existing wire center boundaries, service arrangements, and communities of interest, to minimize subscriber impact, and to avoid disruption of end office toll trunking.[33]

Under the terms of the consent decree, divested BOCs are to provide local exchange traffic, intra-LATA toll traffic, and access by susbcribers to inter-LATA carriers (ICs). To carry out those functions on a nondiscriminatory basis, changes in network configurations, signaling, and transmission plans were required.

13.4.2. Intra-LATA Network Switching Plan

Intra-LATA networks may serve a small geographic area with a high concentration of subscribers, such as a large metropolitan area, or they may extend for up to about 650 kilometers and include several small cities and towns as well as rural areas. Most LATAs employ a two-level hierarchical network configuration. End offices are connected to each other by direct interend office trunks (IOTs) when justified by traffic loads, and tandem offices are used in the second hierarchical level. A LATA may have one or more tandem switching systems, each connected to subtending end offices in a sector. Very large LATAs may have a third-level tandem office or may designate one or more second-level tandems as a principal sector tandem. Fig. 13–11 illus-

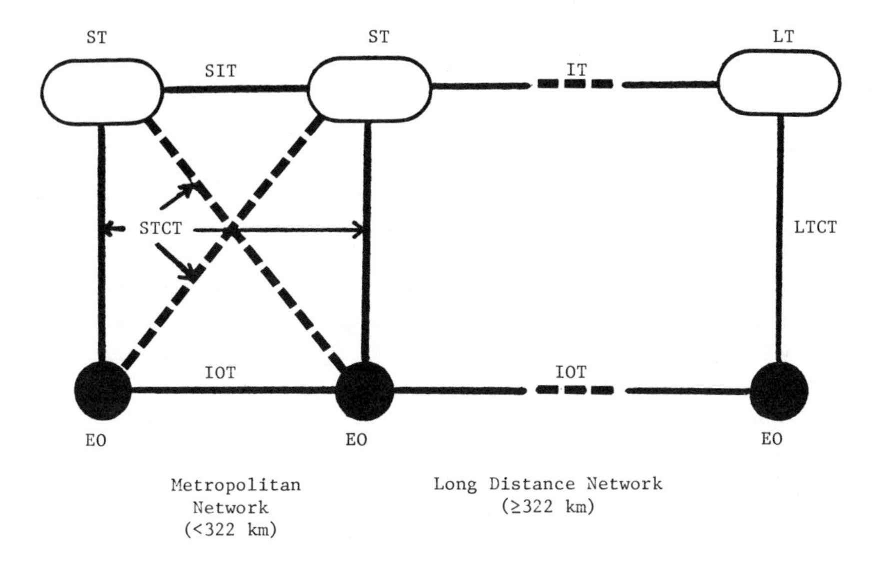

Legend

ST - Sector Tandem

EO - End Office

LT - LATA Tandem

IT - Intertandem Trunk

SIT - Sector Intertandem Trunk

STCT - Sector Tandem Connecting Trunk

LTCT - LATA Tandem Connecting Trunk

IOT - Interend Office Trunk

Fig. 13–11. Intra-LATA network configuration.

trates a skeletonized two-level hierarchy in a large LATA containing both a metropolitan network and an intra-LATA long distance network.[34]

At divestiture, many switching systems performed combined switching of local traffic, intrastate toll traffic, and interstate toll traffic. Those switching systems were assigned to either AT&T Communications or a BOC based upon the dominant function performed. Class 4/5 offices generally were assigned to a BOC. Tandems performing both intrastate and interstate switching functions generally were assigned to AT&T Communications. Continued sharing of switching systems through leasing is permitted during a transition period. As additional switching systems are placed in service, shared trunking also will be eliminated.[35]

Intra-LATA routing employs principles similar to those described for interregional routing in Section 13.2.2. A maximum of four routes can be provided for calls from any end office to any other end office in the same LATA. No more than three trunks are used in any intra-LATA connection. Intra-LATA networks are designed for a blocking probability of B.01.[36]

Inband signaling is the principal signaling method used in the intra-LATA networks. Some stored program control end offices and some tandems, however, are equipped with CCIS and continue to use separate channel signaling for trunk groups so equipped. Indeed, separate channel signaling, using CCITT System No. 7, is planned for introduction in late 1986.

The intra-LATA transmission plan is based upon a combination of trunk loss and connection loss objectives. In metropolitan networks, digital connections via IOTs are designed for a 3-dB net loss end office to end office. Digital tandem connections are designed for a 6-dB net loss. Analog and mixed connections in a metropolitan area may encounter losses between end offices ranging from 3 dB to about 9 dB depending upon the type of facility and office balance parameters.[37]

Intra-LATA long distance transmission standards are similar to those of the national network plan. The digital connection loss objective is 6 dB between end offices whether or not a tandem is used. Loss objectives for intertandem trunks are the same as those for intertoll trunks as shown in Fig. 13–7. Mixed connection loss objectives are shown in Table 13–2.[38] The

Table 13–2. Mixed Connection Trunk Loss Objectives (DB).

	INTEREND OFFICE TRUNKS		LATA TANDEM CONNECTING TRUNKS	
TRUNK TYPE	<322 KM	≥322 KM	<322 KM	≥322 KM
Digital	3	6	3	3
Combination	3	6	3	3
Analog	3	VNL + 6	3	VNL + 2.5

additional loss on connections over 322 kilometers compensates for added echo susceptibility.

13.4.3. Switched Access for Inter-LATA Carriers

The BOCs provide access for inter-Lata carriers (ICs) to LATA subscribers under tariff. Access may be direct from an IC's point of presence (POP) to each end office, or it may be provided via an access tandem (AT) switching system. Access tandems also can perform sector tandem and LATA tandem functions. ICs may have more than one POP in a LATA. In the case of the AT&T Communications network, the POP is a class 4 switching function.

Prior to divestiture of the BOCs, competing long distance carriers other than AT&T could access their subscribers through *exchange network facilities for interstate access* (ENFIA). Three types of ENFIA facilities were available. ENFIA-A provides connections between the switching systems of other common carriers (OCCs) and line terminations on BOC end office switching systems. OCC subscribers dial a 7-digit number, receive a second dial tone or equivalent from the OCC switching system, dial an authorization code (usually 5 digits) for billing identification, and then dial the called number. Since most switching systems do not have the capability of repeating dial pulses on line-to-line connections, OCC subscribers have been required to use DTMF signaling.[39]

ENFIA-B provides trunk rather than line connections, and subscribers dial a 7-digit number in the form 950-10XX, where XX is a 2-digit number identifying the OCC. Second dial tone operation and DTMF signaling also characterize ENFIA-B service. ENFIA-C is similar to ENFIA-B except that the OCC trunk connection is to a local tandem switching system. Answer supervision is provided on ENFIA-B and ENFIA-C connections but not on ENFIA-A connections. Calling party identification is not available with ENFIA-A but is optional with ENFIA-B and ENFIA-C.[40]

The BOCs have defined feature groups and their technical and operational parameters to provide access for all inter-LATA carriers. Feature group A is the direct equivalent of ENFIA-A, and feature group B is the equivalent of ENFIA-B and ENFIA-C for either direct or tandem connections. Dial pulse signaling can be used with feature group B and No. 1/1A ESS switching systems. Feature group C is the type of interconnection enjoyed by AT&T Communications on January 1, 1984. It is provided only on an interim basis until feature group D is available.[41]

Feature group D is a new interconnection arrangement designed to fulfill the requirements of the consent decree that all inter-LATA carriers be accorded equal access to the local networks of the BOCs. Subscribers can presubscribe to the services of a specific inter-LATA carrier, in which case they dial the called number plus a 1 or 0 prefix if applicable. The BOC routes that call to the presubscribed carrier of choice. Inter-LATA carriers can provide their own billing services or contract with the BOCs for billing to their subscribers. Subscribers can use dial pulse or DTMF address signaling. Feature group D employs trunk connections, trunk signaling, calling party identification, and answer supervision. There is no requirement for an authorization code. If a subscriber does not presubscribe to the services of an IC or desires to place a call via a different IC, an access code in the form 10XXX must be dialed preceding the called number address. The XXX digits identify the IC.[42]

Not all switching systems can provide the equal access services of feature group D. No. 1/1A ESS, No. 5 ESS, DMS-10, and DMS-100F can provide equal access service as end offices or as tandem offices. The No. 2B ESS can function as an equal access end office, and the 4-wire No. 1A ESS can function as an equal access tandem office. Other stored program switching systems, which may be used by BOCs in the future, likely will be capable of performing equal access functions. Equal access modifications will not be required for electromechanical switching systems or other systems serving less than 10 000 lines. Table 13–3 shows a comparison of selected characteristics of feature groups A, B, C, and D.[43]

The consent decree required that BOCs begin offering equal access by September 1, 1984; that at least a third of the exchange lines in each BOC serving area be equipped by September 1, 1985; and that equal access be available at all nonexempt offices by September 1, 1986.[44] The first provision

Table 13–3. Inter-LATA Access Feature Group Comparison.

FEATURE CHARACTERISTIC	F.G. A	F.G. B	F.G. C	F.G. D
Access code	NXX–XXXX	950–10XX	1 or None	10XXX or None with Presubscription
Termination	Line	Trunk	Trunk	Trunk
Second dial tone (or equivalent)	Yes	Yes	No	No
Authorization code required	Yes	Yes	No	No
Predivestiture equivalency	OCC ENFIA-A	OCC ENFIA-B/C	AT&T DDD	None
Answer supervision	No	Yes	Yes	Yes
Signaling EO-IC POP	Line	Trunk	Trunk	Trunk
Calling party identification	No	Optional	Yes	Yes
Available to	All IC	All IC	AT&T (Interim)	All IC

of feature group D was in July, 1984, and it appears that most BOCs are ahead of the schedule requirements of the consent decree.

Figure 13–12 shows the network configuration options and transmission objectives for inter-LATA access. Inter-LATA carriers may optionally access LATA end offices in three ways. Direct inter-LATA connecting trunks (DICTs) may be provided between end offices and the IC POP. DICTs are designed for 3-dB fixed loss and have the same quality as predivestiture toll connecting trunks. A BOC access tandem (AT) switching system can

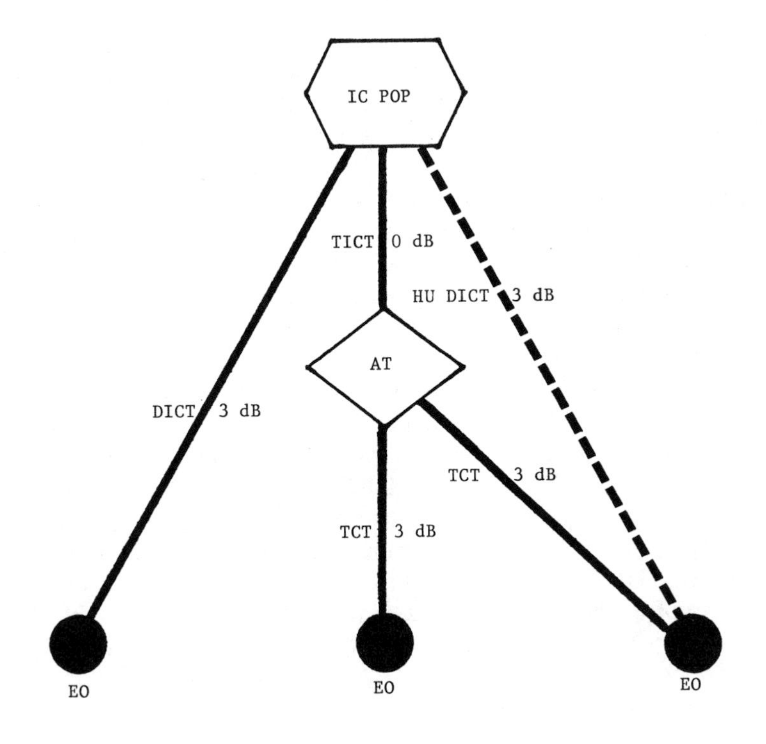

Legend:

IC POP - Inter-LATA Carrier Point of Presence
EO - End office
AT - Access tandem
TCT - Tandem connecting trunk
DICT - Direct inter-LATA connecting trunk
TICT - Tandem inter-LATA connecting trunk
HU - High usage

Fig. 13–12. Inter-LATA network configuration and transmission plan.

be interposed between an end office and IC POP. The AT is connected to end offices by tandem connecting trunks (TCTs) designed to the same standards as DICTs. Tandem inter-LATA connecting trunks (TICTs) between the AT and IC POP are 4-wire, intertoll grade facilities designed for 0-dB loss. If subscriber lines also are served by the AT, it must have 3-dB switchable pads to provide the 3-dB loss between the IC POP and the end switching office.[45]

13.4.4. Equal Access Dialing and Signaling Plan

Under feature group D, subscribers who have presubscribed dial their calls in the same manner as formerly used for AT&T calls. Subscribers who have not presubscribed must first dial an access code in the form 10XXX.

A new inband signaling plan was devised for use with feature group D. Formerly, when an end office transmitted automatic number identification (ANI) of the calling party to a class 4 office for billing control, the ANI information was transmitted after the called number. With feature group D, the sequence is reversed. Calling party identification is transmitted by the equal access end office (EAEO) before the subscriber finishes dialing the called number. When the subscriber dials the last address digit, the called number is transmitted.[46]

An illustration of the originating signaling sequence on a direct connection is shown in Fig. 13–13. Subscriber-dialed numbers in parentheses are dialed if required by the dialing plan and type of call. Following receipt of the dialed NXX code, the EAEO seizes a DICT to the IC POP and, after receiving a wink-start signal, outpulses the identification of the calling party if the IC has arranged to receive ANI information. Otherwise the EAEO outpulses KP + ST. ANI information digits (II) indicate the category of the call. Upon receipt of the last four digits of the called party address, the EAEO outpulses the address of the called party. The IC POP acknowledges the address by transmitting a wink to the EAEO. When the called party answers, answer supervision is forwarded to the EAEO.[47]

The originating signaling sequence for a tandem connection is illustrated in Fig. 13–14. Additional signaling is required for the EAEO to inform the AT the identification of the IC designated to handle the call. That information is transmitted in the form shown. The OZZ code designates one of four possible trunk groups between the AT and IC POP to be used. This is necessary because the IC may desire separate trunk groups to be used for certain categories of calls such as operator-handled calls, information calls, etc. A 3-digit code in the form XXX Identities the IC designated to handle the call. The remaining portion of the signaling sequence is the same as for direct connections except that the AT cuts through a path to the IC POP

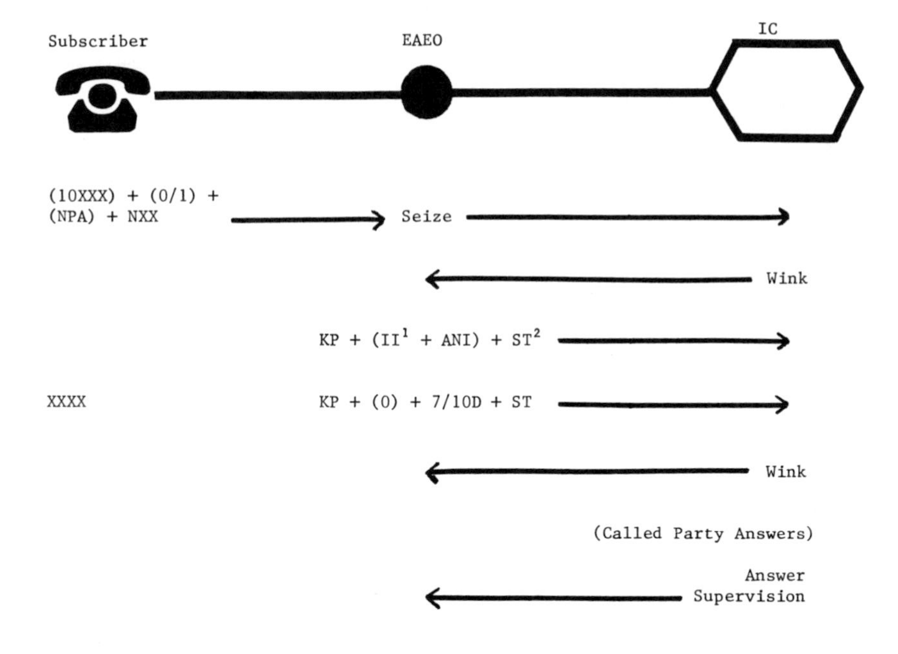

Notes:

1. Information digits to identify type of calling party, e.g., identified direct-dialed call, hotel room call, coin phone, etc.

2. If inter-LATA carrier has not requested ANI, EAEO outpulses KP + ST.

Fig. 13–13. Signaling sequence for a direct connection.

after receipt of the carrier identification and a wink from the IC POP. Acknowledgment winks and answer supervision from the IC POP are repeated by the AT to the EAEO.[48]

The terminating signaling sequence is a straightforward application of wink-start control and MF address signaling in which the IC POP outpulses the called address to either the EAEO or to the AT. Called party answer supervision is returned in the reverse direction.[49]

International calls may be handled directly by international carriers (INCs) through an INC POP in the LATA or through an INC gateway office and the domestic network of an IC. International calls involve multistage outpulsing to the international carrier. Under feature group D, the previous first stage of outpulsing in the form KP + 011 + PCC + ST cannot carry all the information required for international call, carrier, and country identification, and call routing. Therefore, a new 3-stage outpulsing format was designed in the form KP + 1NX + XXX + CCC + ST in the first stage. The three

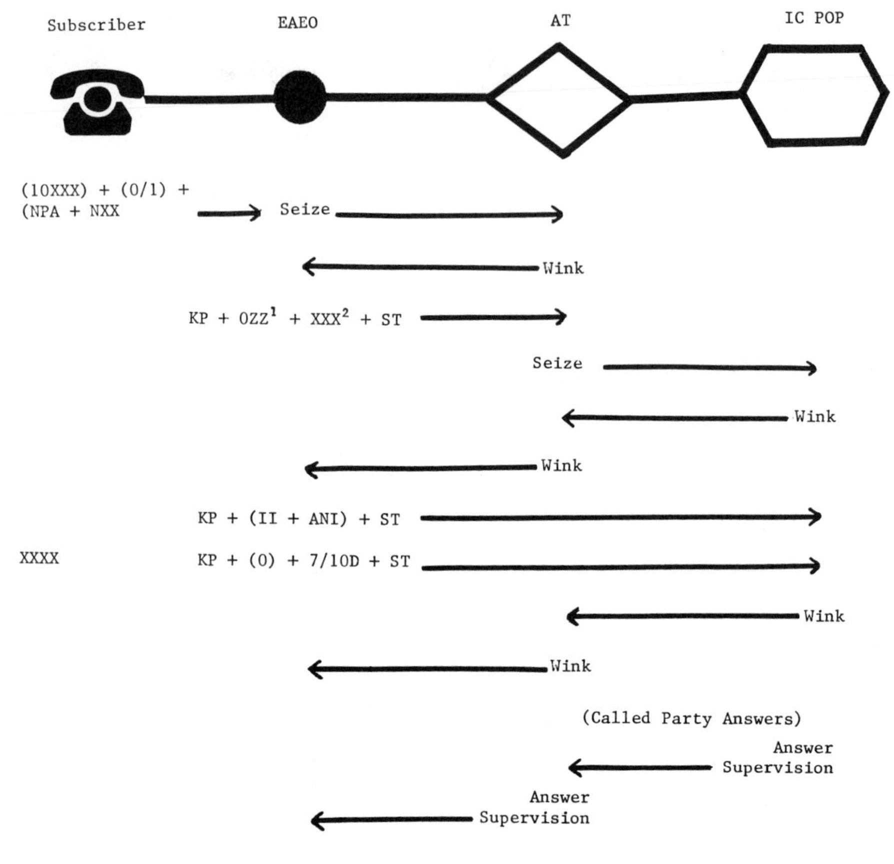

Notes:

1. Code to indicate trunk group routing from AT to IC (based upon type of traffic; e.g., 0+, 1+). Maximum of four codes available.

2. IC identifier.

Fig. 13–14. Signaling sequence for a tandem connection.

3-digit symbols identify call routing, carrier, and country code, respectively. The second stage of outpulsing forwards ANI information if required, and the third forwards the country code and national number of the calling party to the INC directly or through a domestic IC, and possibly through an AT. Winks and answer supervision are the same as for domestic calls.[50]

Some ICs perform combined IC/INC operations. ICs and INCs who do not perform combined operations may negotiate agreements with each other.

Subscribers may select an IC or an INC for international calls. In either case, calls are routed accordingly.

PROBLEMS

13.1 In Fig. 13–1, why is primary center C not permitted to use the high usage trunk group to regional center E as an alternative route for traffic originated at end office A and destined for end office J? Why is the use of the high-usage group from primary center C to regional center N permissible despite the "one-level limit rule"?

13.2 Compare CCIS and CCITT No. 7 signaling systems.

13.3 Why cannot echo be controlled by loss alone?

13.4 Calculate the net loss of a four-wire analog trunk with a microwave route length of 1400 kilometers.

13.5 Why cannot the assigned TLPs at each switching office in a toll connection be used in connection with the net loss of each trunk to calculate the overall loss between the end offices?

13.6 If a standard digital milliwatt is decoded by a −3 DL μ-255 PCM decoder, what is the level of the resulting analog signal?

13.7 Two test tones are transmitted over two separate PCM circuits in the same route. Signal A, at a level of −2 dBm, is encoded by a −3 EL encoder and is decoded at the other end by a −2 DL decoder and put through a 4-dB pad. What analog signal level is measured at the test instrument? Signal B, at a level of zero dBm, is encoded by a 0 EL encoder and is decoded by a −3 DL decoder. What analog pad value is required to measure Signal B at the same level as Signal A?

13.8 What is plesiochronous operation of digital networks?

13.9 The design loss objective for an intra-LATA connection from end office to end office over a digital or combination interend office trunk is 3 dB if the connection is less than 322 kms but 6 dB if the connection is longer. What is the purpose of the additional loss?

13.10 For inter-LATA connections via an access tandem office, the total design loss of 3 dB between the IC POP and the EO is assigned to the tandem connecting trunk while the tandem inter-LATA connecting trunk operates at 0-dB loss. For an inter-LATA connection to a subscriber served by an access tandem office, what is the AT-IC POP loss and how is it accomplished?

REFERENCES

1. *Notes, on the Network,* American Telephone and Telegraph Company, 1980, Section 2 (hereafter referred to as *Notes*).
2. Ibid., Section 3.
3. Ibid., Section 3, Appendix 3.
4. Ibid., Section 3, Appendices 1–3.

5. Pearce, J. Gordon, *Telecommunications Switching* Plenum Press, New York, 1981, pp. 236–240.
6. *Notes,* Section 5, p. 101.
7. Pearce, pp. 257–278.
8. Horing, S., et al., "Overview," *Bell Syst. Tech. J.,* Vol. 61, No. 7, Part 3, pp. 1579–1582 (Sep. 1982).
9. Spring, Peter G., "The Evolving Call Handling, CCIS, Interprocessor Communications Network," *GLOBECOM '82 Record,* p. D1.3.1.
10. Ibid., pp. D1.3.2–D.1.3.4.
11. Pearce, pp. 278–283.
12. R. Williams and A. N. MacDiarmid, "An Operating Telephone Company View of CCIS vs. CCITT #7," *National Telecommunications Conference Record,* pp. 2.4.1–2.4.5 (1980).
13. Schlanger, G. Gary, "Planning for the Appliction of CCITT No. 7 in the CCIS Network," *National Telecommunications Conference Record,* pp. 2.5.1–2.5.4 (1980).
14. *Telecommunications Transmission Engineering,* Vol. 1, pp. 56–58, 413–436 (American Telephone and Telegraph Co., 1977).
15. Ibid., Vol. 1, pp. 104–109, 482–486.
16. Ibid., Vol. 1, pp. 486–490.
17. Ibid., Vol. 1, pp. 596–599.
18. *Notes,* Section 7, pp. 15–18.
19. *Engineering and Operations in the Bell System,* Bell Telephone laboratories, Inc., 1977, p. 183.
20. *Notes,* Section 7, pp. 30–34.
21. *Notes,* Section 7, pp. 15–18.
22. *Telecommunications Transmission Engineering,* Vol. 1, pp. 48–53 (1977).
23. *Local Switching System General Requirements,* PUB48501, American Telephone and Telegraph Company, Basking Ridge, NJ, 1980, Section 7.1.7.3.
24. Ibid., Section 7.1.7.4.
25. Ibid., Section 7.1.4.2.
26. Ibid., Section 7.1.2.
27. *Notes,* Section 7, pp. 21–23.
28. *Engineering and Operations Plan for Synchronization of the Integrated Services Digital Network (EOPS-ISDN),* AT&T Technical Advisory No. 58, Issue 4 (Dec. 8, 1982), pp. 2–4.
29. Ibid., pp., 4–75.
30. W. L. Ross, "Transmission Planning for the Evolving Digital Network," *International Conference on Communications Record,* pp. 26.5–189 to 26.5–192 (1977).
31. *Notes on the BOC Intra-LATA Networks,* American Telephone and Telegraph Company, 1983, Section 1 (hereafter referred to as *BOC Notes*).
32. *The Guide to Communications Services,* Center for Communications Management, Inc., 1984, Section 1.
33. *BOC Notes,* Section 2.
34. Ibid., Section 4.
35. Ibid.
36. Ibid.
37. Ibid., Section 7, pp. 33–35.
38. Ibid., pp. 35–37.
39. Ibid., Section 13, p. 4.
40. Ibid., pp. 4–5.
41. Ibid.
42. Ibid., pp. 5–6.

43. Ibid., Section 6, pp. 192–195.
44. Simon, Samuel A., and Whalen, Michael Jr., *Teleconsumers and the Future: A Manual on the AT&T Divestiture,* Telecommunications Research and Action Center, Washington, 1983, pp 5-10 to 5-11.
45. *BOC Notes,* Section 13, pp. 5, 9–10.
46. Ibid., Section 6, pp. 198–200.
47. Ibid., pp. 201–203.
48. Ibid., pp. 203–205.
49. Ibid., p. 205.
50. Ibid., pp. 205–207.

14
Evolution of the Integrated Services Digital Network (ISDN)

14.1. THE ISDN CONCEPT

In 1972, the plenary assembly of the International Telegraph and Telephone Consultative Committee (CCITT) approved Recommendation G.711 for Pulse Code Modulation (PCM) of Voice Frequencies. This provided for a sampling rate of 8000 per second and 8-bit encoding/decoding with either A-law or μ-law companding (see Chapter 3). In 1980, the plenary assembly approved Recommendation G.705 which recognized that substantial agreement had been reached in studies of integrated digital networds (IDN) for telephony and that studies leading to an integrated services digital network (ISDN) were needed. Some conceptual principles to guide the studies provided for:

- The ISDN to be based upon and evolve from the telephony IDN by adding functions and features;
- A layered functional set of protocols;
- Gradual transition over one or two decades;
- New services to be compatible with the 64-kb/s switched digital connections of the IDN; and
- Arrangements during transition for the interworking of services on the ISDNs and services on other networks.

The CCITT guidelines recognized that countries will develop separate ISDNs but that compatible international connections will be provided. Several study groups are involved. Substantial agreement has been reached in some areas, but there has been pressure from some participants to consider bit rates lower than 64 kb/s for voice services and multiples of 64 kb/s for wideband services over optical fibers and satellites.[1]

The ISDN is a concept for a future digital telecommunications network. It is readily apparent that analog telephony is no longer the state of the art. The state of the art is digital, and digital telephone networks will evolve with or without CCITT standardization. Before the full ISDN becomes a reality, the IDNs, based upon a circuit switched network of 64-kb/s PCM voice channels will evolve. The IDNs will have an inherent capability of handling both voice and digital data. The ISDN is conceived as evolving from the IDN to provide end-to-end digital connectivity to support a wide range of services, including voice, data, sound, and video services, via a variety of communication modes. The key factor is an integrated local access to an apparently transparent network of multiservice capabilities.

Several categories of services are envisioned as the ISDN emerges. Low-bit-rate services could include meter reading, alarms, control signals for energy management, opinion polling, low-speed data, teletext, and videotex. Medium-bit-rate services could provide low-bit-rate voice, medium-to-high-speed data, standard PCM voice, slow-scan video, and facsimile. High-bit-rate services could include wideband music, interactive video, high-speed facsimile, and standard television.[2]

14.1.1. User Perspective of the ISDN

Not every user will need or desire all the services envisioned for the ISDN. As more sophisticated services develop, however, market stimulation will bring more of those services into the home. A primary user requirement is simplicity of operation at reasonable cost. This mitigates against continuation and proliferation of the multiple access arrangements presently necessary for several existing services which require separate access to separate networks. Users will demand a small variety of standard interfaces, differentiated primarily by bit-rate capability. This will require a single access arrangement to a local serving office for all services and standard protocols to access the various services. Advances in microprocessor technology and large-scale demand will permit physical realization of these requirements at reasonable cost.

High on the list of ISDN user demands will be a single billing system for services. This is compatible with the concept of all services passing through a single access arrangement to a local serving office. This does not mean that the provider of the local serving office must provide all services. It means only that the local serving office must provide network interfaces with the different service providers. It also requires a high level of cooperation among users, local service providers, network providers, enhanced service providers, equipment manufacturers, and government entities to develop the technical standards and agreements necessary for economic implementation.

The network should be capable of fast call set-up and disconnect times, a wide range of calling rates and holding times, and variable bit rates for user transmissions. There is increasing demand for cryptographic security for business and industrial transmissions. If the network does not provide such security, it should accommodate user-provided encryption. Efficient directory service for all network services is essential, and users should have the ability to control their costs.

User interface protocols should be oriented toward the standard family of interfaces being developed rather than toward the location or type of local serving office. Flexibility will enable terminals to be connected directly to a user interface at home or at an office, operated behind a PABX, or plugged in anywhere that ISDN standards are used. The only criterion should be service compatibility.[3] The international trend appears to be toward full development and use of the user parts in CCITT Signaling System No. 7.

14.1.2. The User Interfaces

Current studies have resulted in CCITT recommendations which specify three types of user access: basic access, hybrid access, and primary rate access. The principal access and channel structures are shown in Table 14–1.

Basic access includes two B channels at 64 kb/s and one D channel at 16 kb/s (2B + D). The B channel will carry standard PCM voice, digital speech at a subrate of 64 kb/s combined with data to the same destination, or data at rates adapted to 64 kb/s via circuit-switched or packet-switched data networks. Alternatively, a user can switch between voice and data on a B channel, and a B channel can be built up by assembling subrate channels. The D channel will carry signaling information to control the B channel, telemetry, or low-speed packet-switched data. Some users will not require the full basic access, 2B + D; therefore, the user loops can be arranged to support only one B channel and one D channel (B + D) or simply one D channel.

Hybrid access provides for the connection of one analog telephone line in conjunction with a digital interface structure.

Table 14–1. Access Structure and Channel Bit Rates.

ACCESS STRUCTURE	CHANNEL COMBINATIONS	USER LOOP INFORMATION BIT RATES (KB/S)	CHANNEL BIT RATES (KB/S)
Basic	2B + D	144	64 + 64 + 64
	B + D	80	64 + 16
	+ D	16	16
Primary rate	23B + D	1544	64
	30B + D	2048	64
	H_0	1544 or 2048	384
	H_{11}	1544	1536
	H_{12}	2048	1920

The *primary rate access* will support integrated services digital PABXs or wideband digital terminals at bit rates of 1.544 ot 2.048 Mb/s. The bit stream can be arranged as an assembly of 64-kb/s B-channels plus one D channel (or E channel) operating at 64 kb/s.[4]

The primary rate also will support wideband digital channels at the net channel bit rates of the primary access structure or at a specified subrate. In a 24- or 32-channel digital system, the H_0 channel will support channel rates of 384 kb/s. In a 24-channel system, the H_{11} channel will support a rate of 1536 kb/s, while the H_{12} channel in a 32-channel system will support a rate of 1920 kb/s.

A block diagram of the basic ISDN interface is illustrated in Fig. 14–1. Two network terminations, NT1 and NT2, are defined. NT1 terminates the subscriber loop and interfaces the line termination (LT) in the local serving office. NT1 and NT2 may be combined or may be provided separately as determined by the serving common carrier and country regulatory constraints. Three interface reference points are defined. Subscriber ISDN terminal equipment (TE1) connects to the *s* or *t* reference point. Other terminal equipment must connect through the *r* reference point to a terminal adapter (TA).

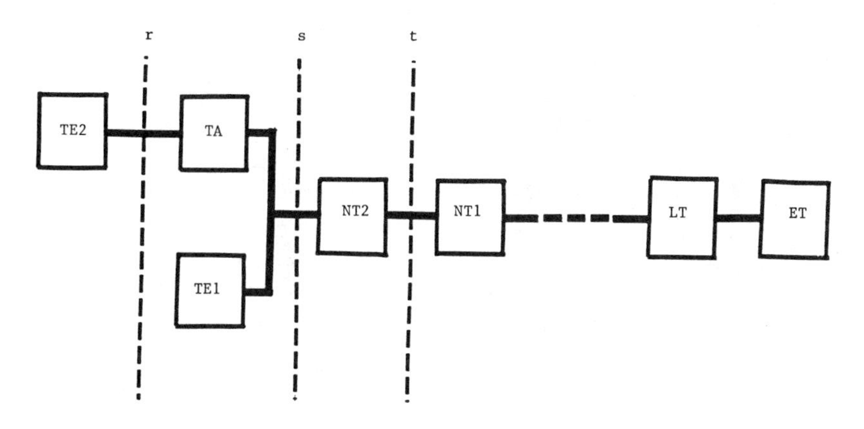

Legend:

ET - Exchange termination
LT - Line termination
NT1 - Network Termination 1
NT2 - Network Termination 2
TE1 - ISDN terminal equipment
TA - Terminal adaptor
TE2 - Non-ISDN terminal equipment

Fig. 14–1. ISDN basic interface.

14.1.3. Subscriber Loop Technology

Section 5.12 discusses techniques for converting loops to digital technology. Virtually all subscriber loops are two-wire facilities with multiple wire sizes and frequent bridged taps (see Section 10.5.1.). Loop interfaces for ISDN terminals require full duplex transmission of synchronous digital signals. Two transmission modes have emerged as leading candidates to support the ISDN basic access structure: hybrid balancing mode and time compression multiplex (TCM) mode. The hybrid balancing mode has better impulse noise immunity and better bandwidth utilization but requires complex echo cancellation and generates more near-end crosstalk compared to the TCM mode. Both are adversely affected by bridged taps, and it appears that bridged taps may need to be removed for either system to achieve acceptable performance over longer loops. Simulations have shown that even one bridged tap anywhere in a loop can impair the transmission eye pattern.[5]

Much research and development has been in progress to develop both the hybrid balancing and TCM concepts. At this time, it appears that TCM is receiving more attention. In time compression multiplexing, a continuous bit stream at a given information rate is divided into equal segments, compressed in time to a higher transmission rate, and transmitted in bursts which are expanded at the other end to the original information bit rate. A sufficient burst interval is provided to allow the burst energy to decay and to enable a burst to be sent in the reverse direction. To allow for bits needed for overhead and synchronization, it appears that the transmission rate for a typical loop length must be about 2.25 times the information rate. Thus the ISDN full basic access structure of 2B + D will require a burst transmission rate of about 324 kb/s to achieve an aggregate information rate of 144 kb/s, and a B + D access will require about a 180-kb/s burst transmission rate for an 80-kb/s information rate. Link bit streams are scrambled to decrease the probability of repetitive zeros, prevent discrete tones from being generated, reduce crosstalk in other systems in the same cable, and simplify extraction of timing information.

14.1.4. Network Topology

At present, the ISDN concept envisions a single network handling all "narrowband" ($\leq$64 kb/s) digital services with special transport capability for wideband services such as standard video transmission. In some countries, competitive ISDNs will be developed. For the transition period of two or more decades, multiple dedicated networks will prevail, and it may be that separate networks may continue to operate longer than initially foreseen. Bulk data are more efficiently transmitted via circuit-switched networks, and interactive data are more efficiently handled by packet-switched networks. In several countries, there already exist such separate networks, and the emerging IDNs will provide efficient transport capability for B channels and subrates thereof. Common channel signaling networks which support the IDNs can provide packet switching for D channels, although messages other than signaling may be diverted to dedicated packet-switched networks by the serving local office. It appears likely that wideband networks will be separate at least until the demand justifies incorporation of wideband hybrid (circuit and packet) switching into the IDNs. The key element is the requirement for a multipurpose local serving office to provide interface transparency to users.

Network performance objectives have not been fully developed for the ISDN. Standards have been approved, however, for 64-kb/s international connections. CCITT Recommendation G.104 described 64-kb/s hypothetical reference connections (HRX), the longest of which includes 25 000-km international portion, 1250-km national portions down to the last toll (transit) office, toll connecting trunks, and end office connections to subscribers, all over terrestrial facilities. Recommendation G.822 specifies an acceptable slip rate for end-to-end digital connections over the longest HRX of not more than one slip in five hours measured over a 24-hour period.[6]

14.1.5. ISDN Signaling Protocols

CCITT Recommendation X.200 has adapted the International Organization for Standardization (ISO) Open System Interconnection (OSI) reference model for ISDN application. Signaling protocols are being standardized for the seven layers of the model. The *physical layer* connects to the subscriber's physical circuit at the *s/t* interface and activates, maintains, and deactivates the physical link to the local serving office. The *data link layer* provides for flow control, error control, and synchronization for information transfer over the physical link. The *network layer* provides a means to establish, maintain, and terminate switched connections between end systems. The *transport layer* provides end-to-end transparent data transfer, optimizes the use of available network service, and provides any additional reliability over that supplied by the network service. It can be said to function as a transition point between the requirements of the upper three layers and the service provided by the lower three layers.

The *session layer* provides a means to transfer data and to control the transfer of data in an orderly manner. The *presentation layer* establishes a common syntax for communication with another end system and performs any conversion which may be necessary. The *application layer* is the highest layer of the model and provides a window through which a user's applications have access to the model.[7]

For switched connections involving the basic ISDN access model, the lower three layers are related as shown in Fig. 14–2. The physical entity supports the D-channel entity and the two B-channel entities in the data link layer. The B-channel entities in the data link layer service the corresponding entities in the network layer. The D-channel entity in the data link layer serves the signaling entity and, where provided, multiple packet-switching and telemetry entities in the network layer.

The application of the ISDN protocol reference model to a circuit-switched connection is shown in Fig. 14–3. Two functional planes are shown. The signaling plane depicts a connection through two ISDN nodes involving the three lower functional levels to establish and maintain the switched network connection via the D channel. Within provider's networks, CCITT system

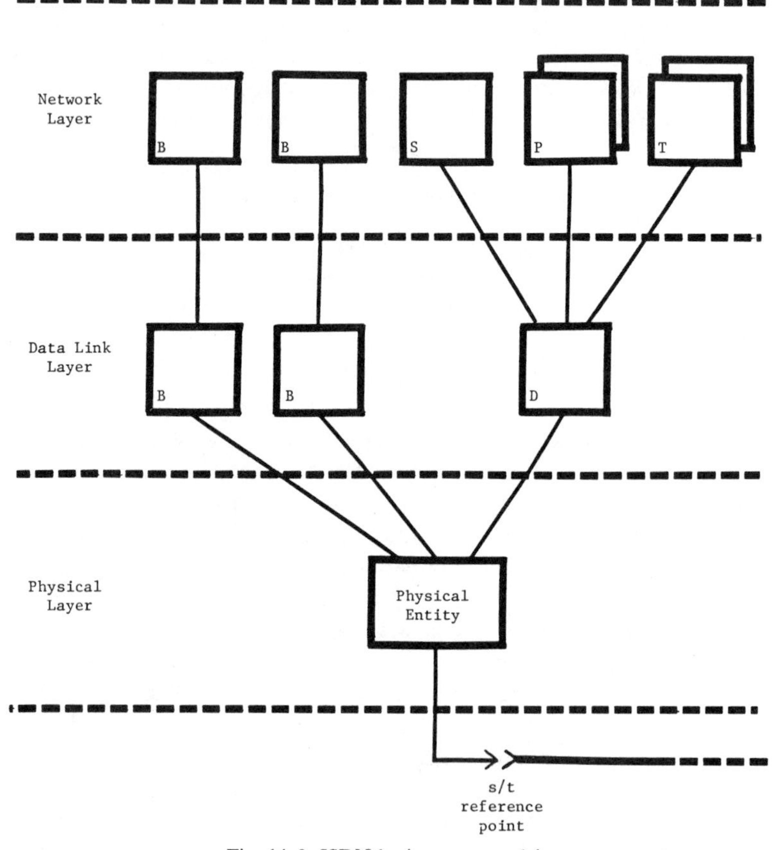

Fig. 14–2. ISDN basic access model.

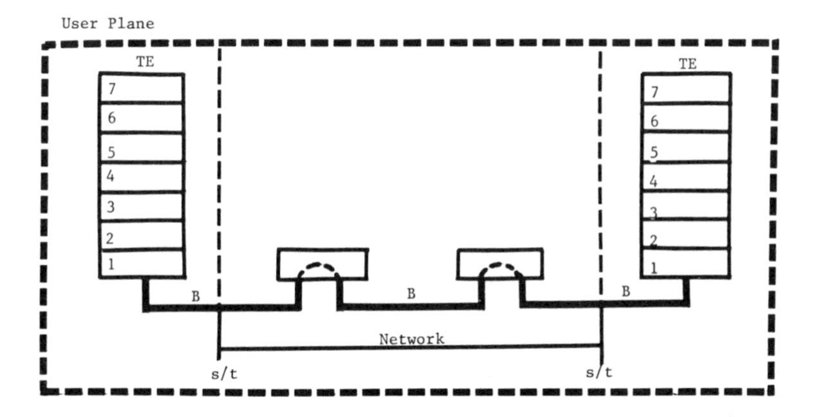

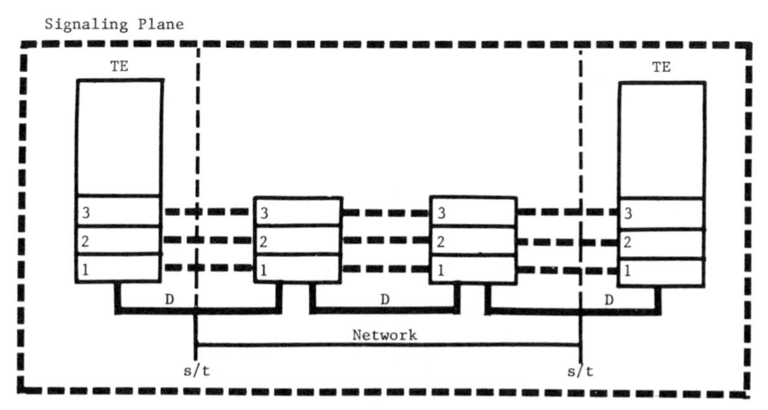

Fig. 14–3. ISDN circuit switched connection.

No. 7 or a modified version of it is expected to be used. The user plane shows the functional call set up and release of a B channel. Only the physical layer is involved at the network nodes in this example.[8]

14.2. ISDN PLANS AND PROGRESS

ISDN planning is progressing in different ways in various countries. Some experimental test beds are being implemented, and others are planned. Plans differ in both concept and technology. One thing is certain: results will be watched closely and carefully evaluated by all. In the following subsections, ISDN plans and related technological developments in five countries are examined. These countries were selected because of the technological interest in their plans and developments, and the omission of events in other countries does not imply a lack of ISDN activity.

14.2.1. ISDN Activity in Japan

Since the early 1970s, the Nippon Telegraph and Telephone Public Corporation (NTTPC) in Japan has been aggressively modernizing its telephone system and constructing new telecommunications networks. Advanced research and development projects have resulted in successful field trials of forward-looking systems. Some of these systems are being placed in commercial service both to satisfy current requirements and to gain further technological experience designed to implement plans to realize an ISDN in Japan at least by the turn of the century.

14.2.1.1. Existing Networks. NTTPC considers that the core of the ISDN in Japan will be digitization of the existing telephone network which consists of about 650 toll switching offices and 6800 local switching centers serving over 58 million telephones. Digital transmission systems, compatible with North American PCM technology, were introduced in 1965, and more advanced systems were placed in service during the middle and late 1970s. Some optical fiber has been placed in service in the telephone network. Integration of digital switching systems into the network has begun, it is expected that about 60% of transit and 30% of local switching systems will use digital technology by the end of this decade.[9]

Digital microwave systems have been under development for several years. A 400-Mb/s system designated 20L-P1, operating in the 20 GHz band, was placed in service in December 1976. The system uses 16-level QAM modulation, and requires repeater spacing of approximately 2.6–3.0 kilometers. Repeaters are mounted in radomes atop 24-meter poles made of steel pipe 1.1 m in diameter. The system reportedly is designed for use over distances up to 2500 km.[10] A 200-Mb/s microwave system, the 5L-D1, operating in the 5 GHz band, was placed in long-haul commercial service in 1982. The system uses 16 QAM modulation and achieves frequency utilization of 5 bits/Hz by using cross-polarization and interleaving the two radio channels with 40 MHz channel separation.[11]

A circuit-switched digital data network (DDX) was placed in service in December 1979 after experimental testing of a prototype system in 1973 and field testing of production equipment in 1976. The system provides seven classes of user data rates ranging from 200 b/s to 48 kb/s. The network employs STS digital switching systems and uses digital line concentrators in cities with insufficient traffic to justify a switching system. The switching system has a capacity of 40 highways operating at 1.544 Mb/s. Common channel signaling is used in the network. A packet-switched data network also is in service in Japan.[12]

The written Japanese language uses Kanji characters (Chinese ideographs) inflected by Hiragana, a syllabic representation of the vocal sounds of the language. All words also can be "spelled out" by using Hiragana sound symbols. Katakana, another syllabic representation, is used for foreign proper names. Finally, Japanese words can be spelled out phonetically by using the English alphabet. Therefore, facsimile has become a very popular means of communication in the Japanese business community not only for transmission of graphics but also for transmission of texts. Since a single Kanji character can require several English letters to phoneticize the word for transmission by teletypewriter text, facsimile transmission actually is much faster than text systems. A store-and-forward digital facsimile network is superimposed on the telephone network. A user establishes a connection through the telephone network to a toll switching system with facsimile facilities and thence to facsimile storage and conversion equipment (STOC), which digitizes, encodes, and temporarily stores the facsimile signal. Transmission between STOCs occurs over 64-kb/s digital channels. The destination STOC initiates a connection through the telephone network to the called party. Common channel signaling is used to set up the connections.[13]

14.2.1.2. Current and Near-Term Activity. Experimentation in fiber optics has been quite extensive in Japan and development of operational systems is occurring at a rapid rate. A system called HI-OVIS (high interactive optical visual interactive system) has been in development and field trial in a new model city near Osaka. HI-OVIS will offer cable TV, videotex, and two-way interactive video services.[14] Initially serving 168 subscribers with two fibers each, the enhanced version serves over 1000 subscribers with one fiber each, using laser diodes and wavelength division multiplexing.

NTTPC is investing a substantial research effort to develop fiber systems for subscriber loops. A field trial of an 7.9-km system was conducted from April 1980 to May 1981 in the Yokosuka area. The primary purpose was to test a system being developed for business use to carry multiplexed PBX signals, high-speed data, and video conferences. NTTPC also is testing other optical fiber systems for CATV distribution, two-way video for home use, and 1.5-Mb/s digital data system for two-way transmission. Several of these tests have involved wavelength division multiplexing (WDM) which NTTPC has concluded is necessary to the economical realization of flexible subscriber loops using optical fiber.[15]

A long-wavelength high-capacity optical fiber system was begun in 1980 over an 80-km route between development laboratories in Atsugi and Musashino. The system uses single-mode fiber cable with transmission in the 1.3-

μm band and involves five cable sections ranging in length from 10 to 20 kilometers. A field trial was completed in 1982. Initial transmission tests indicated that an average transmission line loss of 0.58 dB/km had been achieved.[16] The cable, designated the F-400M system, was designed to carry digital transmission at 400 Mb/s over a distance of 2500 kilometers with a bit error rate of less than 10^{-8}. Maximum repeater spacing is 20 kilometers. NTT planning called for the F-400M cable to be completed by the end of 1984.[17]

A videotex system called CAPTAIN (character and pattern telephone access information network) was placed in initial experimental service in December 1979 with 1000 terminals, 200 information providers, and 100 000 display frames. Based upon evaluation of operation through March 1981, the system was enhanced, and the second phase was placed in service in August 1981 with 2000 user terminals, 250 information providers, and 200 000 display frames. Commercial service was scheduled to be introduced in 1984 with the network partially digitized, providing transmission rates of 8 kb/s and 64 kb/s. It is planned for the network to evolve from an information retrieval service into a transaction and interchange service for graphics. Input is by TV camera, editing terminal, and Kana keyboard. A terminal with a Kana keyboard can input information in Kana (Hiragana or Katakana) character strings which are converted into Kanji character strings by referencing a Kanji dictionary containing about 100 000 Kanji idioms. The commercial system involves digital CAPTAIN switching systems, interconnected by high speed digital transmission links and accessed through the public telephone network. The switching centers will be connected to information centers through the separate switched data network. Future development plans include provision of such features as partially moving graphics, halftone color pictures, high-resolution color graphics, and audio output. It is expected that the evolving network will become part of the ISDN in Japan.[18]

Digital technology has made possible two other new services in Japan which will fit into an ISDN. A voice storage service has been developed to provide automatic answering service, message transfer, and a message mailbox service for closed user groups. The system employs 32-kb/s ADPCM voice coding technology and stores only the voiced parts of messages. When messages are played back, a buffer control unit inserts the silent parts into the messages. This system is being tested for installation in a model plant in a Tokyo suburb.[19] An unlimited vocabularly audio response unit with synthesized voice storage also has been developed. Spoken Japanese is composed of phonemic sequences of the patterns vowel-consonant-vowel (VCV) and vowel-vowel (VV). To produce spoken Japanese, about 900 VCV segments are needed. The VCV segments are read out under processor control and combined at the center of the vowel parts. The primary application is for use in banking for notification of bill payment and for customer inquiry as to bank balance. Eight centers are planned in major cities. Access is through the public telephone network via DTMF telephones.[20]

14.2.1.3. Future ISDN Plans. NTTPC is planning to implement a fully integrated information network system (INS) in four phases. Phase I consists of the present telephone network which is just beginning to be digitized, the digital data (DDX) network, and the facsimile network which is partly digital. Phase II involves a backbone digital telephone network linking the major cities, expansion of the DDX, and expansion and full digitization of the facsimile network by the end of the 1980s. Thus, a thin-route ISDN will be realized. Phase III, to be completed by the end of the 20th century, involves extending the ISDN to smaller major cities and then to much smaller provincial cities. Part of the DDX and most of the facsimile network will be folded into the ISDN. The final integration and realization of the INS will occur at about the beginning of the 21st century.

The telephone network will undergo a radical reconfiguration during this process. The existing four-level network hierarchy will be changed to a two-level hierarchy. Simulation studies have led to a plan to expand the local service areas to approximately the size of the present toll areas, the current second hierarchical level, by introducing digital local switching systems into the toll offices. Concurrently, there will be substantial digitization of subscriber loops by employing digital pair-gain systems, digital concentrators and multiplexers, and optical fiber subscriber loop systems. As the transit (toll) network is digitized, it will evolve into a single-transit-stage configuration with only about 50–60 digital transit offices interconnected by optical fiber.[21]

The user interface structure will support the currently planned CCITT structures (2B + D, etc.). NTT is studying different protocols and has developed a simple D channel protocol for testing in 1984.[22] A new integrated switching system for voice and nonvoice services has been developed to combine circuit switching and packet switching. The D-70 system is capable of providing voice, data, facsimile, and interactive videotex on one subscriber line, using two information channels, at 64 and 16 kb/s, and a separate signaling channel.[23]

The planned Phase II ISDN network configuration is shown in Fig. 14–4, although the plan is subject to modification as the program progresses. Integrated services will access the local serving office through terminal control units which will handle all signaling and transmission protocols. Above the local level, voice and circuit-switched data services will be integrated at the transit level while other services will involve dedicated networks. All services will have access to data bases at information processing centers. Wideband services likely will use channel multiples of 64 kb/s.[24]

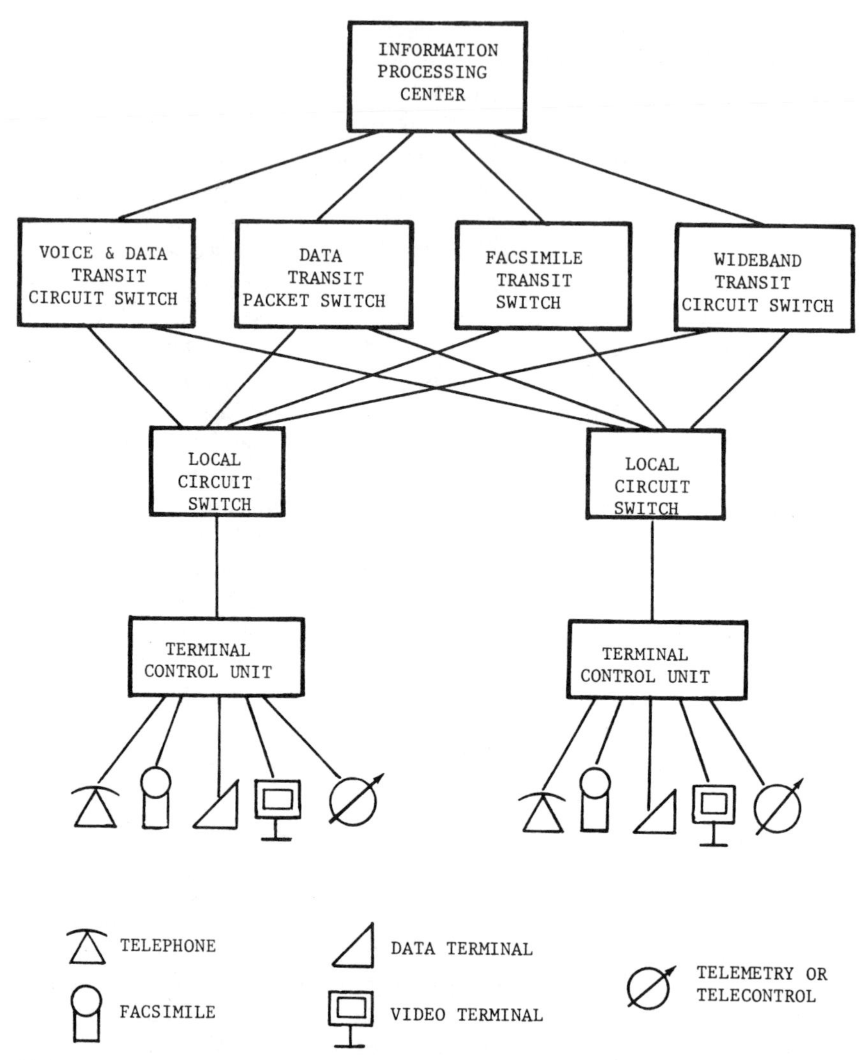

Fig. 14–4. Proposed initial ISDN in Japan. (Iimura, Mizuashi, Kikuchi, "Network Planning Toward the ISDN," *NTC '81,* © 1981 IEEE.)

In preparation for ISDN implementation, NTT has constructed a major model ISDN plant as a test bed for operation in late 1984 with some 1500 digital subscribers.[25] The planned configuration of the model plant encompasses the arrangements shown in Fig. 14–5. An integrated digital local serving office will handle all services except wideband services which will remain separate throughout. A wide variety of user access techniques will be used,

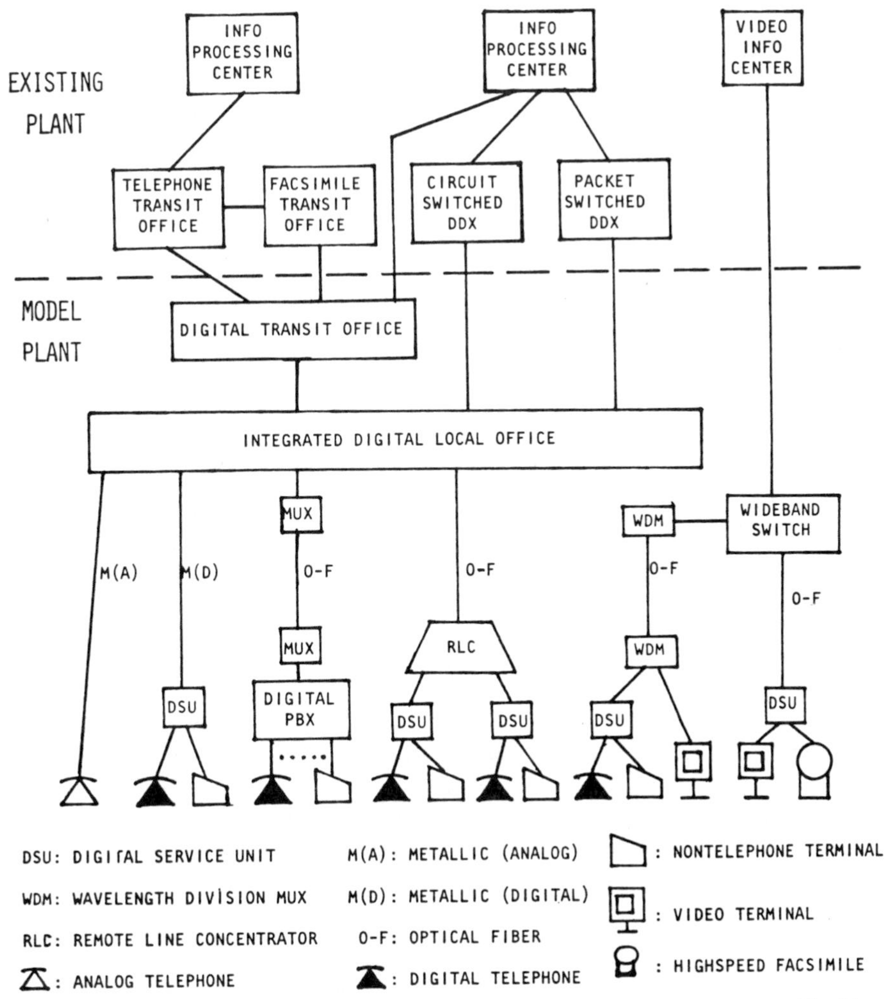

Fig. 14–5. Model ISDN plant configuration. (Based on Fig. 3, Iimura, Mizuashi, Kikuchi, "Network Planning Toward the ISDN," *NTC '81* © 1981 IEEE.)

including metallic pairs and optical fiber cables with varying treatments. Both analog and digital telephone sets will be used. Video terminals will access the wideband switch via a digital service unit (DSU) over optical fiber using wavelength division multiplexing and subscriber microwave radio systems.[26] The performance of the model plant will be watched closely by the international telephony community.

Digital emphasis in Japan is being placed not only on domestic communications but also on international communications. Field tests of submarine optical fiber cable over the past three years have been highly successful. As a result, a submarine cable consisting of six single-mode fibers is planned for commercial use over distances up to 1300 kilometers. It will operate at a wavelength 1.3-μm and will have a capacity of 400 Mb/s with repeater spacing of 30 kilometers. A 300-km cable is planned for service along the Japanese coast and a 1000-km cable between Kyushu and Okinawa by 1985. A Japan-Hawaii cable is planned for completion by 1988 at an initial cost of less than 20% of conventional cables.[27] In anticipation of a 120-Mb/s Time Division Multiple Access (TDMA) system in the INTELSAT network, research and development of a Direct Digital Interface (DDI) has begun. The DDI will be located at satellite ground stations and will perform three major functions: compensation for satellite drift (Doppler effect), slip control, and frame format conversion between plesiochronous networks using different PCM standards. A simulation system has been developed. Two terminals simulate domestic networks, one using 32-channel A-law PCM and the other using 24-channel μ-law PCM. A third terminal simulates an INTELSAT terminal and the periodic delay variation due to satellite drift. Experimental results indicate that slip control well below the marginal level can be achieved.[28]

14.2.2. ISDN Activity in France

Since the world's first digital telephone switching system was installed in France in 1970, the French telecommunications administration (PTT) has undertaken a program to digitize the telephone network. This is being accomplished by overlaying digital transmission systems and by using digital technology for all new switching equipment. In addition, the National Center for Telecommunications Research (CNET) has undertaken several research programs leading to realization of an ISDN in France. (In this discussion, some acronyms derived from the French language do not correlate with their English translations.)

14.2.2.1. Existing Networks. The telephone network in France is a three-level hierarchical network serving some 25 million telephones. At the end of 1980, 12.8% of the telephones were switched by electronically controlled logic, a 123% increase in one year. Some 64% of the total number of telephones terminate main station lines.[29] The number of lines per main exchange area ranges from about 15 000 to about 100 000. Nationally, only about 50–60% of French residences are equipped with telephones. A switched 64-kb/s service was scheduled for 1984.

A packet-switched data network, called Transpac, was introduced in 1978 and interconnection with the networks of other countries began in 1979. A

Telex network has been in service for many years. Digital microwave transmission systems at 8, 34, and 140 Mb/s rates are in service, and about 25% of the long-haul networks and some 50% of the local network is digital.[30] A videotex system, called Teletel, is accessed via the telephone network.

14.2.2.2. Current and Near-Term Activity. A digital subscriber loop system, called PRANA, has been developed. The initial version was installed in an experimental test bed configuration in 1980. PRANA is evolving along with the CCITT ISDN concept. Originally, it provided an 80-kb/s information rate with a TCM channel rate of 176 kb/s. Intermediate transmission equipment (ITE) grouped six subscribers in a 528-kb/s data stream in each direction on a four-wire line. The planned information rate has been expanded to 144 kb/s (2B + D), and the ITE data rate is being increased to 704 kb/s for four subscriber lines at 144 kb/s or 8 subscriber lines at 80 kb/s. Additional research is in progress to develop a hybrid separation method with echo cancellation in addition to TCM. B channel service options include 64-kb/s s voice, 56-kb/s voice plus 8-kb/s data, and 64-kb/s data. The 16-kb/s D channel supports three subrates of 12 kb/s data at a usable rate of 9.6 kb/s, 2 kb/s for signaling and telemetry, and 2 kb/s for multiplex frame synchronization and a system resource of spare bits. Signaling uses CCITT No. 7. Theoretically, PRANA can function on nonrepeatered loops up to 7.5 kilometers, but is expected to be limited to loops of about 5 kilometers which will cover about 88% of the French subscribers. An A-law codec chip incorporating the compression-addition-expansion functions for the subscriber side of the B channel is under development. CNET expects the 144-kb/s (2B + D) rate to evolve as the basic rate with 80-kb/s (B + D) as a lower-grade option.[31]

An electronic directory service has been placed in commercial service in Brittany. In the initial phase, the PTT is installing Minitels, stand-alone videotex terminals, in homes and offices of three million customers by 1986 to replace printed telephone directories. The data base in Brittany comprises about 1 200 000 entries, and the national data base, which was scheduled for completion by the end of 1983, is expected to hold over 23 million entries.[32]

The PTT designed a switched satellite service called Telecom 1 for service in 1984. The system will provide service to as many as 256 earth stations at 2.4 kb/s to 2 Mb/s data rates. It employs time-division, multiple-access (TDMA) technology. Large businesses will install small earth stations on or near their premises.[33]

In expectation of a growing video services market in future years, several projects have been undertaken to investigate techniques for offering such services. Demand for one-way broadcast services, one-way request services, and two-way interactive services is expected to materialize within a few years at most. A SAFO project is investigating the use of digital technology to provide multiservice subscriber connections via a single optical fiber, possibly shared by several subscribers. Planning studies indicate that the economic goals of capital investment not exceeding three or four times that required for telephone service and operation/maintenance costs not exceeding two or three times those for telephone service are reasonable. This will require digitization of the video signals. Research indicates the feasibility of providing initially four, and then six, signals by wavelength division multiplexing, using 96 Mb/s for broadband channels. The concept includes space division switching of the 96-Mb/s digital signals with signal regeneration. It is recognized that service responsibility will be shared by the service provider, the subscriber, and the telecommunications administration.[34]

A major project is in process to provide an optical network in the city of Biarritz. The initial phase, completed in 1983, connected 1500 subscribers to three secondary centers which will function as concentration and distribution centers. Services include picturephone, telephone, two television channels from among 15, and one high-fidelity music channel from among 12. The three secondary centers connect to a principal center in a star configuration. Optical cable with 70 fibers connects the principal center to each secondary center, and cables of diminishing sizes form the distribution network, beginning with 70 fibers, reducing to 10 fibers, and then to 2 fibers to each home. The video network portion initially is analog but ultimately will likely change to digital technology. Speech channels associated with the picturephone service and signaling information use digital techniques at the CCITT standard rates. Separate switching networks, with coordinated control, handle video and voice traffic with the constraint that a single telephone set can be used for both picturephone service and ordinary telephone service. The video switching systems use space division analog switching while the speech networks use time division digital switching. The project is planned for expansion to 5000 subscribers[35] and could reach as many as 1.5 million homes by the end of 1986.

14.2.2.3. Future Plans. The French ISDN concept provides for complete integration of services at the local exchange area, including digitization of the subscriber loop plant, and four separate service networks: a 64-kb/s integrated telephone and data network (ITDN), one or more domestic satellite networks, a packet-switched network, and a wideband (video quality) network. Implementation is planned in three phases.

Phase I involves extension of digital connectivity to subscribers requiring digital services. By 1985, digital switching systems are expected to be in service in all local areas, and they will be interconnected by digital transmission and switching in about 50% of the transit network. A primary development to support subscriber line digitization is ELOISE, French acronym for local integrated service unit. ELOISE will provide three types of subscriber interfaces: basic subscriber access at 144 kb/s (2B + D), multiplexed service at 450–800 kb/s for small PBXs and for multiplexed basic services, and primary 2048 kb/s for larger PBXs, multiplexers, and ELOISE remote units. ELOISE capacity will range from 250 to 1000 terminals depending upon the individual bit rates required. Large integrated services PBXs and local business digital networks can be connected directly to local digital switching systems. Phase I is to be implemented in the 1985–1990 time period.

Phase II (1990–1995) will encompass the introduction of new services and the substantial integration of access procedures for all services. The introduction of integrated services control (CISE) will provide a uniform numbering plan, uniform access procedures for all services, and a charging mechanism. During this phase, most potential business subscribers will be connected to the network.

The final phase (1995–2005) covers the extension of the ITDN to all telephone subscribers. Local exchange areas will be completely digitized, and an integrated services switching system will replace the existing local switching systems and the CISE implemented during Phase II. Telex will disappear, and the four separate transport networks will carry all services, resulting in an ISDN.[36]

14.2.3. ISDN Activity in the Federal Republic of Germany

The Federal Republic of Germany (FRG) is moving rapidly into optical fiber development, which could greatly facilitate ISDN development not only in Germany but also in other countries.

14.2.3.1. Existing Networks. The telephone network operated by the Deutsche Bundespost is almost entirely switched by electromechanical systems. The existing public digital data network is to be expanded and enhanced. The Telex network in Germany is quite extensive, and a videotex system was introduced in 1983. The videotex service, called *Bildschirmtext,* has been developed through field trials involving over 7000 users and more than 35 external data bases. Access is via local telephone calls, and the data transmission is asynchronous at a 1200-baud rate.[37]

14.2.3.2. Current and Near-Term Activity. One project of interest involves the conversion of the existing public digital data network to a 64-kb/s switched service. Digital time division switching systems are planned for introduction at the network nodes beginning in 1985, and a planned model network will serve some 4000 residential and business subscribers.

The enhanced network will be capable of handling both voice and data.[38]

An extensive wideband optical fiber system has been designed and installed to study various techniques of providing integrated services via optical fiber. The HHI broadband communications system services include telephone, data, color television, color videophone, and stereo broadcasts. The system employs both analog and digital transmission technology. The analog system is constructed in a star network configuration centered in a PABX, simulating a central exchange, using pulse amplitude modulation for speech and a separate solid state crosspoint switching network for videophone. Data communication between subscribers uses modems in a TCM mode. From the exchange to the subscriber, the 108-MHz bandwidth provides three standard television programs, one videophone channel, one telephone channel, one data channel, and four stereo broadcast channels. In the opposite direction, about 30 MHz bandwidth carries telephone, videophone, data, and control information. The digital system operates in a multiparty loop arrangement with subscriber loops operating at 10, 17, 140, and 280 Mb/s. The loops are interconnected through switching units by local trunks operating at 280 Mb/s and long distance trunks operating at 560 Mb/s. Distributed switching under subscriber control is used to establish connections.[39]

Operational test results indicated the integration of wideband and narrowband services into one common TDM frame was not completely effective because of the very high-speed switching required at the subscriber station. Other disadvantages of the digital distributed control concept include high power consumption of the extensive electronics required in the subscriber station, network impairment resulting from a single component failure, and lack of privacy. The analog system also has problems. To satisfy the transmission requirements for video and stereo broadcast signals in cable TV systems, overmodulation of about 4 dB and use of phase-coherent carriers were required. Analysis of the results produced the conclusion that future such networks probably should use star topology with central switching and optical transmission with TDM or WDM.[40]

An experimental project of major interest is the broadband integrated glass fiber optical network (BIGFON). The goal of the project is to evaluate concepts and techniques which will enable the telephone network to be converted to a wideband color videophone network in the late-1980s. The Duetsche Bundespost concluded that optical fiber technology will be necessary to achieve that goal and that wideband fiber systems can be made available at costs which closely approximate those of conventional systems as early as 1985. This will require very high production levels which can be realized only by forcing very high demand. Consequently, the plan is to use fiber in newly developed residential areas and to overbuild existing areas with fiber when capacity nears exhaustion. To further stimulate demand, it has been decided to use the conventional television set as the terminal for interactive data services and videophone service.

Subscriber services to be provided by BIGFON include simultaneously a minimum of two 2-Mb/s ISDN primary channels, a group of 24 stereo sound programs or 4 stereo channels selected from 24 channels offered by the BIGFON switching system, 2–4 TV programs selected from the 15 offered by BIGFON, and a bidirectional channel for picture telephony.

Contracts were awarded in September 1981 to six companies to provide ten BIGFON "islands" in seven cities. Eight of the service cells are to have 28 subscribers each, with 6 having videophone service. The other two cells are to have 48 subscribers with 10 of them having videophone service. Half the subscribers to each network can apply for a second digital telephone set, two television receivers, and/or two stereo sound conversion units. Switching equipment for narrowband services will be supplied separately from the BIGFON contracts. A typical BIGFON local access arrangement is shown in Fig. 14–6.

The experimental networks will be integrated with the public telecommunications network for the narrowband services. For the videophone service, a modified digital telephone will be used for both videophone and normal telephone services.

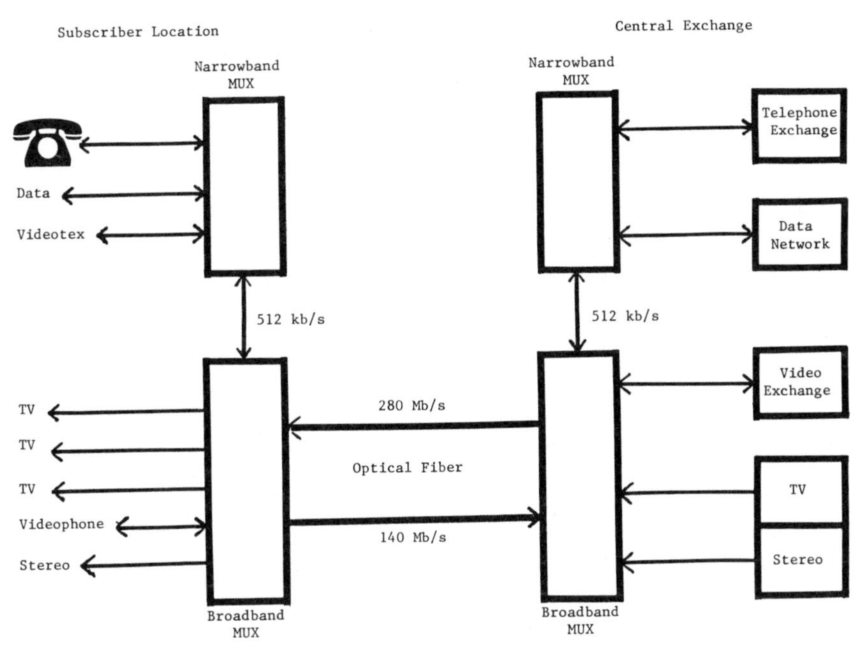

Fig. 14–6. Typical BIGFON local access arrangement.

Most of the BIGFON subscribers were connected by the end of 1983, and the system trials began in 1984 and are to be completed by the end of 1986, after which the Deutsche Bundespost expects to use the results obtained to develop a standard system concept for widespread implementation. Signal processing was specified for only narrowband services. The contractors have wide latitude to leave their own conceptual imprints on the wideband portion of the systems. Most subscribers will be served by two fibers at a maximum bit rate of 280 Mb/s. Some systems will use wavelength division multiplexing at 850 and 1300 nm, and one system will use four wavelengths per wavelength range. Both space division switching and time-space switching will be used for wideband services.

If the demand for wideband services develops as expected by the Deutsche Bundespost, the existing trunk network will have to be converted to an integrated wideband system. Consequently, a plan to interconnect the BIGFON islands with a graded index optical fiber network, equipped initially with a 140-Mb/s system per fiber, was developed.[41]

Much research has been focused on the development of advanced interactive wideband systems to provide information services under the assumption that the ISDN will stimulate demand for widespread use of such services by a very large number of subscribers. Demand is anticipated for information data bases in the areas of general information (news, cultural affairs, sports, consumer affairs, governmental matters, law, health, etc.), education, reservations, electronic mail, entertainment, and computer access. It is believed that some of the services will require interactive audio/visual techniques. The new videotex system could provide the basis for further development of more sophisticated services.[42]

14.2.3.3. Future Plans. Decisions on the exact direction of ISDN implementation appear to depend upon assessment of the results of the BIGFON and other projects. If the Deutsche Bundespost confidence in the early cost-effectiveness of optical fiber technology is substantiated, it would appear that implementation of a major portion of an ISDN with color video services by 1990 could be a realistic expectation.

14.2.4. ISDN Activity in the United Kingdom

The continuing installation of digital transmission and stored program control switching systems in the United Kingdom is preparing the way for early introduction of new digital services and ultimate realization of an ISDN.

14.2.4.1. Existing Networks. The public telephone network operated by British Telecom serves an estimated 20 million main station lines and some 30 million telephones, about 25% of which are switched by electronic switching systems, based upon extrapolation of 1981 reported data.[43] Digital switching systems began to be deployed in 1981. The worldwide Telex network also has many subscribers in the United Kingdom, and a packet-switched data network is in service.

The first public videotex network, called Prestel, was introduced by the British Post Office (now British Telecom) in 1979. The system uses the same display as is used by television broadcasters for teletext. The principal difference between the two systems is that videotex uses the public telephone network and can easily implement interactive services as opposed to the one-way teletext provided by broadcasters. (Interactive teletext, of course, can be provided via cable television systems.) In the spring of 1982, a gateway feature was added, enabling subscribers to access videotex data bases in other countries while remaining logged into Prestel. This enables subscribers to use data bases in other videotex systems which use technology incompatible with Prestel. Communication between Prestel and other videotex networks is via the packet-switched data network.[44]

14.2.4.2. Current and Near-Term Activity. British Telecom is testing a limited ISDN system in London. The system uses the System X digital switching systems which began providing service in 1981. Three types of network terminating equipment (NTE) have been developed for the initial system. The simplest is a desk set containing a digital telephone for the CCITT B channel and a data adaptor for the D channel. Telephone functions are line powered, but the data adaptor is locally powered. The second type of NTE is similar, but the data adaptors (up to three) are external. Only basic telephone service and maintenance service are available if local power fails. The third NTE type is a wall mounted unit with separate terminal equipment which can support up to six adaptor units. If local power fails, only maintenance messages can be handled. Any of three transmission systems can be used with any of the three terminals. Two are TCM systems with different line codes, and the third is hybrid balance system with echo cancelers. The initial information rate is 80 kb/s to provide B + D capability. Plans are to upgrade the transmission links later to 144 kb/s for 2B + D capability.[45]

Along with other research laboratories, British Telecom has been studying long-wavelength optical fiber systems. Studies concluded that single-mode fibers operating at 1.3 and 1.5 μm are needed to provide the data rates required for future services. A prototype repeater section of 31.6 km of fiber was cabled. Mean attenuations before and after cabling were 0.44 and 0.55 dB/km, respectively, for the 1.3 μm wavelength, and 0.32 and 0.51 dB/km, respectively, for the 1.5 μm wavelength. The attenuation for a fully cabled repeater section of 31.6 km with joints at 2-km intervals was 17.5

dB at 1.3 μm and 16 dB at 1.5 μm. British Telecom placed the first 30-km, 140-Mb/s production system in service during 1984.[46]

14.2.4.3. Future Plans. British Telecom's objective is to have a nationwide ISDN in service by 1990.[47] A substantial quantity of PCM transmission systems are in service, and digital local and toll switching systems are being installed in an orderly manner to facilitate digital networking. Replacement of analog switching systems with digital systems as the older systems approach the end of their economical life cycles is a very slow process. For a country the geographic size of the United Kingdom, establishment of a digital network as an overlay network will enable a nationwide ISDN to be realized at an earlier date.[48]

British Telecom expects to replace their 60 analog toll switching systems with digital systems and to achieve limited interconnection by digital transmission systems by 1986. Digital transmission capacity should increase sufficiently by 1988 to carry all trunk traffic. In the same time frame, digital principal local exchanges will concentrate trunk traffic from the remaining analog exchanges still served by analog transmission facilities. Finally, they expect to discontinue the analog trunk network by 1990. Competition in the United Kingdom could aid the early realization of ISDN as the Mercury network begins to offer 64-kb/s business services.[49]

14.2.5. ISDN Activity in North America

The telephone system in the United States and Canada is unique in the world in that the individual geographic area networks of some 1500 telephone companies are amalgamated into a single North American network with unified numbering, compatible signaling, and common routing systems. In addition, subscribers may own their terminal equipment or lease it from outside suppliers, and multiple common carriers provide long distance voice and nonvoice services. Irrespective of this diversity of ownership and interest, progress toward ISDN is occurring at a more rapid rate than perhaps is generally realized.

14.2.5.1. Existing Networks. The North American telephone network is described in substantial detail in Chapters 10 and 13. There are multiple public packet-switched data networks in both Canada and the United States, including Telenet, Tymnet, Datapac, and Infoswitch, and the ACCUNET Packet Service offers virtual call and virtual circuit private line services. There are facsimile networks such as Graphnet, Speedfax, Faxnet, and Fax-Pak, and message-switching services such as Teletypewriter Exchange Service (TWX), and Telex. There are numerous long distance carriers offering satellite communications services, terrestrial microwave services, and resale services involving facilities of other carriers. High Speed Switched Digital Service became available in 1982, providing switched 1.5 Mb/s or 3-Mb/s data rate via satellite between switching centers in five cities for picturephone conferences. Two parallel 1.544-Mb/s T-carrier channels provide the 3-Mb/s data rate. Service has been extended to 18 cities and to Europe.

14.2.5.2. Current and Near-Term Activity. Digital transmission systems are being implemented rapidly by satellite, microwave, coaxial cable, and optical fiber. AT&T and the Bell operating companies have over 130 million circuit-miles of digital systems, mostly in short-haul systems. By 1990, Bell operating companies expect that about 80% of their metropolitan area circuit-mileage will be provided by digital systems. Digital microwave systems are becoming more cost effective. The DRS-8 long-haul system operates in the 8-GHz band, provides 1344 voice channels per radio channel, and provides digital connectivity across most of Canada.[50] The DR 6–30 digital microwave system provides 2016 voice channels per radio channel in the 6-GHz band and can be added to existing TD-2 (4-GHz) microwave routes.[51] An optical fiber system became operational between New York and Washington in February 1983, and is planned to be extended to other cities.[52] Other optical fiber systems have been installed in short-haul configurations, and other long-haul systems are being implemented by several common carriers.

The Manitoba Telephone System conducted an integrated optical fiber field trial which began in October 1981. The system provided telephone, FM stereo, cable television, and videotex services to 150 residences in two nearby rural communities. Telephone and videotex services employed digital technology while analog transmission was used for FM stereo and television services. The system used graded index fibers, and the field trial continued until 1983. Results are reported to be highly satisfactory.[53]

Saskatchewan Telecommunications has installed a wideband optical fiber network totaling 3450 kilometers of fiber, over half of which is long-wavelength fiber to operate at 1.3 μm. The system, completed in 1984, provides integrated services to about 58 communities in Saskatchewan. Transmission is at the DS-3 rate (44.736 Mb/s).[54]

A videotex system called Telidon was introduced in Canada in 1978. The basic design was based upon computer graphics principles to endeavor to avoid hardware constraints on development of services. The British, French, and Canadian videotex systems all use different techniques, and the CCITT approved recommendations covering all three systems. North American groups have developed a unified standard built upon the protocol developed in the United States for the AT&T VIEWTRON system. Standards working groups in both countries have adopted the North American Presentation

Level Protocol Syntax (NAPLPS). The key attribute of the standard is that it frees the coding scheme from the constraints of hardware implementation, which will permit more sophisticated terminals in the future.[55]

Interim end-to-end switched digital networks are planned for both the United States and Canada. The Canadian system, called DATALINK, was introduced in 1982 and uses the DMS-100 digital switching equipment.[56] The American system is called *circuit switched digital capability* (CSDC) and is described here. The CSDC will use No. 4 ESS and No. 1A ESS switching systems interconnected with T-carrier systems or other digital transmission systems. A digital carrier trunk frame can be added to the No. 1A ESS office to directly interface digital transmission systems and to enable the digital data signals to be switched through the reed relay switching network. Full duplex operation across the two-wire switching network of the No. 1A ESS can be achieved by requiring two cross-office connections. The subscriber loop can use time compression multiplexing to provide a full duplex 64-kb/s digital channel over two-wire nonloaded metallic loops.[57] The network connection configuration is shown in Fig. 14–7. The CSDC actually will be an alternate voice/data capability. Calls will be established in the voice mode by dialing an access code and the called number. Plug-in units in the No. 1A ESS can provide analog PAM voice cross-office signals in the voice mode and data buffering in the data mode. Once established, the call can be switched to the data mode to provide a 56-kb/s digital path between the terminals. Seven of the eight bits in the T-carrier time slot will be used for data, and the eighth bit will be used to transmit on hook/off hook and control/data mode information.[58] Synchronization can be achieved by adding a digital office timing supply (DOTS) to the No. 1 ESS.[59] The system requires CCIS to be used within the network to assure that only proper digital facilities and No. 4 ESS switching systems are used for connec-

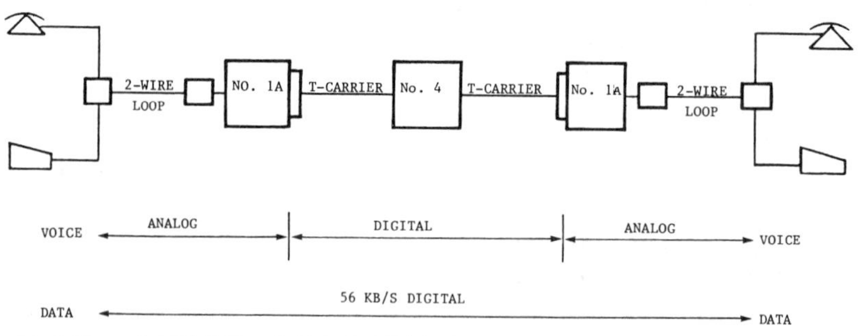

Fig. 14–7. Planned AT&T Circuit Switched Digital Capability (Courtesy Business Communications Review.)

tions. Since echo suppression or cancellation will not be used, additional loss will be used to control echo.[60]

A telecommunications test bed has been developed at Bell Laboratories to conduct advanced experiments in communications networks. An experimental digital switch (XDS) is an intelligent network processor which operates at very high speed to execute a pair of processor instructions during each sample period. Switching is performed by writing "ear" samples and reading "mouth" samples. A commercial multiprocessor minicomputer permitted experiments to involve encryption, electronic mail, word processing, and other programs. Several iterations produced the general-purpose electronic telephone set (GETSET 80), a teleterminal with an 80-character by 24-line display and 59 alphanumeric and function keys. The purpose of the test bed is to provide both technical and human factors research in development of integrated systems and services.[61]

14.2.5.3. Future Plans. Because of the proliferation of commercial companies providing telecommunications services in North America, there is no single unified plan leading toward an ISDN. It is very likely that there will evolve multiple integrated transport networks, each providing a variety of services. Several factors will influence ISDN development in North America more than in most other areas.

The first factor is competition in long distance telecommunications services. While competition encourages innovation, the economics of business mitigate against extremely expensive innovations which promise only long-term recovery of investments. A typical example is the issue of analog versus digital microwave. That digital microwave is needed for an all-digital network is readily apparent. Radio frequency (RF) spectrum is at a premium. Current long-haul digital microwave systems provide 2016 voice channels per radio channel, while the AR6A analog single sideband system provides 6000 voice channels per RF channel. Technological pressure pushes digital systems, but economic constraints favor analog systems. The demise of Datran is a classic reminder of what can happen when technology alone drives investment. Optical fiber systems are providing service on high-density routes and in metropolitan areas, but the long-haul networks may rely on thin-route digital systems for many years.

The second factor is economics, not only in the long-haul plant but also in the local plant. The great majority of telephone plant investment is in local switching and local distribution cable plant, and it is nearly all analog. Although cable TV systems have a substantial amount of coaxial cable plant in place, it is far from being capable of providing universal access even if there were no regulatory constraints. (Recent Federal legislation prohibits large telephone companies from providing cable TV services.) Optical fiber appears to be the solution to digitization of the subscriber loop plant, but economics will prevent wholesale replacement of plant that is serviceable for ordinary telephone service. Time compression multiplexing and some hybrid balance systems will be required for integrated services until fiber is economically attractive compared to copper.

The third factor is local access. The consent decree signed in January 1982 resulted in divestiture of the Bell operating telephone companies by AT&T on January 1, 1984. The long-standing method of apportioning toll revenues to local companies is being replaced by a system of access charges. Local telephone companies are extremely cautious about undertaking risky and costly ventures into new technology until a more stable environment exists.

The fourth factor is transmission. The T-carrier concept of using the least significant bit in every sixth frame for signaling is incompatible with the provision of 64-kb/s clear channels. To assure bit integrity in ISDN B Channels, signaling must be removed. Clear channel capability can be provided over T-carrier by grouping the supervisory signaling for 23 channels into the 24th channel and by employing a B8ZS zero substitution technique. The use of B8ZS is expected to begin in 1986.

Therefore, it appears that multiple ISDNs in North America will evolve slowly and will be confined largely to voice and digital services not exceeding the standard 64-kb/s data rate until optical fiber becomes economical for large-scale installation in the subscriber loop plant and until variable-bit-rate switching and transmission systems are available to handle wideband requirements cost-effectively. That is not likely to occur much, if any, before the year 2000. Undoubtedly some wideband services will be available in some areas during the late 1980s, but universal service likely will occur much later.

14.3. FUTURE TRENDS AND ISSUES

State-of-the-art technology is sufficiently advanced at the present time to permit implementation of the 2B + D subscriber loop concept and network integration of services using bit rates up to the CCITT standard 64-kb/s channel. One major deterrent, however, is the lack of all the standards needed to encourage the necessary investment in operational systems. Standards for the 2B + D and primary rates for the subscriber loop were approved by the CCITT plenary session in 1984. Another major deterrent is signaling. Expansion of CCITT Signaling System No. 7 from a 4-layer protocol model to a full 7-layer protocol model is not complete. A Signaling Connection Control Part (SCCP), and ISDN User Part, and an Operation and Maintenance Application Part (OMAP) have been added. The ISDN User Part provides for control of circuit-switched connections between subscriber line exchange terminations, but the definition of customer and network services to be supported is incomplete.

It appears likely that the first generation ISDN will, in most networks, be limited to services which can be provided over separate specialized networks accessed by digital channels not exceeding a bit rate of 64 kb/s. Future development of a second or third generation ISDN will employ concepts and techniques which will render the first generation as obsolete as it will render the present telecommunications networks.

To be most effective, networks must develop into predominantly transport networks providing connections between terminals with various degrees of sophistication for signaling and communicating after call setup. This could involve variable-bit-rate digital pipes (Fig. 1–1),[62] new switching concepts for service integration such as Bell Laboratories' experimental digital switch,[63] optical fiber loop plant, standards which will permit networks and terminal hardware to evolve independently, and provision for digital speech encoded at less than 64 kb/s—such as 32-kb/s ADPCM, or perhaps as low as 16 kb/s.

PROBLEMS

14.1 What is an ISDN?

14.2 What are the relative advantages of time compression multiplexing and hybrid balancing to provide ISDN access on two-wire subscriber loops?

14.3 In time compression multiplexing, why must the burst rate be more than twice the information rate?

14.4 Considering current network technology, why are specialized transport networks more efficient than one combined network for all services?

14.5 What problems must be solved to enable two national ISDNS to synchronize their operation when operating plesiochronously by satellite?

REFERENCES

1. MacDonald, V. C., "Issues in CCITT Digital Studies for 1984," *The International Conference on Communications Record (1982)*, pp. 1A.1.1–1A.1.5 (hereafter cited as *ICC '82*).
2. Decina, Maurizio, "Managing ISDN Through International Standards Activities," *IEEE Communications Magazine,* pp. 19–24 (Sept. 1982).
3. Gifford, W. S., "The User Perspective of ISDN," *ICC '82,* pp. 6A.6.1–6A.6.4.
4. Decina, Maurizio, "Progress Towards User Access Arrangements in Integrated Services Digital Networks," *IEEE Transactions on Communications,* p. 2177 (Sept. 1982).
5. Ahamed, S. V., "Simulation and Design Studies of Digital Subscriber Lines," *Bell Syst. Techn. J.,* pp. 1024, 1038–1042 (July–Aug. 1981).
6. Decina, Maurizio, "Performance of Integrated Digital Networks: International Standards," *ICC '82,* pp. 2D.1.1–2D.1.6.
7. Day, J. D., and Zimmermann, H., "The OSI Reference Model," *Proceedings of the IEEE,* pp. 1334–1340 (Dec., 1983).

8. Draft CCITT Recommunication I.320, "ISDN Protocol Reference Model" (1984).
9. Iimura, Osamu, et al., "A Comprehensive Study of an ISDN in Japan," *National Telecommunications Conference Record (1980)*, p. 5.2.4.
10. Nakamura, Yoshio, and Yoshikawa, Tatsuo, "20 GHz Digital Radio-Relay System," *International Conference on Communications Record (1977)*, pp. 40.3-88 to 40.3-92.
11. Saito, Yoichi, Morita, Kozo, and Yamamoto, Heiichi, "5L-D1 Digital Radio System," *ICC '82*, pp. 2B.1.1–2B.1.7.
12. Kato, Masao, Tatatsuki, Toshiharu, and Yoshida, Yutaka, "DDX," *IEEE Communications Magazine*, pp. 34–36 (Mar. 1981).
13. Kamae, Takahiko, "Public Facsimile Communication Network," *IEEE Communications Magazine*, pp. 47–51 (Mar. 1982).
14. Dorros, Irwin, "Telephone Nets Go Digital," *IEEE Spectrum*, p. 51 (Apr. 1983).
15. Asatani, Koichi, et al., "A Field Trial of Fiber Optic Subscriber Loop Systems Utilizing Wavelength-Division Multiplexers," *IEEE Transactions on Communications*, pp. 2172–2184 (Sept. 1982).
16. Ishida, Yukinori, et al., "Performance of Field Installed Optical Cables for Use in Long-Wavelength Large-Capacity Transmission System," *ICC '82*, pp. 5D.1.1–5D.1.5.
17. Bell, Trudy E., "Communications," *IEEE Spectrum*, p. 56 (Jan. 1984).
18. Nakajima, Hirohito, and Watanabe, Toshiaki, "Advanced CAPTAIN System and an Approach to Commercial Service," *IEEE Journal on Selected Areas in Communications*, pp. 278–284 (Feb. 1983).
19. Morisawa, Tadakazu, Miyabe, Hiroshi, and Ohyama, Minoru, "Voice Storage Communication Facility for the Information Network System," *GLOBECOM '82 Record*, pp. C3.2.1–C3.2.5.
20. Nurakami, K., and Takeuchi, Seikichi, "An Audio Response Unit with Unlimited vocabulary, and its Application for Public Services," *GLOBECOM '82 Record*, p. F8.4.1–F8.4.5.
21. Iimura, Osamu, and Koide Takaaki, "Network Planning for the Information Network System (INS)," *GLOBECOM '82 Record*, pp. C3.1.1–C3.1.5.
22. Kano, S., Kitami, K., and Ohnishi, H., "ISDN User/Network Protocol—Overall Architecture," *GLOBECOM '82 Record*, pp. D2.2.1–D2.2.5.
23. Kuwabara, Moriji, "Japan tests its model communications system," *IEEE Spectrum*, p. 55 (May 1984).
24. Iimura, Osamu, Mizuashi, Kazunori, and Kikuchi, Shoichi, "Network Planning Toward the ISDN," *National Telecommunications Conference Record (1981)*, pp. G9.7.1–G9.7.4.
25. Kuwabara, pp. 53–56.
26. Ibid.
27. Rutkowshi, A. M., "Optical-Fiber Development and Application in Japan," *Telecommunications*, pp. 70–72 (Dec. 1982).
28. Inagaki, K., et al., "International Connection of Plesiochronous Networks Via TDMA Satellite Link," *ICC '82*, pp. 5H.4.1–5H.4.4.
29. *The World's Telephones, a Statistical Compilation as of January 1, 1981*, pp. 12–13, 96.
30. Ibid.
31. LeGuillou, Jean-Alain, Marcel, Francois, and Schwartz, Arthur J., "PRANA at the Age of Four: Multiservice Loops Reach Out," *IEEE Transactions on Communications*, pp. 2185–2210 (Sept. 1982).
32. "French PTT Installs Minitels for New Electronic Directory Service," *Communications News*, p. 14 (April 1983).
33. Dorros, p. 52.
34. Popovics, Jean-Luc, "Optical Fiber Multiservice Subscriber Connection System: SAFO," *IEEE Transactions on Communications*, pp. 2215–2220 (Sept. 1982).

35. Gravey, P., "The Wired City of Biarritz: A First Step to an Optical Multiservice Network," *ICC '82,* pp. 4D.3.1–4D.3.5.
36. Clost, Marcel, Roche, Alain, and Vomsheid, Alain, "Perspectives of Evolution Towards the Integrated Services Digital Network," *ICC '82,* pp. 3I.5.1–3I.5.8.
37. von Vignau, Ralph A., "*Bildschirmtext* and the CEPT Videotex System," *IEEE Journal on Selected Areas in Communications,* pp. 254–259 (Feb. 1983).
38. Dorros, p. 52.
39. Buenning, Helmut, Kreutzer, Heinrich W., and Schmidt, Fred, "Subscriber Stations in Service Integrated Optical Broad Band Communications Systems," *IEEE Transactions on Communications,* pp. 2163–2169 (Sept. 1982).
40. Ibid., pp. 2169–2171.
41. *Communications Week,* pp. C1–C7 (Mar. 26, 1984).
42. Wilkens, Henning, et al., "Interactive Broad-Band Dialogue Systems in the Integrated Services Digital Network (ISDN)," *IEEE Journal on Selected Areas in Communications,* pp. 295–302 (Feb. 1983).
43. *The World's Telephones, a Statistical Compilation as of January 1981,* pp. 12–13.
44. Childs, Geoff, "United Kingdom Videotex Service and the European Unified Videotex Standard," *IEEE Journal on Selected Areas in Communications,* pp. 245–249 (Feb. 1983).
45. Griffiths, John M., "ISDN Network Terminating Equipment," *IEEE Transactions on Communications,* pp. 2137–2141 (Sept. 1982).
46. Midwinter, J. E., "Development of High Bit Rate Monomode Fibre Systems in the UK," *ICC '82,* pp. 6D.6.1–6D.6.4.
47. Dorros, p. 51.
48. Wedlake, John, "An Introduction to the Integrated Services Digital Network," *National Telecommunications Conference Record (1980),* pp. 5.1.3–5.1.4.
49. Stumpers, F. L. H. M., "The History, Development, and Future of Telecommunications in Europe," *IEEE Communications Magazine,* p. 93 (May 1984).
50. deWitte, G. H. M., "DRS-8: System Design of a Long Haul 91 Mb/s Digital Radio," *National Telecommunications Conference Record (1978),* p. 38.1.1.
51. Kenney, J. J., "Digital Radio for 90-Mb/s, 16-QAM Transmission at 6 and 11 GHz," *Microwave Journal,* p. 71 (Aug. 1982).
52. Dorros, p. 51.
53. Tough, G. A., et al., "An Integrated Fiber Optics Distribution Field Trial in Elie, Manitoba," *GLOBECOM '82 Record,* pp. 4D.4.1–4D.4.5.
54. Levi, Israel, "The FD-3L: a long wavelength fiber trunk system," *Telesis,* No. 4, pp. 13–17 (1982).
55. O'Brien, C. Douglas, and Brown, Herbert G., "A Perspective on the Development of Videotex in North America," *IEEE Journal on Selected Areas in Communications,* pp. 260–265 (Feb. 1983).
56. Vilis, R. A., "Circuit Switched Digital Data Service," *ICC '82,* pp. 6A.4.1–6A.4.5.
57. *Notes on the BOC Intra-LATA Networks,* American Telephone and Telegraph Company, Section 9, pp. 4–8 (1983).
58. Straley, H. D., "Switching System Considerations for SDC," *ICC '82,* pp. 3D.3.1–3D.3.2.
59. *Technical Advisory No. 75,* American Telephone and Telegraph Company, p. 5 (Mar. 1983).
60. Ibid., p. 6 and Appendix 2, p. 3.
61. Bergland, G. D., "Experiments in Communications Technology," *IEEE Communications Magazine,* pp. 4–14 (Nov. 1982).
62. Dorros, p. 49.
63. Bergland, pp. 4–14; Alles, H. G., "The Intelligent Communications Switching Network," *IEEE Transactions on Communications,* pp. 1080–1087 (July 1979).

Glossary

A-Law. A companding characteristic used in European PCM systems.

ABSBH. Average busy season busy hour.

Adaptive Delta Modulation. Delta modulation using a step size that depends on the magnitude of the input.

Adaptive Predictive Coding. Coding based on predicting the present input sample using the previous cycle or pitch period.

Adaptive Pulse Code Modulation. Pulse code modulation using adaptive quantization.

Adaptive Quantization. Quantization in which the step size varies so that it matches the variance of the input signal.

Adaptive Transform Coding. A coding technique in which input subbands are transformed and the resulting coefficients encoded, usually using adaptive PCM.

Address. (1) The destination of a message in a communications system. (2) The group of digits comprising a complete telephone number.

Address Signals. Signals which are used to convey destination information.

ADM. Adaptive delta modulation.

ADPCM. Adaptive DPCM.

Aliasing. A condition in which a frequency Δf above half the sampling rate is reproduced as a frequency Δf below half the sampling rate.

Alternate Routing. The use of a route other than a direct or first-choice route.

Amplitude-Phase Keying (APK). Digital modulation in which both the amplitude and the phase of the carrier are altered to produce the various symbol states.

ANSI. American National Standards Institute.

APC. Adaptive predictive coding.

APCM. Adaptive pulse code modulation.

APD. Avalanche photodiode. A detector used in optical fiber systems.

Array, Switching. A matrix of crosspoints forming part of a switching network.

Asynchronous Communication. A mode of transmission characterized by start/stop operation with undefined time intervals between transmissions.

ATC. Adaptive transform coding.

Availability. The percentage of time during which a system provides its intended service.

Average Busy Season Busy Hour. The three months with the highest average traffic in the busy hour, excluding Mother's Day, Christmas, and extremely high traffic days attributed to unusual events.

b.e.r. Bit error rate.

Block Code. A code in which the redundant bits relate only to the information bits of the same block.

Blocking. The inability of a telecommunications system to establish a connection because of the unavailability of a path.

BORSCHT. An acronym referring to functions performed in, or in connection with, subscriber line circuits in a switching system: Battery, Overload protection, Ringing, Supervision, Coding, Hybrid, and Test access.

BPSK. Bi-phase shift keying.

Break. The open state of relay or switch contacts.

Bridged Tap. A pair of wires branched from a main pair in a telephone cable.

BRZ. Bipolar return to zero. A channel code used for digital transmission.

Busy Hour. The clock hour during which the most traffic is experienced in a switching system or over a group of circuits.

B3ZS. Bipolar with three-zero substitution. A channel code.

B6ZS. Bipolar with six-zero substitution. A channel code.

B8ZS. Bipolar with eight-zero substitution. A channel code.

C-Message Weighting. A noise weighting used in a noise measuring set to measure noise on a line that would be terminated by a device having acoustic properties similar to those of a Western Electric Type 500 telephone set.

Call Progress Signals. Signals sent to a caller by a switching system to provide information relative to the establishment of a connection.

CCIS. Common channel interoffice signaling.

CCITT. International Consultative Committee for Telegraphy and Telephony.

CDM. Continuous delta modulation.

CDMA. Code-division multiple access. A technique involving spread spectrum transmission to provide a degree of protection against jamming.

Central Office. One or more public network switching systems installed at a single location; the term is often used synonomously with *switching system.*

Cepstrum. The Fourier transform of the logarithm of the power spectrum.

Channel Bank. A device which combines a number (e.g., 24) of voice channels together into a digital stream (e.g., at 1.544 Mb/s) based upon sampling each voice channel at a specific rate (e.g., 8000 times per second).

Channel Coder. A device that processes a binary input into a multilevel or modified binary signal.

Characteristic Impedance. The impedance a transmission line would present at its input terminals if it were infinitely long.

Clock. Equipment providing a time base used in a transmission system to control the timing of certain functions such as the control of the duration of signal elements, and the sampling (CCITT).

Coding. Conversion of an analog function to digital form using a specific set of rules.

Combined Switching. In time division switching, the switching of each direction of conversation by a single control.

Common Channel Interoffice Signaling (CCIS). A signaling system used in North America, in which a separate signaling network is used to exchange supervisory and address signals between switching systems.

Common Equipment. Items of like equipment used on a shared basis by a switching system to establish connections.

Communications Quality. Speech quality which is acceptable to military, amateur and citizens band operators, often in a mobile environment.

Compandor. A compressor-expander.

Compressor. A device which reduces the dynamic range of a signal.

Concentration. The process of connecting any of a number of inlets to one of a smaller number of outlets.

Continuous Delta Modulation. Delta modulation using syllabic adaptation of the step size.

Continuously Variable Slope Delta Modulation. Delta modulation using a set of discrete values of slope variations in which the slope changes are done at a syllabic rate.

Convolutional Code. A code in which the redundant bits check the information bits in previous blocks.

Correlation. Multiplication of an incoming signal by a locally generated function and averaging of the result. If the locally generated function is a delayed form of the received signal, the process may be called *autocorrelation.*

Correlative Coding. Coding which uses finite memory to change the baseband digital stream to a form which improves coding efficiency from a spectrum occupancy viewpoint.

Crosspoint. A controlled device used by a switching system to connect one path to another.

Crosstalk. Undesired power coupled to a communications circuit from other communications circuits; may be intelligible or unintelligible.

CSU. Channel service unit.

CVSD. Continuously variable slope delta modulation.

CWR. Continuous word recognition.

Data Under Voice (DUV). Transmission in which a digital stream is sent using baseband frequencies lower than those used for analog transmission.

dBm. Power level in decibels referred to a power of one milliwatt, used in telephony as a measure of absolute power.

dBm0. The power in dBm measured at, or referred to, a point of zero transmission level.

dBrnc. Decibels above reference noise with the reference at −90 dBm using C-message weighting.

DCDM. Digitally controlled delta modulation.

Delay. The amount of time by which a signal or event is retarded, expressed in time or in number of symbols or characters.

Delta Modulation. A one-bit version of DPCM in which the output bits convey only the polarity of the difference signal.

Demand Assignment. The assignment of a channel on demand when needed, for the duration of the communication.

Dial Pulsing. A means of address signaling consisting of regular, momentary interruptions of the direct current path at the sending end, in which the number of interruptions corresponds to the value of the address digit being transmitted.

Differential Pulse Code Modulation. Pulse code modulation in which the quantization is done on a differential waveform produced by subtracting from the input the previous value of the output, or a weighted combination of previous output values.

Digit Synchronization. The condition in which each digit (usually bit) is correctly sampled by the receiver, thus assuring its proper reception.

Digital. Information in the form of one of a discrete number of codes.

Digital Radio. A radio that transmits a signal whose informational content is at least partly digital.

Digital Termination Systems (DTS). Radio local loops provided by competing common carriers in a given metropolitan area.

Digitally Controlled Delta Modulation. Delta modulation in which step size information is derived directly from the bit sequence produced by the sampling and quantization process, with companding at a syllabic rate.

Diphthong. A gliding monosyllabic speech item that starts at or near the articulatory position for one vowel and moves to or toward the position for another.

Dispersion. Pulse broadening in an optical fiber caused by multipath wave propagation. The result is intersymbol interference, corresponding to a limitation on the bandwidth that can be transmitted.

Distribution. The switching of traffic between concentration and expansion portions of a switching system network.

Dithering. A variation of the quantization levels to break up signal-dependent patterns in the quantized result.

Diversity. The use of dual frequencies, paths, or polarizations to minimize fading problems in a microwave radio relay system.

DLQ. Dynamic locking quantizer.

DM. Delta modulation.

DPCM. Differential pulse code modulation.

Dry Contacts. Contacts through which no direct current flows.

DSI. Digital speech interpolation.

DSU. Data service unit.

DTMF. Dual tone multifrequency signaling.

Dual Tone Multifrequency Signaling. A method of transmitting address signals by transmitting a pair of discrete tones from a group of eight tones, usually used on subscriber lines.

Duobinary. A correlative coding technique in which a two-level binary sequence is converted into one that uses three levels. The conversion involves intersymbol interference extending over one bit interval.

Dynamic Locking Quantizer. A quantizer capable of sensing and handling either speech or data inputs.

Dynamic Range. The difference between the overload level and the minimum acceptable signal level.

E & M Signaling. A technique for transferring information between a trunk circuit and a separate signaling circuit over leads designated *E* and *M*.

Echo. A reflection of a transmitted signal having sufficient magnitude to be perceived as separate from the transmitted signal.

EIRP. Effective isotropically radiated power.

Elastic Store. A digital store unit which accepts data under one timing source but outputs it under another. In this manner, jitter related to either the input or the output can be removed.

Erlang. A unit of telephone traffic intensity obtained by multiplying the number of calls by the average length of calls in hours. One erlang equals 60 minutes.

Error Coder. A device which adds redundant bits to a bit stream to provide for error detection and possibly correction, at the receiver.

Expander. A device which increases the dynamic range of a signal.

Expansion. The process of connecting any of a number of inlets to one of a larger number of outlets.

Extended Framing Format. An extension of the multiframe structure for use of the framing bit sequence for signaling, cyclic redundancy check, and a data link.

F-Bit. Framing bit.

Fade Margin. The number of decibels by which a signal can fade before the fade takes the signal's level below the receiver's threshold.

Far-End Crosstalk (FEXT). Crosstalk that is propagated in a disturbed channel in the same direction as the propagation of signals in the disturbing channel.

FDMA. Frequency-division multiple access.

FEXT. Far-end crosstalk.

Footprint. The coverage area on the surface of the earth from a satellite beam.

Formant. A band of speech energy in the frequency spectrum. A resonance frequency of the vocal tract tube.

Four-Wire Circuit. A two-way transmission circuit using separate paths for each direction of transmission.

Frame Synchronization. The condition in which each frame or block of received bits is correctly timed with respect to the received signal for the proper identification of the received bits as well as the individual channels.

Framing. The determination of which groups of bits constitute quantized levels and which quantized levels belong to which channels.

Fresnel Zone. A means of expressing the clearance of a microwave beam over an obstacle. The boundary of the nth Fresnel zone consists of all points from which the reflected wave is delayed $n/2$ wavelengths.

Fricative. A sustained unvoiced sound produced from the random sound pressure that results from turbulent air flow at a constricted point in the vocal system.

G/T. The ratio of a receiving system's gain to its noise temperature.

Glare. The condition resulting from a near-simultaneous seizure of a two-way trunk from both ends, in which the seizure signal appears to each switching system as a signal indicating a readiness to receive address digits.

Graded Index Fiber. An optical fiber which minimizes dispersion effects by providing nearly the same travel time for the various propagating modes.

Ground Start. A method of supervision on subscriber lines by which a seizure is indicated by placing ground potential on one of the conductors.

HDBH. High day busy hour.

High Day Busy Hour. The hour in the one day, among the ten in the ten high days, which has the highest traffic during the busy hour determined from the ten high day busy hour analysis.

Holding Time. The total time that a circuit is held busy, usually expressed in seconds.

HPA. High powered amplifier. The output stage in a transmitting earth station for satellite communication.

Hybrid Coder. A combination waveform and parametric coder. Often a hybrid coder will perform waveform coding of voice pitch but parametric coding of the voice formants.

IC. Integrated circuit. An electronic circuit that consists of many individual circuit elements such as transistors, diodes, resistors, capacitors, inductors, and other active and passive semiconductor devices, formed on a single chip of semiconducting material and mounted on a single piece of substrate material.

Impulse Function. A function that begins and ends within a time so short that it may be regarded mathematically as infinitesimal although the area described by the function remains finite.

Impulse Noise. Intermittent or spasmodic noise consisting of high-level pulses of short duration.

Index of Refraction. The ratio of the velocity of light in vacuum to the velocity of light in a given medium (e.g., an optical fiber).

Integrated Services Digital Network. A switched network, with end-to-end digital connectivity, which supports a wide range of services.

Interdigital Time. The time interval between address digits being transmitted over a circuit.

Interface. The point at which two systems or two parts of one system interconnect.

IRED. Infrared emitting diode. A source used in optical fiber systems.

ISDN. Integrated services digital network.

IWR. Isolated word recognition.

Jitter. Short term variations of the significant instants of a digital signal from their ideal positions in time (CCITT).

Laser Diode. A junction diode consisting of positive and negative carrier regions with a *P–N* transition region (junction) that emits electromagnetic radiation at optical frequencies. The emitted beam is very narrow, allowing the output to be coupled efficiently into single mode fibers usable in long range optical transmission systems.

LDM. Linear delta modulation.

LED. Light-emitting diode. A source used in short range optical fiber systems. A diode that emits electromagnetic radiation at optical frequencies. The emitted beam is broad, allowing the output to be coupled efficiently only into relatively large cross-section multimode fibers. Such fibers normally are used only in short-range optical transmission systems.

Linear Delta Modulation. Delta modulation in which the input time function is approximated by a series of linear segments of constant slope.

Linear Predictive Coding. A parametric coding technique in which the perceptually significant features of speech are extracted from its waveform.

LNA. Low noise amplifier. The input stage of a receiving earth station for satellite communication. Its effective noise temperature usually is lower than the ambient temperature.

Logarithmic Compression. Reduction of the dynamic range of a signal based upon the logarithm of its instantaneous amplitude.

Long Haul. Transmission over a microwave radio relay system to distances in excess of 400 km.

Loop Start. A method of supervision on subscriber lines by which a seizure is indicated by a closure of the two conductors in the subscriber loop.

Loss. (1) Power that is dissipated in a circuit without doing useful work. (2) The drop in power of a signal traversing a circuit or a switched connection.

LPC. Linear predictive coding.

LSI. Large-scale integration of circuitry in semiconductor elements.

MF. Multifrequency signaling.

μ-Law. A companding characteristic used in North American PCM systems.

Make. The closed state of relay or switch contacts.

Metropolitan Area Trunk (MAT). A cable designed to minimize crosstalk where large numbers of circuits are required between central offices.

Modem. A modulator-demodulator. This device is used to convert a digital stream to a quasi-analog form (tones) suitable for transmission on analog facilities, and to reconvert to digital form at the receiving end.

Modified Duobinary. A correlative coding technique in which the intersymbol interference extends over two bit intervals.

Modulation. Variation of the amplitude, frequency, or phase of a carrier wave to convey information.

MSK. Minimum shift keying. A form of frequency shift keying in which the peak frequency deviation equals $\pm$ 0.25 times the bit rate and coherent detection is used.

Muldem. A multiplexer-demultiplexer.

Multiframe. A set of twelve consecutive frames in which the position of each frame can be identified by reference to a multiframe alignment signal for the group of consecutive frames.

Multifrequency Signaling (MF). A signaling system, normally used on telephone trunks, by which address digits are indicated by a pair of discrete tones.

Multimode Fiber. An optical fiber whose diameter is large enough to allow the transmission of multiple propagation modes.

Multiple Access. Techniques allowing variously located earth terminals to use portions of a satellite's transponder on either a frequency- or a code- or a time-division basis.

Near-End Crosstalk (Next). Crosstalk that is propagated in a disturbed channel in the direction opposite to the propagation of signals in the disturbing channel.

Noise. Any unwanted signal or interference on a circuit other than the signal being transmitted.

Noise Figure. The ratio, expressed in decibels, of a system's input signal-to-noise ratio to its output signal-to-noise ratio.

Nonblocking. The ability of a telecommunications system to establish a connection from an inlet to an outlet irrespective of the amount of traffic.

NRZ. Nonreturn to zero. A channel code, in which there are only two states of a signal parameter used to represent data. These are the 0 state and the 1 state.

Off Hook. (1) In line signaling, the condition indicating that a line is in use (line loop: closed). (2) In trunk signaling, the signaling state which exists, in the forward direction, to indicate a seizure of the trunk by the switching equipment, and in the backward direction, to indicate an answered call or an element of signaling protocol.

On Hook. (1) In line signaling, the condition indicating that a line is idle (line loop: open). (2) In trunk signaling, the signaling state which exists, in the forward direction, to indicate that the trunk is not in use and, in the backward direction, to indicate that a call is awaiting an answer, a disconnect signal from the called end, or an element of signaling protocol.

Outpulsing. The transmission of address digits necessary to establish a switched connection.

Packet Transmission. The transmission of a stream of bits which has been divided into packets of a specific length (e.g., 1024 bits). Each packet carries its address. The overall stream is reassembled at the receiving location.

Pad. A resistance or other network inserted into a transmission path to provide a controlled amount of loss in the path.

Parametric Coder. A device which is designed to digitize an input in terms of its parameters, such as frequency bands, amplitudes, periodicities, etc.

Parity Bit. The name given to a redundant bit added to a sequence of information bits so the total sequence adds to either one or zero. If the received sequence does not add to the same number (one or zero), a parity error is said to have occurred.

Partial Response Signaling (PRS). The use of controlled intersymbol interference to increase the transmission rate in a given bandwidth.

PCM. Pulse code modulation.

Peg Count. The count of the number of traffic attempts made on a group of circuits or equipment elements during a given time period.

Phase Shift Keying. A form of digital modulation in which the bits shift the instantaneous phase between predetermined discrete values. It uses 2^m phases to represent m bits of information each.

Phoneme. A distinctive sound within a language.

Pitch. The fundamental or lowest predominant frequency produced by the human voice.

Plosive. A sound resulting from making a complete closure (usually toward the mouth end), building up pressure behind the closure, and abruptly releasing it. The letters B, K, P and T are plosives.

Polarization. The direction of the electric vector of a propagating electromagnetic wave. For circular polarization, this vector rotates at a rate equal to the carrier frequency.

Polybinary. A correlative coding technique in which the intersymbol interference extends over more than two bit intervals.

Prediction. The process of estimating a future value by using weighted sums of past values.

Private Automatic Branch Exchange (PABX). An automatic switching system providing switched telephone communications at a subscriber's premises and connections between the premises and the public switched network.

Protection Channel. A spare channel for use when channel equipment outages occur, or during deep fades on a microwave radio relay system.

Pseudonoise Generator. A generator of a very long periodic digital sequence. The length of the stream is great enough that it appears to be random.

Psophometric Weighting. Selective attenuation of voiceband characteristics based upon the use of a filter recommended by the CCITT and calibrated with an 800 Hz tone at 0 dBm.

Pulse Code Modulation. The use of a code to represent quantized values of instantaneous samples of a waveform.

Pulsing. The generation and transmission of pulses to provide signaling information to a switching system.

QAM. Quadrature amplitude modulation.

QPRS. Quadrature partial response signaling.

QPSK. Quaternary phase shift keying.

Quadrature Amplitude Modulation. The independent amplitude modulation of two orthogonal channels using the same carrier frequency.

Quantization Noise. Noise produced by the error of approximation in the quantization process.

Real Time. Pertaining to the actual time during which a physical process transpires.

Regeneration. The process of recognizing and reconstructing a digital signal so that the amplitude, waveform, and timing are constrained within stated limits (CCITT).

Reliability. The percentage of time during which equipment performs its intended function.

Retrial. An additional attempt to seize a unit of equipment or a circuit, or to find a path through a switching network, after a previous attempt has failed.

Ring. One conductor in a pair of wires, distinguished from the *tip* conductor.

Ringing Signal. A signal sent over a called line or trunk to alert the called party by audible or visual means to the incoming call.

Sampling. The process of sensing a waveform's amplitude at specific instants of time. Sampling usually is done periodically.

SBC. Subband coding.

SCPC. Single channel per carrier. A technique used in thin-route satellite communications.

Scrambler. A device which alters a bit stream using a specific set of rules so that the transmitted stream does not contain long sequences of zeros, but so that the receiver can reconstruct the original stream.

Sender. A device in an electromechanical common control switching system which receives address or routing information and outpulses the correct digits to a trunk or to the local equipment.

Separated Switching. In time division switching, the switching of each direction of conversation by separate controls.

Server. A circuit or item of equipment which provides service to a call attempt.

Service Circuit. Any of several groups of common equipment used on a shared basis within a switching system to establish connections.

Short Haul. Transmission over a microwave radio relay system to distances less than 400 km.

Sidetone. The signal produced in a telephone receiver by one's own voice or by room noise through the telephone transmitter.

Signaling. The transmission of address and other switching information between subscribers and switching systems or between switching systems.

Single Frequency (SF) Signaling. A method for conveying dial pulse and supervisory signals from one end of a trunk to the other end by the presence or absence of a single specified frequency.

Single-Mode Fiber. An optical fiber whose diameter is so small that only a single mode can propagate through it.

SNR. Signal-to-quantizing-noise ratio.

Source Coding. The process of digitizing an analog input using a specific algorithm.

SPADE. *S*ingle-channel per carrier *P*ulse code modulation multiple *A*ccess *D*emand assigned *E*quipment. A SCPC technique used in thin route satellite communication.

SPEC. Speech predictive encoded communications.

Stop. A sound produced by an abrupt release of a pressure built up behind a complete occlusion.

STS. Space-Time-Space switching architecture.

Subband Coding. Frequency-domain coding in which each of several subbands is coded separately.

Subrate. A digital rate less than 64 kb/s.

Supervision . The process of detecting a change of state between busy and idle conditions on a circuit.

Switcher. A switching system.

Switching Array. A matrix of crosspoints forming part of a switching network.

Sync Character. A repetitive bit pattern used by a receiver to establish that synchronization has been achieved.

Synchronization. A means of insuring that both transmitting and receiving stations are operating together (in phase).

Synchronous Communication. A mode of digital transmission in which discrete signal elements (symbols) are sent at a fixed and continuous rate.

Synthetic Quality. The quality of computer-generated speech, which often lacks human naturalness. Synthetic quality speech is intelligible, but the speaker may not be recognizable.

Tandem. (1) A network arrangement in which a trunk from the calling office is connected to a trunk to the called office through an intermediate point called a tandem switching office. (2) To establish a trunk-to-trunk connection through a switching office.

TASI. Time-assignment speech interpolation.

TCM. Time compression multiplexing. Time-domain separation of two directions of transmission.

TDM. Time division multiplexing.

TDMA. Time-division multiple access.

Ten High Day Busy Hour. The ten-day average traffic level for the time-consistent busy hour, excluding Mother's Day, Christmas, and extremely high traffic days attributed to unusual events.

THDBH. Ten high day busy hour.

Time Division Multiplexing. The sharing of a transmission circuit among multiple users by assigning time slots to individual users during which any one of them has the entire circuit's bandwidth.

Time Slot Interchange. A switching system element which switches between circuits by separating signals in time.

Tip. One conductor in a pair of wires used in telephony, distinguished from the *ring* conductor.

Toll Quality. Speech quality based upon a laboratory test in which the signal-to-noise ratio exceeds 30 dB and the harmonic distortion is less than 2%. The bandwidth is 300–3200 Hz.

Traffic. (1) The messages sent and received over one or a group of communication channels. (2) A quantitive measure of the total messages and their length, expressed in specified units.

Transient. A rapid fluctuation of voltage or current in a circuit, usually of short duration, caused by switching, changes in load, momentary crosses, ground, or by lightning surges.

Transmission Level. The power measured at a given point in circuit, usually expressed in dBm, over a given range of frequencies or at a specific frequency.

Transmission Level Point (TLP). The reference level point in a transmission system at which signal strength comparisons are made

Transmultiplexer. A device used to convert TDM signals to FDM, and vice versa, thus serving as an interface device between digital and analog networks.

Trunk. A communication channel provided as a common traffic path between two switching systems.

Trunk Circuit. A network of circuit elements used to connect a switching system to one of its associated trunks.

TSI. Time slot interchange.

TST. Time-space-time switching architecture.

Usage. The intensity of traffic carried by a group of circuits.

VLSI. Very large-scale integration of circuitry in semiconductor elements.

Vocoder. A device which is designed to digitize voice in terms of its pitch, amplitude, voicing and formants.

Waveform Coder. A device which converts samples of an analog waveform to bits.

Wavelength-Division Multiplexing (WDM). The simultaneous transmission of optical carriers of different wavelengths on a given optical fiber.

Wink. A single supervisory pulse used between switching systems to signal a readiness to receive address digits.

Index